K.-J. Reutter · E. Scheuber · P. J. Wigger (Eds.)

Tectonics of the Southern Central Andes

Structure and Evolution of an Active Continental Margin

With 189 Figures

Including 3 Geological and Geophysical Maps

Springer-Verlag
Berlin Heidelberg New York
London Paris Tokyo
Hong Kong Barcelona
Budapest

Professor Dr. Klaus-Joachim Reutter
Dr. Ekkehard Scheuber
Dr. Peter J. Wigger

Institut für Geologie,
Geophysik und Geoinformatik
Freie Universität Berlin
Malteserstr. 74–100
12249 Berlin, Germany

ISBN 3-540-55232-4 Springer-Verlag Berlin Heidelberg New York
ISBN 0-387-55232-4 Springer-Verlag New York Berlin Heidelberg

Library of Congress Cataloging-in-Publication Data. Tectonics of the southern central Andes: structure and evolution of an active continental margin / Klaus-Joachim Reutter, Ekkehard Scheuber, Peter J. Wigger (eds.). p. cm. Most of the papers presented at a workshop held in Berlin, May 23–25, 1990. ISBN 3-540-55232-4 (Berlin: acid-free). – ISBN 0-387-55232-4 (New York: acid-free) 1. Geology, Structural – Andes – Congresses. 2. Earth – Crust – Congresses. I. Reutter, Klaus-Joachim, 1934– II. Scheuber, Ekkehard, 1952– . III. Wigger, Peter J. QE230.T43 1993 551.8′098 – dc20 93-15961 CIP

Typesetting: Camera ready by author
Printing: Weihert-Druck, Darmstadt
Binding: Schäffer, Grünstadt
32/3145-5 4 3 2 1 0 – Printed on acid-free paper

PREFACE

The Central Andes, whose orogenic activity is so impressively documented by recent volcanism and earthquakes, have always attracted the attention of geoscientists. This interest became even more accentuated since, a quarter of a century ago, Plate Tectonics became the basis for the New Global Tectonics concept, in which this huge mountain range was the most spectacular example of an active continental margin. Thus, in addition to the continuing research work by South American and foreign geoscientists dedicated mostly to regional and economic problems, a great number of special research programmes were initiated aiming at a better understanding of the processes acting at a convergent plate margin.

In 1982, the earth science institutes of the Freie Universität and the Technische Universität of Berlin, which have had some tradition in geological work in the Andes since the early 1950s, launched a research programme called "Mobility of Active Continental Margins". Its intention was to obtain more insight into the structures of the Central Andes by an interdisciplinary approach which comprised a variety of mostly geophysical and geological studies. These were concentrated on a segment of the southern Central Andes extending between 20° and 26°S from the Pacific Ocean to the eastern Andean foreland, thus including parts of Argentina, Bolivia and Chile. Fieldwork in South America, additional equipment for laboratory research and furthering of young scientists were generously financed by the Deutsche Forschungsgemeinschaft (German Research Council) within the frame of a "Forschergruppe" (Research Unit), the Freie Universität Berlin and the Technische Universität Berlin.

As in the case of other foreign research programmes, successful work was made possible only through the assistance of the South American colleagues of the three mentioned countries and on the basis of their previous work and field experience. In this respect cooperation programmes with the Universidad Nacional de Salta (Argentina), the Yacimientos Petrolíferos Fiscales Bolivianos, the Universidad Mayor de San Andres in La Paz (Bolivia), the Universidad Catolica del Norte in Antofagasta (Chile), and the Universidad de Chile in Santiago were most important. We seize this opportunity to express our gratitude to our South American colleagues and friends.

The results of the geoscientific research programme of the Berlin universities were presented, together with contributions by invited geoscientists and counterparts from other countries, during a workshop held in Berlin, 23-25 May 1990. A great number of the papers presented at this workshop are included in this volume. While most of the chapters refer regionally to the segment of the southern Andes mentioned above, others treat general aspects or deal with Andean regions farther south, thus showing not only that the structures of this mountain range can be followed to more distant parts but also that there are significant structural variations along strike.

Like other books which originate from workshops and are comprised of contributions from many authors, also this one cannot give a complete and well-balanced view of the scientific subject dealt with, in this case the southern Central Andes. However, we believe that the combination of the results of geophysical studies using different methods and several geological and petrological approaches to different problems gives a broad view of the structures and the tectonic evolution of this active continental margin and points out some of the processes acting on it. The geophysical investigations and petrological research on recent volcanism are contributions to the "Structure and State of the Lithosphere", while geological studies and other petrological work elucidate aspects of the "Geological Evolution". Accordingly, the chapters have been subsumed under these headings. "Maps of the Central Andean Segment Between 20° and 26°S" is the heading of the last chapter which refers to the three maps which are enclosed in this book. The geological maps, at the scale of 1:1 000 000 of the Andean segment between 20° and 26°S, originally intended as an internal basis of discussion between the different working groups, may serve here for a better understanding of the regional context. A direct comparison between geological structure and gravity anomalies has been made possible by also adding a Bouguer anomaly map as well as an isostasy anomaly map at the same scale and of the same area.

Each chapter was reviewed by at least two external referees whose comments were very important for a better and more concise presentation of the chapters. The authors and editors are greatly indebted to them for their patient work.

It should be noted that the Global Geoscience Transect 6 ("Central Andean transect, Nazca plate to Chaco plains, southwestern Pacific Ocean, northern Chile and northern Argentina")[1] runs through the area of the enclosed maps. It was elaborated in cooperation with geoscientists from the Universidad Nacional de Salta (Argentina), the Universidad Católica del Norte in Antofagasta (Chile) and by members of the universities of Berlin, i.e. by many authors who also contributed to the present volume. Therefore, the notes and maps of Transect 6 can be considered as supplementary to the present publication. The same is true for a volume with the title "The Southern Central Andes, contributions to structure and evolution of an active continental margin"[2] which resulted from a first workshop held in Berlin in 1986 on the occasion of the completion of the first half of the research programme and which also refers regionally to the area in the maps presented in this volume.

Klaus-J. Reutter
Ekkehard Scheuber
Peter J. Wigger

[1] Omarini R, Götze H-J (eds) (1991) Am Geophys Union, Int Lithosphere Program, Publ 192, 30 pp, 2 sheets

[2] Bahlburg H, Breitkreuz C, Giese P (eds) (1988) Lecture Notes in Earth Sciences 17, Springer, Berlin Heidelberg New York, 261 pp

Contents

MAPS OF THE CENTRAL ANDEAN SEGMENT BETWEEN 20° AND 26°S

CONTRIBUTORS

ANDRIESSEN, PAUL A. M. Laboratorium voor Isotopengeologie z.w.o., De Boelelaan 1085, 1081 HV Amsterdam, The Netherlands

ARANEDA, MANUEL M. Depto. de Geologia y Geofisica, Universidad de Chile, Casilla 2777, Santiago, Chile

ASCH, GÜNTER Institut für Geologie, Geophysik und Geoinformatik, Freie Universität Berlin
Present address: Geoforschungszentrum Potsdam, Telegrafenberg A6, O-1561 Potsdam, Germany

BAHLBURG, HEINRICH Institut für Geologie und Paläontologie, Technische Universität Berlin
Present address: Geologisch-Paläontologisches Institut, Im Neuenheimer Feld 234, 6900 Heidelberg, Germany

BALDZUHN, SIBYLLE Institut für Geologie, Geophysik und Geoinformatik, Freie Universität Berlin, Malteserstr. 74-100, 1000 Berlin 46, Germany

BARRIENTOS, SERGIO E. Depto. de Geología y Geofisica, Universidad de Chile, Santiago, Chile
Present address: CIRES, Campus Box 216, University of Colorado, Boulder, CO 80309-0216, USA

BOGDANIC C., TOMISLAV Depto. de Ciencias Geologicas, Universidad Catolica del Norte, Antofagasta, Chile
Present address: Compañia Minera Carolina de Michilla S.A., Superintendencia de Geologia, Casilla 898, Antofagasta, Chile

BREITKREUZ, CHRISTOPH Institut für Geologie und Paläontologie, Technische Universität Berlin,
Present address: Department of Geology, University of Kansas, 120 Lindly Hall, Laurence, Kansas 66045-2124

CHARRIER, REYNALDO Depto. de Geologia y Geofisica, Universidad de Chile, Casilla 13518, Correo 21, Santiago, Chile

CHONG D., GUILLERMO Depto. de Ciencias Geologicas, Universidad Catolica del Norte, Av. Angamos 0610, Casilla 1280, Antofagasta, Chile

CLARKE, CHRIS Department of Earth Sciences, The Open University, Walton Hall, Milton Keynes, MK7 6AA, UK

DAMM, KLAUS-WERNER Institut für Geologie, Geophysik und Geoinformatik, Freie Universität Berlin
Present address: Lubrizol GmbH, Bogenallee 10, 2000 Hamburg 13, Germany

DANNAPFEL, MANFRED Abt. Geologie der Universität, Universitätsstr. 30, 8580 Bayreuth, Germany

DAVIDSON, JON P. Department of Earth and Space Sciences, University of California, Los Angeles, CA 90024, USA

ENTENMANN, JÜRGEN Institut für Geowissenschaften, Johannes Gutenberg Universität, Saarstr. 21, 6500 Mainz, Germany

GIESE, PETER Institut für Geologie, Geophysik und Geoinformatik, Freie Universität Berlin, Malteserstr. 74-100, 1000 Berlin 46, Germany

GÖRLER, KONRAD Institut für Geologie, Geophysik und Geoinformatik, Freie Universität Berlin, Malteserstr. 74-100, 1000 Berlin 46, Germany

GÖTZE, HANS-JÜRGEN Institut für Geologie, Geophysik und Geoinformatik, Freie Universität Berlin, Malteserstr. 74-100, 1000 Berlin 46, Germany

HARMON RUSSEL S. NERC Isotope Geosciences Laboratory, British Geological Survey, Keyworth, Nottingham NG12 5GG, UK

HAWKESWORTH, CHRIS Department of Earth Sciences, The Open University, Walton Hall, Milton Keynes, MK7 6AA, UK

HEINSOHN, WOLF-DIETER Institut für Geologie, Geophysik und Geoinformatik, Freie Universität Berlin, Malteserstr. 74-100, 1000 Berlin 46, Germany

HERVÉ, FRANCISCO Depto. de Geologia y Geofisica, Universidad de Chile, Casilla 13518 - Correo 21, Santiago, Chile

v. HILLEBRANDT, AXEL Institut für Geologie und Paläontologie, Technische Universität Berlin, Ernst-Reuter-Platz 1, 1000 Berlin 10, Germany

HORN, SUSANNE Institut für Geowissenschaften, Universität Mainz, Saarstr. 39, 6500 Mainz, Germany

JENSEN I. ARTURO, Depto. de Ciencias Geologicas, Universidad Catolica del Norte, Av. Angamos 0610, Casilla 1280, Antofagasta, Chile

KELLEY, SHARI Dept. of Geological Sciences, Southern Methodist University, Dallas, TX 75205, USA

KLEY, JONAS Institut für Geologie, Geophysik und Geoinformatik, Freie Universität Berlin, Malteserstr. 74-100, 1000 Berlin 46, Germany

KRÜGER, DETLEF Institut für Geologie, Geophysik und Geoinformatik, Freie Universität Berlin, Malteserstr. 74-100, 1000 Berlin 46, Germany

LAHMEYER, BERND Institut für Geophysik, Technische Universität Clausthal
Present address: BEB Erdöl und Erdgas, Riethorst 12, 3000 Hannover 51, Germany

LEHMANN, BERND Institut für Geologie, Geophysik und Geoinformatik, Freie Universität Berlin
Present address: FB Lagerstättenforschung, Technische Universität Clausthal, Adolph-Roemer-Str. 2A, 3392 Clausthal-Zellerfeld, Germany

LOPEZ-ESCOBAR, LEOPOLDO Depto. de Geologia y Geofisica, Universidad de Chile, Casilla 13518, Correo 21, Santiago, Chile

MARTINEZ, ELOY Yacimientos Petroliferos Fiscales Bolivianos (YPFB), GXG, Casilla 1659, Santa Cruz, Bolivia

MASSOW, WINFRID Institut für Geologie, Geophysik und Geoinformatik, Freie Universität Berlin, Malteserstr. 74-100, 1000 Berlin 46, Germany

MOORBATH, STEPHEN Department of Earth Sciences, University of Oxford, Parks Road, Oxford OX1 3PR, UK

MOYA, M. CHRISTINA Facultad de Ciencias Naturales, Universidad Nacional de Salta (UNSa), Buenos Aires 177, 4400 Salta, Argentina

MUÑOZ, NELSON Empresa Nacional del Petróleo (ENAP), Compaña 1085, Santiago-Chile

PICHOWIAK, SIEGFRIED Institut für Geologie, Geophysik und Geoinformatik, Freie Universität Berlin
Present address: HPC Consult, Oraniendamm 64-72, 1000 Berlin 28, Germany

PRINZ, PETER Institut für Geologie und Paläontologie, Technische Universität Berlin, Ernst-Reuter-Platz 1, 1000 Berlin 10, Germany

RAMOS, VICTOR A Depto. de Ciencias Geológicas, Universidad de Buenos Aires, Ciudad Universitaria, Pabellón II, (1428) Buenos Aires, Argentina

RATH, VOLKER Institut für Geologie, Geophysik und Geoinformatik, Freie Universität Berlin, Malteserstr. 74-100, 1000 Berlin 46, Germany

REINHARDT, MARTIN Institut für Geologie, Geophysik und Geoinformatik, Freie Universität Berlin
Present address: MOBIL Erdöl und Erdgas, Burggrafstr. 1, 3100 Celle, Germany

REUTTER, KLAUS-JOACHIM Institut für Geologie, Geophysik und Geoinformatik, Freie Universität Berlin, Malteserstr. 74-100, 1000 Berlin 46, Germany

RICALDI, EDGAR Instituto de Investigaciones Fisicas, Universidad Mayor de San Andres (UMSA), Casilla 8635, La Paz, Bolivia

RÖWER, PETER Institut für Geologie, Geophysik und Geoinformatik, Freie Universität Berlin, Malteserstr. 74-100, 1000 Berlin 46, Germany

SCHÄFER, KARLHEINZ Abt. Geologie der Universität, Universitätsstr. 30, 8580 Bayreuth, Germany

SCHEUBER, EKKEHARD Institut für Geologie, Geophysik und Geoinformatik, Freie Universität Berlin, Malteserstr. 74-100, 1000 Berlin 46, Germany

SCHMIDT, SABINE Institut für Geologie, Geophysik und Geoinformatik, Freie Universität Berlin, Malteserstr. 74-100, 1000 Berlin 46, Germany

SCHMITZ, MICHAEL Institut für Geologie, Geophysik und Geoinformatik, Freie Universität Berlin, Malteserstr. 74-100, 1000 Berlin 46, Germany

SCHWARZ, GERHARD Institut für Geologie, Geophysik und Geoinformatik, Freie Universität Berlin, Malteserstr. 74-100, 1000 Berlin 46, Germany

STRUNK, SYBILL Institut für Geologie, Geophysik und Geoinformatik, Freie Universität Berlin
Present address: RWE-DEA, Überseering 35, 2000 Hamburg 60, Germany

VIRAMONTE, JOSÉ Facultad de Ciencias Naturales, Universidad Nacional de Salta (UNSa), Buenos Aires 177, 4400 Salta, Argentina

WIGGER, PETER J. Institut für Geologie, Geophysik und Geoinformatik, Freie Universität Berlin, Malteserstr. 74-100, 1000 Berlin 46, Germany

WILKE, HANS-GERHARD Institut für Geologie und Paläontologie, Technische Universität Berlin,
Present address: Depto. de Ciencias Geologicas, Universidad Catolica del Norte, Av. Angamos 0610, Casilla 1280, Antofagasta, Chile

WILKES, EBERHARD Institut für Geologie, Geophysik und Geoinformatik, Freie Universität Berlin,
Present address: KRAVAG, Heidenkampweg 100, 2000 Hamburg 1, Germany

WÖRNER, GERHARD Institut für Geowissenschaften, Johannes Gutenberg Universität, Saarstr. 21, 6500 Mainz, Germany

ZEIL, WERNER Institut für Geologie und Paläontologie, Technische Universität Berlin, Ernst-Reuter-Platz 1, 1000 Berlin 10, Germany

STRUCTURE AND STATE OF THE LITHOSPHERE

The Lithospheric Structure of the Central Andes (20-26°S) as Inferred from Interpretation of Regional Gravity

HANS -JÜRGEN GÖTZE, BERND LAHMEYER, SABINE SCHMIDT and SIBYLL STRUNK

Abstract. Isostatic calculations, inversion theory and 3D modelling were applied to the regional gravity database in a segment of the Central Andes between 20° and 26° S. It extends from the Peru Chile Trench in the Pacific Ocean in the west to the NW Argentinean Chaco Plains in the east. Characteristics of the Andean orogen are marked by a negative Bouguer anomaly of more than -450 mGal and great gravity anomalies even in a north-south direction. In this chapter, we present quantitative interpretations from analysis of the regional gravity field along that particular geotraverse which was compiled by the Berlin Research Group "Mobility of Active Continental Margins" and its partner institutions in South America. Tree-dimensional forward modelling as part of an integrated interpretation leads to a crustal depth of less than 70 km in the central part.

1 Introduction

The interpretation of the gravity field in the rugged topography of mountainous regions is extremely difficult, because great differences can exist between heights of even adjacent gravity sites. In the region at hand (Fig. 1), elevations vary by as much as 5000 m. For more detailed information concerning the investigated area in the Central Andes, the gravity data acquisition, data processing and preliminary interpretation of gravity anomalies see Götze et al. (1988, 1990, 1991), Lahmeyer (1990) and Strunk (1990).

1.1 Gravity Field Interpretation in Rugged Topography

There are two terms in the basic Bouguer gravity formula which are particularly affected by topographic masses: (1) the term "free-air" reduction and (2) the "Bouguer slab" reduction, both normally calculated by use of the normal gravity gradient.

On relatively flat terrain, it can be assumed with reasonable accuracy that the term free-air reduction will reduce the height-dependent anomaly at any station to a constant reference level. However, this approximation is not valid in highmountain regions. Since height-dependent, anomalous gradients caused by density inhomogeneities of subsurface structures are not taken into consideration, Bouguer anomalies should thus apply to their station levels due to falsely calculated free-air reductions.

Differences in Bouguer gravity fields calculated at station levels and at the reference level caused by these gradient anomalies can amount to as much as 15 mGal (Götze et al. 1990). Dealing with the interpretation of gravity measurements by 3D modelling (see Sect. 3) here it is supposed to use actual gravity station heights of the relevant survey instead of a common reference level (e.g. normal sea level surface) in model calculations. More extensive calculations become necessary to compensate effects of the local gravity gradient, if a Bouguer gravity field is processed in the wave number domain by means of Fast-Fourier Transformation techniques (FFT). For example, Lahmeyer (1988) describes a combination of both FFT and the collocation method for potential field continuation from a surface of irregularly scattered data to a constant level .

The reduction of the Bouguer slab in the above mentioned basic formula for gravity stations in high mountains leads to another problem. Although it could be expected that reduction would lead to an overall anomaly decrease, it increases for high mountain regions: e.g. the amplitude of free-air anomalies in that region is approx. 550 mGal, yet the derived Bouguer anomaly covers a range of some 750

Correspondence to: H.-J. Götze, Fachrichtung Geophysik, Freie Universität, Malteserstr. 74-100, D-1000 Berlin 46

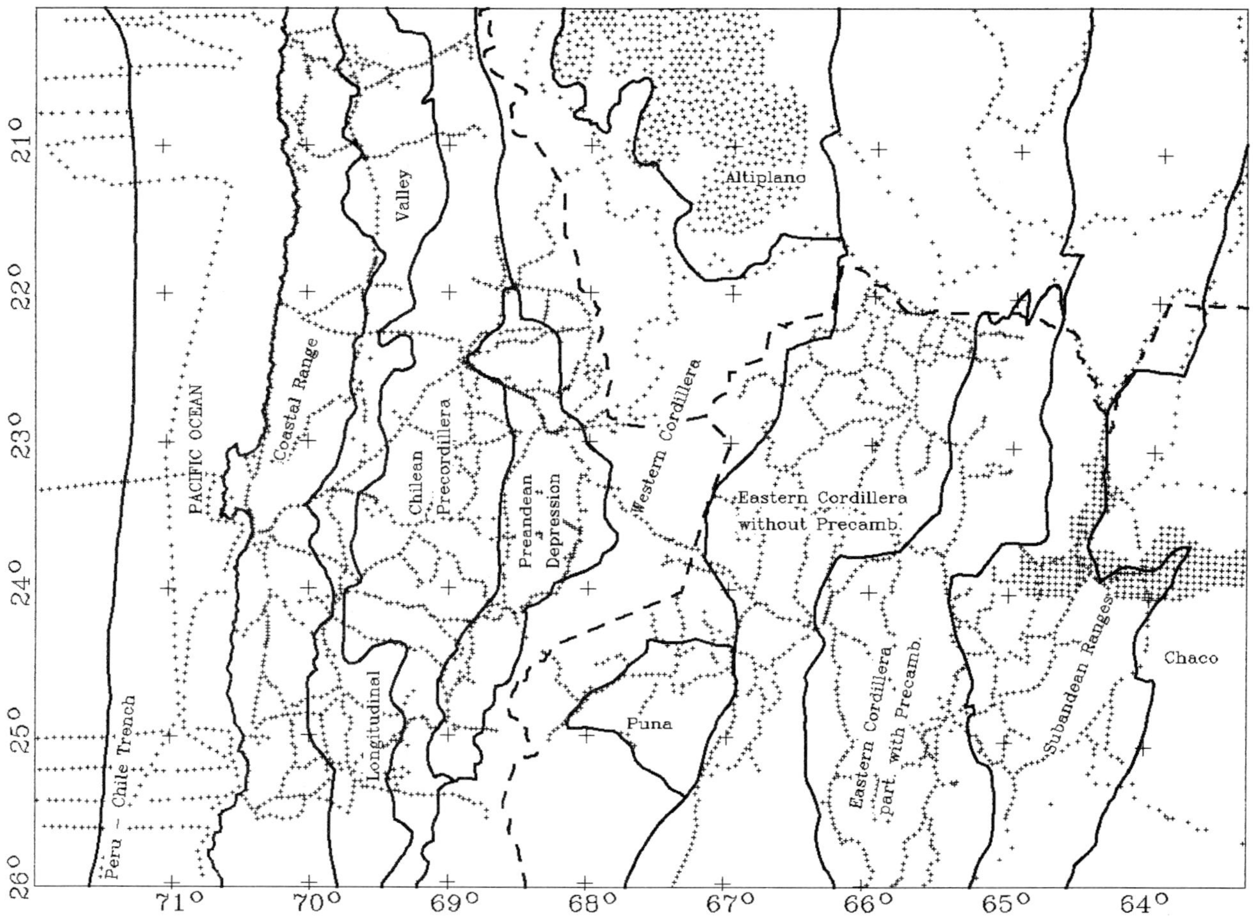

Fig. 1. Location map of gravity stations and geomorphological features (digitized after Reutter et al. 1988) in the Central Andes

mGal. The strong correlation of free-air gravity and local topography on the one hand and of Bouguer gravity and regional topography on the other hand indicates that the strong regional negative Andean gravity anomaly caused by Bouguer slab reduction can largely be interpreted in terms of isostatic crustal thickening. This conceals the small scale anomalies caused by intracrustal density inhomogeneities. To interpret these local anomalies, a regional field must be separated from the Bouguer gravity while taking its already mentioned heightdependence into consideration.

For regional field separation we applied direct methods (FFT, quadratic programming etc.) to a simple model of the crust/mantle interface since crustal thickening apparently causes the regional part of the Andean gravity field. The advantage of this procedure is two-fold: (1) the simple model is ascribed directly to the regional part of gravity, and (2) station dependence can be easily included in that calculation.

While complying with these considerations, we recommend the following concepts for the numerical interpretation of gravity in areas of rough topography:

1. Calculation of isostatic gravity based on a crustal model of regional compensation. This isostatic anomaly serves as a regional field and depends on gravity station heights. To calculate a series of various models without extensive effort, FFT can be applied in the wave number domain of the gravity field.
2. Use upward continuation of the isostatic residual field to a constant reference level at 6 km by FFT procedures. The 6 km level compromises the existence of short wavelengths in the gravity field and its position outside most of the Andean topographic masses and above all gravity stations

of the survey. Downward continuation of the field to sea level would be inhibited by incomplete massreduction between sea level and the topographic surface due to incorrect rock densities.

3. Upward continuation of the Bouguer gravity field to the 6 km reference level.
4. Calculation of the gravity field of the crust/mantle interface by inversion techniques. This leads to a second regional field calculated independently from (1) and (2). By numerical combination of both FFT methods and "quadratic programming" techniques (Lahmeyer 1990) even depth determinations from refraction seismic experiments (Wigger et al. this Vol.) could be used as constraints.
5. Elimination of the gravity effect of the subducted Nazca slab from the measured gravity field by quadratic optimization.
6. Modelling the Bouguer gravity field of lithospheric structures by 3D forward modelling techniques. It has to be considered that model stations have to correspond to stations of the original gravity survey.

The numerical tools related to the concept outlined above will be briefly discussed in the following. Detailed information can be found in Götze and Lahmeyer (1988) and Lahmeyer (1990).

1.2 Numerical Background

Calculating the residual isostatic anomaly

An isostatic Vening-Meinesz model incorporating regional compensation was chosen to calculate the gravity effect of the crust/mantle interface. Calculations were performed in the frequency domain according to the formalism of Banks et al. (1977). The topography model of the Andes (25 x 25 km grid, derived from the TUG 87 elevation model, (Sünkel pers. comm.) served as data input. Formula (1) shows the operator used in the frequency domain on the grid of topography.

The factor D of flexural rigidity controls the size of the regionality of isostatic compensation. For example, D = 0 leads to local isostatic compensation according to the Airy-Heiskanen model (Götze et al. 1990). The gravity at sea level is first calculated by the aid of FFT using a modified formula after Parker (1972). The "chessboard technique" (Cordell 1985) is then applied for upward continuation of the residual isostatic anomaly to stations located at the topographic surface. Due to the fast performance of FFT, numerous models could be tested within a short period. Synthetic models showed that the accuracy of the applied technique was adequate.

$$T^*_i(k_x,k_y) = \underbrace{\frac{\rho_K}{\rho_M-\rho_K}\left(1+\frac{16\pi^4\cdot\left|(k_x,k_y)\right|^4\cdot D}{(\rho_M-\rho_K)\cdot f}\right)^{-1}}_{\text{low passfilter: } TP\left(\left|k_x,k_y\right|\right)}\cdot H(k_x,k_y) \quad (1)$$

$$t_i(x,y) = T_0 - t_i^*(x,y)$$

where

T_0	: normal crustal thickness at h(x, y) = 0
ρ_K	: crustal density
ρ_M	: mantel density
D	: flexural rigidity
$H(k_x, k_y)$	: Fourier transform of topography h(x, y)
$T^*_i(k_x, k_y)$	: Fourier transform of $t^*_i(x, y)$
f	: gravity constant

"optimal" depth parameters for isostatic modelling were found by looking for relatively small standard deviations of the calculated isostatic residual fields. These residual fields depend on their models' crustal thicknesses. These were taken independently from studies of refraction seismics (Wigger et al. this Vol.).

Upward continuation of a gravity field to constant levels

For upward continuation of the isostatic residual field to the 6 km reference level, again we applied iterative inversion according to the chessboard technique. On the basis of synthetic models, we developed a procedure (Lahmeyer 1990) which can be used in the case of irregularly distributed data. This "two step procedure" includes (1) continuation of data by a median gradient to a grid which corresponds to the reference level and (2) interpolation of values at stations by bi-splines from the gridded values. The same technique was applied to continue the Bouguer gravity field from topography to the constant separation.

Constrained inversion of the crust/mantle interface

The method utilized here works interactively, similar to the technique developed by Oldenburg (1974), which is based on Parker's formula (Parker 1972) to solve the "upward" problem. The crust/mantle interface to be inverted was transformed into the frequency domain. Inversion of its Fourier coefficients was done by quadratic programming. Upper and lower limits for crustal thicknesses can be set where information is available. In order to prevent edge effects, both (1) the Bouguer gravity field defined at the 6-km reference level and (2) the gravity of the isostatic model outside the particular surveyed area were taken into account.

Independent information from refraction seismics was used as constraints in this technique. In this case the uncertainties and ambiguity of potential field interpretation are reduced and minimized. The regional isostatic model of western South America served as a starting model for this interactive technique. Topography also provided independent information which could be used to reduce ambiguity of gravity interpretations. The density contrast between crust and mantle was calculated by inversion and determination of the minimal standard deviation of the remaining residual fields of gravity.

Estimation of the gravity effect of the subducted Nazca slab

To estimate the gravity effect of the Nazca slab, a 3D model of the subducted oceanic plate was designed to fit both the distribution of earthquake hypocentres (Buness et al. 1986; Cahill 1990) and the petrophysical behaviour of the plate (Grow and Bowin 1975). The model geometry consists of numerous rectangular prisms for which the number of density contrasts related to their surroundings was calculated by quadratic programming. As mentioned earlier, the upper and lower limits (taken mainly from the literature and rare seismics studies) for model densities were included.

3D forward modelling

Three-dimensional forward modelling was carried out by using the interactive gravity application program, IGAS (Götze and Lahmeyer 1988). Here, underground structures are approximated by arbitrarily shaped polyhedrons. Polyhedrons are very useful elementary bodies employed to construct realistic geological structures, no matter how complicated they are. More information, particularly on the mathematical background and the specific solutions to volume integrals by transferring them first to surface and then to line integrals, can be found in Götze (1976). The actual program version consists of two parts: (1) a batch program to establish and test the data structure for 3D modelling and to create data input for the interactively operating second program; (2) an interactive program which offers the user interactive computer graphics functions to modify geometry and density of the model as well as further auxiliary support.

2 Results of Numerical Data Processing

In this Section, we will discuss the results obtained by applying the numerical techniques introduced in section 1 as they pertain to the regional data base for the Central Andes. Model calculations of the Andean crust will be discussed separately in Section 3.

Crustal Flexural Rigidity

Variation of "normal" crustal thickness, density contrast of crust/mantle interface and flexural rigidity of the crust yielded optimal parameters to calculate of the isostatic model:

Normal crustal thickness:	40 km
Density contrast:	0.4 - 0.5 g/cm^3
Flexural rigidity:	10^{23} Nm

The normal crustal thickness was taken from seismic refraction studies. While the density contrast was not well defined, the given range (0.4 - 0.5 g/cm^3) yielded almost identical results. Consideration of the calculated standard deviations of the isostatic residual field allowed the determination of the flexural rigidity. Figure 2 shows the dependence of the standard deviation of the residual field with regard to flexural rigidity, whereby all other parameters remain fixed. While the S 1 curve applies to the entire investigated area, S 2 applies only to the continental part of the survey. In both curves, the

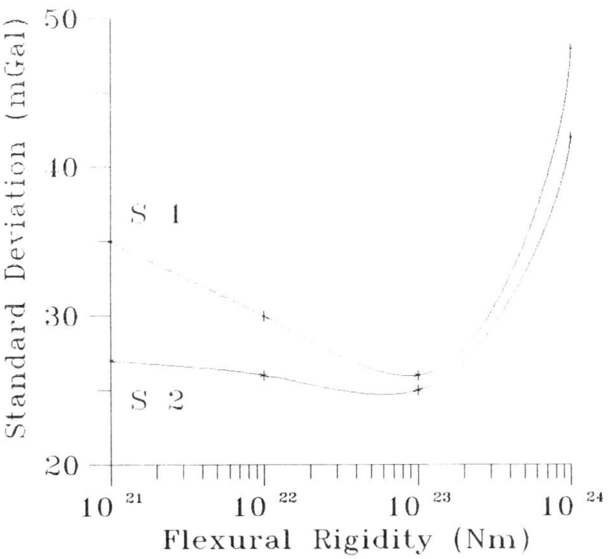

Fig. 2. Standard deviations for isostatic anomalies calculated of four Vening-Meinesz models with different values of flexural rigidity and constant values of crust/mantle density contrast (0.5 g/cm^3) and a normal crustal thickness (35 km). **S 1** standard deviations of residual fields in the entire area covering ocean and continent; **S 2** standard deviations of residual fields on the continent only

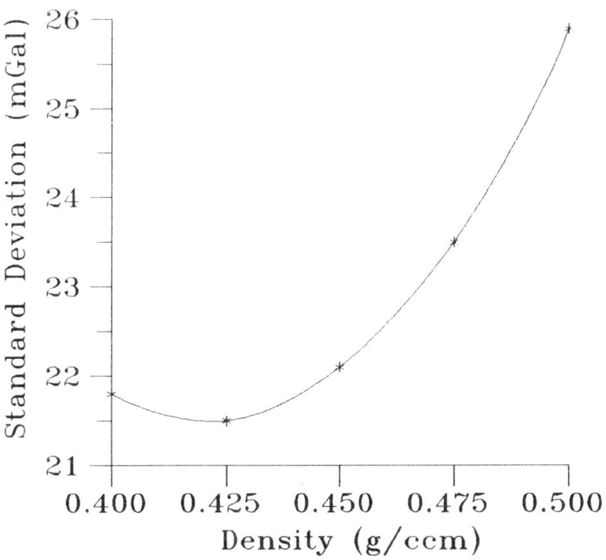

Fig. 3. Plot of standard deviations of residual fields, calculated from the inverted Moho model. Five different density contrasts between crust and mantle were used

minimum value is clearly recognizable at 10^{23} Nm. This value seems to be relatively high, compared with results from, for example the USA, which range from 10^{21} to 10^{22} Nm (Banks et al. 1977).

However, one has to consider that their investigation applies to the entire US territory. The results reflect an average flexural rigidity only. However, we have to emphasize that our results on the rigidity of the Andean crust are only "approximately true" because of uncertain or nonexisting data on crustal temperatures which considerably control flexural rigidity in that complicated crustal environment.

A comparison of the regional and local isostatic residual field of the Central Andes led to a detailed study of rigidity. This study showed that the crust in the W-E segment between 20° and 22°S has a rigidity of 10^{23} Nm, while the crust in the segment between 22° and 26° S reaches only an "optimal" value of 5 x 10^{22} Nm. The crust in the southern part thus appears to be "softer" than in the more northern areas, and, on the whole, it reacts more sensitively to flexural rigidity.

Density Contrast at the Crust/Mantle Interface

Constrained inversion of the crust/mantle interface (Sect. 1.2) allowed the determination of the density contrast at that boundary. To illustrate results, the standard deviations of the residuals, calculated from the gravity effect of the inverted interface were plotted against the density contrast (Fig. 3). The function shows that the contrast must be lower than 0.5 g/cm^3. In the final inverted model 0.425 g/cm^3 was assumed to be a plausible value. This density contrast corresponds to a velocity contrast at the Moho of 1.1 to 1.6 km/s (Gebrande 1982). However, this velocity contrast (calculated for a simple model with a homogeneous crust and mantle density) does not coincide with the extremely low average velocities (see Wigger et al. this Vol.) in the lower parts of the Andean crust. This misfit of both density and velocity might be explained in terms of the existence of small portions of melted material spread over the lower crust which decreases the seismic velocity but does not influence density in the same manner.

Shape of the Crust/Mantle Interface

The contoured map of the depths of the crust/mantle boundary is shown in Figure 4. It documents the results of depth calculations for three different

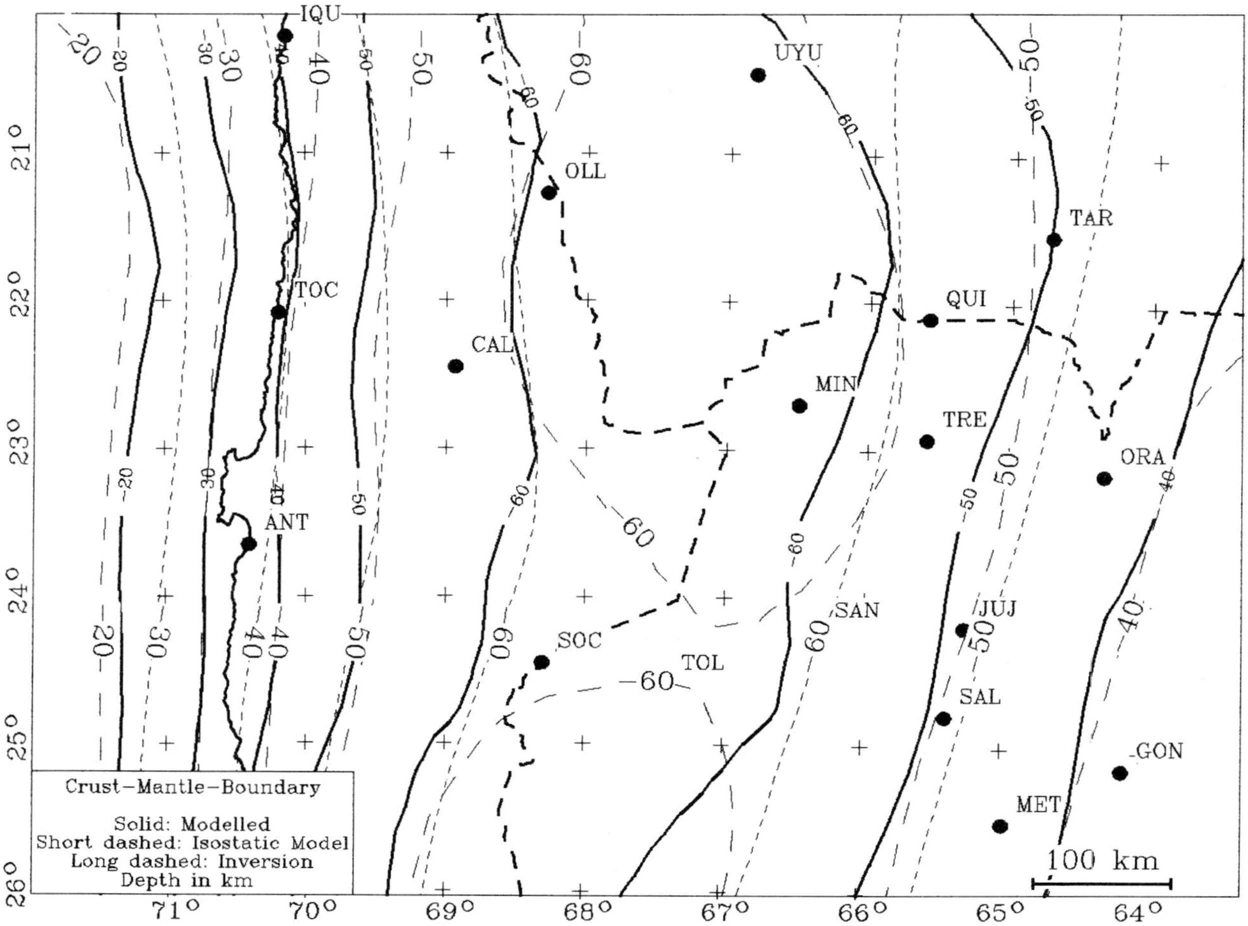

Fig. 4. Moho depths (in km) from the calculus of three independent methods. *Solid lines* 3D forward modelling; *short dashed lines* isostatic modelling; *long dashed lines* inversion with seismic constraints

gravimetric methods: isostatic modelling, constrained inversion, and 3D forward modelling (see Sect. 3). In all methods the crust and the mantle were assumed to be homogeneous. Seismic constraints serving inversion and forward modelling in the oceanic trench area were taken from Fisher and Raitt (1962). It can be assumed that they are representative of the entire oceanic trench. On the continent we took data from the seismic refraction studies carried out by Wigger (1988) and Wigger et al. (this Vol.).

South of Calama, along 69° W a crustal thickness of 55 km was found by gravity-based calculus. However, the velocity-depth function of a seismic profile from Calama towards the south indicates a Moho depth of some 70 km (Wigger 1988). Using this value in a 3D density model as a constraint, extreme positive anomalies of +50 mGal and even up to +170 mGal were obtained in a region 22° to 24° S and 67.5° to 70.5° W. No realistic explanation exists for these positive residuals in modelled gravity. This observation led us to assume that a layer with a thickness of approx. 15 km exists near the Moho. It consists of an anomalous relationship of "low velocity and high density". A very similar structure was observed under Vancouver Island in Canada (Riddihough 1979; Spence et al. 1985). Riddihough (1979) associated this phenomenon with the mantle wedge between the continental crust and the subducted oceanic lithospheric slab. As its components, he named mafic rocks from amphibolite to granulite facies.

Gravimetric Effect of the Subducted Nazca Slab

One of the main contributions to anomalous gravity caused by possible density inhomogeneities at the active continental margin is made by the subducted

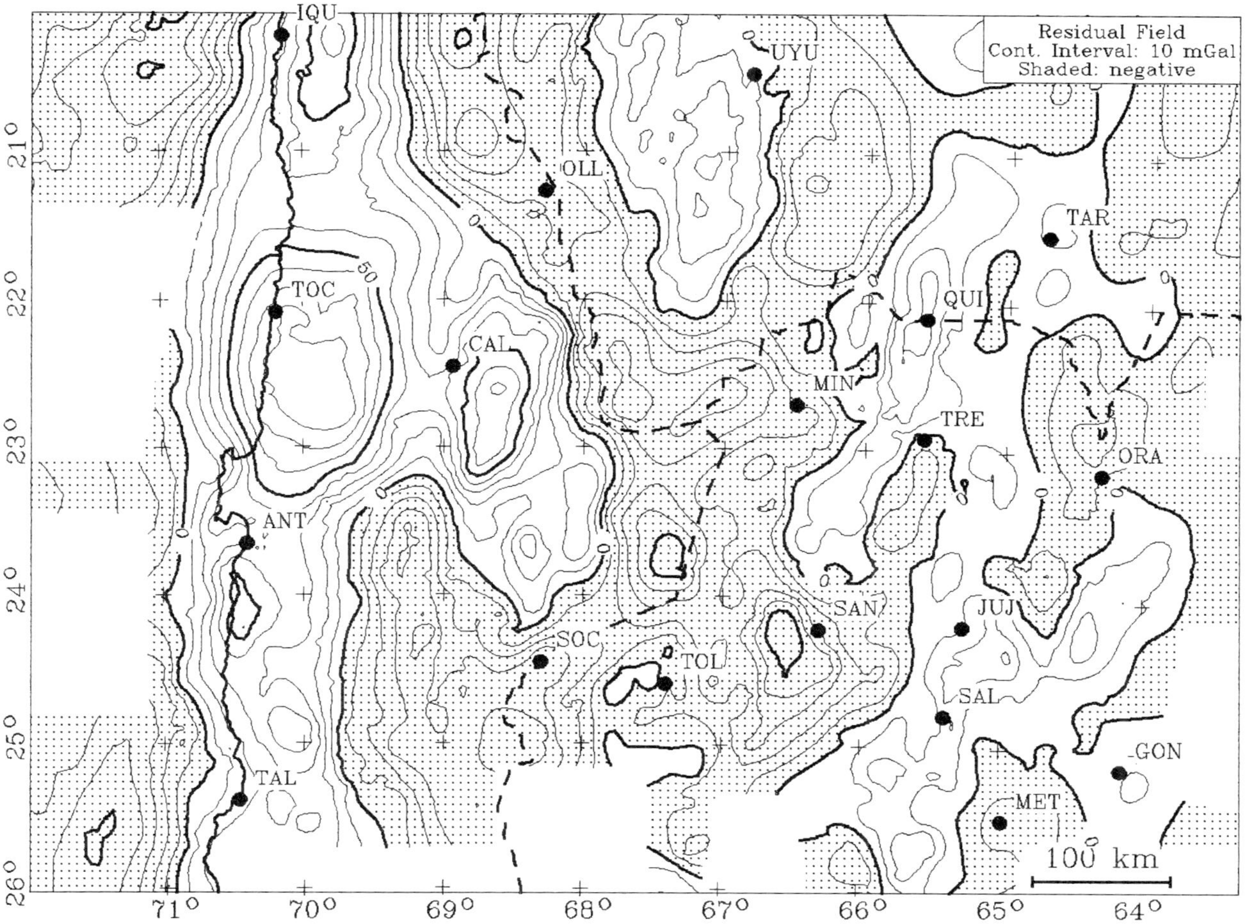

Fig. 5. Residual gravity field as inferred from contouring the differences of the inverted Moho gravity field and the measured Bouguer gravity. Reference level: 6 km a. s.l.

slab. Disregarding both the gravity effect of a possibly existing inhomogeneous mantle wedge and the effect of a possible "A-subduction" under the Eastern Cordillera, which would partially compensate the gravimetric effects of the Nazca slab (Fukao et al. 1989), the maximal gravity effect of the subducted Nazca slab can be deduced. This was done with the aid of (1) quadratic programming and (2) 3D forward modelling of the slab.

1. In the first case, the geometry of the oceanic lithospheric slab was modelled schematically with an angle that flattened out from 24° in the north to 20° in the south. A thickness of 8 km and adepth of 230 km of the oceanic plate were assumed in modelling; the model consists of 13 segments. Modelling such a "thin plate" is in accordance with the concepts of Grow and Bowin (1975) and Wyllie (1981). They assume that important petrophysical processes (dehydration of oceanic crust, phase transitions etc.) with corresponding changes in rock densities take place in the upper part of the slab. The densities of the individual segments were inverted by quadratic programming. Constraints were taken from the literature. Thus, the density contrast to the surroundings was positive and totalled a maximum of 0.5 g/cm^3. Both the isostatic anomaly (see map in Appendix) and the residual gravity (Fig. 5) were involved in this inversion process. Two of the segments located in the uppermost part of the slab beneath the Coastal Cordillera preserved positive density contrasts to the adjacent mantle. However, we know from combined seismic and gravity modelling (Heinsohn 1990; Strunk 1990) that positive gravity effects in the Coastal Cordillera are mainly caused by high density material in the

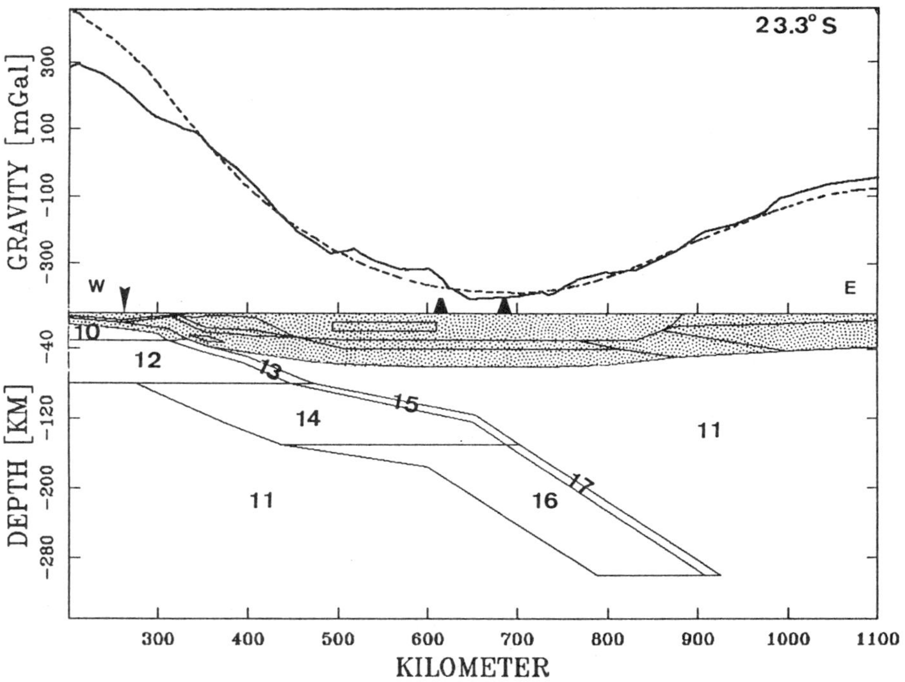

Fig. 6. Vertical cros section at 23.3° S latitude of the 3D forward model at the South American continental margin. For explanations, see text. *Solid line* Bouguer gravity; *dashed line* modelled gravity. *Numbers* refer to densities (see Sect. 3.)

Jurassic arc (for references, see Sect. 3). This means that the gravity effects of the Nazca slab contributed to the entire gravity field would be some 50 mGal.

2. In the second case, 3D gravity forward modelling was applied to combine data from earthquake studies (Buness et al. 1986; Cahill 1990) and seismic refraction surveys (Wigger 1988; Wigger et al. 1991; Wigger et al. this Vol.) with observations from structural geology and tectonics (e.g. Reutter et al. 1988). Figure 6 shows one vertical W-E section of the density model with corresponding gravity profiles. The part of the model under discussion is located between km 300 and 450 of the profile. Again, segmentation and density distribution of the oceanic plate were modelled after Grow and Bowin (1975). Modelled gravity was adjusted to the Bouguer anomaly (see Appendix). Following the results of 3D modelling, the gravity effect of the Nazca slab could be as much as approx. 55 mGal. This value returned from both:

(a) modelled gravity of the 8-km-thick oceanic lithospheric plates (segmentation numbers 13, 15 and 17 in Fig. 6) with density contrasts varying from +0.20 to +0.05 g/cm^3 in relation to mantle density;

(b) modelled gravity of a 75-km-thick plate (segmentation numbers 12 and 13, 14 and 15, and 16 and 17 in Fig. 6) with density contrasts of +0.03 g/cm^3 in the upper part to +0.01 g/cm^3 in the lower part of the slab.

Thus the results obtained contradicted the density model of the Nazca slab by Grow and Bowin (1975). It yielded a regional gravity field of more than +100 mGal or even up to +170 mGal when using the detailed plate geometry model with higher densities. Their density model is based mainly on petrological observations and on the temperature model of Toksöz et al. (1971). Our study showed that the temperature regime within the Nazca plate is cooler than was assumed by Toksöz.

In order to compensate for such a great gravity effect of Grow and Bowin's Nazca slab, the thickness of the continental crust under the Subandean Ranges and Chaco in the east would have to be approx. 10 km thinner. However, this observation is not supported by the latest seismic refraction data (Wigger et al., this Vol.) from the crust/mantle interface under the Bolivian Subandean Ranges.

3 Forward Modelling of Upper Crustal Structures

In this Section 3D forward gravity modelling is used to combine data from geophysical surveys such as seismic refraction (Wigger et al. 1991; Wigger et al. this Vol.), seismology and magnetotellurics (Schwarz et al. 1984, 1991) and observations from geology and tectonics (e.g. Scheuber et al. this Vol. and Omarini

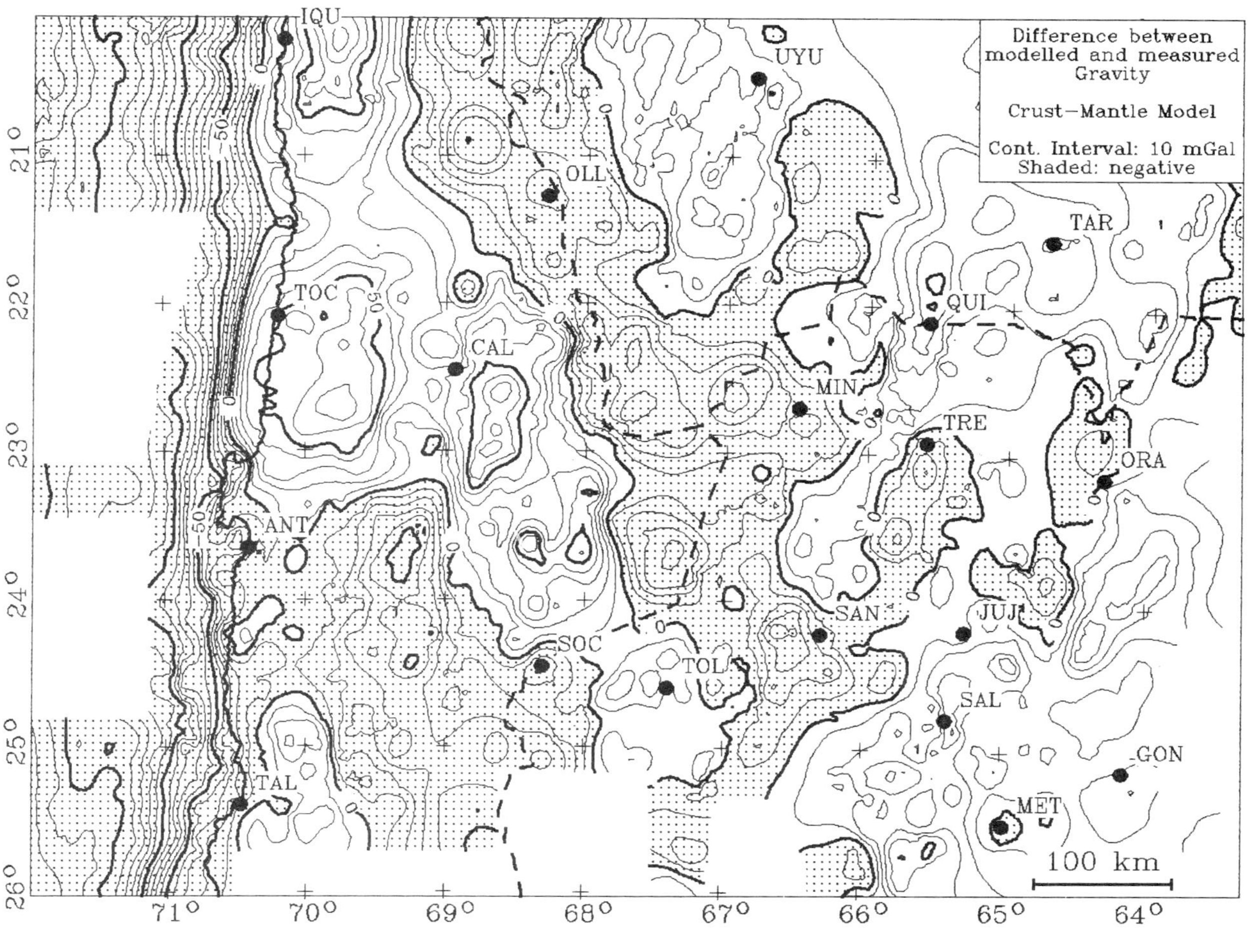

Fig. 7. Residual gravity as inferred from contouring the differences in gravity effects of the crust-mantle model and measured Bouguer gravity

et al. 1991). It must be emphasized that these data and observations are essential to forward modelling due to the ambiguity of potential field interpretations. In particular the results of seismic ray tracing were used as constraints to define layers in the crust, although there were some uncertainties in the results due to different interpretation techniques and personal experience. Here, we concentrate on the main aspects of gravity modelling in the surveyed area, for further information on the methods used and an extended interpretation of results, the reader is referred to Strunk (1990).

The 3D density model covered the entire region between 20° and 25°S as well as from 71° to 64°W. Figure 6 contains one vertical cross section at 23°S latitute to illustrate both the matched gravity curves in the upper box and crustal structures at the bottom. Gravity was calculated at stations which exactly corresponded to station coordinates of the gravity survey (as described in Sect. 1.2). The subsurface model contained a homogeneous crust (shaded area in Fig. 6), a 75-km-thick oceanic lithospheric slab, and a homogeneous mantle density. The differences from measured Bouguer gravity and modelled gravity are shown in Figure 7. Assuming the constraints used here were correct, this residual field should include only inner-crustal density inhomogeneities.

In general, significantly larger and stronger anomalies were found in the region west of the recent volcanic arc (68°W in Fig. 7) compared with those regions to the east. According to studies by Delleske (1989) and Götze et al. (1989), the latter can be largely explained by inhomogeneities in the upper crust of the Puna and Altiplano. However, for the western part of the model strong gravity residuals are

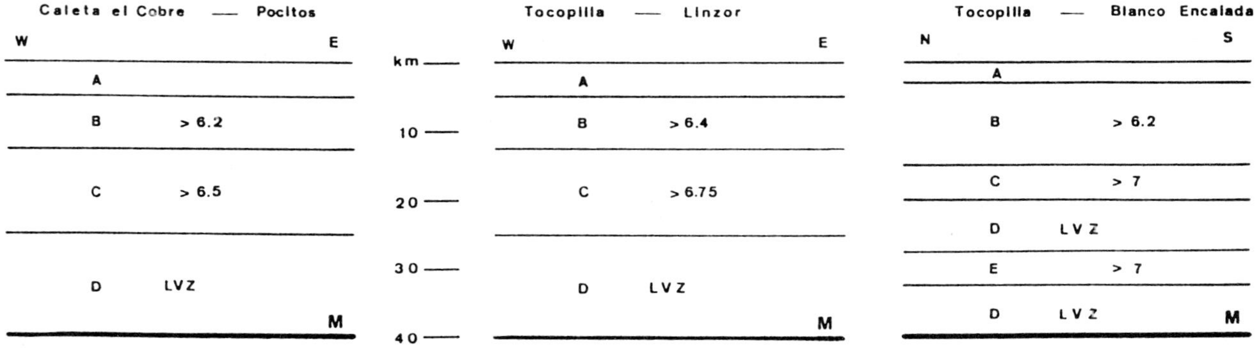

Fig. 8. Sketch of crustal velocities from ray tracing models in Chile. Layers with same *letters* have the same velocities; **A** < 6.0 km/s; **LVZ** low velocity zone. (Redrawn from an earlier stage of seismic data processing; Wigger et al. this Vol.)

caused by inhomogeneities of upper to middle crustal layers.

A few remarks concerning the various anomalies which are relevant for further studies must be made here:

1. From west to east we find in agreement with seismic studies (Fig. 8): an oceanic crust type (with relatively high densities; profile km 200 - 450, Fig. 9), an Andean crust type (with moderate densities; profile km 450 - 850), and a crust type from the Andean foreland (high densities; profile km 850 - 1100).
2. A clearly defined minimum of 50 mGal can be seen along the western margin of the Precordillera (between SOC and ANT in Fig. 7).
3. The Western Cordillera (central part in Fig. 7, following the Chilean-Argentinean/Boliviean border) with recent volcanism can be recognized by a chain of local gravity lows. This area corresponds to a zone of reduced electrical resistivity (Schwarz et al. this Vol.). We assume that fluids and partly molten material in the magma chambers of the active volcanic arc caused both low resistivity and density.
4. Most remarkable is the big gravity high between Calama (CAL) and the southern margin of the investigated area in the Argentinean Puna (TOL). This gravity anomaly causes positive values, even in the area of the volcanic arc.

As mentioned earlier, model densities were calculated from velocities as they result from seismic ray tracing and 1D inversion techniques. Figure 8

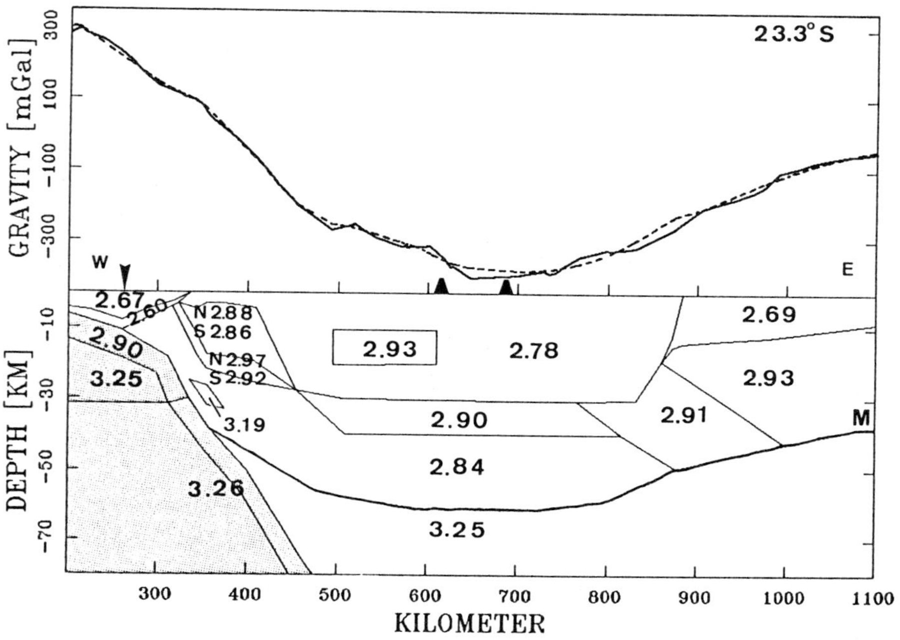

Fig. 9. Cross section through the 3D model of the crust. Densities in g/cm^3; vertical exaggeration: 2.6; *solid line* measured Bouguer gravity; *dashed line* modeled gravity. For further information, see Section 3

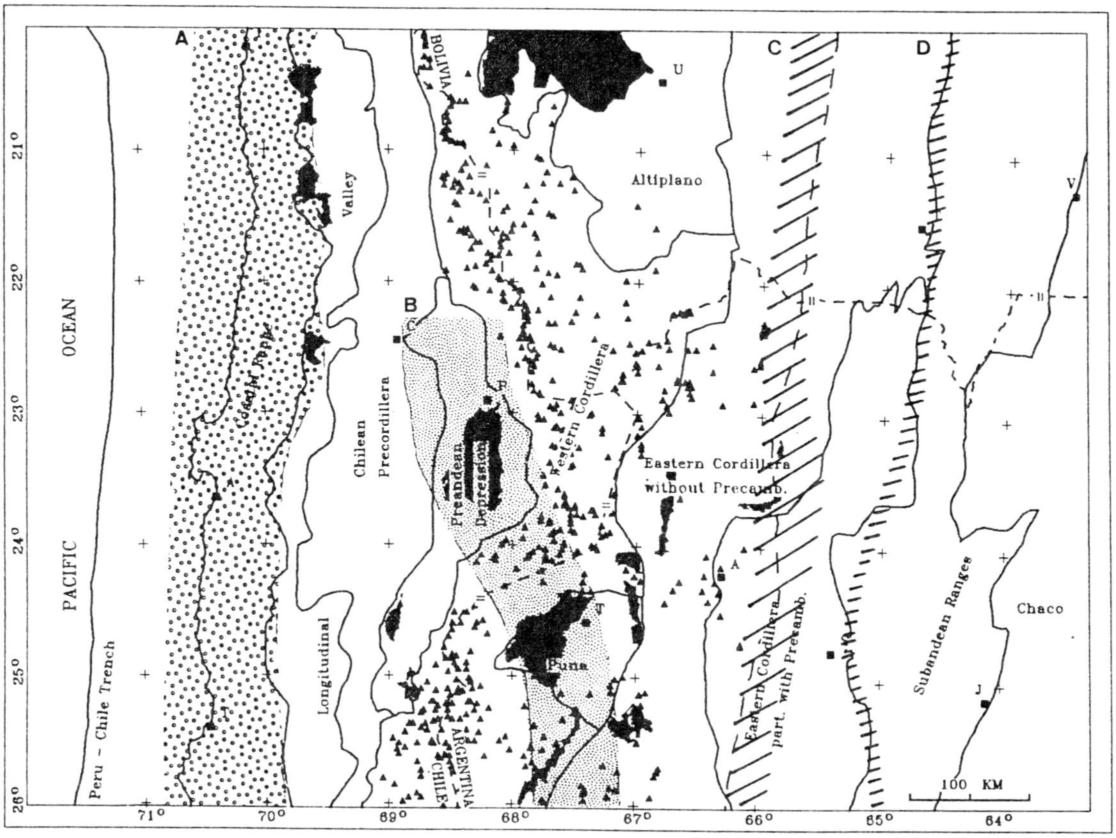

Fig. 10. Important geological units and boundaries and gravimetric features as derived from 3D modelling and data processing. ***A*** Jurassic magmatic arc; ***B*** gravity "terrane"; ***C*** eastern boarder of low-density lower crust; ***D*** transition zone between different crust types of the Eastern Cordillera and the Subandean foreland. *Dashed line* eastern boarder of Neogene to Holocene volcanism. (After Reutter et al. 1988)

gives a general view of the velocity structure used for density modelling. Figure 9 presents a cross section at the same latitude as before (Fig. 7). The measured Bouguer gravity ideally coincides in this profile with the modelled gravity. This model consists of (densities are given in g/cm^3):

- A homogeneous upper crust (2.78) down to a depth of 30 km beneath the Central Andean orogene.
- In near coastal areas structures of high density starting near the surface; these bodies consist of rocks of the Jurassic and/or Cretaceous magmatism. We distinguish in Figure 9 between densities in the northern part of this model (N 2.88) and (N 2.97) and those in the southern part (S 2.86 and S 2.92).
- A small layer of middle crust (2.9).
- A 20-km-thick lower crust with an anomalously low density (2.84), which is in agreement with seismic studies.
- The crust underneath the Andean foreland (Piedmont) which is formed by sediments and a Precambrian basement (2.69). Density information was taken from density logs of the Argentine State Oil Company (YPF pers. comm.) as well as from density determinations by Introcaso et al. (1987). The thickness of the sediments was modelled according to Omarini et al. (1991) and Lyon-Caen et al. (1985). Crust formed in the South American craton consists of 2.91/2.93 g/cm^3).
- Oceanic crust modelled in accordance with the results of seismic refraction studies by Fisher and Raitt (1962). Density values of 2.6 and 2.9 were

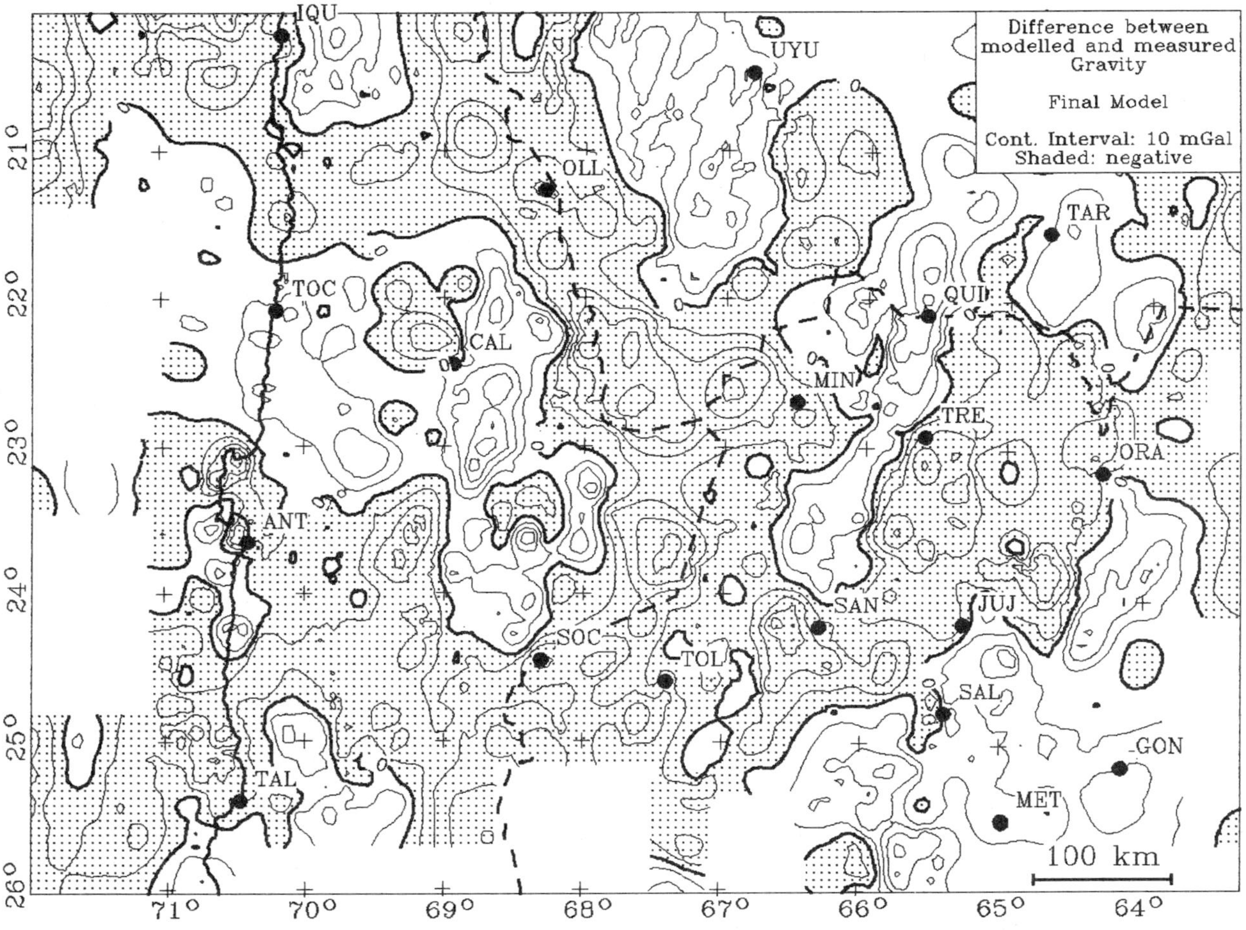

Fig. 11. Residual gravity as inferred from contouring the differences in gravity of the 3D model and the measured Bouguer gravity

used for oceanic sediments and the upper part of the oceanic slab respectively; 2.67 replaces the density for oceanic water in the ocean area.

- Upper mantle modelled by a standard density of 3.25; the downgoing slab has a density of 3.25 and 3.26.
- A density of 2.93 in the upper crustal part (profile km 500 - 600) marking the crustal block (see point 4 above) which causes the gravity high in Figure 7.

4 Discussion and Conclusions

A review of the Central Andes gravity field, which extends from the offshore trench zone to the Argentine Chaco, reveals how geotectonic units of the active South American continental margin have been changed in space and time. The essential results of quantitative gravity field analysis and 3D forward modelling as part of an integrated interpretation together with non-gravimetric studies will conclude this chapter.

The gravity field of the Central Andes consists of numerous local and regional anomalies which are caused by density inhomogeneities at different vertical depth levels and in rather complex horizontal positions. One of the most surprising results of this gravity study is the observation that many of the gravity anomalies do not fit recent geological trends in the region (e.g. the main north-south direction). These anomalies could be caused by Pre-andean remnants in the recent crust. Large anomalies (Figs. 7 and 11) even in the north-south direction point to lateral variation in the crustal structures from north to south. The filtered, long wavelength gravity field (represented in Fig. 4) reflects the gravity effect of flattening of the dropping oceanic slab. Closely linked to seismic ray tracing results, we identified a lower crust with anomalously low densities. However, this low density in the lower crust results in an extreme regional gravity low which does not fit the measured field. To compensate this negative effect, a zone of higher density in the middle crust was included in the model (Fig. 9). The existence of such a zone was found in the west- east refraction profile at 22° and at 24°S (Wigger et al. this Vol.).

Significant differences in the gravity field can be described in relation to the interpretation of filtered gravity fields in Section 2. In general, higher crustal densities characterize the northern part of Chile, lower densities are typical for the crust south approx. 23.5°S latitude. Also, flexural rigidity, estimated from isostatic gravity calculations, changes at nearly the same latitude from a "more rigid" behaviour in the north to a "softer" behaviour in the southern part. High-density rocks and formations at 22° S are located structurally higher and further to the west up to the Precordillera than structures at 24° S, where high densities in the upper crust extend to the eastern boundary of the Longitudinal Valley (Figs. 7 and 11). Another change in crustal density of both the Coastal Cordillera and the Longitudinal Valley from north to south can be identified at depths between 5 and 25 km by forward modelling. High density diminishes significantly in that region.

Figure 10 displays the geomorphological units (digitized after Reutter et al. 1988), volcanoes, salt lakes and some of the most important features and boundaries which can be determined from gravity field interpretation and modelling. For instance the projection of the eastern border of lower crust to surface (Fig. 10, hatched area C) limited the eastern extension of Neogenic to Holocene volcanism (Fig. 10, dashed line) described by Reutter et al. (1988). It seems that the position of extreme eastern volcanism is linked to the eastern extension of lower crust.

The most "mysterious", very puzzling anomaly extends as an extreme gravity high from the northern Cordillera Domeyko, via the Salar de Atacama Basin, to the Salar de Arizaro basin and further to the SSE (Figs. 7 and 10: area B). Its northern boundary seems to be controlled by a refraction profile at 22° S (Wigger et al. this Vol.). There are no indications for anomalous layers or structures that could effect the gravity field density as it was observed. To date its extreme southern extension is still uncertain since no geophysical surveys exist for the southern Puna. In its central part, however, high velocities at depths of approx. 10 km were recorded on the Chuquicamata profiles (Wigger 1988; Wigger et al. 1991). Further, the distribution of electrical resistivity supports the existence of an anomalous crustal structure (Schwarz et al. this Vol.). East of the gravity high a good conductor under the Western Cordillera was separated from an area of less conductivity corresponding to the gravity high area.

Analysing the power spectra of the gravity field, (positive) gravity sources were found near the surface at a depth of approx. 4 km as well as at a maximum depth of 24 km. Using this information as constraints, a homogeneous body extending to a depth of 10 to 20 km was included in the 3D forward model. This body is seen in Figure 9 in the upper crust. Figure 10 (area B) shows its projection on the surface (shaded area). A density contrast of +0.15 g/cm^3 was assumed for modelling and the resulting gravity field was eliminated from the Bouguer anomaly. In the thus modified gravity field (Fig. 11) local anomalies remain which are mainly caused by surface rocks: e.g. the negative anomalies of the Salar de Atacama and the Salar de Arizaro with their thick salt layers, or the ignimbrites along the recent arc which are marked by small lows in gravity. In contrast, the Permo-Triassic volcanic rocks and sediments located at the southern and eastern flanks of the Salar de Atacama give rise to positive anomalies.

The interpretation of this hitherto unknown structure in the Andean gravity field is not easy because no "visible" correlation with geological exposures and large-scaled outcrops of high density rocks exists, and the "gravimetric terrane" that causes the positive residuals is located at a depth of 10 to 20 km.

The complete elucidation of the gravity anomaly is still not possible. However, there are indications that we are dealing with a very old, at least Palaeozoic structure, whose position has not undergone any essential changes since the beginning of the Andean orogenic cycle. The position of recent volcanoes of the present magmatic arc was presumably influenced by this structure. Therefore, the volcanic belt runs first eastward in the direction of the anomaly, before it changes from NW to SW at 24.5°S, then perpendicularly crosses the unknown crustal body (Fig. 10). The trend of the gravimetric contour lines (Fig. 7) correlates well with the general direction of a complex of Palaeozoic granitoid rocks which Palma et al. (1986) described as "faja eruptiva de la Puna occidental". However, granitoids do not have a high density: thus we have to assume that they are linked to a larger structure of regional importance. Therefore, we prefer the interpretation of Breitkreuz et al. (1989) and Bahlburg (pers. comm.), and interpret the gravity high as an Ordovician active continental margin with subduction to the east. The Ordovician magmatic arc is expected to be in the region of the "faja eruptiva de la Puna occidental", described by Palma et al. (1986). Accordingly, large amounts of Ordovician basic intrusives, which are not visible from the surface, could cause the anomaly. A similar direction is seen in the ultramafic exposures extendinging from the south to Salar de Antofalla (67.5°W, 25.5°S), which were interpreted by Allmendinger et al. (1983) as the upper part of an ophiolithic sequence. In contrast to the basic Ordovician intrusives, a density similar to ultramafites would explain the gravity anomaly in this case. Ramos (1988) relates the ophiolithic sequence occurring to the southwest of the Puna with a late Hercynian backarc basin, whereby he postulated that the Arequipa-Antofalla craton was connected with the South American continent. In accordance with this interpretation, the positive gravity anomaly could be explained as a remnant of this craton.

Acknowledgements. First, we would like to thank G. Chong (Antofagasta, Chile), J.G. Viramonte, J.A. Salfity, R.H. Omarini (all of Salta, Argentina), A.A. Cerrato (Buenos Aires), M.Araneda and E. Kausel (Santiago, Chile) and M. Wagener (Scherfede, Germany). Without their scientific and logistic help it would not have been possible to realize the fieldwork under the extreme conditions of the beautiful High Andes. The availability of maps and geodetic information made possible by the following institutions is gratefully acknowledged: Institutos Geográficos Militares in Santiago, Buenos Aires and La Paz, Yacimientos Petrolíferos Fiscales of Argentina and Bolivia and the Servicios Geológicos of Chile and Argentina. This work was financially supported by the Deutsche Forschungsgemeinschaft, which is also gratefully acknowledged.

References

Allmendinger RW, Ramos VA, Jordan TE, Palma MA, Isacks BL (1983) Paleogeography and Andean structural geometry, northwest Argentina. Tectonics 2: 1-16

Banks RJ, Parker RL, Huestis SP (1977) Isostatic compensation on a continental scale: local versus regional mechanism. Geophys J R Astr Soc 51: 431-452

Breitkreuz C, Bahlburg H, Delakowitz B, Pichowiak S (1989) Paleozoic volcanic events in the Central Andes. J S Am Earth Sci 2 (2): 171-189

Buness F, Wetzig E, Wigger P (1986) Seismologische Studien in den zentralen Anden. Berl geowiss Abh (A) 66: 5-29

Cahill T (1990) Earthquakes and tectonics of the Central Andean subduction zone. PhD Thesis, Cornell University, Ithaca, New York

Cordell L (1985) Techniques, applications and problems of analytical continuation of New Mexico aeromagnetic data between arbitrary surfaces of very high relief. Proc Int Meet on Potential fields in rugged topography, Inst de Géophysique Université de Lausanne, Lausanne, pp 96-101

Delleske M (1989) Schwerefeldinterpretation der argentinischen Puna und Ostkordillere mit Hilfe dreidimensionaler Modellrechnungen. Diplomarbeit, FU Berlin (unpubl)

Fisher RL, Raitt RW (1962) Topography and structure of the Peru-Chile trench. Deep-Sea Res 9: 423-443

Fukao Y, Yamamoto A, Kono M (1989) Gravity anomaly across the Peruvian Andes. J Geophys Res 94: 3876-3890

Gebrande H (1982) In: Angenheister G (ed) Landolt-Börnstein, Group 5, vol 1. Physical properties of rocks. Springer, Berlin Heidelberg New York, pp 1-96

Götze H-J (1976) Ein numerisches Verfahren zur Berechnung der gravimetrischen und magnetischen Feldgrößen für dreidimensionale Modellkörper. Dissertation, TU Clausthal, Clausthal-Zellerfeld

Götze H-J, Lahmeyer B (1988) Application of three-dimensional interactive modeling in gravity and magnetics, Geophysics 53 (8): 1096-1108

Götze H-J, Strunk S, Schmidt S (1988) Central Andean gravity field and its relation to crustal structures. In: Bahlburg H, Breitkreutz C, Giese P (eds) The southern central Andes. Lecture Notes in Earth Science 17. Springer, Heidelberg New York, pp 199-208

Götze H-J, Schmidt S, Schuricht B (1989) 3D modeling of YPFB gravity data of the Altiplano area. Interner Bericht, FU Berlin, Berlin (unpubl)

Götze H-J, Lahmeyer B, Schmidt S, Strunk S, Araneda M (1990) A new gravity data base in the Central Andes (20° - 26°S). EOS Trans, Am Geophys Union 71 (16): 401, 406-407

Götze H-J, Lahmeyer B, Schmidt S, Strunk S, Araneda M, Chong G, Viramonte J (1991) The gravity data base of the transect compilation. In: Omarini R, Götze H-J (eds) (1991) Global geoscience transect 6: Central Andean transect, Nazca Plate to Chaco Plains, SW Pacific Ocean, N Chile and N Argentina. Am Geophys Union, Washington D C, pp 20-23

Grow JA, Bowin CO (1975) Evidence for high-density crust and mantle beneath the Chile trench due to the descending lithosphere. J Geophys Res 80: 1449-1458

Heinsohn W-D (1990) Krustenseismische Untersuchungen in der Küstenkordillere von Nordchile und Aufbau und Realisierung einer automatischen Feldapparatur. Diplomarbeit, FU Berlin

Introcaso A, Lion A, Ramos VA (1987) La estructura profunda de las sierras de Cordoba. Rev Asoc Geol Argent XLII (1-2): 177-187

Lahmeyer B (1988) Gravity field continuation of irregularly spaced data using least squares collocation. Geophys J 95: pp 123-134

Lahmeyer B (1990) Anwendungen der schnellen Fouriertransformation und der quadratischen Programmierung bei der Interpretation von Schwerefeldern. Dissertation, FU Berlin

Lyon-Caen H, Molnar P, Suarez G (1985) Gravity anomalies and flexure of the Brazilian Shield beneath the Bolivian Andes. Earth Planet Sci Lett 75: 81-92

Oldenburg DW (1974) The inversion and interpretation of gravity anomalies. Geophysics 39: 526-536

Omarini R, Reutter K, Bogdanic T (1991) Geological development and structures. In: Omarini R, Götze H-J (eds)

Global geoscience transect 6: Central Andean transect, Nazca Plate to Chaco Plains, SW Pacific Ocean, N Chile and N Argentina, Am Geophys Union, Washington D C, pp 5-12

Palma MA, Parica PD, Ramos VA (1986) El Granito Archibarca: su edad y significado tectonico. Rev Asoc Geol Argent 41 (3/4): 414-419

Parker RL (1972) The rapid calculation of potential anomalies. Geophys J R Astr Soc 31: 447-455

Ramos VA (1988) Late Proterozoic - Early Paleozoic of South America - a collisional history. Episodes 11 (3): 168-173

Reutter K J, Giese P, Götze H-J, Scheuber E, Schwab K, Schwarz G, Wigger P (1988) Structures and crustal development of the Central Andes between 21° and 25°. In: Bahlburg H, Breitkreutz C, Giese P (eds) The southern central Andes. Lecture Notes in Earth Science, 17. Springer, Berlin Heidelberg New York, pp 231-261

Riddihough RP (1979) Gravity and structure of an active margin - British Columbia and Washington. Can J Earth Sci 16: 350-363

Schwarz G, Haak V, Martinez E, Bannister J (1984) The electrical conductivity of the Andean crust in northern Chile and southern Bolivia as inferred from magnetotelluric measurements. J Geophys 55: 169-178

Schwarz G, Rath V, Krüger D (1991) A cross-section of electrical resistivity structure along the Central Andean transect. In: Omarini R, Götze H-J (eds.) Global geoscience transect 6: Central Andean transect, Nazca Plate to Chaco Plains, SW Pacific Ocean, N Chile and N Argentina. Am Geophys Union, Washington D C, pp 24-25

Spence GD, Clowes RM, Ellis RM (1985) Seismic structure across the active subduction zone of western Canada. J Geophys Res 90: 6754-6772

Strunk S (1990) Analyse und Interpretation des Schwerefeldes des aktiven Kontinentalrandes der zentralen Anden (20°-26° S). Dissertation, FU Berlin

Toksöz MN, Minear JW, Julian BR (1971) Temperature field and geophysical effects of a downgoing slab. J Geophys Res 76: 1113-1138

Wigger P (1988) Seismicity and crustal structure of the central Andes. In: Bahlburg H, Breitkreutz C, Giese P (eds) The southern central Andes. Lecture Notes in Earth Science 17. Springer Berlin Heidelberg New York, pp 209-229

Wigger P, Araneda M, Giese P, Heinsohn W-D, Röwer P, Scmitz M, Viramonte J G (1991) The crustal structure along the Central Andean transect derived from seismic refraction investigations. In: Omarini R, Götze H-J (eds) Global geoscience transect 6: Central Andean transect, Nazca Plate to Chaco Plains, SW Pacific Ocean, N Chile and N Argentina. Am Geophys Union, Washington D C, pp 13-19

Wyllie PJ (1981) Plate tectonics and magma genesis. Geol Rundsch 70 (1): 128-152

Variation in the Crustal Structure of the Southern Central Andes Deduced from Seismic Refraction Investigations

PETER J. WIGGER, MICHAEL SCHMITZ,
MANUEL ARANEDA, GÜNTHER ASCH, SIBYLLE BALDZUHN, PETER GIESE,
WOLF-DIETER HEINSOHN, ELOY MARTÍNEZ, EDGAR RICALDI,
PETER RÖWER and JOSÉ VIRAMONTE.

Abstract. A net of mainly reversed seismic refraction profiles has been measured in the years 1987 and 1989 in northern Chile, northern Argentina and southern Bolivia to investigate the crustal structure beneath the Andes from the coastal range to the Andean foreland. Regular blasts of the Chuquicamata copper mine, sea shots in the Pacific Ocean and land shots in Chile, Argentina and Bolivia were used as energy sources. One- and two-dimensional model calculations were applied to the data. Strong west-east as well as north-south variations in the crustal structure are observed which allow one to distinguish mainly three different crustal blocks.

The forearc block (Coastal Cordillera to Precordillera) increases in thickness from 40 km at the coast to 70 km under the Precordillera. A discontinuity between 20 km (west) and 35 km (east) depth is interpreted as the base of the Jurassic crust. The Moho of the oceanic crust is proven at 40 km depth under the Coastal Cordillera. The depth range from 20-30 (west) to 35-70 km (east) characterized by alternation of high and low velocities is interpreted as a mixture of continental crustal and mantle material and Nazca plate derivates. The average velocities of 6.6 km/s beneath the Coastal Cordillera to 6.2 km/s (Precordillera) suggest a rigid behaviour for this forearc block.

For the Western Cordillera and the western Altiplano the data indicate a strong, vertically structured, about 70-km-thick crust. Only at the southern W-E profile are clear signals observed from the lower crust and uppermost mantle. At the other related profiles signals from this depth range are missing. These data indicate a weakened crust with, in part, strongly reduced velocities, also suggesting partially melted zones.

The backarc crust (eastern part of the Altiplano, the Eastern Cordillera and the Subandean Ranges) shows a rigid behaviour. Here, continental crust is overthrusted over its foreland. The data evidence the base of the crust in a rather shallow position at 25-30 km depth beneath the Eastern Cordillera. The underlying upper and middle crusts are characterized by a low velocity zone. Total crustal thickness decreases from 70 km under the Altiplano to 40 km at the Andean foreland, already representing shield conditions.

1 Introduction

The first investigation of the crustal structure of the Central Andes by means of seismic refraction measurements was carried out by the Carnegie group during the "International Geophysical Year" 1957 using blasts of the Toquepala copper mine in southern Peru and the Chuquicamata copper mine in northern Chile (Tatel and Tuve 1958). The overall crustal thickness was computed at 56 km for the Precordillera and 70 km for the southern Bolivian Altiplano, indicating an average P-wave velocity of 6.6 km/s (Woollard 1960). In 1968 a second experiment by the same group took place with two shot points in the Langui y Layo Lake in Peru and Talacocha Lake in Bolivia. From this a crustal thickness beneath the Bolivian Altiplano of 76 km with an average velocity of 6.4 km/s (Ocola et al. 1971) was derived. Low-velocity zones at about 10 and 30-40 km depth are described by Ocola and Meyer (1972) and explained as sources of acidic and basic lavas. For the lower crust between the coast and the Brazilian Shield Ocola and Meyer (1973) presented P-wave velocity values (v_P) between 6.8 and 7.0 km/s and Fisher and Raitt (1962) found velocities of 6.6 to 7.0 km/s for the oceanic crust off Antofagasta.

Correspondence to: P. Wigger, Fachrichtung Geophysik, Freie Universität Berlin, Malteserstr. 74-100, D-1000 Berlin 46

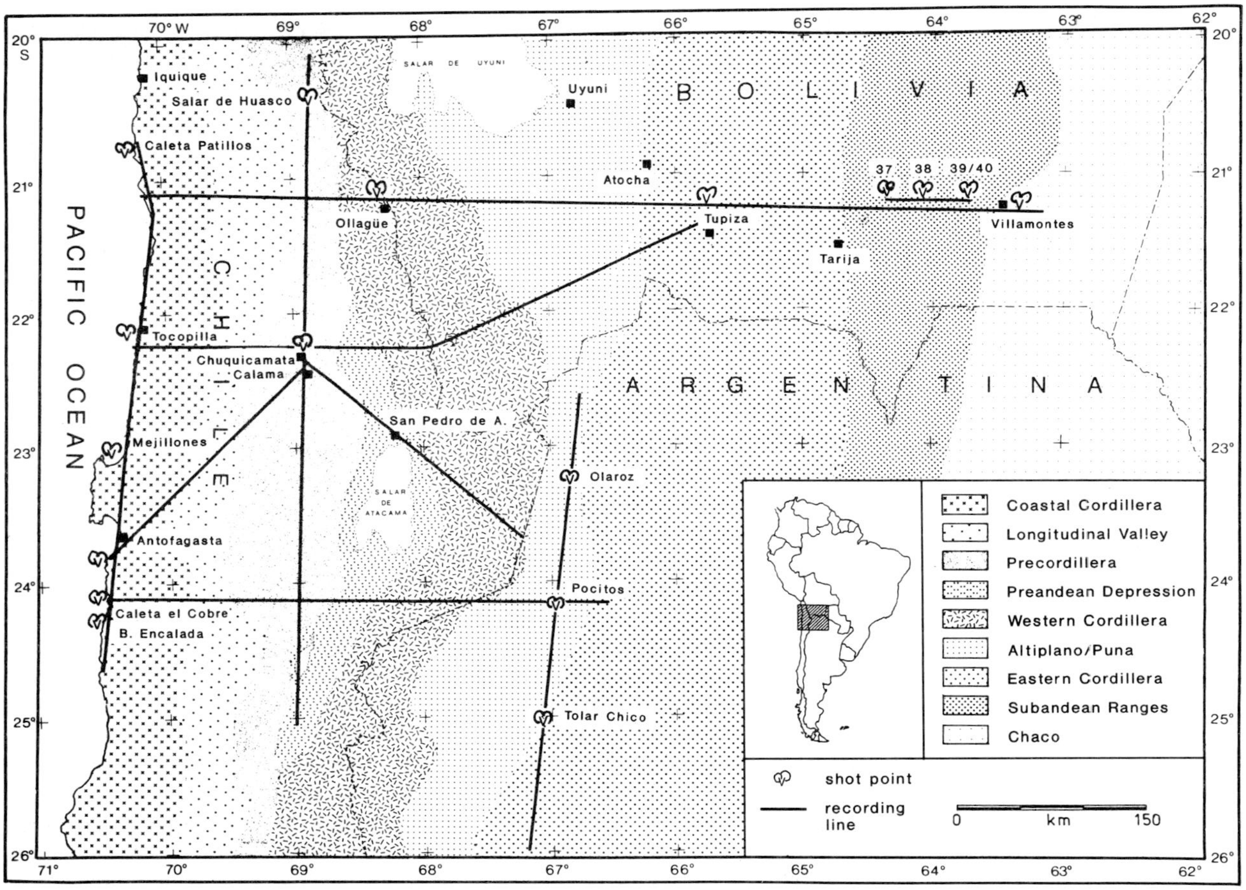

Fig. 1. Position map of the crustal seismic recording lines and shot points in the southern Central Andes observed until 1989. The morphostructural units are drawn after Reutter et al. (1988)

The results of the first seismic measurements by the Berlin research group "Mobility of Active Continental Margins" in northern Chile and southern Bolivia in 1982 and 1984, indicated a discontinuity at about 30 km depth with a velocity of 7.3 km/s below the Chilean Precordillera. This discontinuity was interpreted as palaeo-Moho of Jurassic age (Wigger 1986, 1988). A velocity of 6.5 km/s was calculated in the lower crust (30-40 km depth) beneath the Precordillera with a thickness of up to 40 km. The data were obtained by unreversed profiles, using the blasts of the Chuquicamata copper mine. Thus the observed velocities could be affected by dip influences.

The aim of the seismic experiment, carried out in 1987 and 1989 and presented in this chapter, was to observe a net of reversed profiles which cover all the main morpho-structural units between the Pacific Ocean and the Andean foreland in the east to obtain a reliable, rough model of the crustal structure and its variations in the southern Central Andes. The distribution of the recording lines and the shot points is given in the position map (Fig. 1), which also indicates the main morpho-structural units of the Andes. A description of the covered morpho-structural units, including the Coastal Cordillera, Longitudinal Valley (Pampa del Tamarugal), Precordillera, Western Cordillera, Altiplano and Puna, Eastern Cordillera and Subandean Ranges (in west-east order), the tectonic setting and the development of this region is given e.g. by Reutter et al. (1988) and Scheuber et al. (this Vol.).

The profile net covers one of the strongest, negative Bouguer anomalies on earth (-440 mGal) and a positive residual anomaly in the area of the Pre- and Western Cordillera at about 23°S (Götze et al. 1988, this Vol.; Strunk 1990). The gravity data already indicate an increasing thickness of the continental crust from the Coastal Cordillera to the Western Cordillera. The above-mentioned research group also carried out the gravity investigations as well as the magnetotelluric (MT) studies. The MT studies (Schwarz et al. 1984, 1986, this Vol.) included the study of layers and zones of extremely high electrical conductivities at depths of about 10 km beneath the Pre- and Western Cordillera and at 30-40 km beneath the Altiplano.

2 Data Acquisition and Interpretation Methods

The seismic refraction experiments were carried out in two periods, from August to October 1987 and from April to June 1989. The total number of observed record sections is 31 and the sum of maximum recording distances of the 31 record sections amounts to more than 7000 km of observed lines. As signal sources sea shots off the Chilean coast, regular blasts of mines and specially prepared borehole shots were recorded in Chile, Bolivia and Argentina (Fig. 1).

2.1 The Shot Points

The 1987 shots in the Pacific Ocean, close to the Chilean coast, were fired on the sea bottom from ships of the Chilean Navy at depths between 57 to 80 m. The charges varied between 50 and 250 kg of explosives (Pentolita) for the coast-parallel profiles in Chile and the reversed profiles to Chuquicamata and Pocitos (see Table 1; Wigger et al. 1991). The land shots located in salt lakes on the Argentine Puna (Pocitos, Olaroz and Tolar Chico) were fired in a pattern of 10 to 20 holes of 30 m depth, the charges varied between 500 and 1100 kg of explosive material (Pegagel). The profiles referring to Chuquicamata used the regular blasts of the copper mine with charges of up to 2 tons of explosives for the first hole. The total charges of the blasts could reach more than 200 tons each, ordered in lines with time delays of up to 35 ms between each line of boreholes.

The land shots in Chile (1989) were fired in salt lakes of the Western Cordillera (Salar de Huasco and Ollagüe); the shot points in Bolivia were located in the Eastern Cordillera (Tupiza) and the Chaco (Villamontes) and also fired in boreholes; the charges varied between 700 and 2000 kg Pentolita for the land shots and 250 to 450 kg for the sea shots in the

Table 1. Shot point data of the seismic experiment in 1989 and of the profile Tocopilla-Chuquicamata recorded in 1987.

Shot number, location, recording direction (type of shot)	Coordinates (latitudes, longitudes)	Elevation /depth charge	Date, time LT (h:m:s,ms)
02/05 Chuquicamata - east/west (mine, first hole)	22°16'46",4 S 68°54'16",7 W	2780 m 1000 kg	28.8.1987 13:00:01,567
02/05 Chuquicamata - east/west (mine, first hole)	22°17'19",0 S 68°54'06",7 W	2463 m 800 kg	29.8.1987 13:02:22,617
02/05 Chuquicamata - east/west (mine, first hole)	22°16'47",4 S 68°54'08",2 W	2619 m 950 kg	10.9.1987 13:02:44,673
06 Tocopilla - east (sea)	22°09'11",3 S 70°13'54",9 W	-57 m 200 kg	29.8.1987 13:16:59,714
18 Chuquicamata - north (mine, first hole)	22°16'35",8 S 68°53'44",8 W	2645 m 1400 kg	18.4.1989 12:57:22,490
19 Chuquicamata - north (mine, first hole)	22°17'30",0 S 68°54'00",0 W	2400 m 1200 kg	18.4.1089 12:57:54,850
20 Chuquicamata - north (mine, first hole)	22°16'36",2 S 68°54'08",5 W	2800 m 2100 kg	15.5.1989 12:59:27,560
21 Salar de Huasco - south (drill holes)	20°17'38",5 S 68°49'09",4 W	3794 m 800 kg	22.4.1989 13:15:00,765
22 Ollagüe - west (drill holes)	21°12'39",5 S 68°16'28",0 W	3716 m 700 kg	30.4.1989 11:31:59,580
25 Caleta Patillos - south (sea)	20°54'55",2 S 70°09'42",5 W	-62 m 250 kg	08.5.1989 14:32:00,800
27 Tupiza - west (drill holes)	21°19'51",1 S 65°36'08",0 W	3480 m 2000 kg	17.5.1989 11:01:00,030
28 Chuquicamata - north (fan) (mine, first hole)	22°17'29",3 S 68°54'13",8 W	2500 m 1200 kg	17.5.1989 12:49:52,175
29 Chuquicamata - north (fan) (mine, first hole)	22°17'12",1 S 68°54'04",1 W	2411 m 300 kg	17.5.1989 12:52:20,510
30 Tupiza west/east (drill holes)	21°19'50",2 S 65°36'07",7 W	3477 m 100 kg	30.5.1989 16:15:59,948
32 Tupiza - west (drill holes)	21°19'36",1 S 65°36'06",0 W	3474 m 1200 kg	04.6.1989 11:01:00,000
33 Ollagüe - east (drill holes)	21°12'37",6 S 68°16'22",8 W	3716 m 1050 kg	05.6.1989 11:01:00,400
34 Tupiza - east (drill holes)	21°19'25",6 S 65°36'13",5 W	3465 m 1100 kg	11.6.1989 11:01:00,045
35 Caleta Patillos - east (sea)	20°42'45",2 S 70°13'23",7 W	-70 m 450 kg	12.6.1989 11:30:59,913
36 Villamontes - west (drill holes)	21°19'22",6 S 63°17'49",6 W	346 m 1200 kg	13.6.1989 11:02:00,955
37 Narvaez - east (drill holes)	21°27'33",0 S 64°20'26",2 W	2100 m 200 kg	18.6.1989 08:01:59,930
38 Tacuarandi west/east (drill holes)	21°26'29",6 S 64°00'39",7 W	900 m 100 kg	18.6.1989 13:02:00,000
39 Palos Blancos - west (drill holes)	21°25'13",0 S 63°45'29",6 W	730 m 100 kg	18.6.1989 17:02:00,130
40 Palos Blancos - west (drill holes)	21°25'13",0 S 63°45'29",6 W	730 m 100 kg	18.6.1989 17.13:59,870

Pacific Ocean (Caleta Patillos, Tocopilla, Mejillones, Antofagasta, Caleta El Cobre and Blanco Encalada). Charges of the blasts in the Subandean Ranges for the short profiles up to 70 km length (shot point numbers

Table 2. List of participants of the seismic refraction experiments, 1987 and 1989.

Fachrichtung Geophysik im Institut für Geologie, Geophysik und Geoinformatik, Freie Universität Berlin:

GÜNTHER ASCH; JOACHIM BARTZ; SIBYLLE BALDZUHN; HERMANN BUNESS; WOLF-DIETER HEINSOHN; SEBASTIAN IBBEKEN; UWE KASTENS; CARYL MICHAELSON; JOHANNES PALMER; VOLKER RATH; PETER RÖWER; MICHAEL SCHMITZ; KARL-GEORG SCHÜTTE; GERHARD SCHWARZ; PETER J. WIGGER.

Institut für Geodäsie und Photogrammetrie, Technische Universität Berlin:

JÜRGEN KLOTZ.

Lehrgebiet Angewandte Geophysik, RWTH Aachen:

NORBERT OCHMANN.

Departamento de Geología y Geofísica, Universidad de Chile, Santiago:

MANUEL ARANEDA; ANIBAL DAVANZO; OMAR QUEZADA; HEKTOR ZUÑIGA;

Departamento de Geología, Departamento de Física, Universidad Catolica del Norte, Antofagasta:

JORGE ARAYA; ALEXIS CORREA; JUÁN CATALAN FLORES; MARCELO GOMEZ FLORES; MARKO GOMEZ DEL VALLE; MIGUEL GUERRERO; ENRIQUE GUTIERREZ; GONZALO MENDOZA; MANUEL OLCAY GONZALES; JORGE LUIS OSORIO; MARIA ISABEL RAQUEO MORALES.

Departamento de Geología, Universidad Nacional de Salta:

RAMÓN EDUARDO BARRELIER; LUIS ALBERTO FAVA; JOSÉ IGNACIO FERETTI; RAÚL EDUARDO GONZALEZ; MARCELO RAMÓN OLAÑETA; CARLOS MARCELO PERALTA; JORGE GENARO TORRES.

Instituto de Investigaciónes Físicas, Universidad Mayor de San Andrés, La Paz:

JOSÉ LUIS ANDO BUSTAMANTE; JHONNY EVER VICTORIA PESTANAS; EDGAR RICALDI YARVI; VICTOR RICARDO TARQUINO FLORES; HERNÁN URIBE ZEBALLOS; GERMÁN UZQUIANO ESPINOZA; FELIX ZENÓN ALIAGA QUISBERT.

Observatorio San Calixto, La Paz:

RENÉ RODOLFO AYALA SÁNCHEZ.

Instituto Cultural Boliviano-Aleman, Tarija:

THOMAS FRANK.

Servicio Nacional de Geología y Minería, Santiago de Chile:

BRIGADA DE PERFORACIONES (YOYO).

Armada de Chile:

SHIPS "RANA KAO", "LAUTARO" AND "RANCAGUA" AND THEIR TRIPULATIONS.

Yacimientos Petrolíferos Fiscáles Bolivianos, Santa Cruz:

NINO GAITE AND THE BRIGADA 22.

Yacimientos Petrolíferos Fiscáles, San Pedro, Argentina:

BRIGADA DE PERFORACIONES.

37-40) varied between 100 and 200 kg Pentolita. All specific shot point data of the 1989 experiment are listed in Table 1.

For the time-break recordings two independent instruments with a high-precision quartz clock were used. The clocks were adjusted to the radio time signal WWV. The time signals were recorded together with an electrical impulse of the blast (inductive current transformator) on magnetic tape. Furthermore, the shot times were controlled by recordings by 2-Hz geophones or a hydrophone.

2.2 Equipment and Recordings

For the recordings along the profiles, 35 instruments mainly of type MARS 66 were available. Twenty-one of them, one-component MARS stations with 2-Hz vertical geophones (MARK L4), were able to record in an automatic operation mode as well. The 14 three-component instruments with one vertical and two horizontal components had to be operated manually. The seismic signals and the time signal were frequency modulated (FM) and stored on 1/4" magnetic tapes. Some of the three-component stations which were not type MARS 66 operated with pulse code modulation (PCM). The automatic stations were equipped with a microprocessor-controlled quartz clock (ACT) which also provided the time signal (Heinsohn 1990). Before and after recordings the clocks were compared with external time signals (WWV). The controlled time deviations of the clocks varied between 30 and 50 ms/day; a time error of about 100 ms was observed for some recordings due to fieldwork conditions.

The recordings were carried out by seven groups, each of them including one South American and one German colleague (see Table 2), who operated three automatic and two manual stations. Observation point intervals mainly varied between 5 and 10 km; the spacing for the short profile in the Subandean Ranges was only 2 km. Field playback instrumentation was used to check the recording instruments and data quality regularly after each shot.

Geographical coordinates could be selected from topographic maps (scale 1:50 000) and partly by observations with a Global Positioning System (GPS) receiver (MX 4400). The accuracy of most site coordinates is better than 50 m. An error of 100 to 200 m for only a few locations must be taken into account. Thus, the equivalent time error amounts maximally to 20-30 ms based on an average crustal velocity of 6 km/s.

2.3 Data Processing and Corrections

The field data of FM and PCM type were converted to digital data with a sample rate of 100 Hz/channel. Corrections were applied with respect to time deviations observed on the clock control records and with regard to the extremely varying topography severely disturbing one-dimensional model calculations. To improve the identification and correlation of the particular wave groups, different band-pass filters were applied to the seismograms, and record sections with different reduction velocities were plotted. However, all the record sections presented here are without topographic corrections and have a reduction velocity of 6 km/s; the amplitudes are trace normalized. A complete set of the southern Central Andean record sections was compiled by Schmitz et al. (1990).

2.4 Interpretation Methods

The first and most important step concerning the interpretation of the record sections is the picking and correlation of the seismic arrivals. This paper deals with P-waves only. S-waves were generated mainly by the shots in the Pacific Ocean and the Chuquicamata copper mine. An example of S-wave arrivals is shown in Fig. 3. Due to the recording intervals of more than 5 km, group correlation was applied. The principles of group correlation are given by Giese (1976a).

The correlated travel-time curves were inverted to a one-dimensional (1D) velocity-depth function v(z). For this procedure the Herglotz-Wiechert method (Wiechert 1910) with an extension for low-velocity zones (Gerver and Markushevich 1966) was applied. The correlations, used for 1D modelling, are shown in some of the presented record sections as dotted lines. The existence of low-velocity zones does not allow a unique solution for the detailed velocity distribution within them (Giese 1976b). For the low-velocity zones in this paper v(z)-functions with constant velocity gradient were chosen and its actual value calculated.

The v(z)-functions were the basis for the construction of the two-dimensional (2D) velocity models. A ray-tracing method (Spence et al. 1984) was used to calculate the ray paths and the travel times for the 2D models. In the figures with the 2D model below, the resulting travel times from ray-tracing are plotted in the record sections as dotted lines.

3 Record Sections and Model Calculations

Representative record sections for the different morpho-structural units of the Andean crust will be presented in this section. The observed travel time curves and the derived 1D and 2D models will be discussed starting at the modern forearc.

3.1 The Coastal Cordillera

The structure beneath the Coastal Cordillera can be described by a set of reversed record sections of the four shot points, Blanco Encalada, Mejillones, Tocopilla and Iquique, recorded in a north-south direction parallel to the coastline. The data and first models for the part south of Tocopilla were presented by Wigger et al. (1988) and Heinsohn (1990). For example, the record sections Blanco Encalada to the north and Tocopilla to the south are shown in Fig. 2a and b. In all record sections of the coastal profile clear first arrivals can be observed at a recording distance of up to 300 km . A very prominent feature of these record sections is represented by the early arrivals between 50 and 180 km below the t=0 - axis, which are caused by high velocities in the uppermost crust.

South of Tocopilla, velocities of 5.8 - 6.0 km/s are observed just below the surface. The early, first arrivals at a recording distance of up to 100 km lead to a velocity of 6.7 km/s at 7 to 10 km depth, and from the first arrivals of the same wave group at recording distances up to 180 km a velocity of 7.0 km/s and a penetration depth of 20 km is derived. The average velocity for the upper 20 km is 6.5-6.7 km/s. The deeper structure displays at least two low velocity zones (LVZ) characterized by velocities between 6.2 and 6.4 km/s intercalated by intracrustal discontinuities.These discontinuities are confirmed by weak onsets only. In the record section Tocopilla-south this retrograde phase can be observed at a recording distance between 120 and 210 km (Fig. 2a). In the 2D model for this phases a high velocity structure (7.6) at about 30 km depth is modeled and the resulting traveltimes are plotted in the sections. A well-pronounced discontinuity at a depth between 38 km in the south and 43 km near Tocopilla was derived from the large amplitudes of a retrograde phase at a recording distance between 110 and 170 km (e.g. section Blanco Encalada-north; Fig. 2b). The velocity at this depth, verified by a related, progressive wave group and reversed profiles, is 8.3 km/s. The average velocity shows values between 6.5

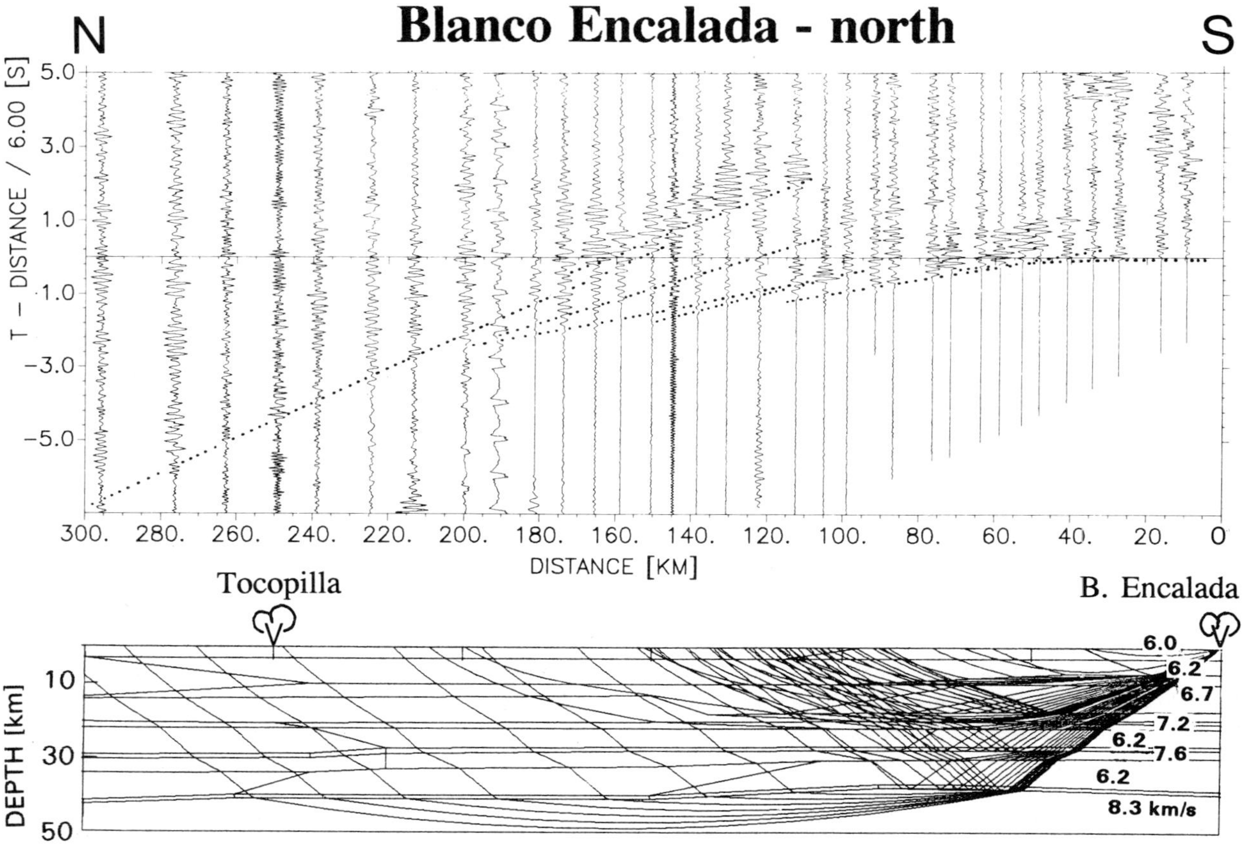

Fig. 2a. Record section and 2D model of the Coastal Cordillera. The records of shot point Blanco Encalada (SP 11) to the north are band-pass (BP) filtered (2-12 Hz) and the record section is plotted with a reduction velocity of 6 km/s as all the following ones in this chapter. The 2D velocity model displays the calculated rays from SP 11. The given velocity values represent the average velocity of the indicated layers. The *dotted lines* in the record section show the calculated traveltimes for the correlated wave groups. The two most important features of the record section are the early first arrivals up to 170 km, being expression of a high velocity upper crust (average velocity 6.5 - 6.7 km/s), and the retrograde phase in the distance range 170 - 110 km, indicating a strong discontinuity at about 40 km depth where the P-wave velocity jumps to 8.3 km/s

and 6.7 km/s for the upper part of the crust, the stack at 20 to 40 km depth, and for the total crust.

For the shot point Caleta Patillos-south (Fig. 3) the first arrivals are more delayed than in the southern part; the velocity varies between 5.0 and 5.6 km/s for the first 3.5 km of depth. At approx. 15 km depth the velocity increases to 6.6 km/s and the 15 to 23 km depth range shows velocities of 6.9 to 7.1 km/s. The deeper structure again is modeled with alternating high (7.4 km/s, 32-36 km depth) and low (6.2-6.4 km/s) velocities down to a discontinuity at 42 km, where a velocity of 8.2 km/s is reached. The discontinuity at 32 km depth is confirmed by weak onsets only and must be considered as one of possible interpretations. The discontinuity at 42 km is less clear than in the southern sections.

In an early stage of the interpretation for the southern part of the profile the derived velocity model between Tocopilla and Blanco Encalada was converted into density values (Strunk 1990) and a good fit of the gravity data was achieved by optimizing the density values and varying slightly the geometry. For the complete coastal 2D velocity model between Caleta Patillos and Blanco Encalada the velocity structure was simplified and then changed into a density model by converting the velocities after Ludwig et al. (1970) into density values without changing the interfaces. The density

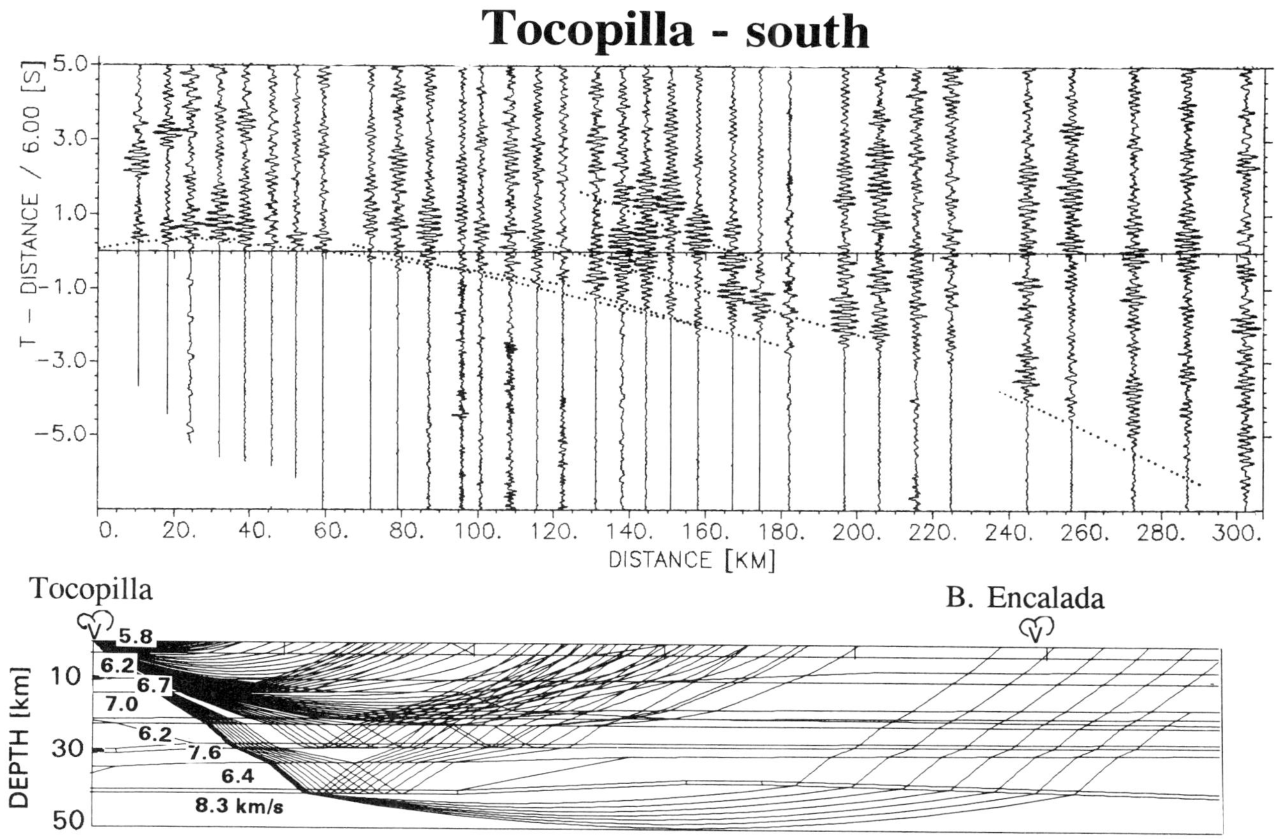

Fig. 2b. Record section of shot point Tocopilla (SP 8) to the south (BP: 2-12 Hz) with the same 2D model and the respective rays. Early first arrivals are observed up to 180 km recording distance as in the reversed section. A retrograde phase can be observed between 120 and 210 km recording distance. A high velocity structure (7.6 km/s) at about 30 km depth is modeled for this phase. The deeper discontinuity with the velocity jump to 8.3 km/s is calculated to be at 43 km depth

values were optimized by a program based on an evolution strategy (Asch et al., in prep.). For the optimization steps only a variation in the density values up to 2% was allowed according to the velocity-density relation published by Barton (1986). The resulting model gravity principally fits the observed gravity values (Fig. 4), thus demonstrating the reliability of the derived seismic model and its compatibility with the observed gravity data.

3.2 The Forearc Crust Between the Jurassic and the Modern Magmatic Arc

The crustal structure of the modern forearc is derived from recordings of the Chuquicamata blasts to the south and north, the latter reversed by recordings from Salar de Huasco to the south, by E-W observations at about 21°, 22° and 24°S, most of them reversed, and the lines Chuquicamata to Antofagasta (reversed) and to the southeast (see Fig. 1). A detailed description of the E-W profile at 24°S is given in Wigger et al. (1991) and for the observations Chuquicamata to the southwest, south and southeast by Wigger (1986, 1988).

Generally, a rapid change in structure as well as in crustal thickness is evident from the Chilean coast to the east. In the western part the W-E profile from Caleta El Cobre at the Pacific Ocean to the east (24°S) shows a velocity distribution similar to the profile of the Coastal Cordillera. The high velocity upper crust (20 km) found under the Coastal Cordillera dips down to the east and can be followed up to the Precordillera (Fig. 1), where a strong

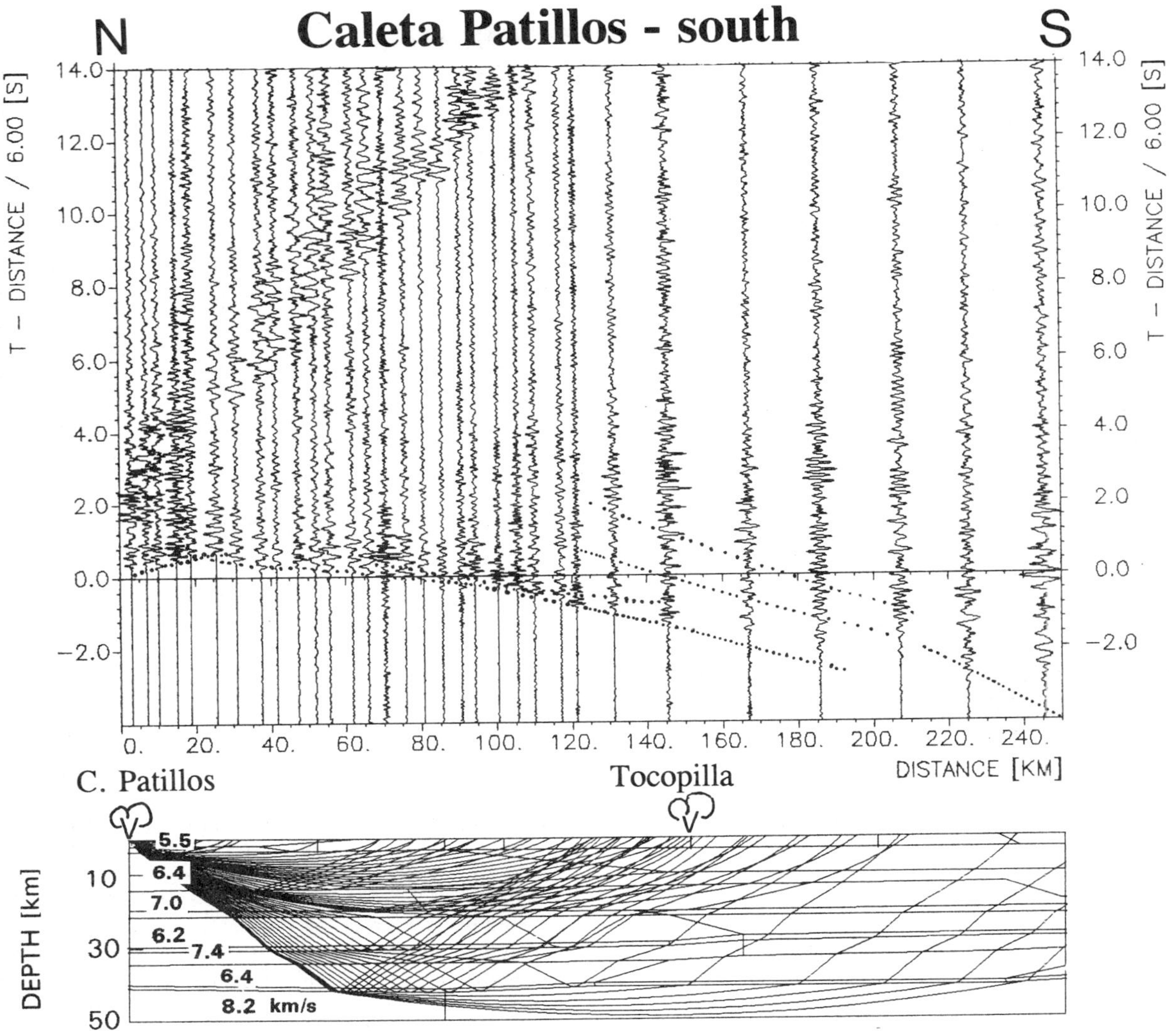

Fig. 3. Record section Caleta Patillos to the south (SP 25, BP. 2-12 Hz) and 2D model of the northern part of the coastal profile. The structure is similar to that of the southern part but the average velocity is lower (6.5 km/s). The time range between 4 and 14 s is shown to demonstrate the quality of the S-waves generated by the sea shots

change in crustal structure is observed. In this area the velocity increases only slightly with depth and the average velocity decreases from the Coastal Cordillera (6.6 km/s) to about 6.3 km/s beneath the Precordillera, approx. 150 km farther east.

The record section Tocopilla-east (Fig. 5a) from the E-W profile at 22°S is observed at a recording distance of up to 240 km, thus entering the Western Cordillera for about 50 km up to the Chile/Bolivia border. For the first 150 km recording distance, the record section resembles the N-S sections of the Coastal Cordillera. Derived from the wave group of these first arrivals, the velocity increases from 5.9 to 6.7 km/s at 15 km depth and to 6.9 to 7.0 km/s at 15-20 km depth. The next phase, a retrograde phase, is observed with a time delay of about 1 s between 240 and 150 km. This phase leads to a slight velocity reduction to 6.7 km/s and to a discontinuity at about 30-40 km depth, where the velocity amounts to 7.4 km/s. Signals from deeper levels are not very strong. Poor indications of a later retrograde phase suggest a gradient zone in the lower crust with a possible velocity increase to 8.2 km/s at about 50 - 55 km depth with vertices between the Coastal Cordillera and Precordillera.

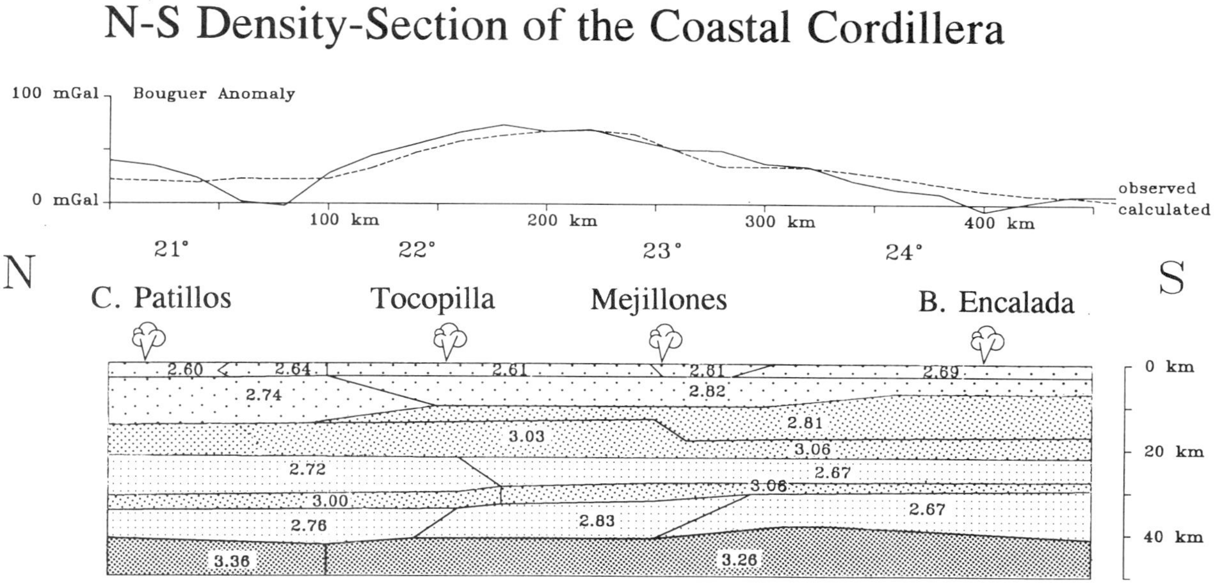

Fig. 4. N-S cross section of a 2D gravity model through the Coastal Cordillera indicating layers of high *(densely dotted areas)* and low *(sparsely dotted areas)* velocities. The density values of the principal units are in g/cm^3. The model gravity optimized by an evolution strategy program without modifying the structural lines fits the observed gravity values shown in the upper portion of the figure

The reversed profile of Chuquicamata to the west (Fig. 5b) ends at the coast at a recording distance of less than 140 km. For this reason signals from the lower crust and the discontinuity at 50-55 km depth are not recorded. The rays through the lower crust of the 2D model show that they just start at that distance. The reversed phases between 130 and 80 km distance in this record section prove the discontinuities at 15-20 and 30-40 km depth where the velocities increase to 6.9-7.0 and 7.2-7.4 km/s respectively (Figs. 5b and 6).

In comparison with the record section Tocopilla-east, the recordings at Chuquicamata-west and -east show a time delay of about 1 s for the first arrivals, and compared to near-surface velocities of 5.9 km/s for the coastal sector, lower velocities farther east (5.0-5.5 km/s) indicate a thickening of the uppermost crust which is characterized by typical velocities for sediments and basement rocks.

The first arrivals of the record section Chuquicamata-east penetrate to a maximum depth of 10 km, reaching a velocity of 6.1 km/s. Below, an LVZ must be assumed due to missing seismic signals at recording distances between 100 and 470 km east of Chuquicamata in prolongation of this profile (Wigger 1986). The depth of this LVZ cannot be determined from the present data. Schwarz et al. (1984) observed for this crustal level below the Precordillera a zone of extremely high electrical conductivity. The lower boundary of the high conductivity zone could not be detected.

The velocity structure of the forearc region at about 22°S Latitude, based on the interpreted observations, is summarized in Fig. 6. Those parts of the velocity lines, which are proven by vertices, are shown by heavy lines. From the coast to the east the actual total crustal thickness and the upper crust increase. The high velocity layers starting at shallow depths at the coast dip down to the east analogously and cannot be seen east of the Precordillera.

For the profiles Chuquicamata to Antofagasta, to the south and southeast the main results will be reported shortly. The data and interpretations are given by Wigger (1986, 1988) and Schmitz et al. (1990). From the reversed profile Chuquicamata-Antofagasta a velocity increase to 6.4 km/s is observed at 11 km depth and at a discontinuity at about 30 km depth the velocity jumps to 7.2 km/s. An additional deeper discontinuity, interpreted as the modern Moho, with a velocity jump to 8.1 km/s, was found for the depth range 50-55 km beneath the area between the Coastal Cordillera and the Precordillera (Wigger 1988).

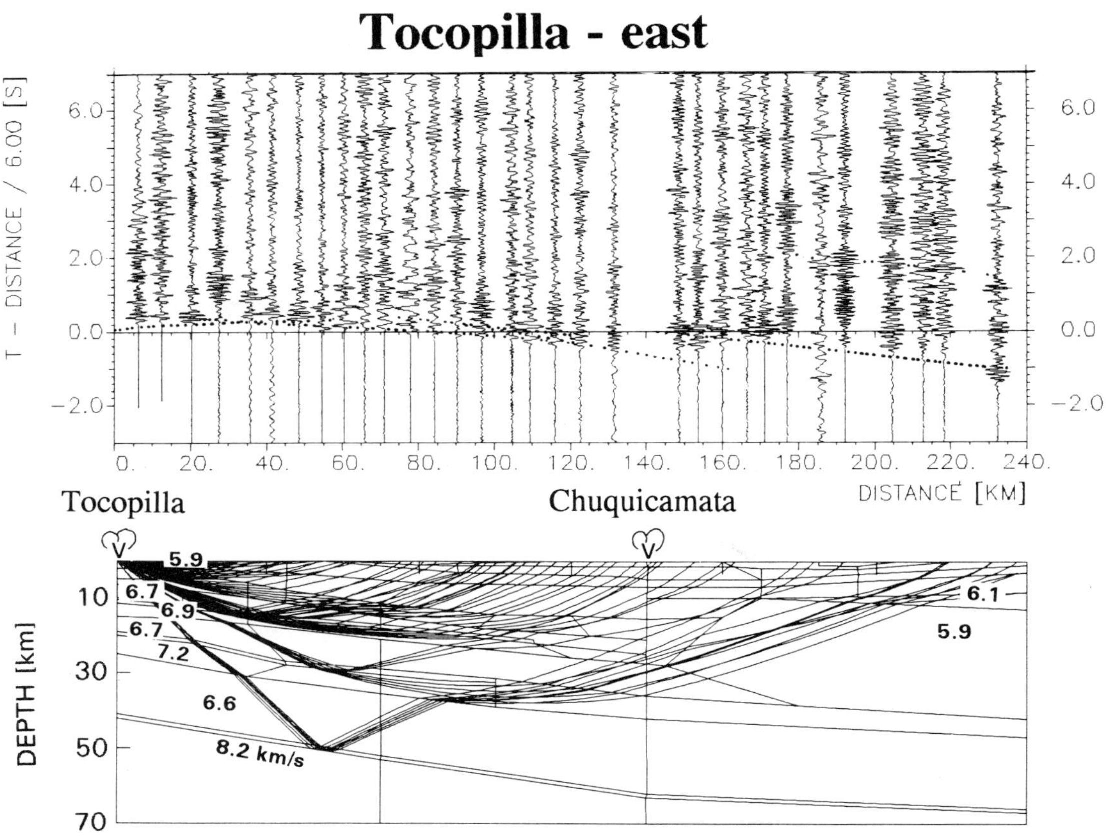

Fig. 5a. Record section of the observation Tocopilla-east (SP 6) and the related 2D model with calculated rays

The record section from Chuquicamata-south (Fig. 7) runs parallel to the strike of the Precordillera. A set of travel-time curves up to the maximum recording distance of nearly 300 km indicates a very strongly structured crust. The velocity increases to 6.4 km/s at 13 km depth and a retrograde phase between 110 and 180 km recording distance leads to a velocity of 6.6 km/s at about 35 km depth. The 1D inversion (Fig. 8, right) of the correlated phases shows a total crustal thickness of already 70 km for the Precordillera and an approx. 30 km thick lower crust with alternating high and low velocities, as indicated by large amplitudes of three late retrograde phases between 180 km and the maximum recording distance. The average velocity of the upper and middle crust down to 35 km is 5.9 km/s and for the whole crust it is 6.2 km/s.

In the record section Chuquicamata-southeast running from the Precordillera in a SE direction with an azimuth of 120°, we can observe similar wave groups for the upper and middle crust as recorded on the line Chuquicamata-south. A velocity of 6.5 km/s is observed down to a depth of 8 km and a prominent discontinuity was found at a depth of about 35 km where the velocity jumps to more than 7 km/s. These relatively high velocities for the upper and middle crust may be related to the high gravity values presented by Götze et al. (1988, this Vol.). Thus, the area of the Precordillera south of Chuquicamata is characterized by a positive residual gravity anomaly with a magnitude of more than 50 mGal and a total width of about 100 km that continues in a SSE direction up to the Puna.

The seismic refraction data of the forearc region north of the line Tocopilla-Chuquicamata and its 2D model calculation and interpretation are reported by Schmitz et al. (1991). The reversed N-S observation in the Precordillera suggests again a strongly structurized crust. A few kilometres of typical, sedimentary velocities are underlain by an upper and middle crust down to about 35 km, characterized by alternating high and low velocities between 6.0 and

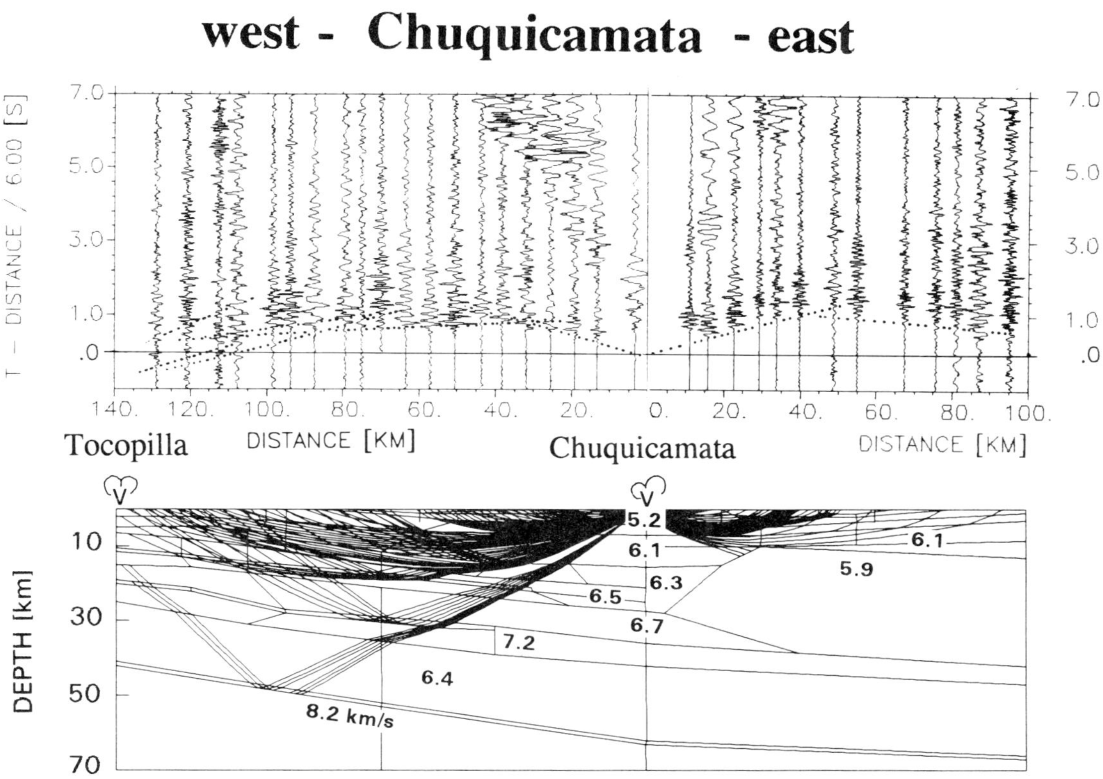

Fig. 5b. The record sections Chuquicamata-west and -east (SP 5 and SP 2, all BP: 2-12 Hz) with the same 2D model and the respective rays

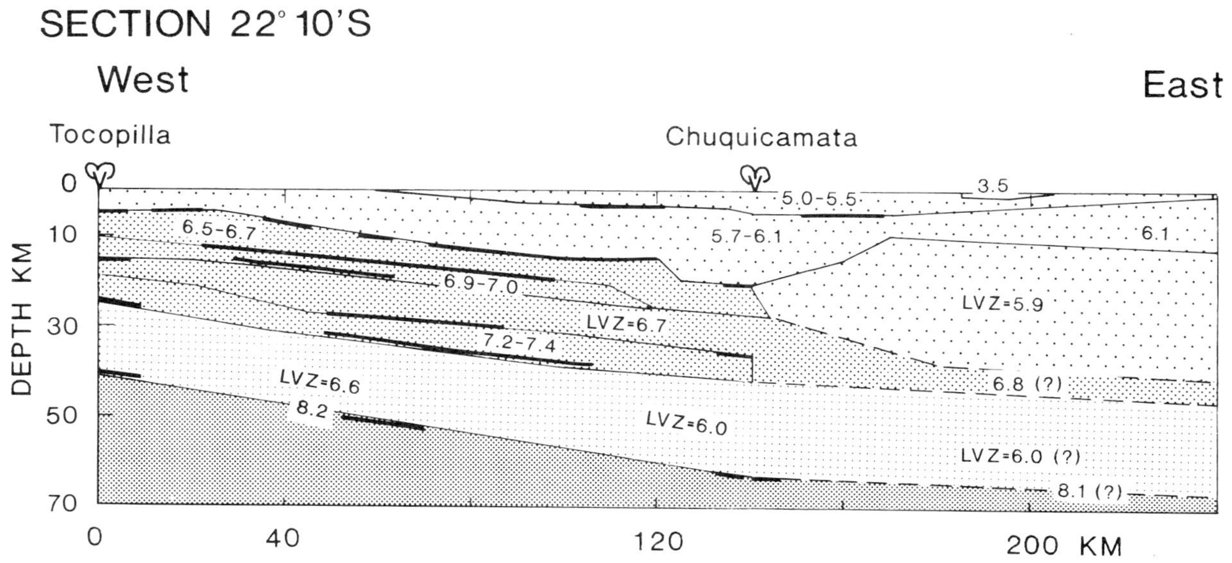

Fig. 6. The W-E cross section at about 22°S between Tocopilla and Linzor as derived by 2D model calculations. The surface of the section represents a straight line between sea level in the west and 4000-m elevation at the eastern border. All velocities (v_p) are given in km/s. The thickened portions of the lines of equal velocities (isotachs) are proven by vertices. Increasing crustal thickness from W to E caused mainly by a thickening of the upper crust and an extended LVZ in the eastern part are the main features of this section

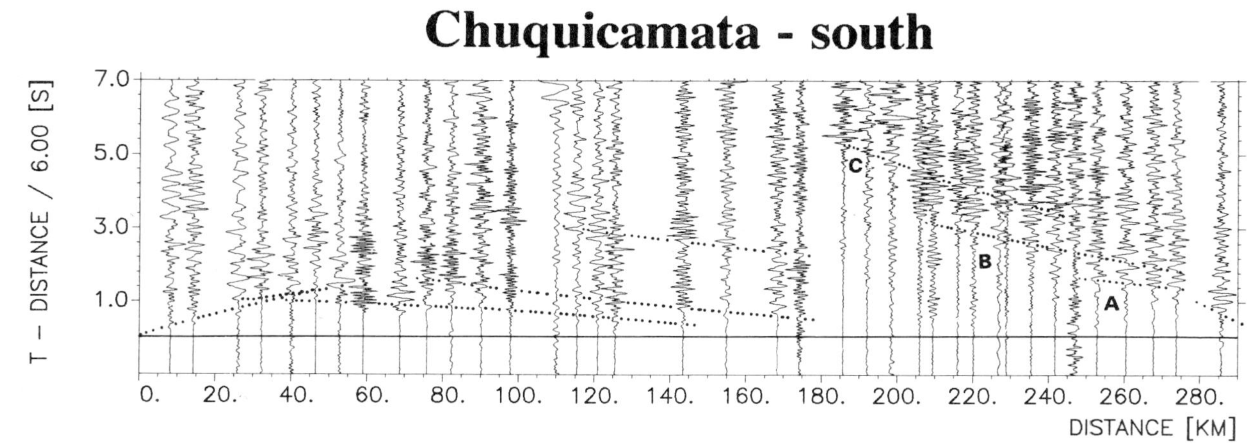

Fig. 7. Record section Chuquicamata-south (SP 4, BP: 2-12 Hz). The *dotted lines* represent correlations compatible with the velocity-depth model in Fig. 8. The early phases up to the recording distance of 180 km describe the velocities in the upper and middle crust, whereas the three late reversed travel-time curves (phases A to C) represent a 30-km-thick lower crust

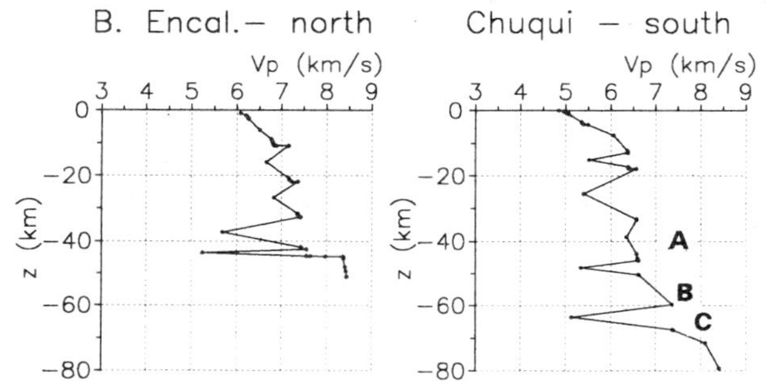

Fig. 8. Typical velocity-depth functions for the Coastal Cordillera *(left)* and the Precordillera *(right)*. They demonstrate a strongly structured crust in both cases, but show different crustal thickness and average velocities resp.: 6.6 km/s for the Coastal Cordillera and 6.2 km/s below the Precordillera. The letters A to C refer to the indicated phases in Fig. 7

6.8 km/s. From retrograde phases at recording distances of 200 to 300 km in the section Salar de Huasco-south and from seismograms with large amplitudes but without clear arrivals between 160 and 240 km recording distance in the section Chuquicamata-north an approx. 30 km thick lower crust with high (7.1 km/s) and low (6.2-6.5 km/s) velocities was deduced. Modern total crustal thickness is calculated to be 65 km and its average velocity varies quite narrowly between 6.2 and 6.3 km/s.

The record section Ollagüe-west shows a first progressive wave group with an increasing velocity from 6.0 to 6.3 km/s. With a time delay of about 1 s phases follow which indicate strong velocity reductions in the middle crust (Fig. 11a). The vertices of these phases indicate that they do not belong to the forearc structure with its relatively high velocities. The mentioned phases must be attributed to the transition of the forearc crust to the crustal structure of the modern magmatic arc which will be described in the next section.

3.3 The Western Cordillera, the Western Altiplano and the Puna

Information about the structure of the Western Cordillera, which represents the modern magmatic arc and about the Altiplano/Puna region, is provided by the recordings of the land shots in salt lakes of the Western Cordillera and the Puna, the copper mine in Chuquicamata, and by the land shot in the Eastern Cordillera (Tupiza).

For the E-W profile at 24°S the waves of the recordings from Pocitos-west (Fig. 9) have passed the Western Cordillera. A description of this record section and 1D and 2D model calculations are given in Wigger et al. (1991). For the first arrivals at a recording distance of up to about 200 km, a time delay of about 1 s to the zero axis is observed, evidencing a relatively monotonous structure of the upper 35 km with a subsurface velocity of 3.5 km/s for the salt lake sediments and velocities between 5.3 and 6.5 km/s for the upper 35 km. The average velocity of this crustal portion is 6.0 km/s. A strongly imbricated reversed phase in the distance range of 330-140 km at 1-7 s traveltime with vertices under the Western Cordillera documents a lower crustal level with alternating high and low velocities at depths between 35 and 60 km (Fig. 10). The different branches of the traveltime curve cannot be explained by a simple reversed phase, which would lead to an extremely high apparent velocity. The latest reversed phase is considered the seismic crustal base. Once again, first arrivals can be observed between 280 and 350 km, which are related to the subcrustal layer with a P-wave velocity of 8.1 km/s. The average velocity of the crust of the modern magmatic arc region at this latitude is calculated to be 5.9 km/s.

Perpendicular to this E-W profile a N-S profile is observed in the eastern, adjacent Puna which has its central shot location at Pocitos, observed to the north and south, and reversed recording lines from Olaroz to the south and Tolar Chico to the north (Fig. 1). This profile, too, is shown in Wigger et al. (1991). The crustal structure is represented by clear wave groups at depths of up to 25-30 km. A young sedimentary cover of less than 4 km with velocities of 4.0 to 5.6 km/s was determined. At a depth of 25 to 30 km a discontinuity with a velocity increase of 6.6 to 6.8 km/s is indicated (Fig. 10) and the average velocity for the upper 30 km is 6.0 km/s. The quality of data does not allow a detailed modelling of deeper structures. There are no indications for a structured lower crust as in the record section of Pocitos-west. In the Tolar Chico-north record section (Fig. 9), weak evidence of a deeper reflection in the distance range 260-150 km at 2-5 s traveltime indicates a depth of 60 km, but the signal/noise ratio does not allow a conclusion as to whether or not there is a Moho reflection.

Following the Western Cordillera to the north, the observation lines of Chuquicamata to the southeast and east run into the Western Cordillera or cross the Cordillera. The line to the southeast does not provide information about the structure under the Western Cordillera due to the maximal recording distance of 240 km. The intracrustal discontinuity at about 35 km depth can be traced from the forearc crust up to the Western Cordillera only. The existing recordings at distances greater than 170 km are very noisy and do not indicate signals from deeper structures. The absence of signals at distances greater than 100 km up to 470 km at the profile Chuquicamata-east (as mentioned earlier) must lead to the assumption that the middle and lower crust consists of material with no strong velocity increase and/or prominent discontinuities. The absence of signals at the most distant recordings might also reflect insufficient seismic source energy and a high attenuation of seismic waves under the Western Cordillera. All indications lead to the assumption that there is no seismic crust/mantle boundary.

The shot point Ollagüe, 150 km NNE of Chuquicamata, lies in the Western Cordillera and is recorded to the west and to the east. As mentioned above, there is a time delay of 1 s for the first arrivals in the Ollagüe-west record section. With a time delay of 2-3 s with respect to the first arrivals a retrograde phase with possibly subcritical reflections can be observed in the distance range of 100-50 km recording distance. A velocity jump to 6.9 km/s at about 20 km depth is derived and a velocity reduction to 4.5 km/s is calculated for the 10-20 km depth range (Fig. 11a). Later arrivals of a deeper reflector between 190 and 160 km recording distance and 3-5 s traveltime can be correlated with great caution.

For the Ollagüe-east record section the vertices of the recorded waves penetrating to a depth of 20-25 km refer to the Western Cordillera, the deeper penetrating ones to the eastern adjacent Altiplano. Clear seismic P-waves are observed only up to 120 km. The results are shown in Fig. 10 (1D) and in Fig. 11b (2D). The first arrivals with a time delay of more than 1 s at a recording distance of up to 120 km represent an increasing velocity of 2.7-4.8 km/s down to 4 km depth, which is interpreted as young sediments and young volcanic rocks and an increase to 5.9 km/s at 4-10 km depth, which is considered to represent basement rocks. A time delay of 2 s with respect to the next retrograde phase indicates a prominent LVZ at the depth range of 10-20 km with an average velocity of 5.2 km/s. The depth range of this LVZ corresponds to that one observed about 100 km to the south on the profile Chuquicamata-east (see also Fig. 6b) on the western side of the modern magmatic arc and the high electrical conductivity zone described by Schwarz et al. (1984). At the discontinuity at 20-25 km depth the velocity jumps to 6.4 km/s. At recording distances greater than 120 km there are no clear wave groups. Only one more wave group might be correlated with great caution, which is characterized by signals with a low signal/noise ratio. It is regarded as a delayed retrograde phase between 180 and 140 km distance at about 4 s traveltime and indicates a moderate velocity increase

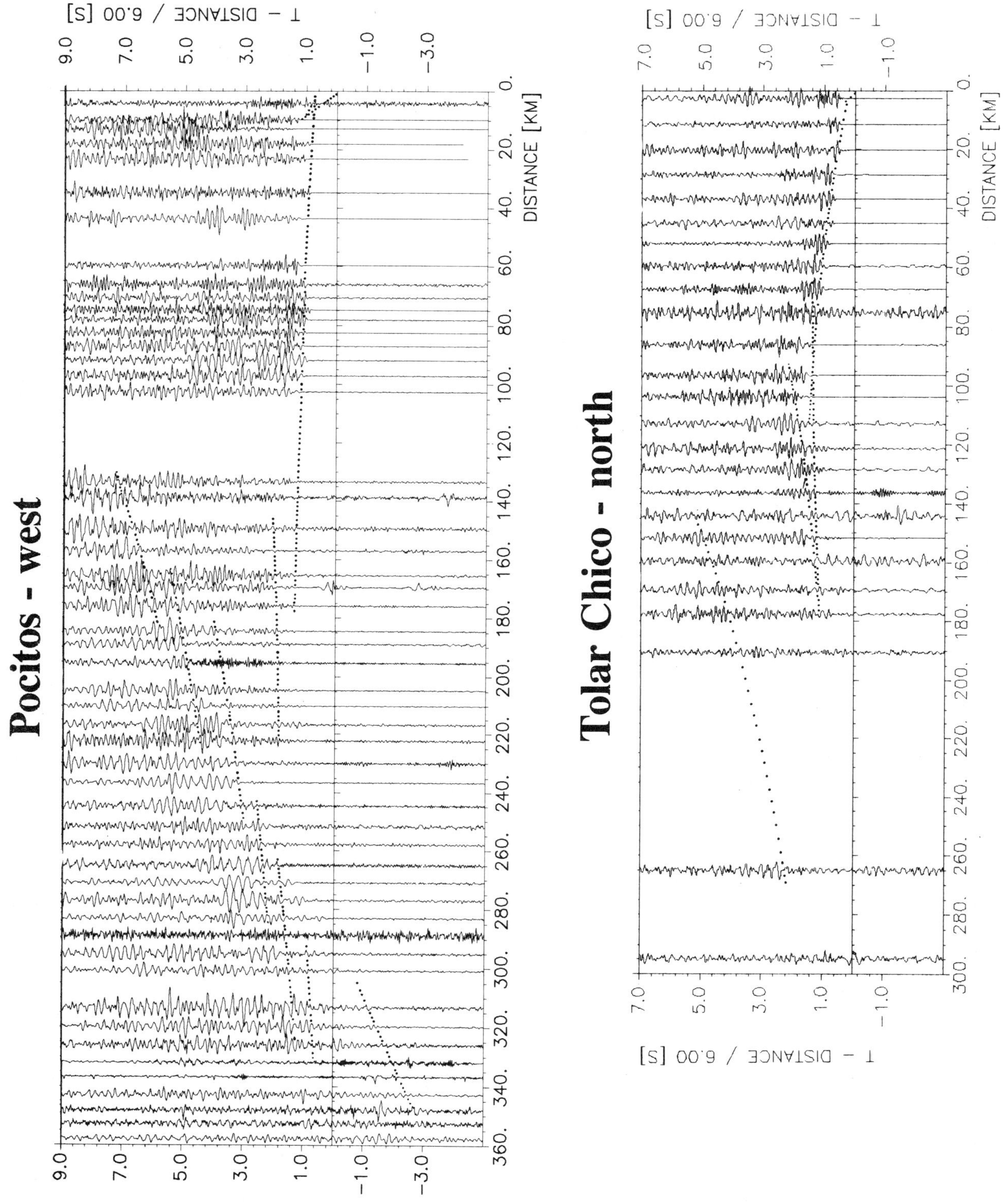

Fig. 9. *Above* Record section of the profile Pocitos-west (SP 13, BP: 2-20 Hz) and *below* Tolar Chico-north (SP 16, BP: 2-12 Hz). The imbricated reversed phases between 330 and 140 km in the record section Pocitos-west indicate layering of high and low velocities in the lower crust. In the Tolar Chico-north record section weak evidence of a deeper reflection in the distance range 260-150 km at 2-5 s traveltime indicates a depth of 60 km. The *dotted lines* refer to the corresponding 1D model calculations (see Fig. 10, *left*)

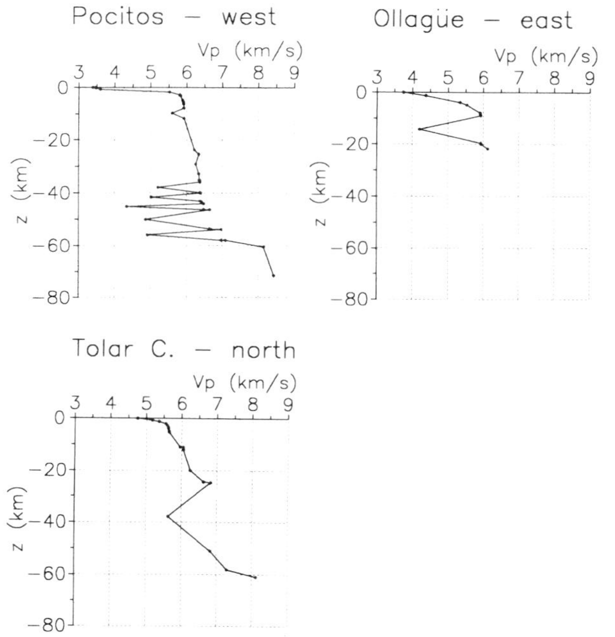

Fig. 10. Velocity-depth functions for the record sections Pocitos-west, Ollagüe-east and Tolar Chico-north, referring to the structures of the Western Cordillera and the Puna

to approx. 6.6 km/s in the depth range of 35-40 km at the most western margin of the Altiplano. There is no evidence for a structured lower crust or a Moho reflection. The distance range of 300 to 470 km on the Tupiza-west profile (Fig. 12), which extends to the Pacific coast, is suitable to obtain information on the deep crustal or lithospheric structure beneath the Western Cordillera. However, the lack of any wave group in this distance range suggests low velocities and high attenuation for the crustal base and its footwall.

3.4 The Eastern Altiplano, the Eastern Cordillera and the Subandean Ranges

The crustal structure further east, up to the Andean foreland, can be described by the observation Tupiza-west, the reversed profile Tupiza-Villamontes, and by a 70-km profile with three shot points (37, 38, 39/40) in the Subandean Ranges (Fig. 1).

The record section Tupiza-west (Fig. 12) extends to the Pacific coast at a recording distance of 470 km. Clear first arrivals are observed up to 140 km offset. Furthermore, in the range 170-270 km, a delayed retrograde phase seems to appear faintly. Thus correlation was done with great caution. For distances greater than 270 km no wave group at all can be detected. The latter point was discussed previously. The first wave groups show velocities of 4.3-5.3 km/s, penetrating to a depth of 4 km. The velocity increases slightly down to a discontinuity at about 13 km depth where the velocity reaches 6.0 km/s (Figs. 12 and 13). A retrograde group with a small time delay with regard to the previous one, which results from a slight velocity inversion, can be correlated between 150 and 70 km at about 2.5-3.5 s traveltime. This correlation shows a velocity jump to 6.8 km/s at 25 km depth. For the deeper crustal structure one can find a retrograde phase between 6-8 s at a distance of 270 and 170 km. Using these correlations for model calculations, a depth of 70 km (or 77 km in 1D calculations) results. The deeper discontinuity is interpreted as the crust/mantle boundary. An important consequence of this calculation is the resulting 30-km-thick LVZ at a depth range of 25-60 km, in the Eastern Cordillera, with an average velocity of 6.1 km/s. The presented model yields an average velocity for the total crust of 6.0 km/s.

Figure 14 shows the record sections and derived 2D models of the segment from the Eastern Cordillera, the Subandean Ranges up to the Andean foreland, the Chaco. In contrast to recordings from the Altiplano and the Western Cordillera we can see very clear wave groups in the record sections of this reversed profile, which provide detailed information about the whole crust. However, the two reversed record sections are distinctly different from each other, thus indicating strong lateral variations. The Tupiza-east section shows a more differentiated picture than Villamontes-west, which has its shot point in the foreland (Figs. 13, 14 and 17). Whereas in the Tupiza-east record section the time delay for the first arrivals up to 80 km does not exceed 1 s, in the Villamontes-west section a time delay of up to 3.5 s for the first wave groups at a recording distance of up to 100 km is observed. The resulting velocities vary between 4.1 and 6.0 km/s for the Eastern Cordillera, penetrating to a depth of 8 km. In the reversed recording the first arrivals up to 100 km form wave groups which show velocities from 2.5 km/s near the surface down to a discontinuity at 12-15 km depth dipping to the west where the velocity jumps from 5.7 to 6.1 km/s. This value is interpreted as the top of the basement.

To obtain more information about the upper crustal structure a reversed profile, including a central shot point with a total length of 70 km and a seismometer

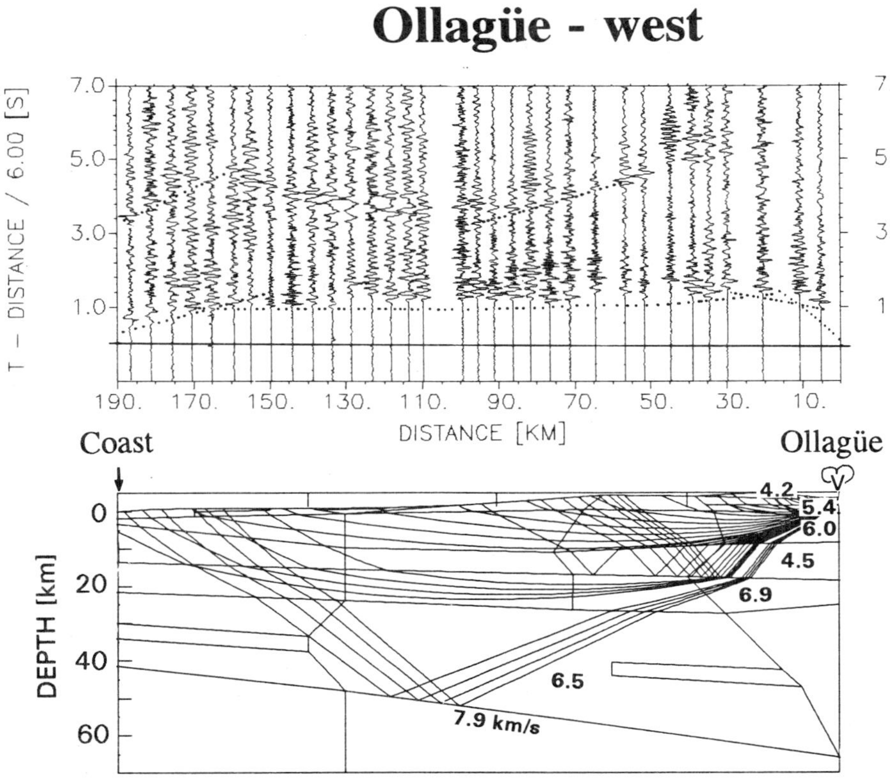

Fig. 11a. Record section Ollagüe-west (SP 22, BP 2-20 Hz) recorded up to the Pacific coast

spacing of 2 km, was observed in the Subandean Ranges (SP 37 = Narvaez, SP 38 = Tacuarandi, SP 39/40 = Palos Blancos). Examples of a record section and velocity depth functions are given in Figs. 15 and 16. For the uppermost layers down to 4 km depth the velocities vary in lateral and vertical directions; the values are between 2.8 and 5.0 km/s. These strong variations must be correlated to the fold and thrust tectonics bearing repetitions of different lithological units of Cenozoic, Mesozoic and Palaeozoic age, which now are predominantly in a steep position in this Andean segment (Kley and Reinhardt this Vol.). The next depth range from 4-10 km is characterized by a smaller velocity gradient with velocity values up to 5.6 km/s. It is assumed that this pile consists mostly of Palaeozoic sediments, here in a more horizontal position. A basement velocity of about 6.0 km/s is observed here and the derived depth is 10 km. Deeper structures could not be detected from these short profiles. The 2D model of this special upper crustal study is incorporated into the 2D model Tupiza-Villamontes (Fig. 14). Its detailed description is given in Baldzuhn (1992).

Going back to the record section Tupiza-east (Fig. 14a), the second wave group, which can very clearly be correlated as a retrograde phase between 140 and 60 km at about 2 s, implies a discontinuity at 20 km depth where the velocity jumps to 6.8 km/s. This discontinuity can be traced by vertices to the east up to the border between the Eastern Cordillera and the Subandean belt (in the 2D model at km 80). In the reversed profile this very prominent wave group is not present. However, it must be recalled that a similar wave group was found in Tupiza-west (Fig. 12) which indicates a depth of 24 km for this discontinuity. A narrow LVZ in the uppermost crust is found in both profiles related to the Eastern Cordillera.

A further prominent feature of the record section Tupiza-east is represented by the late arrivals which must be regarded as signals from the crust/mantle boundary. These large amplitudes form a retrograde travel time curve in the distance range 210-140 km at 3.0-6.2 s, which represents a velocity increase from 7.3 to 8.2 km/s. By model calculation this phase leads to an inclined crustal/mantle boundary between

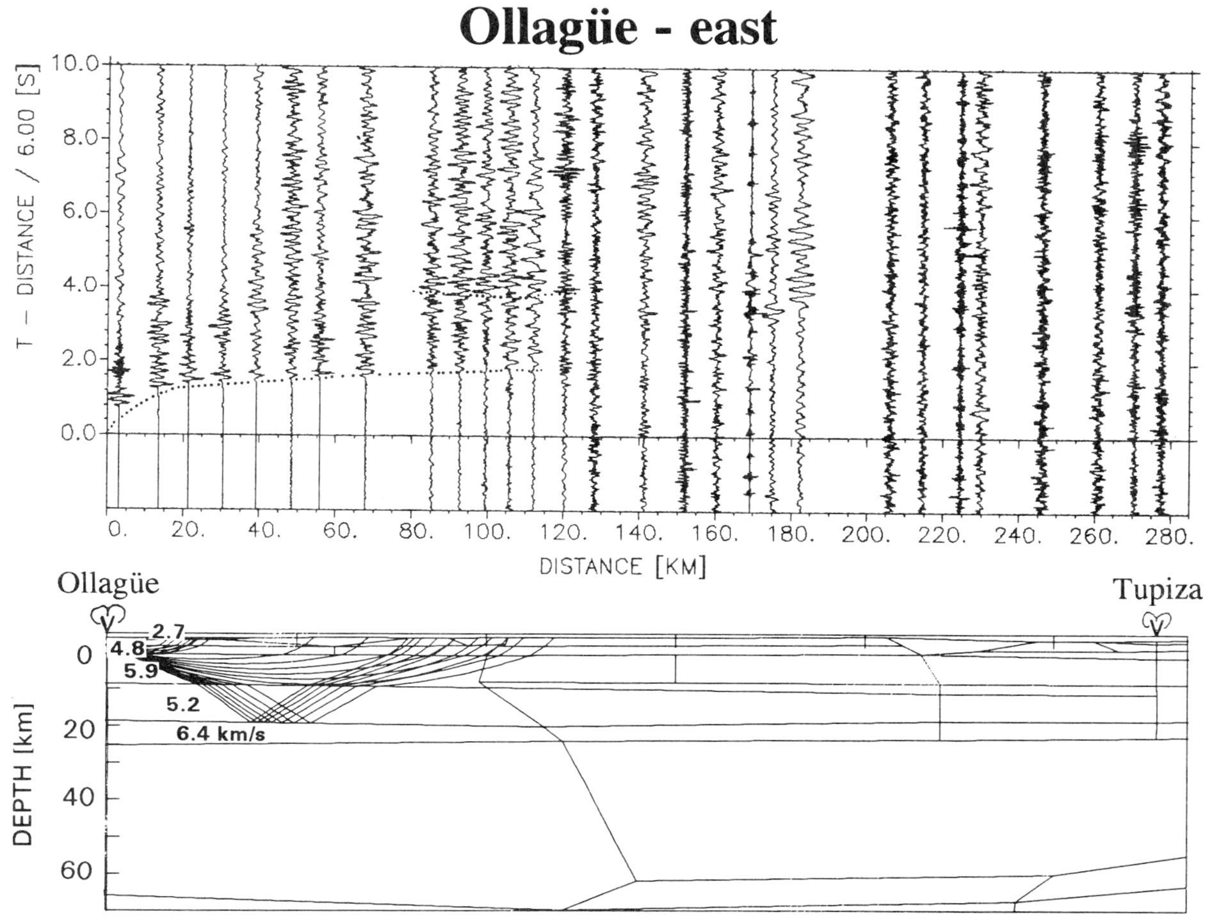

Fig. 11b. Record section Ollagüe-east (SP 33, BP 2-20 Hz). The 2D model in the lower part of the figure is developed from the shown record section and the reversed one as well: Tupiza-west (Fig. 12). The zero line in the 2D model represents sea level. However, ray calculation is performed until the topographic surface is reached (also for the following figures). No wave penetrating deeper than 25 km is visible in the record section

50 (east) and 60 (west) km depth. From both record sections of the Tupiza shots a broad LVZ for the 25-60 km depth range results, which is characterized by an average velocity of 6.1 km/s.

The internal crustal structure of the eastern part of the Subandean Ranges derived from Villamontes-west is simple. As shown in the 1D and 2D models (Figs. 17 and 14b), the velocity increases continously down to the crust/mantle boundary. The data base for these model calculations is established by the very well-pronounced retrograde phase between 210 and 90 km at 3-6 s. The velocity at the Moho as indicated also by a P_n-phase amounts to 8.2 km/s and the average velocity for the total crust is 5.9 km/s, subtracting the sedimentary cover (for v_p < 5.7 km/s) its value increases to 6.3 km/s. Crustal thickness beneath the border between foreland and Subandean belt is 40 km dipping down towards the west. The crustal parameters are very similar to those described for a profile observed on the Brazilian Shield (Giese and Schütte 1980).

4 Conclusions

The presented seismic data prove variations in crustal structure and in average velocities perpendicular to the strike of the Andes as well as in the N-S direction. The results are summarized in an 800 km long W-E section at 21°S from the Pacific coast to

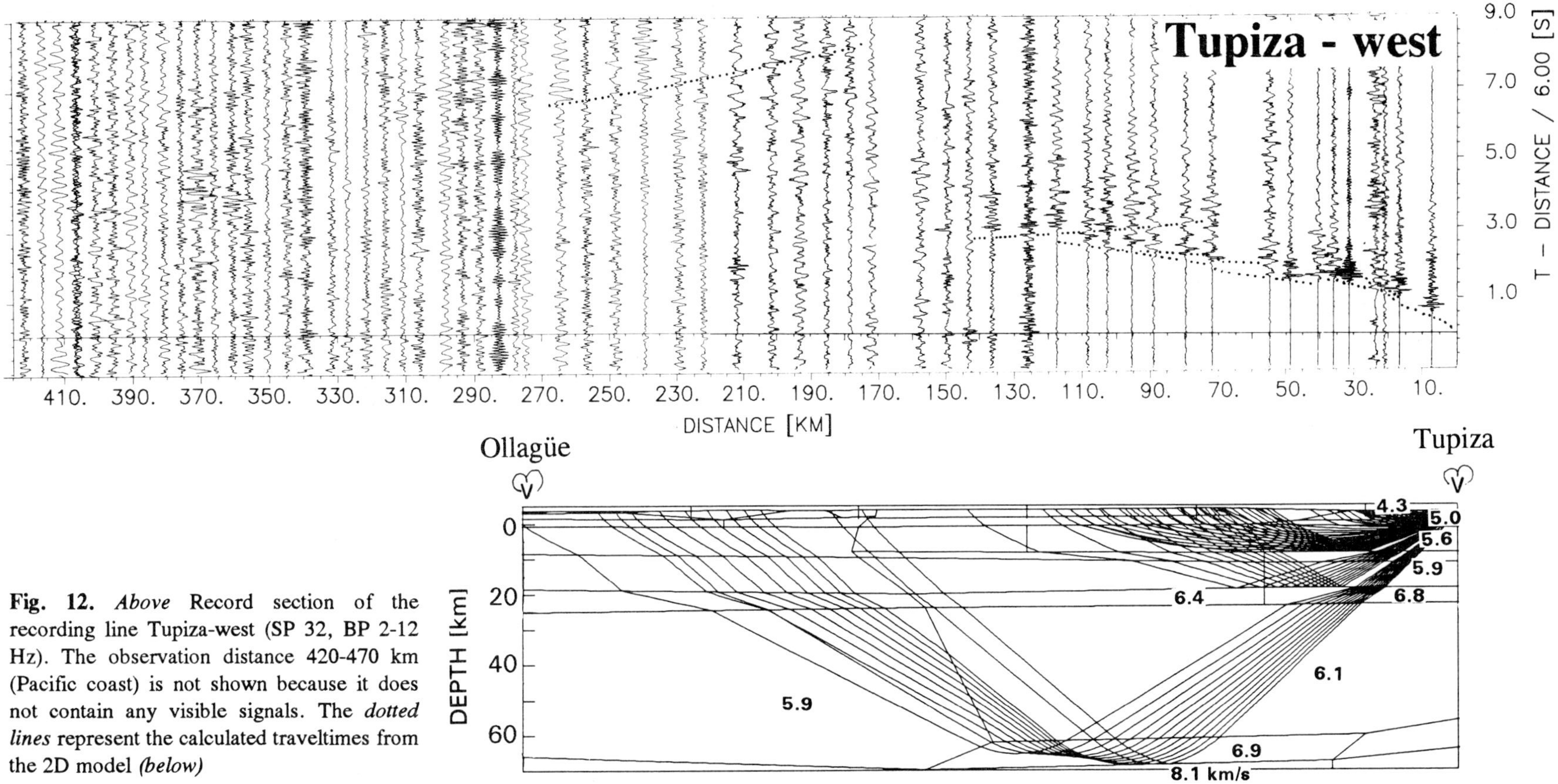

Fig. 12. *Above* Record section of the recording line Tupiza-west (SP 32, BP 2-12 Hz). The observation distance 420-470 km (Pacific coast) is not shown because it does not contain any visible signals. The *dotted lines* represent the calculated traveltimes from the 2D model *(below)*

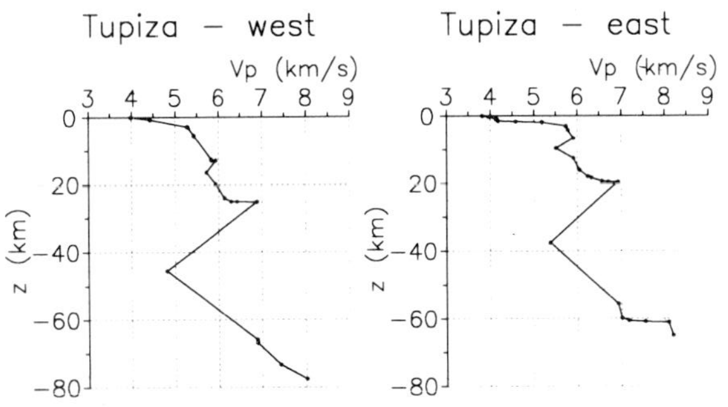

Fig. 13. Velocity-depth functions for the Tupiza-west and the Tupiza-east record sections. Note the generally similar structure of the graphs and the broad LVZ in the 20-55 km depth range but decreasing total crustal thickness to the east

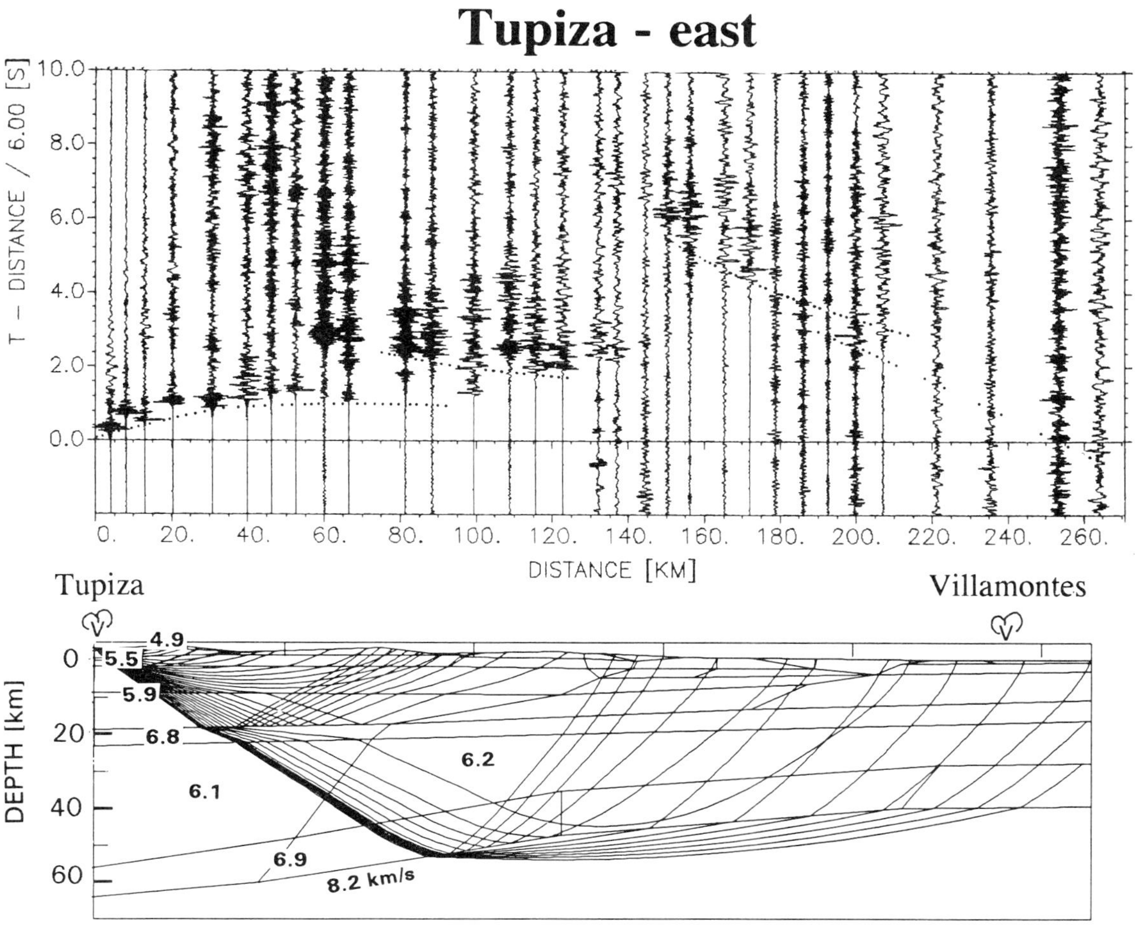

Fig. 14a. Record section Tupiza-east (SP 34, BP: 2-20 Hz). The western part of the profile is characterized by a pronounced intracrustal discontinuity at 20 km depth and a broad LVZ in the lower crust

the Andean foreland (Fig. 18). The discussion will focus along this line, but the results from the profiles situated further south, will be included. The upper part of Fig. 18 shows the main velocity partitions of the 2D ray tracing model only, while the section in the middle part is restricted to the main discontinuities with the corresponding velocity jumps, which are proven by vertices. In the lower part the calculated average crustal velocities are given for the related crustal segments and a rough relation of these values to a rigid or ductile crust was tried. After this classification three main crustal types must be distinguished. The crust of the forearc from the Coastal Cordillera up to the Precordillera can be characterized as rigid. The Western Cordillera and the deeper part of the Altiplano are considered as ductile and the backarc from the Eastern Cordillera to the Chaco is characterized by a rigid crust again.

4.1 The Rigid Block of the Andean Forearc Region

The characterization of the modern forearc as a rigid block is based on the high velocities of the entire crust. We have to distinguish between an upper level down to 20 km beneath the Coastal Cordillera dipping to the east, a level down to 35 km under the Precordillera, and below, a 20 to 30-km-thick zone with alternating high and low velocities. The upper crustal level, with high average velocities and maximum velocities of 7.2-7.4 km/s at its base, is interpreted as the former Jurassic arc and backarc crust of the continental margin. The Coastal Cordillera represents the Jurassic arc and its crust was strongly modified by magmatic activity (Reutter et al. 1988; Pichowiak this Vol.). Later, the crust was uplifted to its actual position accompanied by partial erosion of the upper crust. Rössling (1989) described

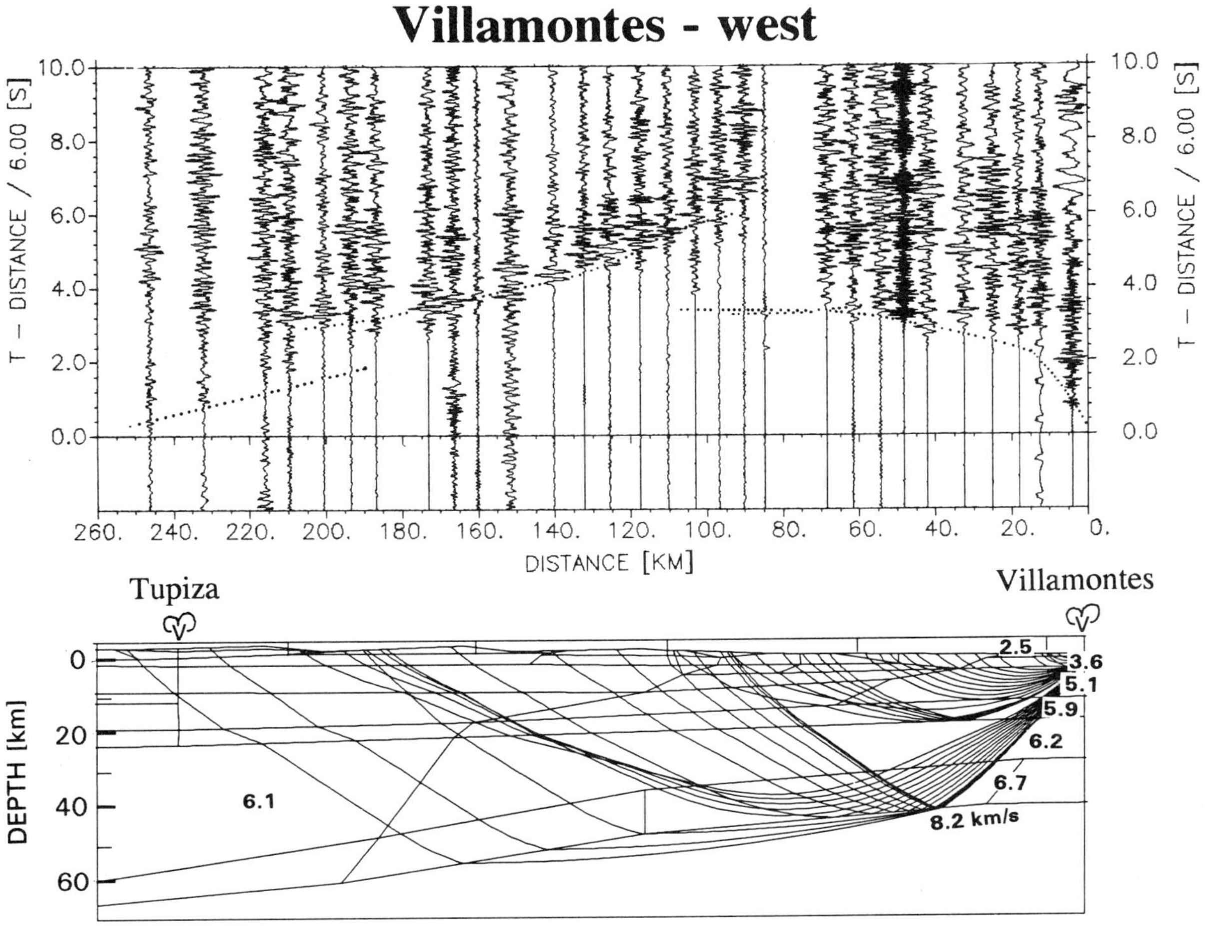

Fig. 14b. Record section Villamontes-west (SP 36, BP: 2-12 Hz). The eastern part of the profile shows a simple crustal structure already representing shield conditions

lower crustal complexes outcropping in the Coastal Cordillera.

The geological situation of the modern continental margin and the main seismic results are represented in Fig. 19. It is assumed that the crustal thickness was 30-35 km in Jurassic to early Cretaceous times, its base proven by the high seismic velocities at the modern depth of 20 km. The velocities of the upper 10 km are relatively high and must be related to the La Negra volcanites of Jurassic age and Jurassic to early Cretaceous dioritic to granodioritic intrusions intruded into the lower crustal level at that time. The decreasing average velocities to the east must be explained by a less intensive magmatic modification of the crust during the eastward migration of the magmatic arc systems (Scheuber et al. this Vol.).

For the deeper level of the forearc crust there is no uniform interpretation. It seems to be clear that the discontinuity at 40 km depth below the Coastal Cordillera, on the basis of the most prominent retrograde phase with a related progressive one, marks the border between the oceanic crust and the mantle of the subducted Nazca plate. Its downdipping slab defines a well-developed Wadati-Bennioff zone. If we consider the oceanic Moho at 40 km depth, consequently, the 30-40 km depth range must be interpreted as the oceanic crust. From this directly the question arises on the character of the depth range of 20-30 km. What is the origin and the nature of this level with an average velocity of 6.6 km/s and a maximum P-wave velocity of 7.6 km/s? It is supposed that the ultrabasic, uppermost Jurassic-

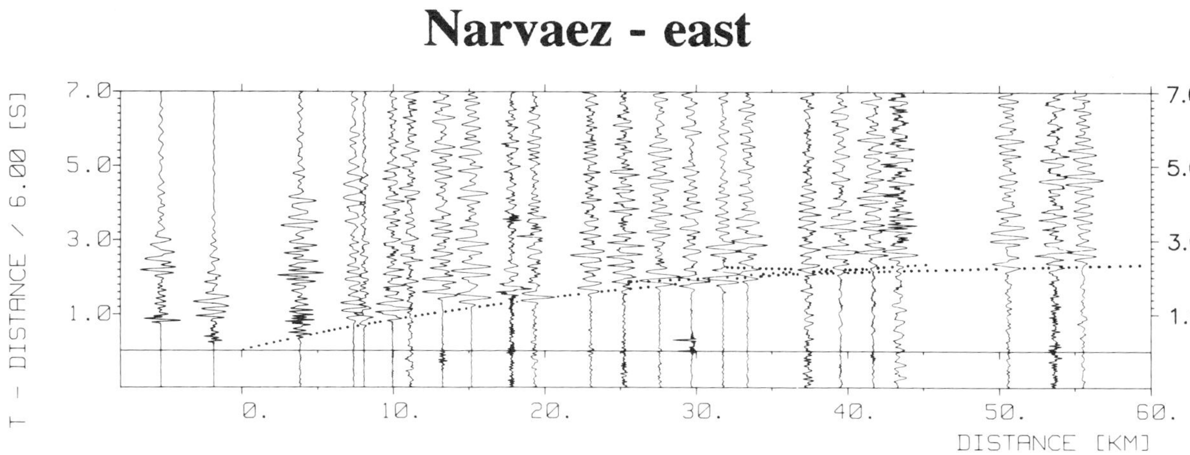

Fig. 15. The record section Narvaez-east (SP 37, BP: 2-12 Hz) as an example of the reversed 70 km profile in the Subandean Ranges. The *dotted lines* refer to the corresponding 1D model calculations (Fig. 16)

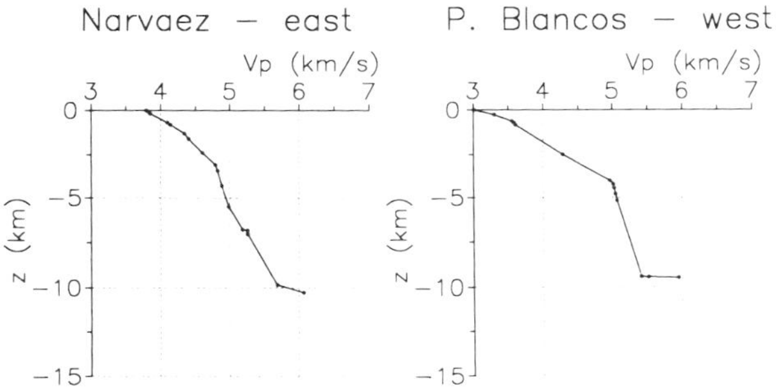

Fig. 16. Velocity-depth functions derived from the Narvaez-east (SP 37) and the reversed Palos Blancos-west (SP 39) record section in the Subandean Ranges. The upper parts of the functions down to 4 km represent steeply dipping sediments of the fold and thrust belt, the 4-10 km range shows older sediments in a more horizontal position; typical basement velocities are reached at 10 km depth

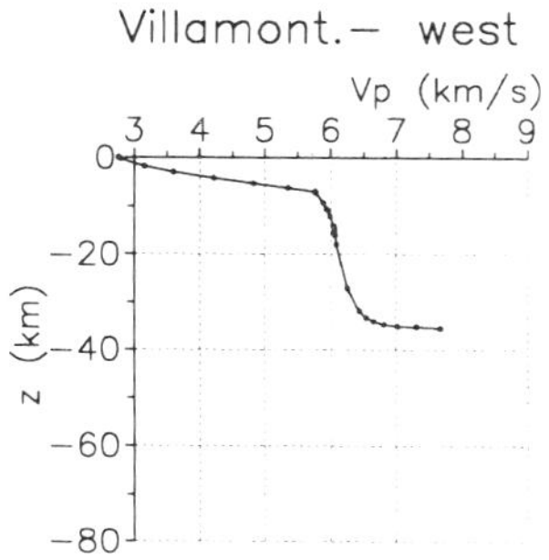

Fig. 17. Velocity-depth function for the section Villamontes-west (SP 36). The graph already represents the typical structure of the Brazilian Shield

Cretaceous mantle has been partly serpentinized by the introduction of water from the descending slab, thus inducing the reduction in mantle velocities (Fig. 19). Another interpretation of this crustal stack follows from the assumption that the Jurassic arc developed upon a continental crust with the forearc positioned west of today`s Coastal Cordillera. Considering the actual distance of 300 km between the trench and the magmatic arc, it can be concluded that 200 km of the former forearc was consumed by tectonic erosion (see also von Huene and Lallemand 1990). This crustal material, too, could have formed this depth level under consideration, or at least contributed, together with the above discussed process, to its genesis. Further contributions to this crustal level might come from magmatic addition and, finally, a combination of all these processes should be taken into account for the 20-30 km depth level under the Coastal Cordillera. This crustal level increases up to 30 km in thickness to the east under the Precordillera, while its average velocity decreases (Fig. 20).

4.2 The Thickened and Weakened Arc Region and the Altiplano Basin

Between the Precordillera and the Western Cordillera a rapid change in crustal structure must be noted, especially for the middle and lower crust. As shown in Fig. 18, this crustal segment is considered ductile, at least in its middle and lower part. LVZs distributed in the entire crust of this region and a related,

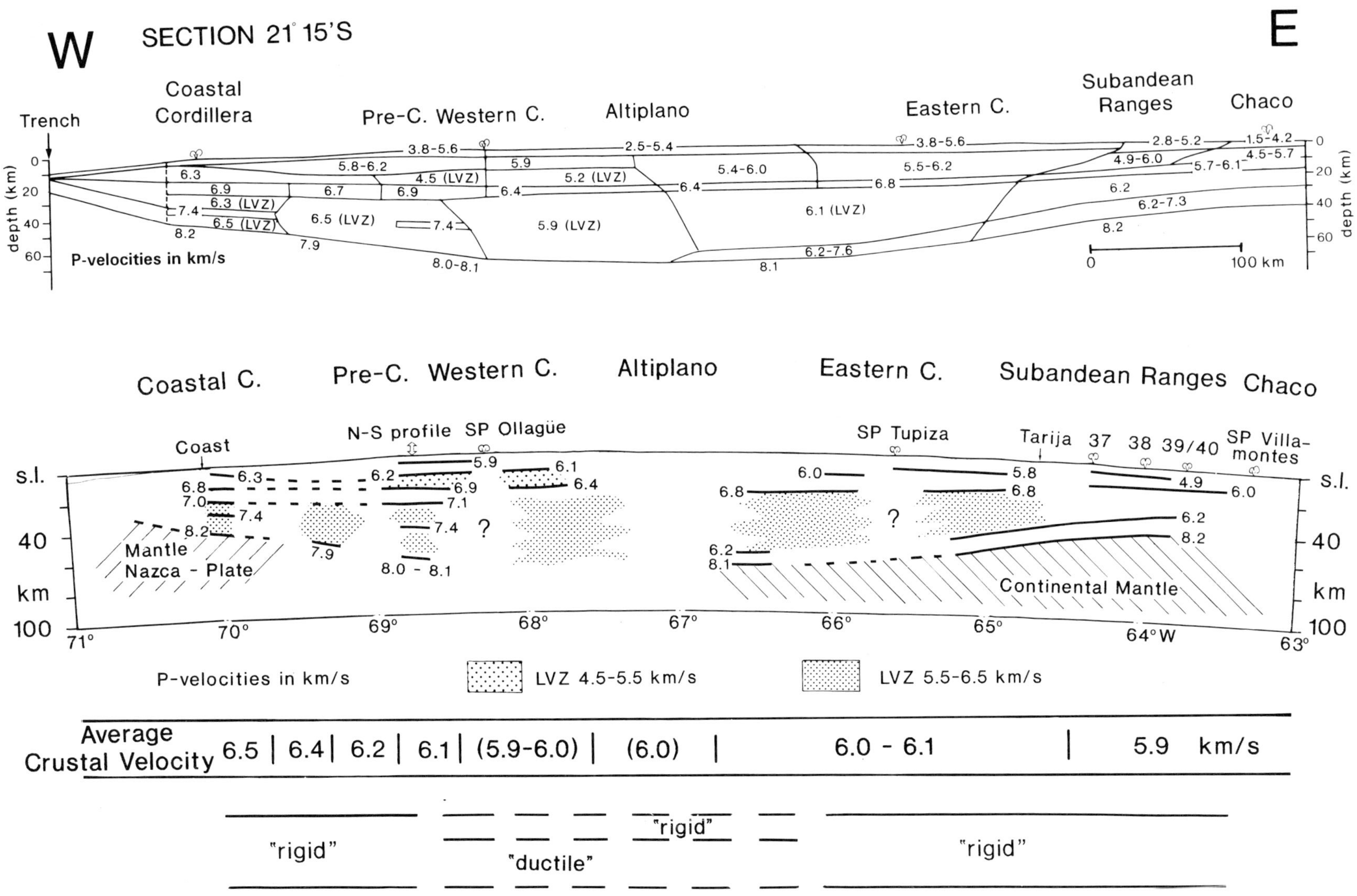

Fig. 18. Cross sections at 21°15'S showing seismic discontinuities and P-wave velocities. The *upper part* of the figure shows the major velocity domains of the 2D ray-tracing model, whereas the section in the *middle part* shows the velocity lines only in those parts which are proven by vertices, thus indicating the most reliable parts of the model. The average crustal velocities *(below)* indicate roughly rigid and ductile crustal segments. Note that the values in the eastern part are considerably influenced by sediments. If this influence is eliminated, the values increase for the area of the Eastern Cordillera to 6.2 km/s and for the Subandean Ranges to the Chaco to 6.3 km/s

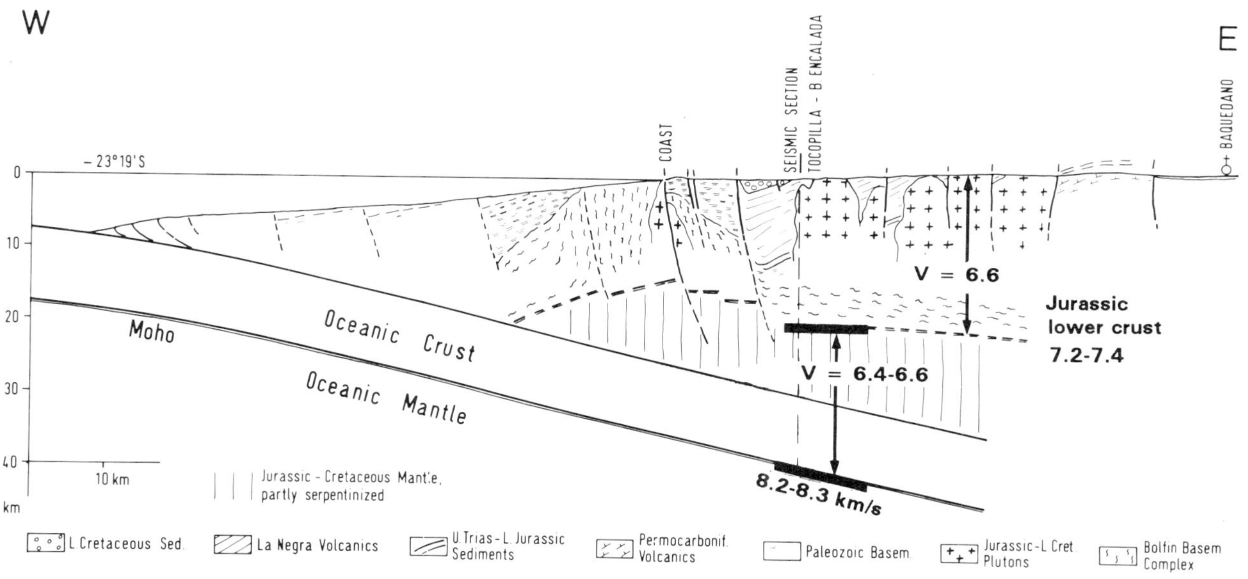

Fig. 19. Cross section through the Coastal Cordillera near Antofagasta (after Reutter et al. 1991). The discontinuity at 20 km depth is interpreted as the base of the Jurassic crust and the very clearly detected discontinuity at 40 km depth is interpreted as the Moho of the Nazca plate

relatively low average velocity evidence this character.

The seismic data from the western margin of the modern magmatic arc point to an at least 70-km-thick crust. Gravity data show a Bouguer anomaly of less than -400 mGal, indicating a crustal thickness in the same order (Strunk 1990; Götze et al. this Vol.). However, as seismic data do not show a crust/mantle boundary below the Western Cordillera and the adjacent eastern Altiplano (except the southern profile at 24°S), it is assumed that the lower crust and uppermost mantle are characterized by drastically reduced velocities, which may represent partially melted zones, caused by ascending hot fluids of the dehydration process of the Nazca plate. Haak and Giese (1986) relate this dehydration process to the earthquake cluster at about 100 km depth (Fig. 20). They argue that metastable phase transitions from gabbro to eclogite could initiate the earthquakes and release the fluids which ascend through the mantle wedge.

One consequence might be the absence of seismic signals in the deeper crust of the arc as discussed above. Another possibility of crustal response at this depth level can be observed on the southern W-E profile at 24°S (Figs. 9 and 10). The imbricated structure could be explained by zones of partial melting induced by mafic intrusions. In comparison with the absence of signals in the northern area this type may represent a less progressive stage in the weakening of the crust. Another explanation might be due to different temperature and viscosity constraints. As observerd often in seismic reflection data this might be the reason for the existance of so called lamellae or missing seismic signature, resp., in the lower crust (Meissner 1986).

Strong correlations between LVZs and zones of high electrical conductivity (HCZ), introduced by Schwarz et al. (1984, 1986, this Vol.), are given at different depth levels. Below the area of the largest geothermal anomaly of South America, El Tatio/Laguna Colorada (east of Chuquicamata, at the Chile/Bolivia border), extremely high electrical conductivity starting from 10 km depth down to at least 30 km as well as low velocities and a scattering of seismic waves (Chinn et al. 1980) are present, pointing to partially melted zones or magma chambers in the upper crust. In the eastern part of the modern magmatic arc and in the western Altiplano the HCZ continues below 40 km depth. This observation agrees well with the absence of seismic signals. The distribution of LVZs and HCZs obviously divides the crust at 40 km depth, thus the upper part might be related to the Jurassic/Cretaceous continental crust and the lower part to the actual lower crust generated in Cenozoic times.

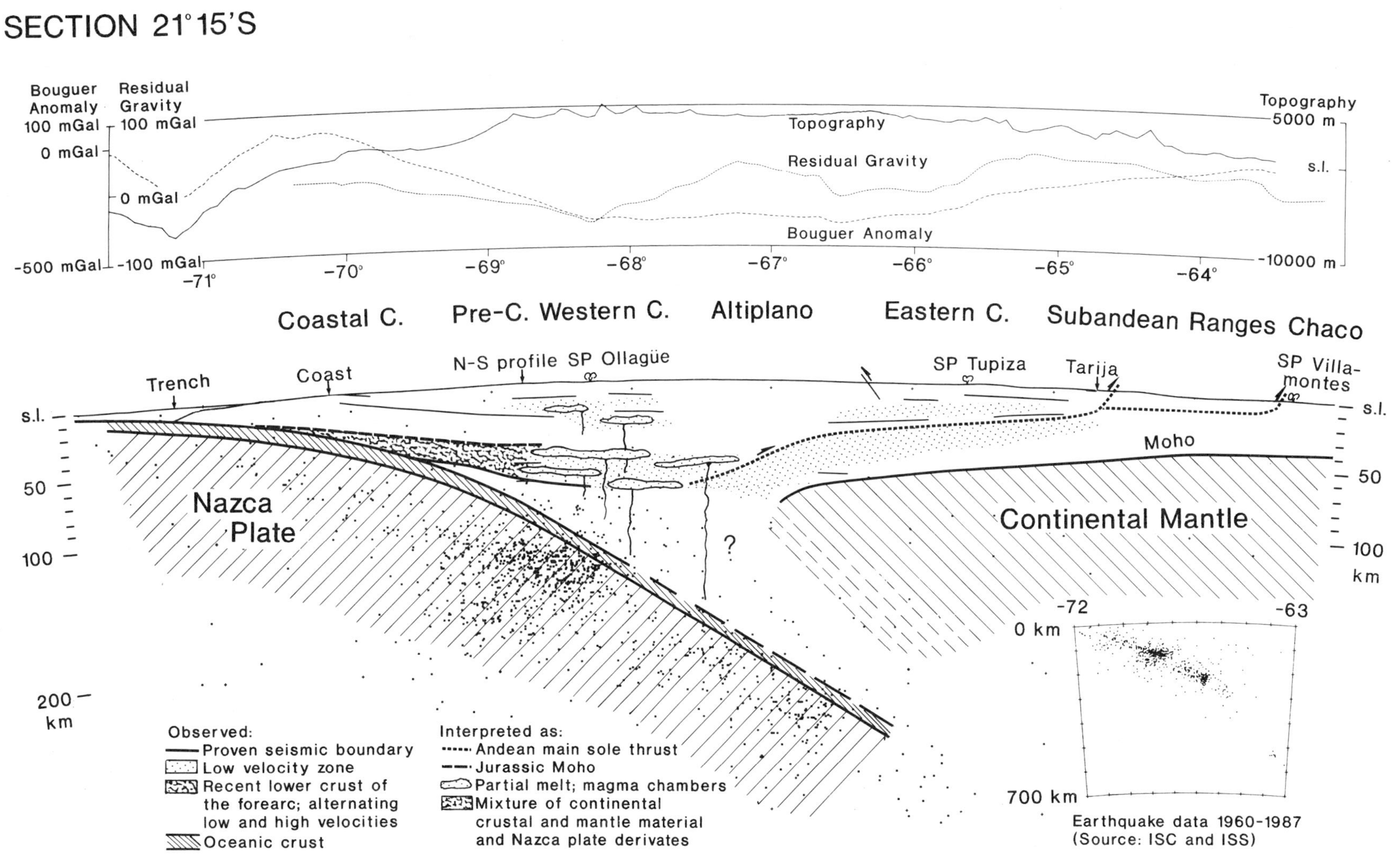

Fig. 20. The interpretative cross section at 21°15'S bases on the main seismic discontinuities and low velocity zones found in the presented data (see Fig. 18). Earthquake hypocenter data describe the subducted Nazca plate as a Wadati-Benioff zone. Related gravity data shown *above* are from Götze et al. (this Vol.). The assumed main Andean sole thrust is indicated by a *hatched line*. The continental backarc crust is overthrusted to the east, thus generating crustal doubling and strongly thickened actual crust. For the forearc and modern arc magmatic addition to the crust must be considered as a contribution to crustal thickening. This process is also responsible for the zones of partial melts in the crust of the Western Cordillera and western Altiplano. For a more detailed explanation see legend and text

4.3 The Crustal Doubling in the Backarc

The tectonic units of the backarc consist of a high plateau, the Altiplano, the Eastern Cordillera and the Subandean Ranges, a typical fold and thrust belt which is the easternmost Andean structure. The amount of crustal shortening in the backarc is not known for all units in detail. For the Subandean belt a minimum of shortening of 140 km was calculated by Kley and Reinhardt (this Vol.). The values of the other units are assumed to be smaller but in general a minimum of up to 230 km can be considered for the Central Andean backarc region (Isacks 1988; Roeder 1988; Sheffels 1990). Our seismic data presented here unambigously do support such a crustal model with crustal shortening effects of this order.

The discovered discontinuities beneath the Eastern Cordillera at 20-25 km depth (see Fig. 18) are interpreted as the base of a lower continental crust, formerly situated further to the west which was overthrust on the foreland crust (Fig. 20). The LVZ observed below the mentioned discontinuities is related to the upper and middle parts of the overridden foreland crust. In contrast to the Eastern Cordillera, there are no indications that the crystalline crust of the Subandean Ranges is involved in this folding and thrusting. Lower crust and modern Moho are also confirmed by the seismic data at about 60 km beneath the Eastern Cordillera ascending to the foreland. There, a total crustal thickness of 40 km represents shield conditions.

Finally, it must be recalled that the total Andean crust is thickened and different processes must be considered to explain its growth. Doubling of two rigid crusts occurred in the backarc. Thickening of the forearc is thought as a modification of the former underlying mantle, probably accompanied by accretion processes. The cause of thickening in the ductile crust of the modern magmatic arc must be seen primarily in the addition of magmatic material at levels immediately below or within the crust. Tectonic thickening is considered to play the dominant role in crustal shortening in the Central Andes.

Acknowledgements. Geophysical measurements in desert and mountain areas demanded greatest efforts from all field participants - Argentinean, Bolivian, Chilean and German colleagues, staff and students. Many thanks to all of them for supporting our field survey! Of great importance was the logistic support given by numerous persons and institutions which cannot be mentioned all. However, without close cooperation with the following institutions, our work could not have been realized: Programa Sismológico, Universidad de Chile, Antofagasta; Universidad Católica del Norte, Antofagasta; Universidad de Chile, Santiago; Universidad Nacional de Salta; Universidad Mayor de San Andrés and Observatorio San Calixto, both at La Paz; Technische Universität Berlin and RWTH Aachen. The Chilean Navy supported us by providing ships and crews to carry out the blasts in the Pacific Ocean. The national oil companies of Argentina YPF (Salta and San Pedro, Prov. de Jujuy), and of Bolivia YPFB (Santa Cruz, La Paz and Tarija) and the geological service of Chile SERNAGEOMIN were engaged in the drilling to prepare the blasts on the different land sites. YPFB once again provided valuable logistic support. The same must be noted for the Chilean national copper company, CODELCO, division Chuquicamata, which enabled us to record the explosion times inside the mine and made available specific data of the regular blasts. We are grateful to these institutions and all the persons who stand behind these names. We thank David Crossley from McGill University Montreal for surrendering the ray tracing program "RAYAMP". The main financial support for the experiments was provided by the Deutsche Forschungsgemeinschaft and the Freie Universität Berlin.

References

Asch G, Blümecke T, Buness H (1992) An application of evolutionary strategy on geophysical modelling. Subm to Geophys J Int

Baldzuhn S (1992) Tiefenseismische Untersuchungen in der Ostkordillere und im Subandin Südboliviens. Dipl Thesis, Freie Universität Berlin, FB Geowiss

Barton PJ (1986) The relationship between seismic velocity and density in the continental crust - a useful constraint? Geophys J R Astr Soc 87: 195-208

Chinn DS, Isacks BL, Barazangi M (1980) High-frequency seismic wave propagation in western South America along the continental margin, in the Nazca plate and across the Altiplano. Geophys J R Astr Soc 60: 209-244

Fisher RL, Raitt RW (1962) Topography and structure of the Peru-Chile trench. Deep Sea Res 9: 423-443

Gerver M, Markushevich V (1966) Determination of a seismic wave velocity from the travel-time curve. Geophys J R Astr Soc 11: 165-173

Giese P (1976a) General remarks on travel time data and principles of correlation. In: Giese P, Prodehl C, Stein A (eds) Explosion seismology in Central Europe, data and results. Springer, Berlin Heidelberg New York, pp 130-136

Giese P (1976b) Depth calculation. In: Giese P, Prodehl C , Stein A (eds) Explosion seismology in Central Europe, data and results. Springer, Berlin Heidelberg New York, pp 146-161

Giese P, Schütte K-G (1980) Resultados das medidas de sísmica de refraçao a leste da Serra do Espinhaçao, M.G., Brasil. In: Nuevos Resultados de la Investigación Geocientífica Alemana en Latinoamerica. Dt Forschungsgemeinschaft y el Inst de la Colab Científica, Tübingen, pp 44-50

Götze H-J, Schmidt S, Strunk S (1988) Central Andean gravity field and its relation to crustal structures. In: Bahlburg H, Breitkreuz C , Giese P (eds) The Southern Central Andes. Lecture Notes in Earth Sciences 17. Springer, Berlin Heidelberg New York, pp 199-208

Haak V, Giese P (1986) Subduction induced petrological processes inferred from magnetotelluric, seismological and seismic observations in N-Chile and S-Bolivia. Berl Geowiss Abh (A) 66: 231-246

Heinsohn W-D (1990) Krustenseismische Untersuchungen in der Küstenkordillere Nordchiles und Aufbau und Realisierung einer automatischen Registrier-Apparatur. Dipl Thesis, Freie Universität Berlin, FB Geowiss, 58 pp

Isacks BL (1988) Uplift of the central Andean plateau and bending of the Bolivian orocline. J Geophys Res 93: 3211-3231

Ludwig WJ, Nafe JE, Drake CL (1970) Seismic refraction. In: Maxwell AE (ed) The sea 4, Wiley, New York, pp 53-84

Meissner R (1986) The continental crust. Academic Press, Orlando USA,426pp

Ocola LC, Meyer RP (1972) Crustal low-velocity zones under the Peru-Bolivia Altiplano. Geophys J R Astr Soc 30: 199-209

Ocola LC, Meyer RP (1973) Crustal structure from the Pacific Basin to the Brazilian Shield between 12° and 30° south latitude. Geol Soc Am Bull 84: 3387-3404

Ocola LC, Meyer RP, Aldrich LT (1971) Gross crustal structure under Peru-Bolivia Altiplano. Earthquake Notes XLII, 3-4: 33-48

Reutter K-J, Giese P, Götze H-J, Scheuber E, Schwab K, Schwarz G, Wigger P (1988) Structures and crustal development of the Central Andes between 21° and 25°S. In: Bahlburg H, Breitkreuz C, Giese P (eds) The southern Central Andes. Lecture Notes in Earth Sciences 17. Springer, Berlin Heidelberg New York, pp 231-261

Reutter K-J, Heinsohn W-D, Scheuber E, Wigger P (1991) Crustal structure of the Coastal Cordillera near Antofagasta, northern Chile. VI. Congr Geol Chil, Viña del Mar, Actas. SERNAGEOMIN Santiago, pp 862-866.

Roeder D (1988) Andean-age structure of Eastern Cordillera (Province of La Paz, Bolivia). Tectonics 7 (1): 23-39

Rössling R (1989) Petrologie in einem tiefen Krustenstockwerk des jurassischen magmatischen Bogens in der nordchilenischen Küstenkordillere südlich von Antofagasta. Berl Geowiss Abh (A) 112, 73 pp

Schmitz M, Wigger P, Heinsohn W-D, Baldzuhn S, Rudloff A (1990) Data compilation from the Central Andes refraction seismic experiments of the research group "Mobility of Active Continental Margins" in the years 1982, 1984, 1987 and 1989. Open file report, Institute of Geophysical Sciences, Free University, Berlin

Schmitz M, Wigger P, Araneda M (1991) La estructura Preandina entre el Salar de Huasco y Chuquicamata mediante sísmica de refracción. VI Congr Geol Chil, Viña del Mar, Actas,. SERNAGEOMIN Santiago, pp 882-886

Schwarz G, Haak V, Martínez E, Bannister J (1984) The electrical conductivity of the Andean crust in northern Chile and southern Bolivia as inferred from magnetotelluric measurements. J Geophys 55: 169-178

Schwarz G, Martínez E, Bannister J (1986) Untersuchungen zur elektrischen Leitfähigkeit in den Zentralen Anden. Berl Geowiss Abh (A) 66: 49-72

Sheffels B (1990) Lower bound on the amount of crustal shortening in the central Bolivian Andes. Geology 18: 812-815

Spence GD, Whittall KP, Clowes RM (1984) Practical synthetic seismograms for laterally varying media calculated by asymptotic ray theory. Bull Seismol Soc Am 74 (4): 1209-1223

Strunk S (1990) Analyse und Interpretation des Schwerefeldes des aktiven Kontinentalrandes der zentralen Anden (20°-26°S). Berl Geowiss Abh (B) 17: 135 pp

Tatel HE, Tuve MA (1958) Seismic studies in the Andes. Am Geophys Union Trans 39: 580-582

von Huene R, Lallemand S (1990) Tectonic erosion along the Japan and Peru convergent margins. Geol Soc Am Bull 102: 704-720

Wiechert E (1910) Bestimmung des Weges der Erdbebenwellen im Erdinnern: 1. Theoretisches. Phys Z 11: 294-304

Wigger P (1986) Krustenseismische Untersuchungen in Nord-Chile und Süd-Bolivien. Berl Geowiss Abh (A) 66: 31-48.

Wigger P (1988) Seismicity and crustal structure of the Central Andes. In: Bahlburg H, Breitkreuz C , Giese P (eds) The southern Central Andes. Lecture Notes in Earth Sciences 17. Springer, Berlin Heidelberg New York, pp 209-229

Wigger P, Araneda M, Röwer P (1988) Investigaciones sísmicas de refracciones en el norte de Chile. V Congr Geol Chil, Actas II, Santiago, F185-F202

Wigger P, Araneda M, Giese P, Heinsohn W-D, Röwer P, Schmitz M, Viramonte J (1991) The crustal structure along the Central Andean transect derived from seismic refraction investigations. In: Omarini R, Götze H-J (compilers) Global geoscience transect 6, Central Andean transect, Nazca Plate to Chaco Plains. ICL and AGU, Washington D.C.: pp 13-19.

Woollard GP (1960) Seismic crustal studies during the I.G.Y. Part II. Continental program. Am Geophys Union Trans 41: 351-355

Crustal High Conductivity Zones in the Southern Central Andes

GERHARD SCHWARZ, GUILLERMO CHONG DIAZ, DETLEF KRÜGER, ELOY MARTINEZ, WINFRID MASSOW, VOLKER RATH and JOSÉ VIRAMONTE

Abstract. Magnetotelluric (MT) and geomagnetic deep sounding (GDS) experiments were performed in the southern Central Andes from 1982 to 1989. There is evidence for high conductivity zones (HCZ) at crustal depth beneath parts of the Central Andes. MT data give quantitative results: a zone of extremely high electrical conductivity (total conductance of 20000 to 30000 Siemens) at shallow depth beneath the volcanic belt of northern Chile strikes about NNW-SSE, running into northwestern Argentina. The eastern border of the HCZ is found in the Eastern Cordillera of southern Bolivia and probably northwestern Argentina, where total conductance of the upper crust reaches more than 10000 S. The area in between both cordilleras, the Altiplano, shows besides a well-conducting cover an increase in conductivity at much greater depth, i.e., at about 40 km in its western part and at about 20 km in the east (total conductance $>$ 10000 S). The HCZs beneath the Andes have quite different origins and do not strictly correlate with the known structural units. They can be classified not only in their E-W but also in their N-S extent. The HCZs correlate with some other geophysical and geological features and may reflect the intense tectonic and magmatic evolution of this mountain belt.

1 Introduction

The Andes of South America are classical examples of plate tectonics: the process of mountain building initiated by the subduction of oceanic lithosphere under the continent can be studied in a wide sense. Although geophysical investigations in the Andes started as early as in the 18th century, e.g. by von Humboldt, the real data base for describing the structure and the physical state of - and processes within - the crust and upper mantle is relatively scarce. The history of the Andes still remains a mystery. One of the reasons for this is obviously related to the large N-S extent of this mountain range of more than 7000 km as well as to the difficult access and the rough conditions which are met there.

The first integrated Andean studies were done during the International Geophysical Year (IGY) 1957-1958: Forbush and Casaverde (1961) conducted a detailed latitude survey of diurnal variations in the equatorial electrojet (EEJ) region of southern Peru and northern Chile with simultaneous registrations of all three magnetic field components. Parallel to the EEJ program, a seismic refraction experiment was done in 1957 (Tatel and Tuve 1958) which led to the discovery of very high attenuation of seismic waves that travel across the Andean mountain range. Could this attenuation possibly be connected to the volcanic structure of the Andes, namely to the anomalous physical state of the crust caused by partly molten material at depth? As there exists a direct relation between high temperature and high electrical conductivity, it seemed promising to investigate the crustal electrical conductivity structure by geomagnetic deep soundings (GDS), in the region where seismic high attenuation had occurred. The first GDS surveys, undertaken by the Carnegie Institution of Washington, the Instituto Geofísico del Peru, and the Instituto Geofísico Boliviano in 1963 and 1965 revealed the existence of a high conductivity zone (HCZ) at crustal depth in southern Peru and Bolivia (Schmucker et al. 1964, 1971, Schmucker 1973). This HCZ became known later as the Andean conductivity anomaly and follows roughly the strike of the Eastern Cordillera. According to 2D modelling the anomalous structure is situated about 20 km under the Cordillera, and extends laterally about 400 km E-W. The picture of the classical subduction regime seemed to be perfect: magma rising from the depth of the Wadati-Benioff zone to the upper crust, where it is stored in magma

Correspondence to: Gerhard Schwarz, Fachrichtung Geophysik, Freie Universität Berlin, Malteserstr. 74-100, D-1000 Berlin 46

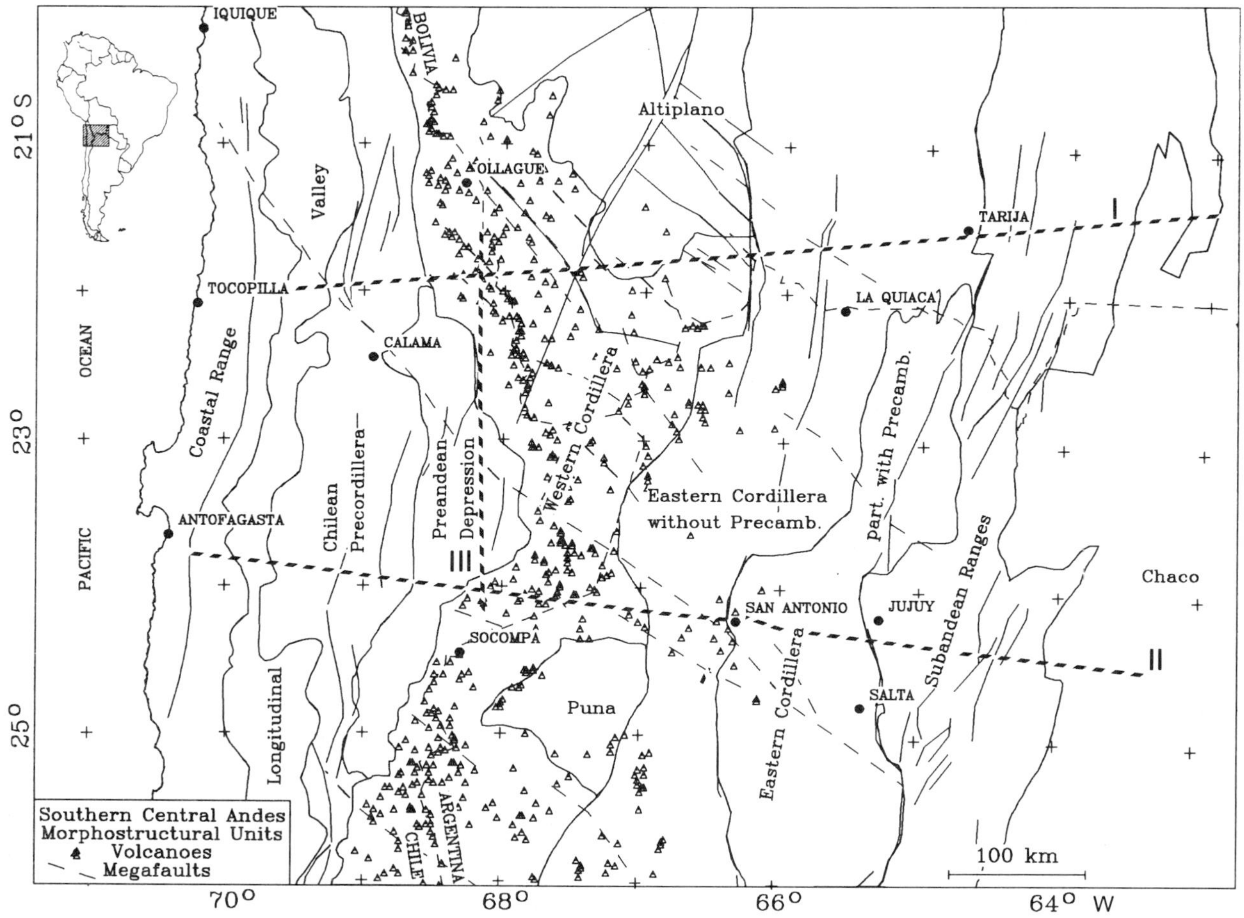

Fig. 1. Morphostructural sketch map of the southern Central Andes according to Reutter et al. (1988). Also shown is the distribution of volcanoes (*open triangles*) and the main large-scale fault systems. Dashed lines mark the measured traverses: traverse I runs between lat 21° and 22°S from the Pacific Coast into the lowland plains of southern Bolivia; II starts in the Coastal Cordillera at lat 24°S, while its easternmost site is located in the Argentine Chaco. Traverse III is parallel to long 68°W, roughly in the strike direction of the Western Cordillera

chambers and detectable by its high electrical conductivity.

In the following years the Carnegie Institution together with national institutions extended their studies southwards into Chile and Bolivia. Aldrich et al. (1975) detected a large-scale HCZ under the Western Cordillera and the southern Bolivian Altiplano in the region of lat 22°S, but quantitative deductions were not made.

Here, at about lat 22°S the Central Andes show their largest width and the migration of the tectonic and magmatic activity starting in the Jurassic and being active until the Quaternary is clearly evidenced. In the early 1980s a group of geoscientists from both universities of Berlin (West) focused their research activities on this region (Fig. 1). Among the methods to probe the crustal structure of the Andes were magnetotelluric and geomagnetic deep soundings. They give the distribution of electrical conductivity with depth, which depends on a wide range of petrological and physical parameters, e.g., fluids and volatiles, conducting minerals (like graphite), or partial melts due to enhanced temperatures. Because electrical conductivity is sensitive to high temperatures, i.e., those exceeding 700°C, it can also provide information on processes in the lithosphere. As temperature gradient is an important controlling parameter in subduction regimes, the interpretation of electrical conductivity data can provide constraints on other physical parameters. Electromagnetic methods

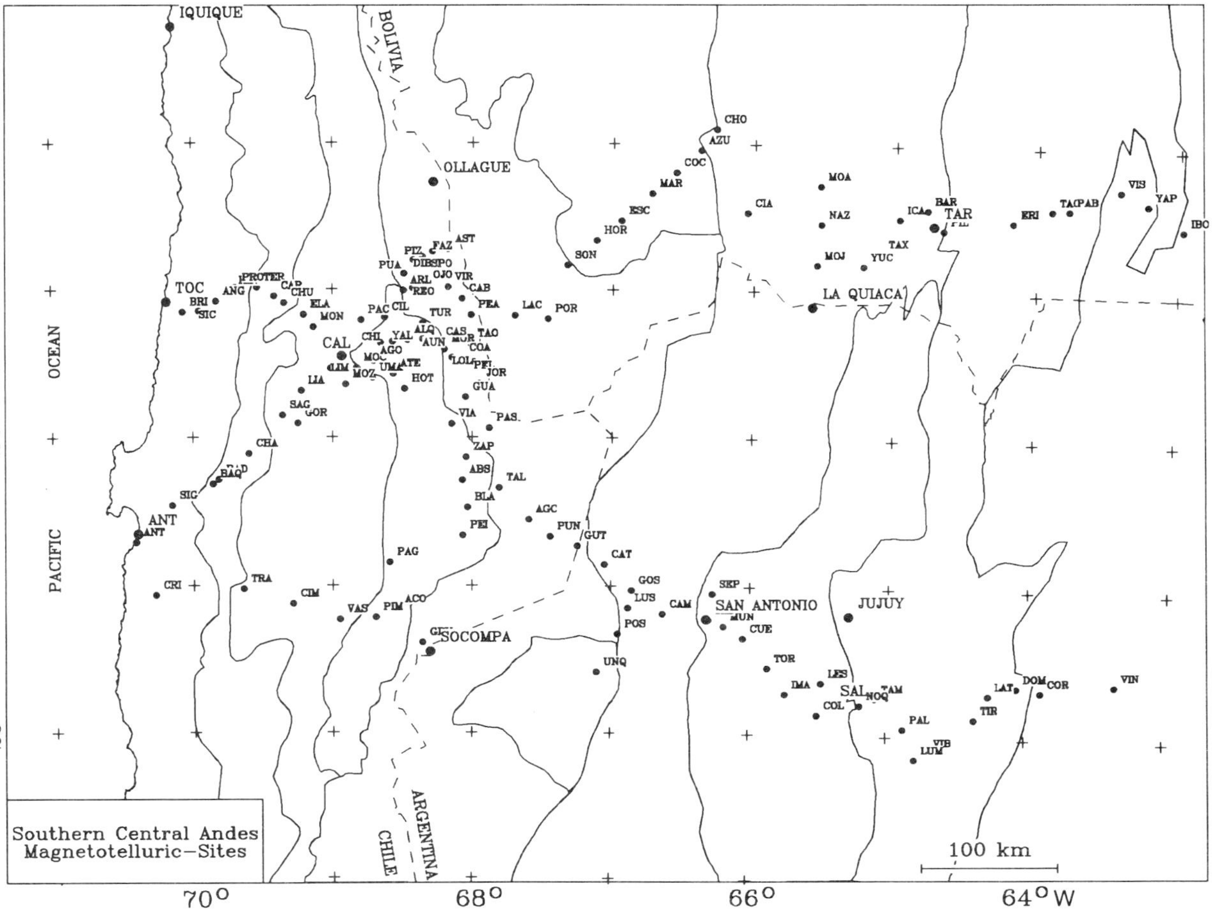

Fig. 2. Map showing the position of magnetotelluric and geomagnetic deep soundings in the southern Central Andes. Structural units according to Fig. 1

have also been successfully used elsewhere (e.g. Emslab-Group 1988, Wannamaker et al. 1989) for probing the deep structures and processes of subduction regimes. In the following we report on and discuss the extended studies of the Andean electrical conductivity structure.

2 Field Experiments

The magnetotelluric and geomagnetic deep soundings were done in the southern Central Andes from 1982 to 1989 (Schwarz et al. 1984, 1986, 1990, Krüger et al. 1990, W. Massow in prep.). Seven field campaigns between the Pacific coast of northern Chile and the Andean lowland plains of southern Bolivia and northwestern Argentina were performed with a total duration of more than 24 months.

The first measurements were pilot studies giving the rough conductivity structures at depth. Two main profiles were measured crossing all Andean structural units - one at about lat 22°S and one at about lat 24°S (Figs. 1, 2). A third and shorter profile connects both. As the area of the Western Cordillera of northern Chile showed an extreme increase in conductivity at shallow crustal depth, the depth of investigation was limited. Therefore this area received special attention: in 1984 it was investigated by long period soundings to resolve lower crustal conductivity structures, and in 1989 the Western Cordillera was covered by a very dense profile of sounding sites. Additionally, seismological measurements were done at some of the sites and seismic refraction experiments were carried out parallel to some sections of the magnetotelluric profiles (see Wigger et al. this Vol., also for further references regarding the seismological work).

Soundings were carried out at more than 120 sites, of which about one-third are situated in the Western Cordillera. The instrumentation included five-component magnetotelluric stations (two electric and three magnetic channels) with either induction coil or flux-gate magnetometers and telluric devices operating in the period range from 10 to 3000 s or 40 to 20000 s. The electromagnetic data were recorded digitally into solid state memory with 12-bit resolution (Beblo and Liebig 1986). A paper chart recorder was used for visual inspection. The equipment was totally solar powered and buffered with lead batteries to allow self-supporting operation in the remote area under investigation. Total measuring time per site varied typically between two and three weeks, except for a long period survey in northern Chile in 1984 where the observation time was three months. Up to eight stations were operated simultaneously. Due to the remoteness of the area under investigation - ranging from the Atacama Desert of exceptionally arid climate in the west to the subtropical Chaco lowland plains in the east - and its harsh environment, special precautions had to be taken both for the field crew and the instruments.

3 Data Analysis

The magnetotelluric method aims at determining period (T) dependent transfer functions (Z) between the horizontal electric field (*E*) and the magnetic field (*B*), measured simultaneously at the earth's surface and following the linear equations

$$E_x = z_{xx} B_x + z_{xy} B_y,$$

$$E_y = z_{yx} B_y + z_{yy} B_x, \qquad \text{with } T = \text{const};$$

while geomagnetic deep sounding only uses the magnetic field (*B*) to calculate a transfer function, relating the vertical to the horizontal magnetic field

$$B_z = z_x B_x + z_y B_y.$$

The complex-valued coefficients of Z are calculated by a least squares algorithm. Processing procedures for data evaluation, e.g. data selection, trend removal, frequency analysis and correction for instrumental transfer functions are described in some detail elsewhere (Schwarz et al. 1986; Krüger in prep., Massow in prep.).

The magnetotelluric transfer function Z is used to calculate period dependent apparent resistivities, phases between electric and magnetic fields and skewness. The GDS transfer functions are displayed in the form of the so-called induction arrows (Schmucker 1970). For more clarity only the 'real' arrow is presented in this work.

In the case of a homogeneous electrical structure, i.e. a 1D-earth, the calculation of the magnetotelluric transfer function elements is easy, as the elements z_{xy} and z_{yx} are equal. Over a strictly 2D-earth with the direction of resistivity structures striking along the x-axis of an orthogonal coordinate system, the elements z_{xy} and z_{yx} represent the E- and B-polarization cases and the electric field is polarized in the strike direction. In this special case the Schmucker induction arrows are orthogonal to the strike of the conductive structures and point in the direction of increasing resistivities. The normal procedure in MT data analysis is to rotate the coordinate system to separate the modes of the E- and B-polarization. In this study, strong anisotropy of apparent resistivities at some sites and the fact that the B-polarization data are much more affected by topography made this a crucial point in data evaluation and interpretation. Rotation criteria have been discussed earlier by Schwarz et al. (1984, 1986). After analysis of the preference directions of the electric field, of the induction vectors, and of the continuity of the MT transfer functions, the approximate strike direction of Andean structures was found to represent the E-polarization. Further analysis for all sites then included one-dimensional modelling of the E-polarization transfer functions to provide a resistivity structure, which served as the basis for subsequent two-dimensional modelling. Two-dimensional model calculations have been extensive and often difficult considering the complexity of the data and the large area under study. These are still in progress (Krüger in prep.). As shown later, in the discussion of the data, conductive structures are, at least in certain areas, three-dimensional. Because of lack of reasonable modelling algorithms (e.g. Cerv and Pek 1990), this strongly limits the interpretation possibilities.

4 Main Features as Inferred from Geomagnetic Deep Soundings

The transfer functions of the geomagnetic field, as introduced in the previous section are used to calculate geomagnetic induction arrows. The geomagnetic deep sounding (GDS) technique may be used to define the lateral extension of the Andean orogen according to the electrical structure of its crust, but

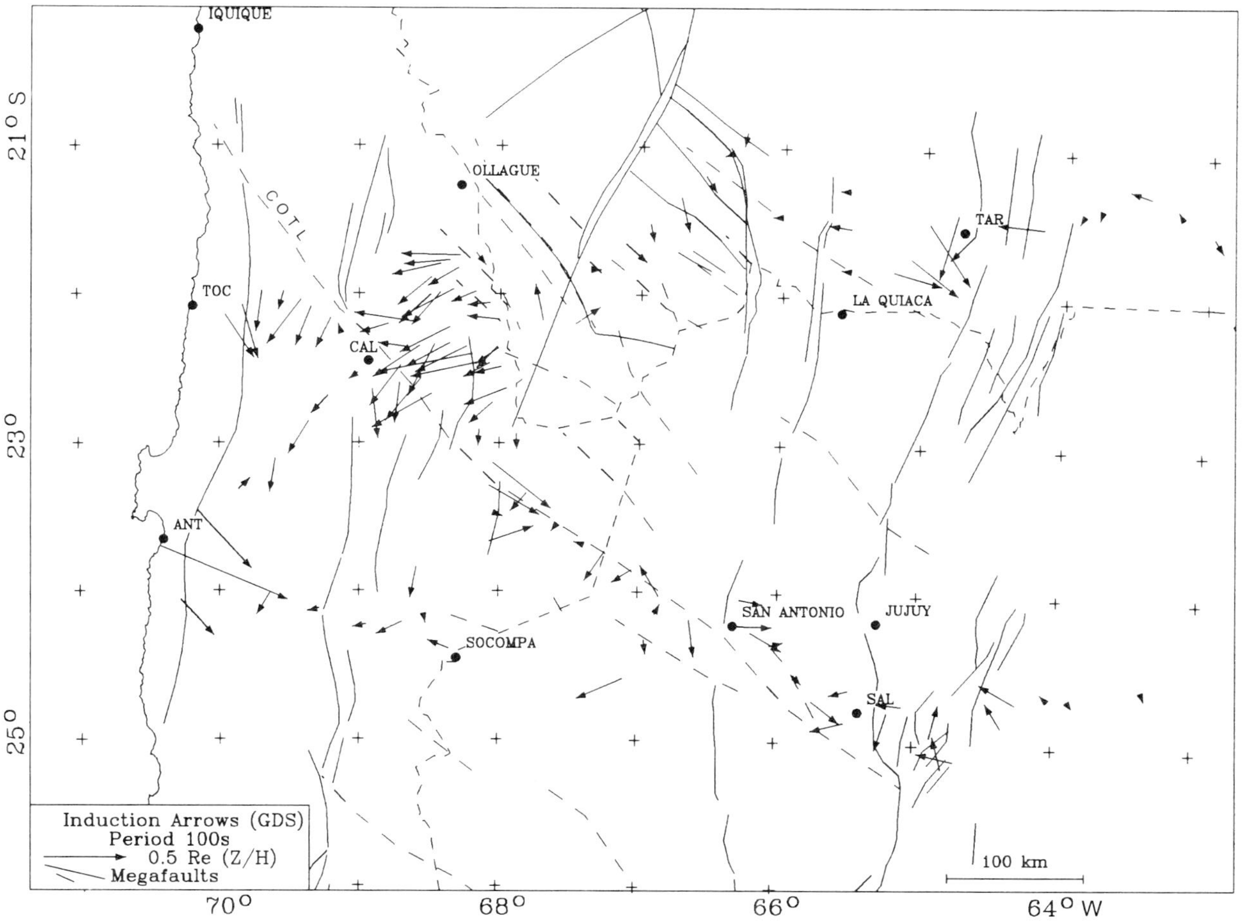

Fig. 3. Real induction arrows for a period of 100 s with the main large-scale fault systems marked. Induction arrows point away from well conducting structures and are longest where the gradient in conductivity is largest. Although the induction arrows represent only small-scale well conducting structures at many sites, some large-scale HCZ may also be identified, e.g. the Western and the Eastern Cordillera, and the Subandean Belt. Abbreviations of some towns: *TOC* Tocopilla; *ANT* Antofagasta; *CAL* Calama; *TAR* Tarija; *SAL* Salta. The Calama-Olacapato-El Toro lineament mentioned in the text is denoted by *COTL*

the depth resolution of GDS is rather poor because of the equivalence of distinct models. This can be seen, e.g., in the interpretation of data from the Eastern Cordillera of southern Peru (Schmucker et al. 1964, Schmucker 1973; Tarits and Menvielle 1986). The induction arrow gives the direction of the gradient in electrical conductivity and its length is proportional to the gradient. The arrows point away from zones of low resistivity.

Figures 3 to 6 show real induction arrows for periods from 100 to 4000 s, which are well suited for resolving shallow as well as deeper seated structures, i.e., from the upper to the lower crust. Either large-scale fault systems, volcanoes and structural units, or residual gravity (after Götze et al. 1988) are added to each of the figures to visualize a possible correlation to observed HCZs. In addition to the expected variations of electrical conductivity in the E-W direction, the data show clear changes along the geological strike direction, i.e. N-S. This was not expected with such clarity. According to the GDS data we can identify several zones with high contrast in electrical conductivity.

For all period ranges given, the Coastal Cordillera and the Longitudinal Valley show large amplitudes of induction arrows. This is mainly due to the strong inductive effect of the Pacific Ocean with a total conductance (conductivity-thickness product) in the

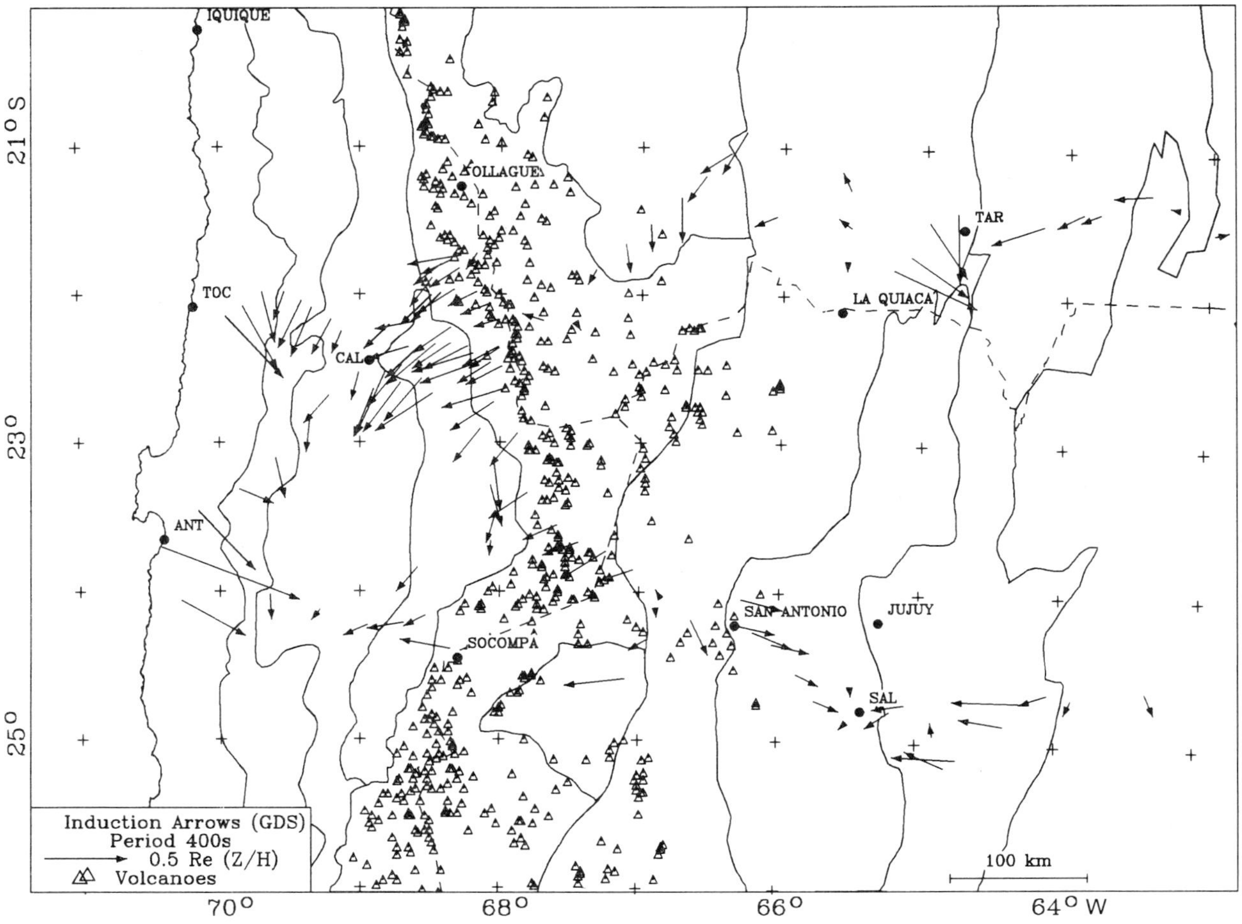

Fig. 4. Real induction arrows for a period of 400 s. The distribution of volcanoes (*open triangles*) and main structural units are also shown. According to the induction vectors, the HCZ under the Western Cordillera at about lat 22°S does not follow the strike of the volcanic belt but cuts it and extends into northwestern Argentina. North of lat 22°S it bends westward

deep sea of more than 25000 S. However, E-W striking conducting structures must also be inferred. At the site ANT, i.e. closest to the coast, the induction arrow reaches its maximum in length at a period of 400 s (Fig. 4) - but it points directly away from the coast only for periods of more than 4000 s (Fig. 6). In a strictly two-dimensional structural setting, with the ocean as the HCZ, the induction arrows should point directly E. The deviation in direction is possibly caused by deep reaching fault systems, which may serve as pathways for fluids if they are still active or which may be mineralized. Total conductance must be much greater in the north (at the latitude of Tocopilla: TOC) than in the south. Here again, northeast of Tocopilla, the inductive effect of the ocean dominates the HCZ only for very long periods (4000 s and more). The ocean effect can be observed as far inland as 69°W.

East of Calama, a prominent large-scale conductive structure is observed within the Western Cordillera striking about NNW-SSE into northwestern Argentina, i.e. not strictly parallel to the mountain belt. This trend is evidenced by the maximum in length of the induction arrows at the individual sites, which indicates the maximum in lateral contrast of electrical conductivity (see esp. Figs. 5 and 6). It cannot be determined, whether this HCZ bends towards the west close to lat 22°S or whether the HCZ NW of Calama has a different source. Induction vectors reach their maximum length at periods of about 1000 s but maintain a stable direction even for the longest periods observed. Deflexions in the general trend of the induction vectors at certain periods point to small-scale inductive structures: e.g. for periods

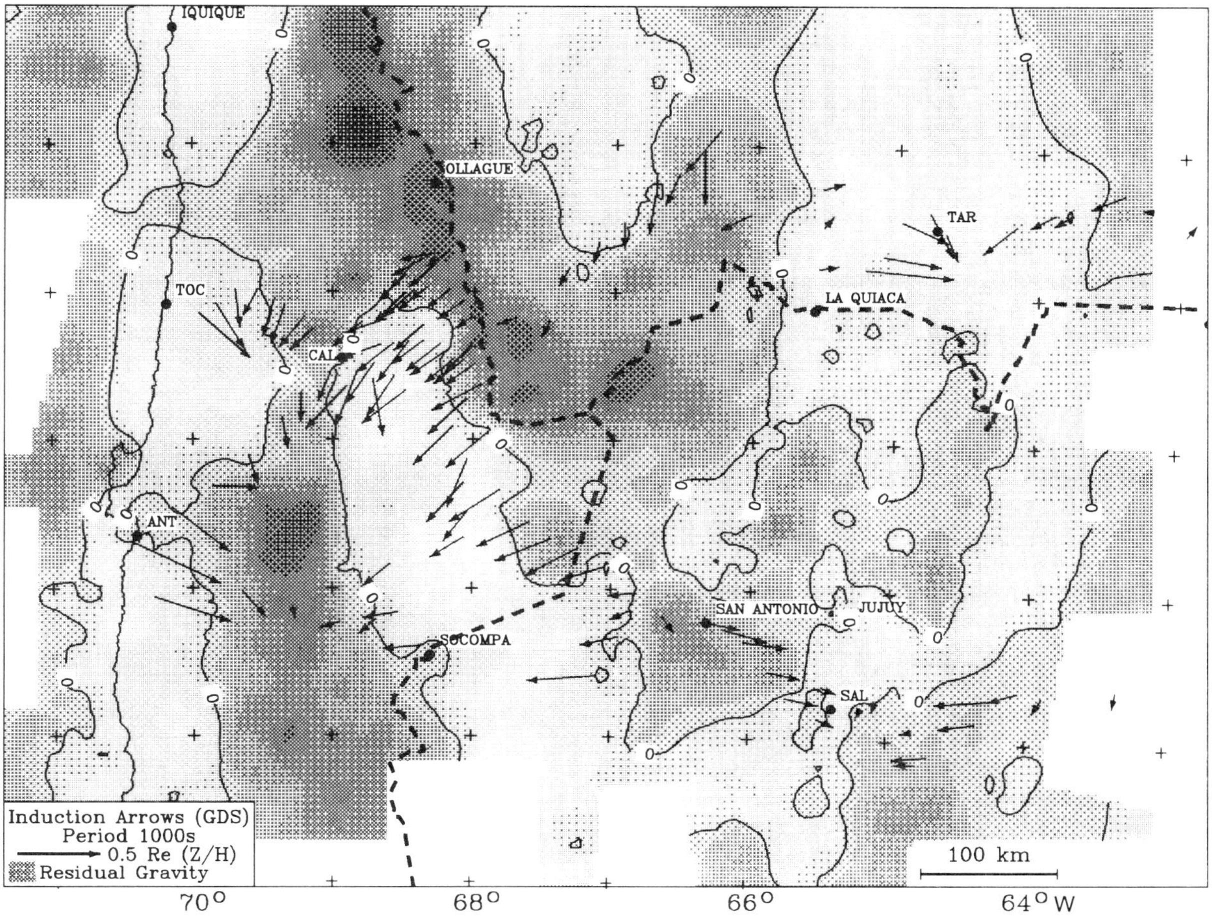

Fig. 5. Real induction arrows for a period of 1000 s, with residual gravity (after Götze et al., this Vol.). A correlation seems to exist between residual gravity and induction vectors: the arrows point away from areas with minima in gravity (darkest in shading), i.e. a deficit in masses seems to correspond to high electrical conductivity

much less than 1000 s the Salar de Atacama, with its eastern border at about long 68°W and lat 23° to 23.5°S, can be identified as a local HCZ.

Further east in southern Bolivia and northwestern Argentina, across the Western Cordillera, induction arrows do not indicate any eastern edge to the observed high conductivity crustal structures. One may expect to observe this, if one considers only the volcanic belt itself as the reason for the HCZ at depth. The Altiplano of southern Bolivia ends at about lat. 22°S and so does an HCZ, as indicated by induction vectors pointing southwards for almost all periods. This suggests that the thick sediments themselves are the conductors with the total conductance decreasing southwards. The decrease may be caused by the Cordillera de Lipez which borders the Altiplano to the south. Farther south on the Argentine Puna and in the westernmost Eastern Cordillera conductive structures are only poorly resolved.

In the Eastern Cordillera of Bolivia the total conductance first increases dramatically and then drops abruptly somewhere W of Tarija. This is best documented for periods of 4000 s (Fig. 6). For shorter periods a superposition of induction vectors was observed, leading even to arrows pointing almost southwards close to Tarija. This effect is due to an increase of conductivity again east of Tarija. The HCZ west of Tarija strikes about N-S for longer periods (i.e. at greater depth). It should have a rather complex structure because of the quite complicated period dependence of the induction vectors, i.e. an increase of total conductance westwards. The HCZ seems to be identical with the southwards striking

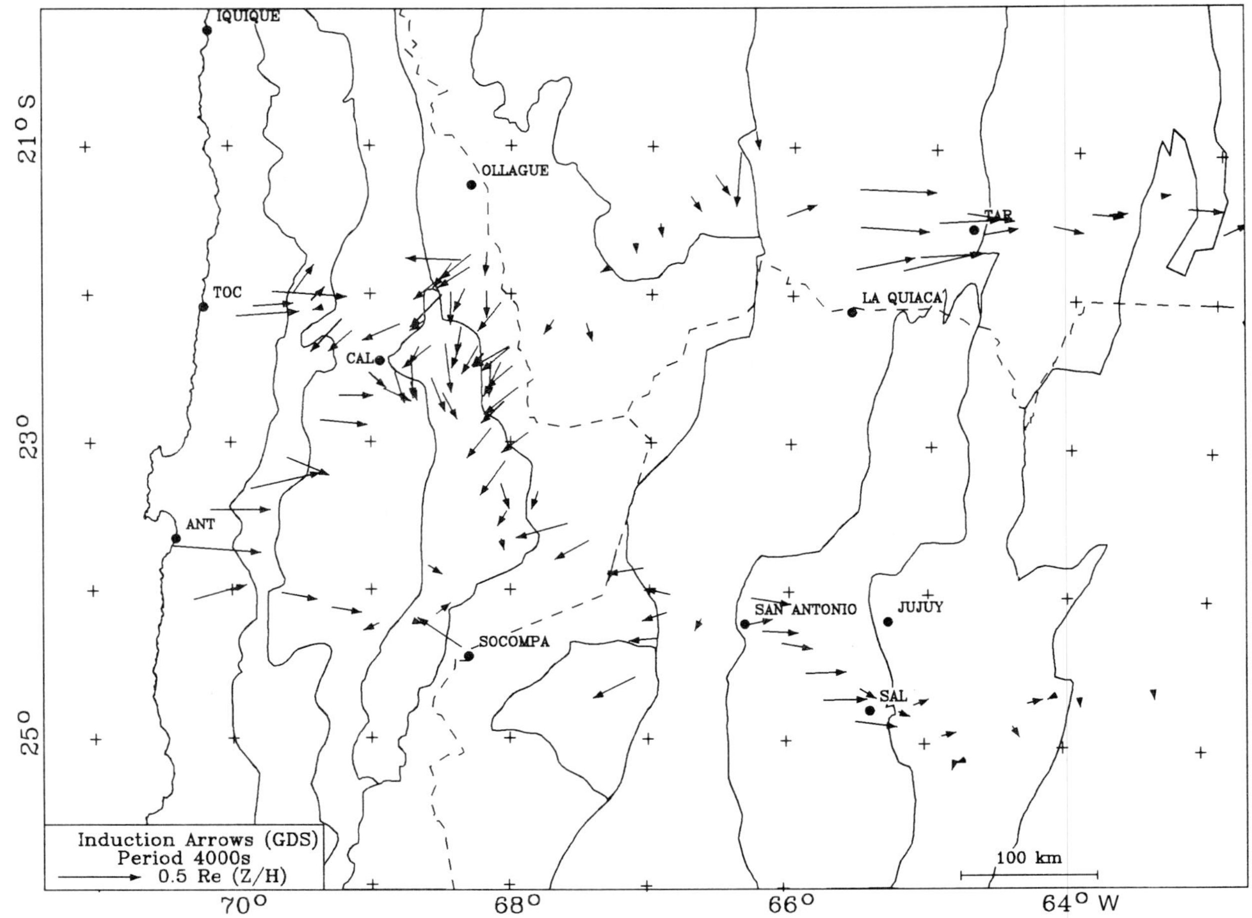

Fig. 6. Real induction arrows for a period of 4000 s, indicating deep seated and large-scale, highly conductive structures, which are best documented in the area between the Western Cordillera of northern Chile and the Eastern Cordillera of southern Bolivia and northwestern Argentina. Furthermore they show the coast effect, i.e. the Pacific Ocean

zone observed by Schmucker et al. (1964) in southern Peru. It is difficult to follow it precisely into northwestern Argentina, but it appears to exist again W of San Antonio de los Cobres, although the total conductance appears to be much lower there according to the length of the vectors.

Summarizing the GDS results we can identify the area stretching from the Western Cordillera across the Bolivian Altiplano to the Eastern Cordillera as being a large HCZ. It forms a triangle with its southern corner close to San Antonio in northwestern Argentina. While in its strike this large-scale HCZ seems to follow the Eastern Cordillera, its western border seems to cut the volcanic belt rather than to follow it. As mentioned already a definite depth scaling of conductive structures is not possible with GDS data alone. Therefore we will try to infer depth ranges of the observed HCZs with the help of magnetotelluric soundings, and finally we will include GDS data again in the two-dimensional model calculations of conductive structures.

5 Magnetotelluric Soundings

Analysis of the horizontal components of the electric and magnetic fields yields period-dependent apparent resistivities and phases. Examples of MT sounding results, typical of sites within the individual structural units of the southern traverse, are displayed in Fig. 7. Apparent resistivity values seldom exceed 100 ohm m, and normally give a three-layered structure. The resistivity curves for the two modes of polarization show marked differences, although the

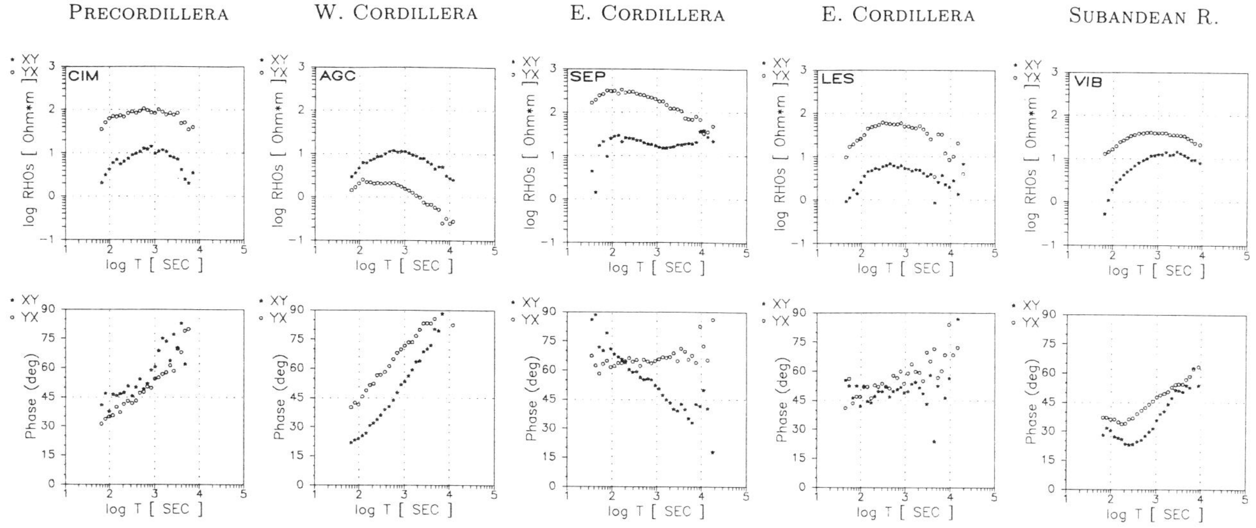

Fig. 7. Magnetotelluric results (period dependent apparent resistivities and phases) from sites within individual structural units (Precordillera to Subandes) crossed by the southern traverse (II)

phases are often comparable and even continuous from site to site. Therefore static shift effects cannot be excluded though they are difficult to handle, as 'normal resistivities' at depth are improbable in an active orogen.

Site CIM of Fig. 7, lying in the Chilean Precordillera, shows a three-layered crust with a distinct high conductivity zone at lower crustal depth. The comparable phases and parallel shifted curves of apparent resistivity may be due to a static shift effect. The situation at AGC in the Western Cordillera is similar, but the AGC data reveal an HCZ at probably shallower depth and with even higher conductivity. Comparable resistivity curves were obtained on the Bolivian Altiplano (e.g., site ESC), leading to three-layered structures with a pronounced HCZ at mid to lower crustal depth. In the Eastern Cordillera of northwestern Argentina the electrical resistivity structure is far from uniform. At SEP the response curve is rather flat with a broad minimum, indicating a good conductor in between more resistive ones while the LES sounding curves resemble those from the Western Cordillera with an HCZ indicated at crustal depth. Together comparison of the LES and SEP results demonstrates the variability of data from this region. Data from site VIB within the Subandean Ranges are very different. The resistivity curves rise for periods of up to 1000 s obviously due to the thick sedimentary cover resting on a more resistive basement. This type of data is characteristic of the lowland plains of northwestern Argentina and southern Bolivia.

The very high electrical conductivity values beneath the Western Cordillera of northern Chile lead to a strong shielding effect, limiting the depth of MT soundings to a maximum of 20 km even for periods of 4000 s. This was the reason for extending the soundings to much longer periods for investigating lower crust to upper mantle conductivities. The special techniques of these soundings are described by Schwarz et al. (1986). Even for sounding periods of 43200 s an investigation depth of only about 30 km was reached, e.g., at the site El Tatio (TAT/TAO, see Fig. 8). These measurements were made at two sites close to each other and gave further proof of the correct separation of the E- and B-polarization modes: the minimum apparent resistivity curves for sites TAT and TAO (which represent the E-polarization case) are continuous and follow each other in amplitude. The maximum curves are separated from each other due to the influence of topography on the B-polarization electrical field (for a review see, e.g., Jiracek 1990). These shallow conductive structures have been studied in more detail with dense MT measurements on a profile from the Longitudinal Valley to the Western Cordillera in northern Chile, i.e., W and E of Calama, sites TER to PEL (see map of Fig. 2). Station spacing was only 10 to 15 km. The MT response curves change very smoothly from site to site and indicate only weak lateral conductivity variations.

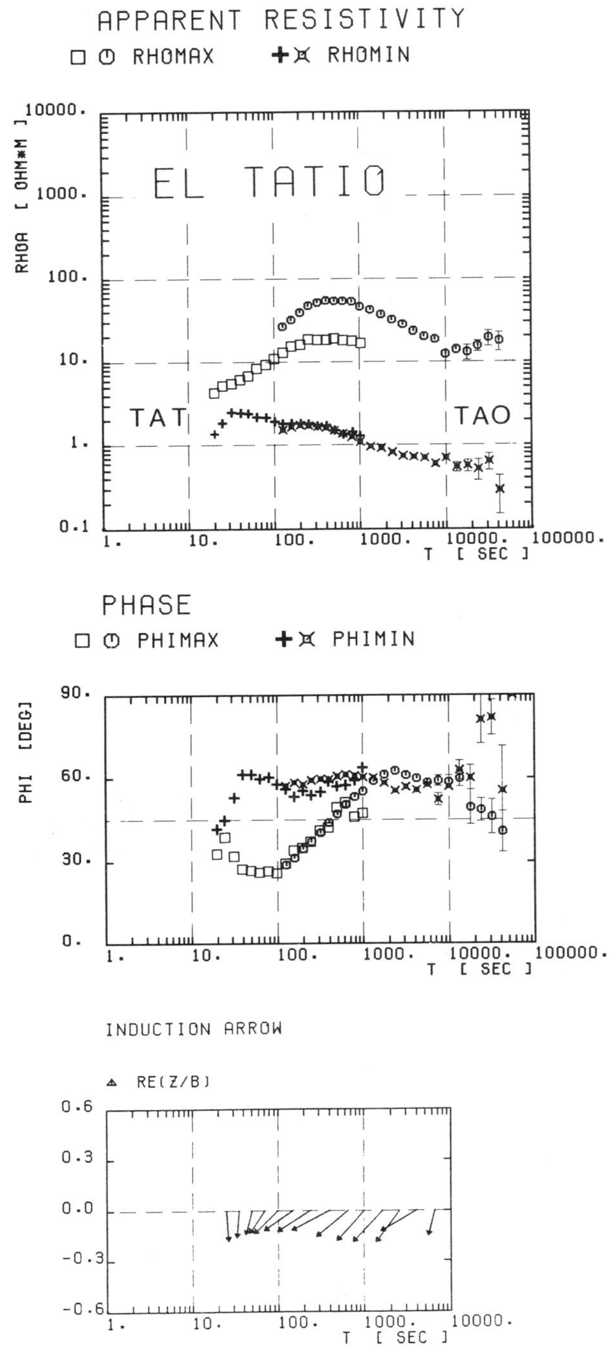

Fig. 8. Long period apparent resistivities and phases for sites *TAT* and *TAO* (El Tatio), lying in the Western Cordillera about 1.3 km from each other. The minimum curves of apparent resistivity represent the E-polarization case while the maximum curves correspond to the B-polarization mode. The latter is biased by the effect of topography on the electric field. Electrical resistivity drops below 1 ohm m at crustal depth

6 Cross Sections of Electrical Resistivity

One-dimensional resistivity models were calculated from the apparent resistivity and phase data and used to construct composite resistivity cross sections for certain parts of the profiles. These are now discussed. Two-dimensional modelling of structures is still in progress and will be presented in forthcoming publications (Krüger in prep.; Massow in prep.).

Figure 9 displays a composite resistivity cross section for the Western Cordillera and the Longitudinal Valley. Starting at site TER in the W at about lat 22°S, within the Longitudinal Valley, the section displays a conductive overburden of up to 10 km thickness. With the present data base we cannot answer the question about the origin of the HCZ north of this profile, which was detected by geomagnetic deep soundings. The overburden alone, even with increased thickness can not explain the GDS data. The conductive overburden stretches about 60 km into the Precordillera where more resistive material is found at shallow depth. Lower crustal HCZs were found at some sites but do not give a really consistent picture. East of Calama, i.e., within the Preandean Depression a conductive zone reaches upper crustal level. It is divided from another shallow and very thick HCZ below the volcanic belt (e.g., El Tatio: TAT) by a broad region of relatively resistive material. At a depth of more than 30 km the good conductors may be combined. The HCZ reaches a total conductance of 20000 to 30000 S. Its lower bound is not known although very long period MT data were analysed, e.g., at site TAO. Figure 10 presents a N-S section of electrical resistivity in the Western Cordillera (sites AST to GRU), showing that the conductive structures change along the strike direction of the volcanic belt. The HCZ between sites TUR (lat 22°14'S) and ABS (lat 23°17'S), also visible in the E-W section of Fig. 9, is the dominant feature. Relatively good conductors are found at greater crustal depth also north- and southwards.

Resistivity sections of the Bolivian Altiplano and the lowland plains farther E, adjacent to the profile just presented, have been described earlier (Schwarz et al. 1984, 1986) and are therefore only summarized in brief: the Altiplano of southern Bolivia has a thick conductive cover due to sediments; the moderately conductive crust is underlain by very conductive material (1 to 4 ohm m) at a depth of ca. 40 km in the west. This HCZ can be followed eastwards, its upper bound being found at a depth of about 20 km at the western border of the E-Cordillera. The lower bound of the HCZ under the Altiplano has not yet been

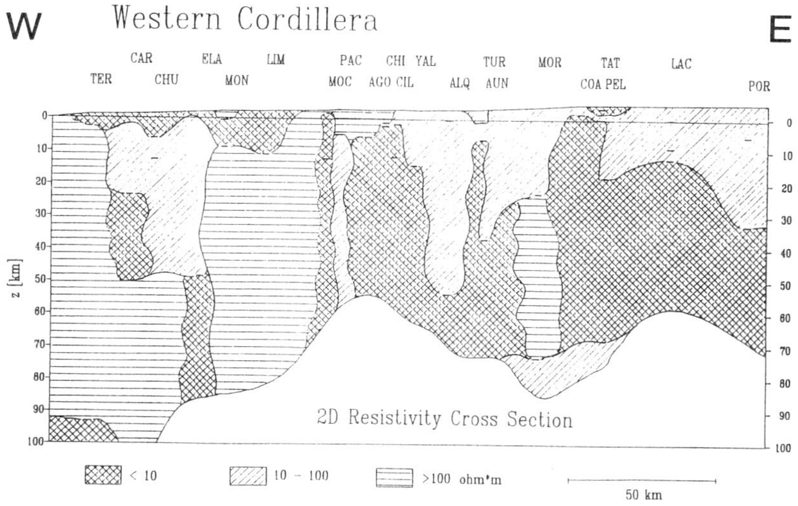

Fig. 9. A resistivity cross section for the western part of traverse I (see Fig. 1), compiled from one-dimensional resistivity models. Data from site *TER* (Longitudinal Valley) to *POR* (Western Cordillera) were used

detected because of screening by the HCZ itself. Based on model assumptions we infer a total conductance of at least 12000 S, which leads to a thickness of the HCZ of more than 12 km when assuming a conductivity of $(1\ \text{ohm m})^{-1}$. Within the Eastern Cordillera an HCZ is found at shallow crustal depth. It is obviously related to thick sedimentary layers providing a total conductance of more than 10000 S. They are bordered to the E, close to Tarija, by uprising and more resistive basement structures.

Figures 11 and 12 present data from the southern profile: on the NW-SE section W-Cordillera-Subandean Ranges two HCZs have been identified at middle to upper crustal levels. One, E of the Salar de Atacama, extends towards the Chilean-Argentine border (site GUT); the other is found at the longitude of San Antonio de los Cobres. Data from the other sites do not give a particularly consistent picture of the electrical crustal structures. The reason for this can be seen in the distortion of the telluric fields by near-surface inhomogeneities (e.g., static shift). The section following to the E reaches from the Subandean ranges to the Chaco plain (Fig. 12). It exhibits a continuously increasing thickness of highly conducting sediments overlying a resistive basement, though its depth is not very well resolved. Similar structures are found in the Bolivian lowland plains, where the total conductance of the overburden varies between 700 and 1300 S. However, the data reveal basement structures which are inhomogeneous and strongly anisotropic, a result which seems to be valid for the southern profile as well.

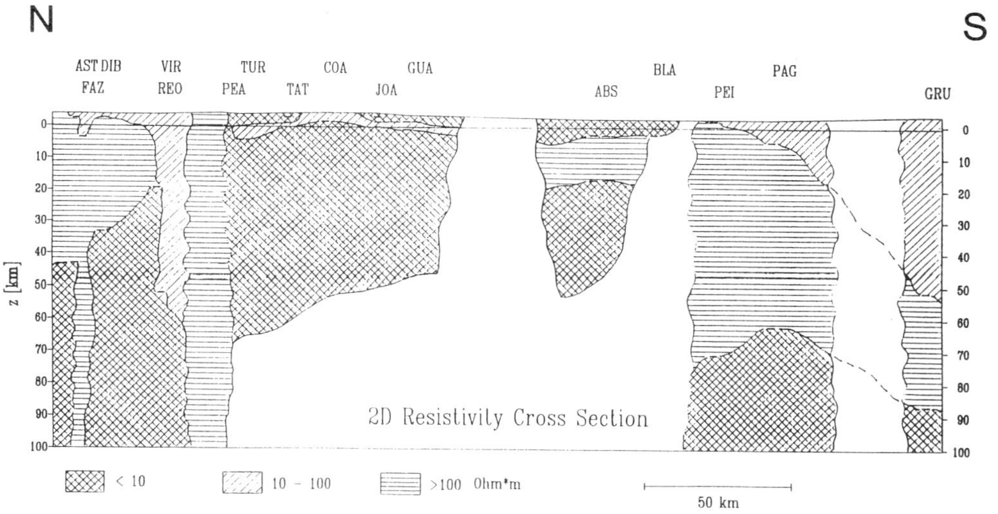

Fig. 10. A cross section of electrical resistivity in N-S direction of the Western Cordillera; from site *AST* (Ascotan) to *GRU* (Socompa), almost parallel to longitude 68°W

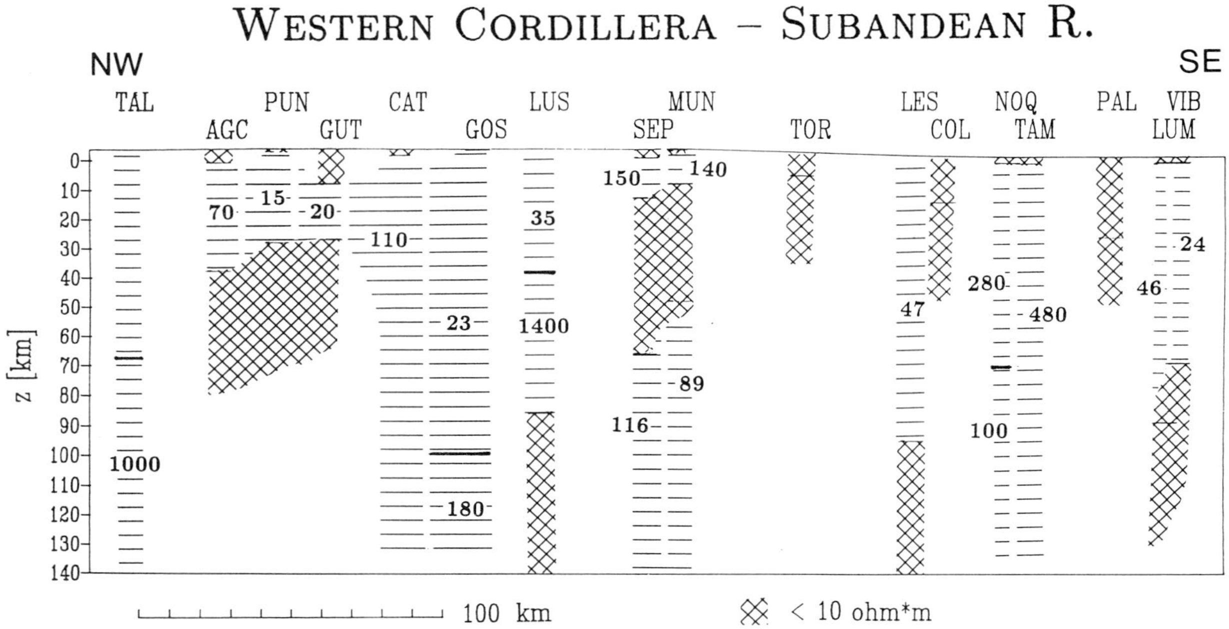

Fig. 11. One-dimensional electrical resistivity models from the Western Cordillera (site *TAL* at about long 68°W) of northern Chile to the Subandean Ranges (site *VIB* at about long 65°W) of northwestern Argentina along traverse II

7 On the Origin of the Observed High Conductivity Zones

The process of Andean mountain building is still debated. The idea of magmatic underplating (e.g., Thorpe et al. 1981, de Silva 1989), accounting for the Andean uplift, opposes the idea of mainly tectonic thickening (Roeder 1988, Sheffels 1990), or of some combination of crustal shortening and thickening with lithospheric thinning (Isacks 1988, Kono et al. 1989). Further questions arise regarding the origin of the magmas which are believed to result from either the mantle lithosphere or large-scale crustal involvement of the continental plate (see Wörner et al. this Vol.). It is well accepted that fluid phases and partial melts play dominant roles within a subduction regime. They are derived either from dehydration of subducted oceanic crust or from melting of oceanic crust or continental upper mantle. The slab loses water at all depths down to approximately 100 km and melting starts at a depth of 100 to 150 km - or even deeper, depending on the actual temperature.

Electrical conductivity is related to temperature, composition, and state of the material. Physical and chemical mechanisms increasing electrical conductivity in the lithosphere are still a matter of discussion (for an overview see, e.g., Haak and Hutton 1986; Hyndman 1988; Schwarz 1990). In an active orogen we can consider (1) fluids and volatiles (with C-O-H-S as main components); (2) rock melts connected to high temperatures; (3) graphite or other conducting minerals as being the source for increased conductivity. Constraints for differentiating the origin of an HCZ may come from additional geological and geophysical data, which have been provided within this research project (Giese et al. 1990).

The measured electrical resistivities for the crust and uppermost mantle of the southern Central Andes are all lower than those in a 'normal' geological setting almost everywhere (Schwarz 1991). Even lower values have been found at depth in both the Cordilleras, the Altiplano and also at a few sites W of the volcanic belt. Figure 13 summarizes qualitatively all HCZs, giving also their depth, as found within the present study. Clearly visible are both the Cordilleras, bordering a large-scale HCZ, the origin of which we will now try to infer.

It seems justified to divide the whole area from west to east into several units, deserving different explanations. Electrical resistivities in the Coastal Cordillera of northern Chile and adjacent areas eastwards are influenced in total by the Pacific Ocean and are generally high, except for indications of HCZs at depths greater than 40 km in the Longitudinal Valley.

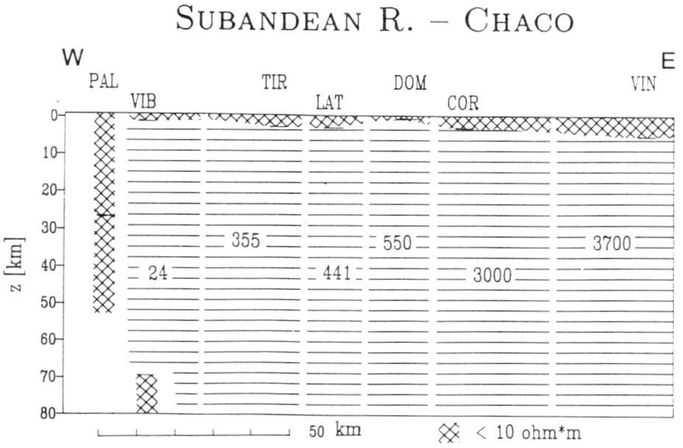

Fig. 12. One-dimensional resistivity models for the Subandean Ranges and the Chaco of northwestern Argentina. Upper crustal electrical resistivities are below 10 ohm m; the thickness of these structures increases from W to E; the basement is highly resistive and electrically not well resolved

However, better conducting structures striking roughly E-W must exist as the GDS data provide southeastwards deflected induction arrows. One reason could be crustal fluids liberated from the descending oceanic plate and rising permanently through cracks and faults to the surface. HCZs situated farther east will now be discussed in some more detail.

7.1 The Pre- and Western Cordillera

The most remarkable features in the western (Chilean) section of the traverse are (1) the extremely high conductivities as high as $(0.5\ \text{ohm m})^{-1}$ at depths greater than 10 km in the Western Cordillera from lat 22° to 23°S and at depths greater than 20 km further north and south; (2) a total conductance higher in the north than in the south, i.e., north of lat 22.3°S (according to GDS data); and (3) the strike direction of the HCZ cutting across that of the volcanic belt into northwestern Argentina.

The HCZ below the Western Cordillera must have a thickness of more than 20 km, although its depth resolution is limited due to the screening effect, and a total conductance of at least 20000 to 30000 S. The lateral extent of the HCZ seems to correlate with an area of minimal residual gravity found by Götze et al. (1988, this Vol.). In this area a high attenuation and/or scattering of seismic waves is observed, and earthquake activity is rather low at a crustal depth above 30 km (Molnar and Oliver 1969, Buness et al. 1986). The lower crust has layers with alternating high and low velocities of seismic waves resulting in a mean velocity of only 5.9 km/s. Total crustal thickness was found to be 60-65 km under the volcanic arc (James 1971, Wigger et al. this Vol.). The depth range from 80 to 100 km at the western border of the Western Cordillera from lat. 23° northwards is characterized by a high release of seismic energy (Buness et al. 1986). Further eastwards at the depth of the upper mantle high attenuation of seismic waves is observed again (Sacks 1968, Chinn et al. 1980, Whitman et al. 1992). This zone follows the Andean strike northwards and also extends further eastwards south of lat 22°. High attenuation and low velocities of seismic waves seem to be a more general feature above subduction zones and may indicate the location of crustal magma reservoirs. The observed physical parameters under the Western Cordillera may be unified in a hypothesis of a partly ductile crust. Strike-slip tectonics may have weakened the crust providing an influx of fluids and initiating partial melting: in explaining the observed HCZ, Schwarz et al. (1984) calculated the maximum amount of melts and fluid phases as 33% by volume. This value is rather high and may be valid in its extreme for crustal conditions, e.g., to be found within the geothermal anomaly of El Tatio (sites TAT, PEL). The partial melts are assumed to be SiO_2-rich, existing at a temperature of 600-700°C and a pressure of about 2.5 kbar, constrained in the crust because of their increasing viscosity with decreasing lithostatic pressure. The water is liberated at depth by dehydration processes in the downgoing oceanic slab, which reaches a depth of about 100 km under the Western Cordillera. Heat is carried upwards by a steady flow of rising fluids. Excess fluids depress the solidus temperature of crustal rocks to 600-700°C. The rocks begin to melt, convect and differentiate so that, finally, superposed on less acid rocks, silicic melts form large magma chambers in the upper crust. Ignimbrites are considered as the typical product of Late Miocene silicic magmatism at depth. Large-scale crustal melting, proposed already by Pichler and Zeil (1972), can account for the giant ignimbrite flare-up (de Silva 1989), which has occurred in the volcanic complex between lat 21° and 24°S - a major volcanic-tectonic province - less than 10.6 Ma ago. A consequence of the model (de Silva 1989) is that a large batholith should be present below the volcanic complex with the shallow magma chambers above. In contrast to the model given by Schwarz et al. (1984) and Haak and Giese (1986), the model of de Silva accounts for

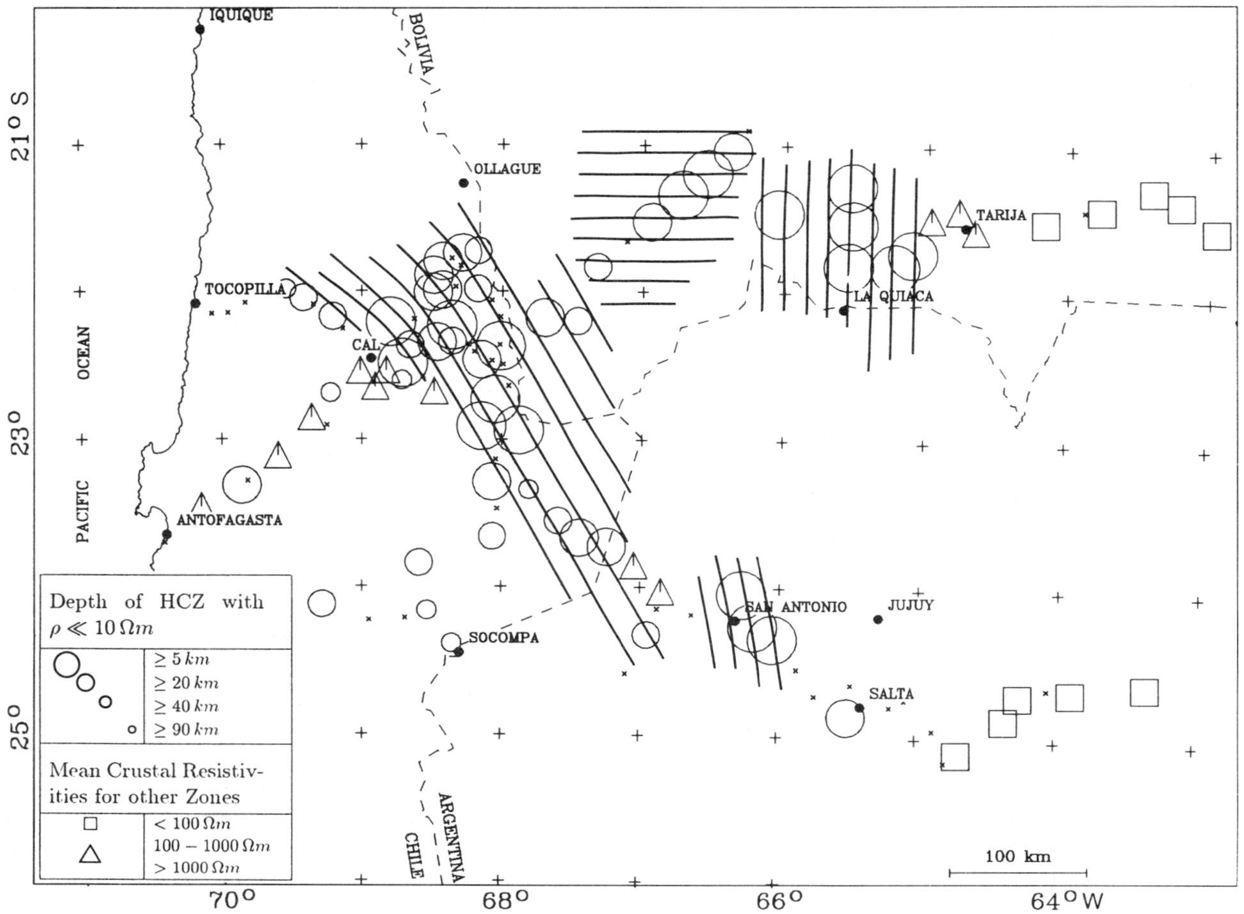

Fig. 13. A map displaying all the southern Central Andes HCZs (marked by *circles*) found within the whole study area. Their diameter is inverse to the depth at which they are seated. Superficial structures above a resistive basement are symbolized by *squares*. Normal to low conductive electrical structures are marked by *triangles*. *Dashed areas* give some regional correlations which have been inferred from MT and GDS results. Clearly visible are both the Cordilleras, bordering a large-scale HCZ. Its origin differs in the different areas. Note that the strike of the western HCZ does not correspond to the strike of the volcanic belt

subduction related basaltic magmas to provide the heat for crustal melting. This seems questionable as temperatures of more than 1300°C would be needed even at a depth between 100 and 130 km to melt basalts from the mantle lithosphere. However the differentiation of magma in the upper crust seems to be no longer questionable (e.g., Wörner et al. this Vol.) as can be seen in studies of volcanic centres.

Within the Andean subduction regime the volume of plutonic rocks may exceed that of volcanic rocks by a factor of 10 (Thorpe et al. 1981). A number of hydrothermal ore deposits (e.g., Cu, Ag, Au) exist in the area. These are connected to plutonic events and even give evidence of increased fluid circulation during the Andean orogeny. Most remarkably is the copper belt at the western border of the Western Cordillera (almost parallel to long 69°W). Hydrothermal ore deposits are associated with an intense plutonism of acidic composition, which has occurred in the upper crust in this area since Jurassic. The intrusive ages decrease eastwards, i.e., the loci of the arc magmatism shift eastwards (Breitkreuz and Berg 1983, Andriessen and Reutter this Vol.). The lateral migration of plutonism and of the trench provides evidence for subduction erosion (Ziegler et al. 1981, von Huene and Scholl 1991). Stern (1991) explained the Andean magmatism by subduction erosion due to the steeply dipping Nazca plate. However the amount of subducted continental crust cannot be estimated precisely: rates are within the range of 50-500 km^3/m.y. of crust per km of trench scraped off the margin of the southern Central Andes.

While the occurrence of acid intrusions in the upper crust is justified by the observations, the origin of the strike direction of the HCZ (trending from NNW-SSE to NW-SE) is still unclear. It does not coincide with the strike of the present volcanic belt. Long-period GDS data (again Figs. 5 and 6) also suggest a westward bending of the HCZ north of Calama, but the present data base does not provide a definite answer to this question. One may also assume a separate HCZ northwest of Calama. At a first glance the HCZ, indicated by the GDS induction arrows, seems to be controlled by major tectonic lineaments, especially the Calama-Olacapato-El Toro system (Calama to Salta, cf. Fig. 3). With more careful inspection, however, one recognizes that induction arrows still point westwards even at sites east of the lineament: therefore, this region cannot possess much higher electrical conductivity. At any rate this lineament seems to coincide in strike with the HCZ and to limit it to the west. Other fault-systems (under local extension?) further eastwards (see again Fig. 3) may be associated with the influx of fluids and melts and thus directly related to the HCZ on a regional scale. Another hint may come from the area south of lat 22°, where the volcanic chain retreats to the east and becomes irregular and wider: the line of active volcanoes follows the contour depth line of 125 km of the oceanic slab. Under northwestern Argentina the 100 and 125 km contours start to diverge; a bench forms which broadens southwards (Cahill et al. 1992). At lat 24°S where the crustal conductivity is still high but at much greater depth (see again Fig. 13), the volcanic chain again moves to the west and follows now the 100 km contour line of the slab. Here, although the electrical structure is not very clear in its extent, its strike is identified to be about N-S.

7.2 The Eastern Cordillera

Another pronounced HCZ is found within the western part of the Eastern Cordillera of southern Bolivia. This was earlier detected by Schmucker et al. (1964) who operated a GDS-station close to Tarija. The old data did not allow adequate quantitative results; the depth range of the detected HCZ was especially questionable. Recognizing that the Andean HCZ could be the result of shallow crustal structures (< 10 km), Aldrich et al. (1972) arranged direct current resistivity soundings in southern Peru. High electrical conductivity structures were found close to the surface, but the data were not consistent from one area to the other. With the present MT data, we can calculate the total conductance of the HCZ to at least 10000 S. The HCZ has its origin in the upper crust and explanations are ambiguous. In Bolivia the anomaly correlates roughly with the 'tin belt', which shows a broad variety of ore minerals (e.g., Sn, W, Bi, Ag, Pb, Zn). It stretches from southern Peru (lat 14°S) to northwestern Argentina, lat 23°S (see Lehmann this Vol). Young volcanism and subvolcanism of Mio- to Pliocene age is known in the area. It is mostly acid in composition. Hydrothermal circulation within thick sedimentary layers (exceeding 10000 m) of Paleozoic age or within a fractured basement is probable. Hydrothermal circulation is also manifested by high terrestrial heat flow reaching 100 mW/m^2 (Henry 1981) and several geothermal springs in the western part of the Eastern Cordillera. Carbon is found within the sedimentary layers (Kley and Reinhardt this Vol.), but it does not seem to be sufficiently coalified (Jödicke 1990) and electrically connected to cause a considerable increase of electrical conductivity. The less conductive Precambrian basement rises and is even close to the surface at some places, e.g., SW of Tarija, and thus provides the lateral contrast in conductivity necessary for increasing GDS induction vectors in amplitude. The latter indicate by their period dependence a much more complicated structure of the HCZ, i.e., a very large increase in conductance westwards. This can be well explained by an increase in thickness of the Paleozoic sediments (see Kley and Reinhardt this Vol.). Seismic refraction investigations east of Tupiza detected high velocities at a depth of about 20 km. They are, in spite of their depth, interpreted to represent lower crustal material (S. Baldzuhn pers. comm.), serving as the base for overlying sediments. We have not been able to resolve this structure properly by means of MT soundings because of too large station spacings, but may see a correlation to an HCZ at a depth of approximately 20 km under the easternmost Altiplano. A relationship is likely between the HCZ and large overthrusting events, i.e., in the form of thin- and thick-skinned tectonics (Reutter et al. 1988, Giese et al. 1990). This even affects the Altiplano. Though the shear zone proposed by these authors in the Eastern Cordillera can not be identified clearly from the MT data, tectonics may be responsible for the overall conditions of the mid to lower crust.

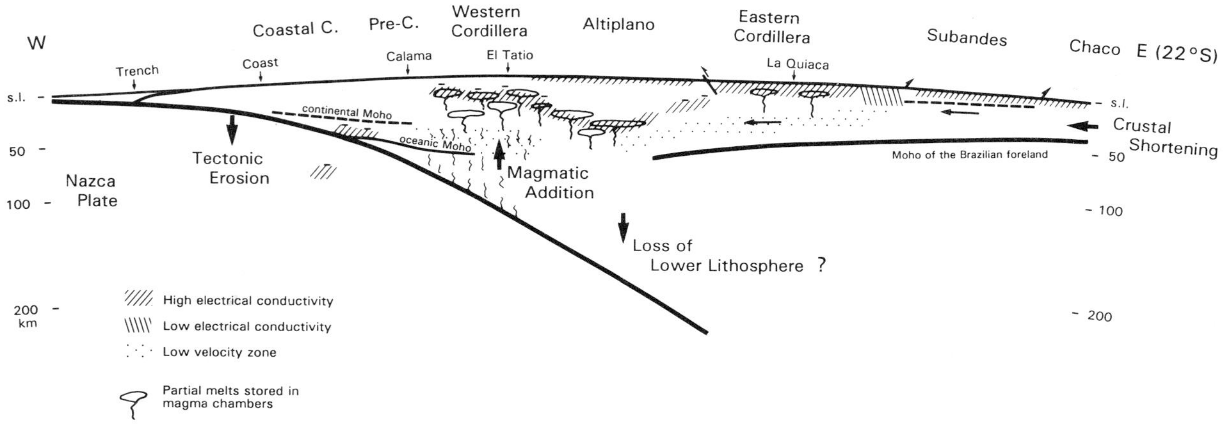

Fig. 14. Cross section (1:1) of the southern Central Andes showing schematically the observed HCZ at different depth ranges at lat 22°S. Seismic interfaces are taken from Wigger et al. (this Vol.) and some tectonic features from Reutter et al. (1988), Sheffels (1990), and Giese et al. (1990). Under the Western Cordillera and westernmost Altiplano, the HCZs may originate from magmatic underplating, i.e. silicic melts stored in shallow magma chambers. The HCZs farther west may relate to subduction-induced fluid flow from the oceanic slab and erosion of the continental lithosphere. The HCZs under the eastern Altiplano, the Eastern Cordillera and the Subandes may be due to large-scale overthrusting, i.e. thin- and thick-skinned tectonics. Crustal shortening or thickening is mainly due to tectonic processes. Magmatic events only dominate the western section

With the present data we cannot clearly decide whether the HCZ of southern Bolivia extends into northwestern Argentina, where another, but less conductive HCZ is found around San Antonio. This strikes approximately N-S, reaching the maximum in total conductance just west of San Antonio. Here, surface expressions of subduction induced processes may be seen in a cluster of Quaternary volcanoes of basaltic composition and low residual gravity (Götze et al. 1988). The descending oceanic slab is characterized by a nest of earthquakes at a depth of ca. 220 km (Buness et al. 1986). A relationship between the HCZ and these observation may be inferred. In total, conductive structures within the crust of northwestern Argentina seem to be more heterogeneous than in southern Bolivia and even three- dimensional (Krüger et al. 1990). Here, zones of very high conductivity were identified at shallow crustal depth (cf. Figs. 9, 13). Most of these structures are directly connected to geothermal anomalies (e.g., Febrer et al. 1981), but thick sedimentary layers are also involved as well as in southern Bolivia.

In contrast to the observed HCZ of the Eastern Cordillera, the Subandean ranges and the Chaco lowland plains of southern Bolivia and northwestern Argentina show normal crustal conductivities. Upper sedimentary layers may reach a depth of 10 km and the electrical structures are either strongly anisotropic or the basement is highly structured.

7.3 The Altiplano-Puna Plateau

MT data from the Altiplano of southern Bolivia show a very consistent electrical picture. From the Puna of Argentina (for structural definitions, see Fig. 1) only one data set is available. Therefore, an 'electrical comparison' of these two plateaus is not really justified. The Altiplano has a thick conductive cover due to sediments. The moderately conductive middle crust is underlain by very conductive material (1-4 ohm m). Total crustal thickness calculated from gravity reaches about 60 km. A discontinuity in seismic velocity is found at a depth of 40 km (see Wigger et al. this Vol.).

In the west this corresponds to the upper bound of the highly conductive structure. Comparable conductive structures were found under the northern Altiplano of Bolivia (Ocola and Meyer 1972, Ritz et al. 1991), which may suggest that this large-scale structural unit is governed by physical regimes at depth similar to those in the south. Farther east the lower crustal conductive structures rise to a depth of ca. 20 km (sites MAR to CHO: see Fig. 2). Geophysical and geological data of the Eastern

Cordillera have implied decollement type tectonics as being responsible for Andean mountain building. If this exists, it might also be responsible for the deeper (40 km) HCZ beneath the Altiplano. A sole thrust, having developed under the compressive stress of convergent oceanic and continental plates, should then mark the upper bound of the HCZ. This zone may be enriched in fluids and/or in graphite. However, the question of how to trap these fluids is still open. Magmatic involvement, however, can not be excluded. Schwarz et al. (1984) and Haak and Giese (1986) suggested such an underplating of basaltic melts. These melts may also serve as the base of underthrusting as the crust is thermally weakened. These partial melts originate and ascend from the subducting slab. Partial melts of 8% by volume are required to explain the observed electrical conductivities. Considering both magmatic and tectonic involvement, the Altiplano may represent a zone of tectonically and thermally weakened crust.

8 Concluding Remarks

In their earliest paper on Andean research, Schmucker et al. (1964) concluded that the crust of southern Peru as a whole appears to have anomalously high values of electrical conductivity. With our present knowledge we can extend this statement to almost the entire region of the Central Andes: electrical conductivities of the whole Andean crust and upper mantle appear to be higher than in 'normal' orogenic settings. Within this electrically anomalous region we find structures with even higher conductivities [> $(1\ \text{ohm m})^{-1}$] at different depth ranges, e.g., in the Western and Eastern Cordillera and on the plateau in between, the Altiplano (Fig. 14). The observed HCZs reflect the intense tectonic and magmatic evolution of this mountain belt. Summarizing the present geological and geophysical knowledge, its origin, the Andean orogeny, is seen in both processes: tectonic underplating was initiated in the east under the Eastern Cordillera in the Oligocene, and only in the Pleistocene under the Altiplano, while the process of magmatic underplating started in the west, i.e., under the Western Cordillera and westernmost Altiplano. Farther west of the present volcanic belt, tectonic and magmatic underplating can account for Andean mountain building. Tectonic underplating would be caused by subduction erosion while magmatic underplating occurred beneath the ancient arc systems. However, the key question concerning which process or mechanism initiated the tectonic uplift of the Andes in the east still remains a challenge for further integrated studies of the mountain belt.

The magnetotelluric and geomagnetic deep soundings now undertaken give new information about Andean conductivity structures but also raise new questions. These structures should be classified not only by their E-W but also by their N-S extent, although the Andean units seem to be structured mainly E-W. The question remains open as to why the structures identified on the northern traverse are different from those of the south. Some answers to this complex of problems may be expected from further studies carried out parallel to the Andean strike and studies of the transition to low-angle subduction of the Pacific slab.

Acknowledgements. A project like the one presented here needs many helpful hands to be successful. The following institutions and their members gave valuable support: Programa Sismológico, Universidad del Norte, both from Antofagasta; Dir. de Riego, Embalse de Conchi; Universidad de Chile, Santiago; Yacimientos Petrolíferos Fiscales Bolivianos (La Paz, Sta. Cruz, Tarija), Universidad Mayor de San Andres, Comibol, Geobol, all from La Paz; Universidad Nacional de Salta; Embassies of the Federal Republic of Germany in Santiago de Chile and La Paz; Geophysical Institutes of Berlin, Göttingen and München Universities. Special mention is made of all co-workers in the Andes project; we would like to express our deep thanks to all of them. The Deutsche Forschungsgemeinschaft (DFG), Deutscher Akademischer Austauschdienst (DAAD), FONEM and YPFB, La Paz and Freie Universität Berlin gave main financial support.

References

Aldrich LT et al (1972) Electrical Conductivity Studies in the Andean Cordillera. Carnegie Inst Wash Yearb 71: 317-320

Aldrich LT et al (1975) Electrical Conductivity Studies in South America. Carnegie Inst Wash Yearb 74: 291-293

Beblo M, Liebig V (1986) Mobile Datenerfassung mit CMOS-Halbleiterspeicher. In: Haak V, Homilius J (eds) Prot Koll Elektromagn Tiefenforschung. Lerbach, pp 357-361

Breitkreuz C, Berg K (1983) Magmatite in der Küstenkordillere südöstlich von Chanaral/Nordchile. Zbl Geol Paläont 1 (3/4): 387-401

Buness F, Wetzig E, Wigger P (1986) Seismologische Studien in den Zentralen Anden. Berliner geowiss Abh (A) 66: 5-33

Cahill T, Isacks BL, Whitman D, Chatelain JL, Perez A, Chiu JM (1992) Seismicity and tectonics in Jujuy Province, northwestern Argentina. Tectonics (submitted)

Cerv V, Pek J (1990) Modelling and Analysis of Electromagnetic Fields in 3D Inhomogeneous Media. Surv Geophys 11: 205-229

Chinn DS, Isacks BL, Barazangi M (1980) High frequency seismic wave propagation in western South America along the continental margin, in the Nasca plate and across the Altiplano. Geophys J R astr Soc 60: 209-244

de Silva SL (1989) Altiplano-Puna volcanic complex of the central Andes. Geology 17: 1102-1106

Emslab-Group (1988) The EMSLAB electromagnetic sounding experiment. EOS 69: 89,98,99

Febrer JM, Gasco JC, Pomposiello MC, Mamani M, Baldis B, Fournier H (1981) Magnetotelluric measurements defining a continental plume in a zone of the Andean belt southeast of the Altiplano in Argentina. Paper, IV IAGA Sci Assemb, Edinburgh

Forbush SE, Casaverde M (1961) Equatorial Electrojet in Peru. Carnegie Inst Wash Publ 620

Giese P et al (1990) Crustal Evolution of the Central Andes. Proc Worksh Structure and Evolution of the Central Andes in Northern Chile, Southern Bolivia and Northwestern Argentina, FU Berlin, May 23-25, 1990, pp 65 - 66a

Götze HJ, Schmidt S, Strunk S (1988) Central Andean gravity and its relation to crustal structures. In: Bahlburg H, Breitkreuz C, Giese P (eds) The southern Central Andes. Lecture Notes in Earth Sciences 17. Springer, Berlin, pp 199-208

Haak V, Giese P (1986) Subduction induced petrological processes as inferred from magnetotelluric, seismological and seismic observations in N-Chile and S-Bolivia. Berliner geowiss Abh (A) 66: 231-246

Haak V, Hutton R (1986) Electrical Resistivity in Continental Lower Crust. In: Dawson JB et al (eds) The Nature of the Lower Continental Crust. Geol Soc Spec Publ 24: 35-49

Henry SG (1981) Terrestrial heat flow overlying the Andean subduction zone. Ph D Thesis, Univ Michigan

Hyndman RD (1988) Dipping seismic reflectors, electrically conductive zones, and trapped water in the crust over a subducting plate. J Geophys Res 93: 13,391-13,405

Isacks BL (1988) Uplift of the Central Andean Plateau and Bending of the Bolivian Orocline. J Geophys Res 93: 3211-3231

James, DE (1971) Andean crustal structures. Carnegie Inst Wash Yearb 69: 447-460

Jiracek GR (1990) Near-Surface and Topographic Distortions in Electromagnetic Induction. Surv Geophys 11: 163-203

Jödicke H (1990) Zonen hoher elektrischer Krustenleitfähigkeit im Rhenoherzynikum und seinem nördlichen Vorland. Lit, Hamburg

Kono M, Fukao Y, Yamamoto A (1989) Mountain Building in the Central Andes. J Geophys Res 94: 3891-3905

Krüger D, Massow W, Rath V, Schwarz G (1990) Neues von der Andengeotraverse. In: Haak V, Homilius J (eds) Prot Koll Elektromagn Tiefenforschung, Hornburg, pp 267-278

Molnar P, Oliver J (1969) Lateral variations of attenuation in the upper mantle and discontinuities in the lithosphere. J Geophys Res 74: 2648-2682

Ocola LC, Meyer RP (1972) Crustal low-velocity zones under the Peru-Bolivia Altiplano. Geophys J R astr Soc 30: 199-209

Pichler H, Zeil W (1972) The Cenozoic rhyolite-andesite association of Chilean Andes. Bull Volcanol 35: 424-452

Reutter KJ, Giese P, Götze HJ, Scheuber E, Schwab K, Schwarz G, Wigger P (1988) Structures and crustal development of the Central Andes between 21° and 25°. In: Bahlburg H, Breitkreuz C, Giese P (eds) The southern Central Andes. Lecture Notes in Earth Sciences 17. Springer, Berlin, pp 231-261

Ritz M, Bondoux F, Herail G, Sempere T (1991) A Magnetotelluric Survey in the Northern Bolivian Altiplano. Geophys Res Let 18: 475-478

Roeder D (1988) Andean-Age Structure of Eastern Cordillera (Province of La Paz, Bolivia). Tectonics 7: 23-39

Sacks JS (1968) Distribution of Absorption of Shear Waves in South America and Its Tectonic Significance. Carnegie Inst Wash Yearb 67: 339-344

Schmucker U (1970) Anomalies of geomagnetic variations in the southwestern United States. Bull Sripps Inst Ocean Univ Calif 13

Schmucker U (1973) Regional induction studies: a review of methods and results. Phys Earth Planet Inter 7: 365-378

Schmucker U, Hartmann O, Giesecke AA, Casaverde M, Forbush SE (1964) Electrical conductivity anomaly in the earth's crust in Peru. Carnegie Inst Wash Yearb 63: 354-362

Schmucker U, Forbush SE, Aldrich T, Hartmann O, Giesecke AA, Casaverde M, Castillo J, Salgueiro R, del Pozo S (1971) Anomalias de la conductividad electrica debajo los Andes. Geofis Panam 1: 51-69

Schwarz G (1990) Electrical Conductivity of the Earth's Crust and Upper Mantle. Surv Geophys 11: 133-161

Schwarz G (1991) Electromagnetic Soundings of the Earth's Crust and Upper Mantle in Western South America. Münchner geophys Mitt 5: 309-318

Schwarz G, Haak V, Martinez E, Bannister J (1984) The electrical conductivity of the Andean crust in northern Chile and southern Bolivia as inferred from magnetotelluric measurements. J Geophys 55: 169-178

Schwarz G, Martinez E, Bannister J (1986) Untersuchungen zur elektrischen Leitfähigkeit in den Zentralen Anden. Berliner geowiss Abh (A) 66: 49-72

Schwarz G, Chong D G, Krüger D, Martinez E, Massow W, Rath V, Viramonte J (1990) Crustal electrical resistivity structure in the southern Central Andes. Proc Int Symp Andean Geodynamics, Grenoble, pp 41-43

Sheffels BM (1990) Lower bound on the amount of crustal shortening in the central Bolivian Andes. Geology 18: 812-815

Stern CR (1991) Role of subduction erosion in the generation of Andean magmas. Geology 19: 78-81

Tarits P, Menvielle M (1986) The Andean Conductivity Anomaly reexamined. Ann Geophysicae 4B: 63-70

Tatel HE, Tuve MA (1958) Seismic studies in the Andes. Trans Am Geophys Union 39: 580-582

Thorpe RS, Francis PW, Harmon RS (1981) Andean andesites and crustal growth. Phil Trans R Soc London 301: 305-320

von Huene R, Scholl DW (1991) Observations at convergent margins concerning sediment subduction, subduction erosion, and the growth of continental crust. Rev Geophys 29: 279-316

Wannamaker PE et al (1989) Resistivity Cross Section Through the Juan de Fuca Subduction System and its Tectonic Implications. J Geophys Res 94: 14,127-14,144

Whitman D, Isacks BL, Chatelain JL, Chiu JM, Perez A (1992) Attenuation of high frequency seismic waves beneath the Central Andean Plateau: evidence for along-strike changes in lithospheric thickness. J Geophys Res (submitted)

Wigger P (1988) Seismicity and crustal structure of the Central Andes. In: Bahlburg H, Breitkreuz C, Giese P (eds) The Springer, Berlin, pp 209-229

Ziegler AM, Barrett SF, Scotese CR (1981) Paleoclimate, sedimentation and continental accretion. Phil Trans R Soc London (A) 301: 253-264

Geothermal Structure of the Central Andean Crust - Implications for Heat Transport and Rheology

PETER GIESE

Abstract. This study aims to determine the central Andean geothermal structure by downward continuation of the surfcace temperature field. The heat flow density values along a trans-Andean profile show a minimum of 40 mW/m² in the region of the Coastal Cordillera and Chaco and a maximum of 80 mW/m² in the Western Cordillera and the western Altiplano. The necessary structural data are provided by deep seismic sounding measurements. The petrophysical parameters used for this study are taken from the literature. Steady state and one-dimensional conditions and pure conductive heat transport are assumed. For the region with heat flow values between 40 and 50 mW/m² the temperature at the base of the crust does not exceed 700 °C. But in the zones with high heat flow density values of 60-80 mW/m² the temperature in the lower crust reaches values exceeding 1000 °C. Such high temperatures seem to be unrealistic. In order to lower the temperature gradient an additional advective heat transfer is postulated. For the advecting material (fluids and volatiles) a Darcy velocity of about 1 mm/year results. For comparison the amount of upstreaming fluids and volatiles released by the downgoing oceanic crust is determined. This rough estimation yields the amount of fluids pervading the hanging lithosphere and is in the order of 0.3 mm/year. Because most parameters used for these calculations are only estimates the result may vary by a factor of 0.2-5 Finally the rheological response has been determined as the maximum possible stress as a function of depth for the different heat flux locations and a strain rate of $7 \cdot 10^{-15}$ s^{-1} .

1 Introduction

In order to understand the tectonic and magmatic processes going on in active orogenic systems it is necessary to have a knowledge of the temperature distribution and the way in which heat is transferred through the lithosphere. There are two different approaches to determining the geothermal field in a colliding system.

One approach is based on calculations that attampt to model how the temperature distribution develops in a convergent ocean/continent system. The first model calculations were carried out by Andrews and Sleep (1974) and Toksöz et al (1971). Although there are differences between these starting models, the geothermal field and heat transfer values obtained, the results are broadly similar. The thermal effects of a subducting slab on the overlying crustal and mantle wedge can be summarised as follows (Fowler 1990): (1) an influx of upward moving volatiles from the descending slab, (2) some rising melt from the deeper parts of the oceanic slab, and (3) the driving of convective flow in the wedge, the flow lines of which show movement of mantle material from the distant part of the wedge to its edge. By this conductive and convective heat transfer an upwelling of the isotherms is generated in the upper plate causing an increase in the surficial heat flow density.

The other method to determine the geothermal field uses the heat flow density observed at the surface and its downward continuation; certain assumptions are made concerning the structure, the depth distribution of heat sources and the kind of thermal heat transfer. In this study the second approach has been applied. The resulting temperature distribution implies some consequences with respect to the problem of conductive and convective heat transfer and the rheological behaviour of the Andean crust.

Any temperature-depth calculation requires a knowledge of geothermal and structural parameters, including boundary and starting conditions. This chapter outlines the necessary and available basic data for this study.

Correspondence to: Peter Giese, FR Geophysik, Freie Universität Berlin, Malteserstr. 74-100, D-1000 Berlin 46

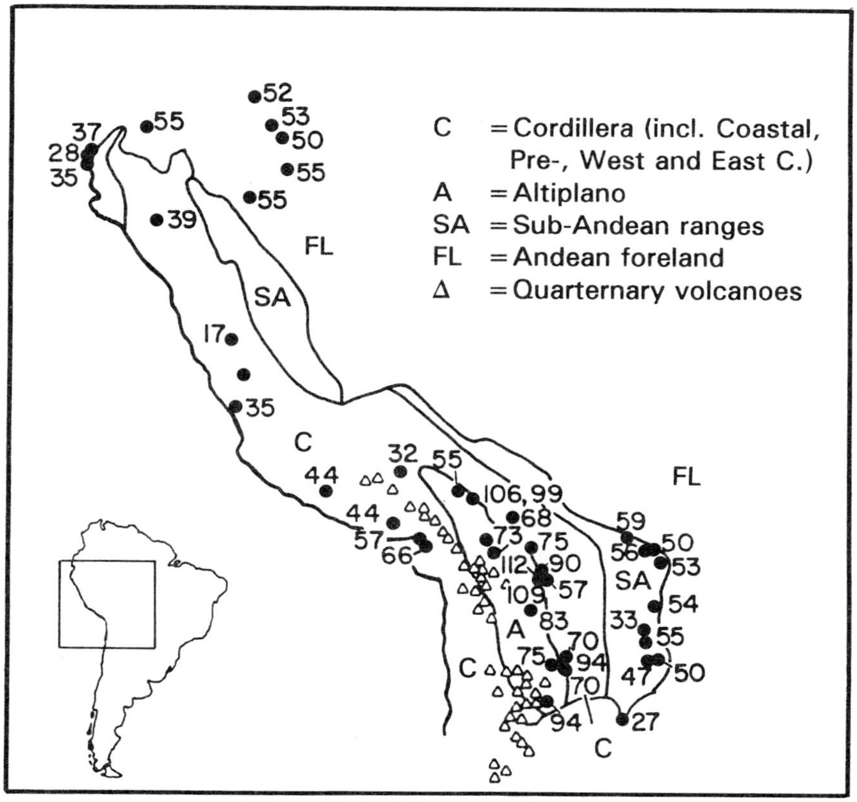

Fig. 1. Locations of the heat flow sites in Peru and Bolivia (Henry and Pollack 1988). The numbers are the heat flow density values in mW/m²

2 Heat Flow Density Data

This study is based on a compilation of heat flow values obtained from measurements in the Central Andes and presented by Henry (1981) and Henry and Pollack (1988). These authors reported on 35 new and 9 revised heat flow measurements overlying the Andean subduction zone in Bolivia and Peru (Fig. 1). On the basis of the data Henry divided the area under study into five geothermal regions, characterised by their different tectonic and magmatic history which, in turn, is related to the position with respect to the magmatic arc and the angle of subduction.

Region 1 (Northern and Central Peru)
In northern and central Peru the heat flow is low (20-40 mW/m²), as is also the angle of subduction (10°-15°). Here the major period of volcanism occurred in the mid-Miocene (10-15 Ma ago).

Region 2 (Southern Peru and Northern Bolivia)
In this region heat flow density values are between 50 and 80 mW/m², with higher values of between 65 and 80 mW/m² towards the recent magmatic arc in the Western Cordillera.

Region 3 (Northern Chile and Central and Southern Bolivia)
Region 3 is associated with high heat flow density values between 80 and 100 mW/m². The region immediately east of the active volcanic arc has been identified as the recent backarc region in a continental environment.

Region 4 (Chile Between 22° and 26° S)
From region 4 no heat flow data are available. Because the geological and tectonic history of this region is comparable with that of southern Peru, similar heat flow values (20-40 mW/m²) can be expected here.

Region 5 (Central and Southern Chile)
Uncorrected values have been collected in this region and all values are as low (20-40 mW/m²) as in region 1.

Region 6 (Brazilian Shield)
The Precambrian shield in Brazil, with low heat flow density values (30-60 mW/m²), is introduced as region 6 (Vitorello et al. 1980) and is added to the regions proposed by Henry.

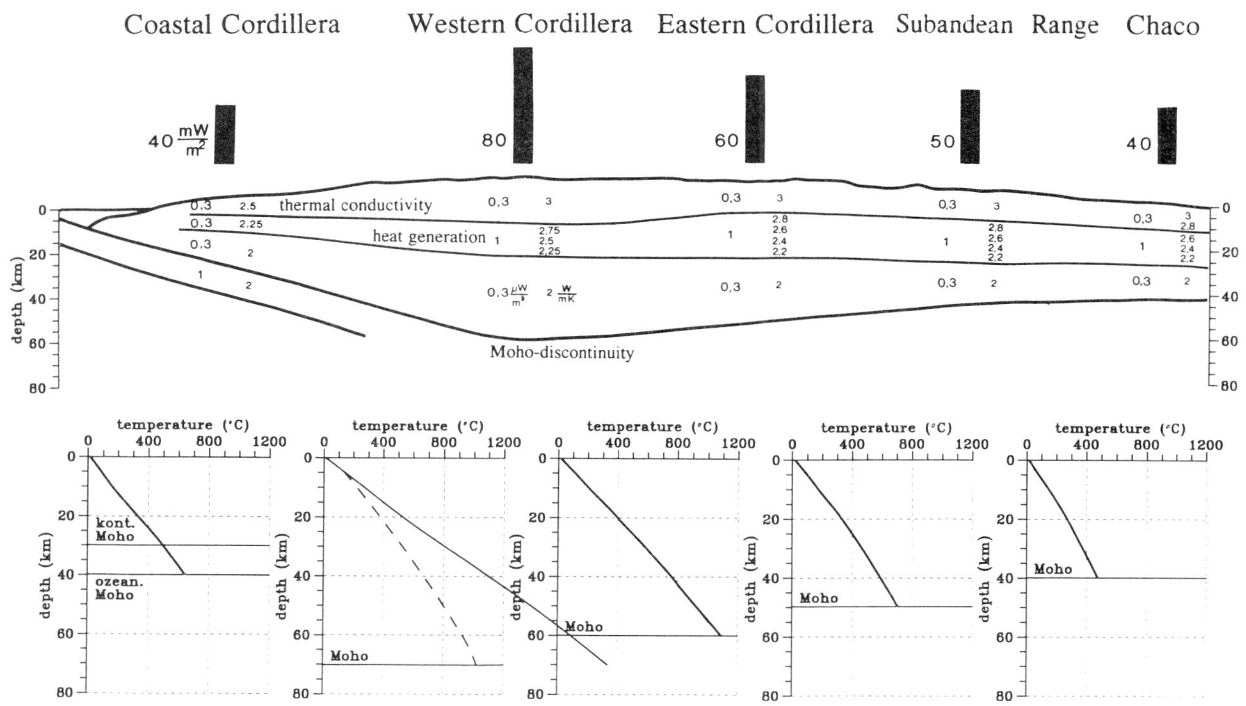

Fig. 2. *Above* The W-E crustal profile between the Pacific coast and the Chaco. The *columns* on top show the heat flow density values used. The values inside the crustal profile indicate the thermal conductivity and the heat generation. *Below* The temperature-depth functions under steady state and one-dimensional conditions

2.1 Heat Flow Density Profile

Based on these values a heat flow density profile running from the Pacific coast near Antofagasta up to Villamontes at the eastern margin of the Andes, has been compiled (Fig. 2, upper part). For the Coastal Cordillera (region 4) a heat flow density of 30 mW/m² is assumed, and of 80 mW/m² for the recent magmatic arc of the Western Cordillera and the adjacent western Altiplano region. Local anomalies of higher values, up to 100 mW/m² or even more, may exist in the vicinity of volcanic or geothermal activity. The Eastern Cordillera is associated with a heat flow density of 60 mW/m²; the Subandean range is characterised by 55 mW/m². The easternmost end of the profile, situated on the western margin of the Brazilian Shield, has a value of 40 mW/m². Generally, a variation of ± 10 mW/m² must be attributed to these heat flow density values.

3 Structural Model and Rock Geothermal Parameter

The necessary structural data are taken from the seismic model proposed by Wigger et al. (this Vol.). They are displayed in Fig. 2 along with the geothermal parameters used for the calculation. Thevalues for the geothermal parameters are mainly based on data published by Cermak and Bodri (1986) for the purpose of crustal thermal modelling.

Steady state conditions have been assumed. As mentioned above; local heat flow density values higher than 100 mW/m² are reported in the Western Cordillera; these can probably be attributed to young volcanic activities. In such local anomalies unsteady state conditions must prevail, but such localities are outside the scope of this study.

The five heat flow density values are distributed along a profile of 600 km length. Thus, the average spacing is 150 km, which is two to four times larger than the crustal thickness. Because of this relation it is possible to simplify the solution of the general heat transfer equation to the one-dimensional case.

Figure 2 (lower part) displays the temperature-depth curves obtained from the one-dimensional calculations for the five selected sites. As mentioned above the heat flow density values are associated with an error of ± 10 mW/m², which corresponds to variation of the geothermal gradient by ± 3.3 °C/km and by ± 165 °C of the temperature at 50 km depth.

Although the one-dimensional steady state temperature calculations must be regarded as rough and simplified estimates, they do allow two important statements to be made. First, the temperature distributions in the Chaco, the Subandean range and the Coastal Cordillera seem to be realistic, because the calculated curves do not exceed the melting temperature of lower crustal rocks; this temperature is assumed to be between 600 and 800 °C. Second, the picture is quite different and unusual for the Eastern Cordillera and especially so for the Western Cordillera. In these zones the temperatures are totally unrealistic at regions deeper than 20-30 km, because here the temperatures exceed those required to melt lower crustal rocks by several hundred degrees. To conclude, in the Central Andean region other geothermal models must be considered for the middle and deeper crustal zones.

There are some possibilities that the geothermal gradient in the crust may be lowered. Increasing contents of heat-producing minerals reduce the temperature and thus the gradient as well. In order to produce an average geothermal gradient of about 15 °C/km, heat generation of 3-4 μW/m³ between 30 and 70 km must be required, and such values are unknown. The geothermal gradient could also be lowered by an increase in thermal conductivity. In this case values rising from 4 W/(m·K) at 30 km depth to more than 7 W/(m·K) at 70 km depth would be necessary in order to obtain a temperature of about 1000 °C at the base of the crust. But measurements of heat conductivity on samples at high temperatures clearly demonstrate that the thermal conductivity decreases strongly with rising temperature. Values between 2.0 and 2.5 W/(m·K) are realistic. These two explanations must be rejected.

A third possibility considers tectonic thickening and stacking of the crust of the Central Andes which has progressed over the last 10-15 Million years. An "instantaneous" crustal doubling produces a thickened crust with a zigzag temperature distribution. But Shi and Wang (1987) demonstrated, by two-dimensional model calculations, that substantial amounts of lateral heat transfer can take place during the overthrusting event and there is a smoothing of the zigzag curve. Nevertheless, a lowering of the geothermal gradient, as demonstrated by the geothermal considerations outlined in the previous section, is achieved by crustal stacking.

An increase in heat conductivity in the middle and lower crust can be achieved by introducing convective heat transfer. The role of such heat transport will now be discussed.

3.1 Convective Heat Transfer Model

Another way of lowering the temperature gradient is by introducing upward directed convective heat transfer between the top of the mantle and the base of the upper crust. This convective heat transfer through the hanging lithosphere could be realised by fluids released by dehydration processes occuring in the depth range 50-100 km in the subducted oceanic crust (Peacock 1987, 1989).

In order to estimate the amount of ascending fluids necessary to produce a reasonable temperature distribution and gradient in the middle and deeper crust the following calculations are carried out. These assume that the upgoing fluids do not undergo any chemical reactions causing additional heat production or loss. The author is well aware that this is an oversimplification.

Upper Crust

In the central region of the Andes the surface heat flow density is 80 mW/m². Between the surface and 5-km-depth level normal conductive heat transfer occurs. The value of 5 km is more or less arbitrary. The heat generation in the upper 5 km crust may amount to 0.5 μW/m³. Thus, here remains a heat flux of 77 mW/m² passing through the 5-km-depth level from below. With a heat conductivity of 3 W/m·K, a temperature of 141 °C results at the depth of 5 km.

Middle and Deeper Crust

Heat generation in the depth range of 5-70 km is assumed to be 0.20 μW/m³, giving a total radiogenic heat production of 13 mW/m² for the middle and deeper crustal column. Thus, the heat flux through the base of the crust is 64 mW/m².

It is supposed that between the base of the crust (70 km depth) and 5 km depth, a temperature difference of $\Delta T = (1041\text{-}141)$ °C = 900 °C may be realistic. A temperature of 1050 °C is just below the melting temperature of ultramafic rocks.

The amount of convectively transported heat Q_{conv} is given by

$$Q_{conv} = d \cdot c_f \cdot v_D \cdot \Delta T$$

where

d: is the density of the fluid,

c_f: is the volumetric heat capacity of the fluid,

v_D: is the Darcy velocity, that is the amount of fluid equal to the hight of a column per m² and per year, and

ΔT: is the Temperature difference.

For d a value of 1200 kg/m³ is used. For crustal conditions with temperatures up to 800 °C the volumetric specific heat capacity for H_2O varies between 2 and 4 J/(cm³·K), in general this parameter varies between 2.6 and 4.2 J/(cm³·K). Thus, this parameter shows practically the same value range for both types of material (Peacock 1989). In the following calculation an average value of $1 \cdot 10^3$ J/(kg·K) is used. As mentioned above, the temperature difference is ΔT= 900 °C.

A problem arises with the quantity Q_{conv}. As a first approximation the value of 36.5 mW/m² is used. It will be shown that this value will satisfy the boundary condition that the total heat flux at a depth of 5 km must amount to 77 mW/m². The remaining conductive heat flux amounts to about 40 mW/m², which is the average surface heat flux in normal crustal areas.

With these values the Darcy velocity v_D becomes

$$v_D = 1.07 \text{ mm/a}.$$

The temperature distribution between 5 and 70 km depth is calculated assuming a homogeneous layer with a constant thermal conductivity (2.5 W/(m·K) and heat production (0.2 μW/m³), composed of vertically advecting material. A value of 0.01 has been introduced for the porosity of the material. The resulting temperature distribution is shown in Fig.2 (dashed curve of "Western Cordillera"). The conductive/convective part of the temperature-depth distribution starts with 141 °C at the depth of 5 km and ends with 1041 °C at the depth of 70 km, as defined by the assumed temperature difference.

The temperature gradient between 5 and 10 km depth is 16.2 °C/km. With k (heat conductivity) at 2.5 W/(m·K) the conductive portion of heat transport is 40.5 mW/m². As mentioned above, at the depth of 5 km a convective portion of the heat flux of 36.5 mW/m² has been used in order to determine the Darcy velocity. The sum of these two values is 77 mW/m².

To determine the partition of conductive and convective heat flux at the base of the crust the following calculation is applied: The temperature gradient near the base of the crust is 11.4 °C/km. With k=2.5 W/(m·K) a conductive heat flux of 28.5 mW/m² results. As outlined above the total heat flux through the crust/mantle boundary is about 64 mW/m², and thus the convective heat flux must be 35.5 mW/m². This value agrees rather well with that determined as the convective portion at a depth of 5 km.

The heat transport equation gives a Darcy velocity of 1.06 mm/a. The actual velocity of the upgoing fluid depends on the porosity of the pervaded rocks. Assuming a porosity of 1% an actual velocity of about 10 cm/a is achieved. If the pervaded layer is 65 km thick the fluid needs 0.65 million years, or roughly 1 Ma, to pass the crust. This value is one order of magnitude smaller than the assumed time of 10-15 Ma for the young thickening of the Andean crust.

The result means that the recent temperature field of the Andean crust is mainly controlled by the ascending fluids and not by a temperature disturbance caused by the stacking process. A similar logic has been applied to the temperature curve of the Eastern Cordillera. Here again the conductive crustal heat flux is assumed to about 40 mW/m². The resulting temperature distribution is plotted as dotted curve in Fig. 2 as well.

Finally it must be reemphazised that these calculations are only very rough estimations. Many of the parameters used in this calculation are only approximations, and with respect to the Darcy velocity an error factor of between 0.2 and 5 must be taken into account. Nevertheless the magnitude of the values obtained does seem to be realistic.

3.2 The Effect of Dehydration on Convective Heat Transfer

These results can be checked by a different approach which is based on the study of the fate of fluids in subducting slabs. It is generally accepted that hydrothermal circulation of sea water takes place by convective flow, probably through the whole of the oceanic crust (e.g. Fyfe and Lonsdale 1981). There is a dispute over how much H_2O is captured in the oceanic crust. Peacock (1987) proposes 2 wt %

whereas Anderson et al. (1978) assumes 6 wt % of fluids. In the following a value of 5 wt % is used for fluids which are completely released and recycled in the hanging plate. The up-streaming fluids can react with the overlying continental lithosphere in various ways. The released fluids can: (1) transport significant amounts of heat, (2) cause retrograde endothermic hydration reactions in the descending oceanic slab, (3) transport dissolved species such as alkalis and silica, (4) promote melting, and (5) affect the rheology of infiltrated rocks (Peacock 1987).

The following simplified estimate should help with an understanding of the possible contribution of upstreaming fluids to advective heat transport. In this estimation it is again assumed that the fluids pass through the rocks without causing any chemical reactions. A 6-km-thick oceanic crust with a porosity of 5% contains a fluid column of 300 m. The oceanic crust is subducted and it is assumed that the fluids are released in a depth range of 50-100 km over a length of 100 km. The subducting velocity is taken as 10 cm/a. From these values a Darcy velocity for the up streaming fluids of 0.3 mm/a results. This value is smaller than the 1 mm/a determined from thermal considerations. The discrepancy amounts to a factor of 3, which is within the limits of uncertainty set by the poorly constrained physical parameters used in this model calculation.

Peacock (1989) also studied in detail the possible contribution on mass transfer of released fluids in subducting slabs. His results of numerical heat transfer modelling of subducting zones place constraints on the nature of dehydration processes and the amount of fluid generated in subducted oceanic crust. Three dehydration models are discussed: a pressure sensitive dehydration, a temperature sensitive dehydration, and the dehydration of amphibole. In all these dehydration reaction models release of 2 wt % volatiles and consumption of 50 kJ/kg of rock are assumed.

In order to estimate the amount of ascending fluid the following calculation has been carried out by Peacock (1989). As a first approximation, the effective width (W) of the fluid channel is equal to twice the characteristic diffusive length (D) given by

$$D = \sqrt{k \cdot t},$$

where k is the thermal diffusivity of 10^{-6} m²/s and t is time of 10 Ma. From this the effective width of the pervaded hanging lithosphere is

$$W = 35 \text{ km}.$$

The total amount of fluid subducted each year must be equal to the amount released by the dehydration models. For a convergence rate of 10 cm/a there results 1,3 kg fluid/($m^2 \cdot a$) with a fluid density f=1200 kg/m³, the effective volumetric fluid flux is $1.1 \cdot 103$ m³ fluid/($m^2 \cdot a$). This fluid flux corresponds to a vertical mass transfer of 1 mm/a. Peacock uses an effective width of 35 km, while the model used by the author had a width of 100 km. In narrowing that length of the slab where the release of the fluids takes place from 100 to 34 km, a value of 1 mm/a is obtained for the Darcy velocity. In conclusion, the quantitative study on the input side of the fluids gives a result which agrees with the order of magnitude of the Darcy velocity.

4 Rheological Behaviour of the Andean Crust

The strength of the lithosphere controls its deformation in both compression and extension regimes. The response of the lithosphere to an applied force is dependent on the vertical distribution of both brittle and ductile strength, which, in turn, depends on variation in rheology with depth. While brittle strength is controlled primarily by lithostatic pressure and increases with depth, the ductile strength is governed by temperature and decreases with depth. The variation in strength is therefore critically important in defining the value of the force which must be applied to the lithosphere in order to produce significant deformation (Park 1988).

The strong temperature dependence of crustal rheology means that the geothermal gradient is very important in determining the variation of strength with depth. In a previous section the results of temperature determination for five sites along a trans-Andean profile were presented.

Following Park (1988) the ductile deformation of quartz corresponds to that of dislocation creep. The continental upper crust is assumed to deform according to a wet quartz rheology with 50% quartz. For the lower crust it is assumed that rheology is controlled by plagioclase which amounts to 40-50%. The creep rates for these materials are given by the following equations:

wet quartz:

$$\dot{\varepsilon} = 4.36 \cdot (\sigma_1 - \sigma_3)^{2.44} \cdot \exp(-19332/T) \qquad (s^{-1})$$

and

plagioclase:

$$\dot{\varepsilon} = 8200 \cdot (\sigma_1 - \sigma_3)^{3.2} \cdot \exp(-28788/T) \qquad (s^{-1}),$$

where

$\dot{\varepsilon}$ (creep strain rate) is $7 \cdot 10^{-15}$ (s^{-1}),

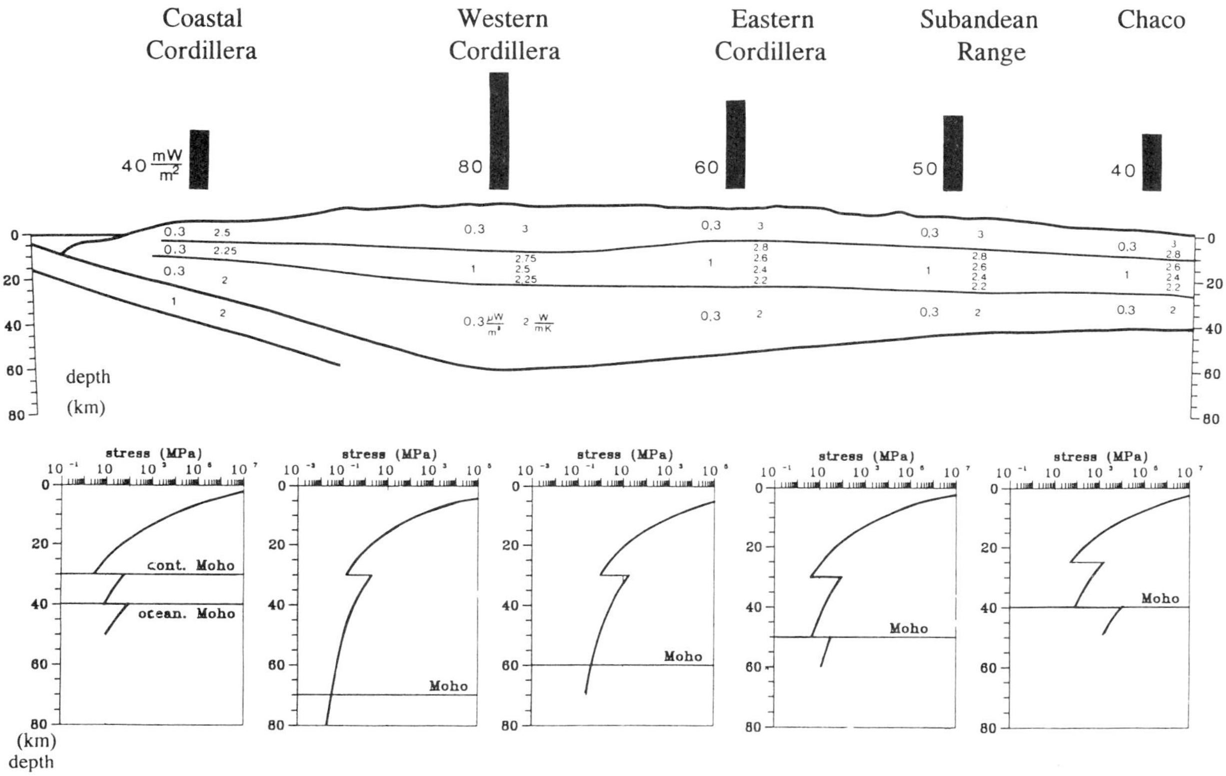

Fig. 3. *Above* See legend for Fig. 2. *Below* The maximum stress profiles derived from the temperature profiles in Fig. 2. Further explanation is given in the text

$(\sigma_1 - \sigma_3)$ is twice the maximum shear stress, (kbar), and

T temperature (K).

Figure 3 shows a simplified structural subdivision of the Andean crust which was used for the calculation of the stress-depth relationships for the five locations under study. These relationships are presented in the lower part of Fig. 3. For the determination of the strength-depth curve for the Western Cordillera the convective temperature curve was used. Due to the petrologic division in the upper and lower crust the curves show typical rheological layering with decreasing strength toward the base of the corresponding layer. It is possible to define the rigid zones of the crust for which their lower part shows a strength which is below a reasonable threshold level of about 50 MPa. Such a value is assumed to be a realistic strength in a crust under compressional stress.

Through the whole crust in the Chaco the shear strength is higher than the assumed active crustal stress of 50 MPa. For the Subandean curve the active stress is somewhat larger than the calculated shear stress. The situation changes significantly in the Eastern Cordillera and even more so in the Western Cordillera. In these locations beneath the depth range 20-30 km the assumed shear stress of 50 MPa distinctly exceeds the calculated stress. The crust of the Coastal Cordillera again behaves as a rigid layer.

During extensional or compressional deformation the main rheological boundaries in mid and deep crustal levels can form zones of low ductile strength providing horizons of décollement. The rheological curves described above show that beneath the Eastern Cordillera, the Altiplano and the Western Cordillera a basal detachment horizon, independent of the crustal rock type, must be assumed.

References

Anderson RN, DeLong SE, Schwarz WM (1978) Thermal model for subduction with dehydration in the downgoing slab. J Geol 86: 731-739

Andrews DJ, Sleep NH (1974) Numerical modelling of tectonic flow behind island arcs. Geophys J R Astr Soc 38: 237 257

Cermak V, Bodri L (1986) Two-dimensional temperature modelling along five East-European geotraverses. J Geodyn 5: 133-163

Fowler CMR (1990) The solid earth, an introduction to global geophysics. Cambridge University Press, Cambridge, 472 pp

Fyfe WS, Lonsdale P (1981) Ocean floor hydrothermal activity. In: Emiliani C (ed) The oceanic lithosphere, the sea 7. Wiley, New York, pp 589-638

Henry SG (1981) Terrestrial heat flow overlying the Andean subduction zone. PhD Thesis, The University of Michigan, 194 pp

Henry SG, Pollack HN (1988) Terrestrial heat flow above the Andean subduction zone in Bolivia and Peru. J Geophys Res 93 (12): 15153-15162

Park RG (1988) Geological structures and moving plates. Blackie, Glasgow, 337 pp

Peacock SM (1987) Thermal effects of metamorphic fluids in subduction zones. Geology 15: 1057-1060

Peacock SM (1989) Thermal modelling of metamorphic pressure-temperarture-time paths: A forward approach. In: Spear S S, Peacock S M (eds) Metamorphic pressure-temperature-time paths. Short Course Presented at the 28th Int Geol Congr, American Geophys Union, Washington DC, pp 57-102

Shi Y, Wang CY (1987) Two-dimensional modeling of the P-T-t paths of regional metamorphism in simple overthrust terrains. Geology 15: 1048-1051

Toksöz MN, Minear JW, Julian BR (1971) Temperature field and geophysical effects of a downgoing slab. J Geophys Res 76: 1113-1138

Uyeda S, Watanabe T (1982) Terrestrial heat flow in Western South America. In: Gupta ML (ed) Terrestrial heat flow. Tectonophysics 83: 63-70

Vitorello I, Hamza VM, Pollack HN (1980) Terrestrial heat flow in the Brazilian highlands. J Geophys 85 (B7): 3778-3788

Large- and Fine-Scale Geochemical Variations Along the Andean Arc of Northern Chile (17.5°- 22°S)

GERHARD WÖRNER, STEPHEN MOORBATH, SUSANNE HORN,
JÜRGEN ENTENMANN, RUSSEL S. HARMON,
JON P. DAVIDSON, and LEOPOLDO LOPEZ-ESCOBAR

Abstract. Geochemical data from 37 volcanic centres from the active volcanic front in the Central Andes between 17.5° and 22°S of northern Chile provide constraints on crustal contributions to arc magma genesis in that region. Crustal thickness, distance from the trench, height above the seismically active subduction zone, and sediment supply to the trench are all constant along this segment of the arc. The only significant variable along the current arc segment is in mean crustal age (Palaeozoic in the south to Proterozoic in the north). In addition, the crustal thickness has varied through time from around 40 km in the Lower Miocene to about 70 km today. Variations along the N-S chain of the volcanic front include: (1) regional segmentation into zones of abundant and voluminous volcanic complexes that evolve from rare basaltic andesites to rhyodacites and rare rhyolites; (2) northern centres with higher incompatible trace element contents than southern ones; (3) constant isotopic composition with differentiation for most centres, with the exception of V. Ollague where AFC trends are observed; (4) Sr- and, in particular, Pb - isotopic compositions which differ north and south of the Alto de Pica area at about 19.5°S, and reflect interaction of magmas with domains of different crustal type and mean crustal age; and (5) interruption of the volcanic chain north and south of the Alto de Pica area where recent volcanism is absent over 180 km in a N-S direction, except for a single rhyodacite mega-dome (Co. Porquesa). The Alto de Pica sector is also unusual in that it lacks the tectonic horst of the Western Cordillera and may correlate with a major transverse crustal domain boundary. There is also a striking correlation between the variation in isotopic composition of erupted magmas through time and the crustal uplift/crustal thickening history of this region. Together these are interpreted as clear evidence that isotopic (and thus bulk chemical) compositions are strongly affected by magma-crust interaction rather than by changes in the mantle source. Crust-magma interaction is viewed as a MASH-type process (melting, assimilation, storage, homogenization) in lower crustal melting regions. Homogenization of deeply-derived contaminated magmas produced typical "MASH-baseline values", i.e. chemical and isotopic characteristics that are constant throughout the entire compositional range of an individual volcanic structure. However, for a number of unusual basalts and basaltic andesites in small and rare isolated monogenetic centres, MASH homogenization is not indicated and is in conflict with chemical characteristics.

In order to account for variation of magma composition in time and space, we envisage a transitional MASH zone which has responded to crustal thickening since the Upper Miocene, involving breakdown of mineral phases such as plagioclase and amphibole, and/or variable degrees of partial melting of the crust. In contrast, small-volume basaltic andesites show regional geochemical domain boundaries that are distinct from the much larger stratovolcanoes (andesites to rhyodacites). Their trace element and isotopic signature is regarded as being derived from the deepest crust, with the northern crustal domain extending further south at depth rather than nearer the surface.

1 Introduction

At subduction zones new additions to the continental crust by magmatism may result in the growth of continents. It is therefore important to identify details of the bulk composition of magmas derived from the mantle above subduction zones and to understand processes and rates of interaction between these mantle-derived arc magmas and continental crust, in particular where continental crust above a subduction zone is thick and likely to contribute to subduction zone magmatism. In the subduction zone of the Central Andes of South America (Fig. 1) the continental crust is thicker than anywhere else on

Correspondence to: Gerhard Wörner, Institut für Geowissenschaften, Universität Mainz, Saarstr. 21, D-6500 Mainz

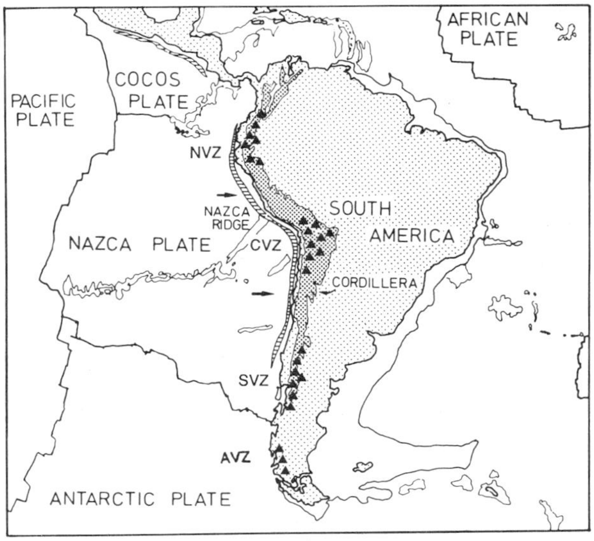

Fig. 1. Plate tectonic setting of South America and Pacific plates and location of currently active volcanism (*filled triangles*) in the South American Andes. *NVZ, CVZ, SVZ, AVZ* northern, central, southern, and austral volcanic zones respectively. *Large arrows* indicate direction of movement of the Nazca plate towards the Peru-Chile trench (horizontal bar signature). *Sparse dots* South American continent; *dense dots* South American Cordillera (i.e. >3000 m a.s.l.)

earth in such a setting.

With regard to the Central Andes it is debated whether the magmas from the mantle wedge are derived from either an asthenospheric (geochemically depleted) reservoir which is subsequently contaminated by crustal melt, or - at least in part - from an old, enriched lithospheric mantle source. Such discussions are based on (1) isotopic and trace element variations in magmas from the Central Andes from a single transect across the arc that spans 180 million years in subduction history from the Jurassic to the present (Rogers and Hawkesworth 1989), and (2) variations in magmas that were erupted in the High Andes along the volcanic chain between 17.5° and 22°S during the past 10 Ma (this study). These data thus document the chemical and isotopic changes in Central Andean magmas both in time and space.

Addressing the question of origin of Andean magmas is fundamental for understanding how the continental crust forms along active continental margins through geological time: If the "enriched" geochemical character of Andean magmas results from enriched mantle lithosphere, then the magmas represent new addition to the crust. On the other hand, if crustal assimilation before or during differentiation has caused the enriched magma isotopic and chemical composition, then this would correspond to variable intra-crustal recycling and, correspondingly, reduced bulk crustal growth (Davidson et al. 1990, 1991).

The possible contribution of upper crustal rocks through contamination in shallow magma chambers by combined assimilation and fractional crystallization (AFC) is widely accepted. Several examples from the Central Andes, e.g. at the Cerro Galan Caldera in western Argentina (Francis et al. 1989), Sierra de Lipez (Harmon et al. 1984) or Ollague (this study), have documented such upper crustal contamination processes parallel to magma evolution by crystal fractionation (AFC, DePaolo 1981). We note that the first two centres are located slightly behind the active volcanic front and that Cerro Galan is clearly controlled by major regional faults (Viramonte pers. commun. 1990). We also emphasize that typical AFC trends appear to be the exception in the Central Volcanic Zone (CVZ) rather than the rule. In most Andean volcanic complexes studied in sufficient detail magma compositions change systematically with differentiation while retaining constant isotopic compositions, suggesting evolution in a closed system (e.g. Nevados de Payachata, Wörner et al. 1988; Davidson et al. 1990) by crystal fractionation, mixing and back-mixing. However, some workers have stressed that assimilation of up to 10-15 % of crust may remain undetected if the isotopic contrast between magma and (young, igneous) crust is small or if parental magmas are insensitive to contamination, e.g. high Sr and Nd contents (Davidson et al. 1990). Careful investigation of other chemical tracers is necessary in such cases to identify both components and processes (e.g. Davidson et al. 1987, 1989).

Case studies from the Central Andean Volcanic Zone have been extremely valuable in placing constraints on possible processes at individual centres. These will now be put into a regional framework in order to identify variations in chemical and isotopic compositions of the magmas which may be related to types and age of underlying continental crust.

Such an attempt is of particular importance between 17° and 22°S, because volcanic centres in this part of the front have been sampled and geochemically analysed only very sparsely. Moreover, at 18°S Wörner et al. (1988) documented highly unusual magma compositions at the Nevados de Payachata complex; compared with all other data in the literature on Central Andean volcanic rocks these are enriched in Ba, Sr and other incompatible trace elements by a factor of

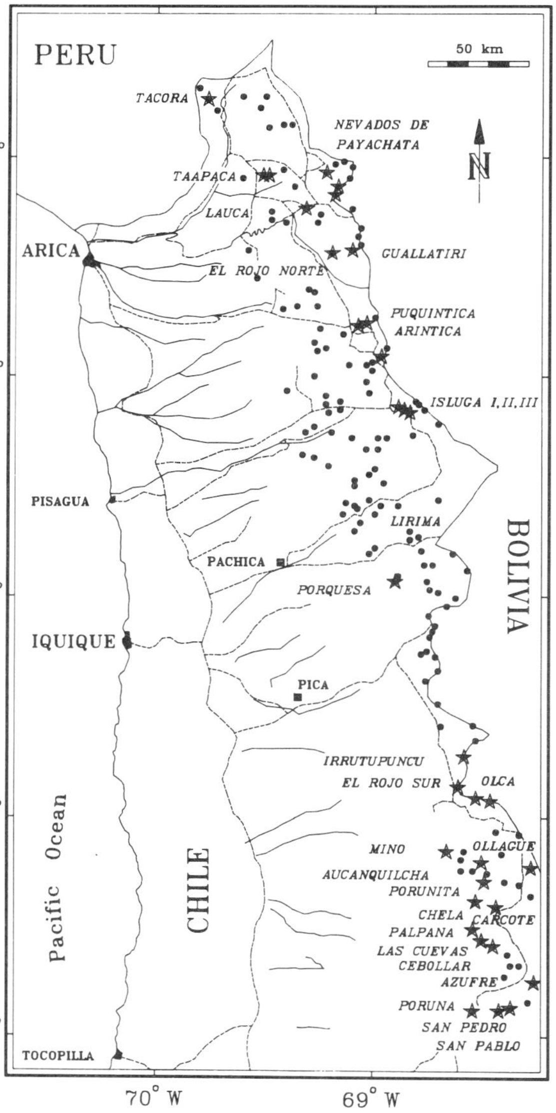

Fig. 2. The Central Volcanic Zone (CVZ) of northern Chile between 17° and 22°S with Plio-Pleistocene volcanoes (*stars*) sampled in this study and older volcanic centres (*dots*) of the active volcanic front

up to 2. It is important to know whether this anomaly is a local feature or represents a broader geochemical signal in the CVZ. In addition, Wörner et al. (1988) demonstrated Pb-isotopic compositions from centres in the northern CVZ to be different from those in the southern CVZ. Moreover, the Nevados de Payachata lavas are distinct in their Pb isotopes which differ from those in all other centres of the CVZ. It is an important goal of this work to trace this unusual Pb-isotopic signature within the CVZ volcanic chain.

We have now sampled 37 volcanic complexes along a 550 km-long segment of the volcanic front between 17.5° and 22°S (Fig. 2). For most of these the sampling and geological information is sufficient to give a full coverage of the compositional spectrum with related stratigraphic control on sample localities. In total, we have collected 530 rock specimens and selected 273 for XRF (X-ray fluorescence) major and trace element analysis. A large representative number was also analysed for Sr (61), Nd (10), Pb (59), and oxygen (44) isotopes. Sampling is restricted to Plio-Pleistocene stratovolcanoes, and older ignimbrites and intrusives are not considered here.

The aim of this contribution is to: (1) present geological information on all volcanic centres sampled, (2) discuss the petrography and petrographic variations in the rocks from different complexes, (3) document the chemical and isotopic database, from which (4) we can place constraints on crustal contributions to arc magmatism in the CVZ of the Andes.

2 The South American Subduction Zone

Currently active subduction-related volcanism is separated into four main volcanic zones in South America (Fig. 1): (1) the Northern Volcanic Zone in Colombia and Ecuador (NVZ), (2) the Central Volcanic Zone in southern Peru and northern Chile (CVZ), (3) the Southern Volcanic Zone in south-central Chile (SVZ), and the Austral Volcanic Zone (AVZ) in southernmost Chile. Intervening zones of shallow subduction have been without volcanic activity in the recent geological past.

The Peru-Chile trench in the CVZ sector of the Andean arc is relatively deep (ca. 7000 m) and almost devoid of sediments (Thornburg and Kulm 1987). The direction of plate convergence changes from oblique (75°) in the northern CVZ to nearly perpendicular in the southern CVZ (Pilger 1981), the transition taking place at the Arica bend.

Upper Miocene to Recent volcanoes (de Silva and Francis 1990) of the CVZ form an almost continuous N-S aligned volcanic zone which features the highest volcanoes on Earth with summits between 5000 and 7000 m (Fig. 2). The northern segment of the CVZ between 16° and 22°S is characterized by a fairly regular chain of evenly spaced volcanoes (with one exception, see below), while further south (22° - 28°S) the volcanic chain steps back to the east, becomes more irregular and wider, and features volcanic gaps related to geochemical variations and aseismic windows within the subduction zone (S.M. Kay pers. commun. 1990).

The crust of the CVZ reaches a maximum thickness of around 75 km (James 1971a,b). Most of the upper crust of the Altiplano is made up of Mesozoic igneous and marine sedimentary rocks. These are covered by Cretaceous continental sediments and younger Miocene to Recent volcanics, intruded by young plutonic rocks. In the Belen (18°20' S) and Quebrada Choja (21°5' S) areas, this crustal section contains N-S oriented lenses 10 km in length of rocks from the deeper crustal basement (gneisses, amphibolites, and serpentinites) representing Proterozoic (~1.5 Ga) protoliths which were metamorphosed in the Cambrian and Precambrian (Damm and Todt 1986). These older rocks possibly correlate with the Arequipa Massif in southern Peru and/or the stable South American continental craton. The nature of the middle and lower crust, however, is unknown. Geophysical measurements on the Andean crust in the CVZ (Schwarz et al. 1986; Wigger 1986) suggest a zone of low seismic wave velocity and high electrical conductivity at mid-crustal levels (10-50 km). This anomaly probably relates to crustal fluids and partial melts liberated from the underthrust crustal segment due to crustal thickening since the Miocene (Benjamin et al. 1987; Isacks 1988).

Magmatism has been intense during the Cenozoic and older geological history of the Andean crust. Even so, the cumulative effect of igneous activity through time, including igneous-derived sedimentary rocks, is probably insufficient to account for the 30 to 45 km of crustal thickening in the central CVZ during the past 15 - 20 million years (Isacks 1988). Thus, foreland thrusting and crustal imbrication have been advocated as the main modes of crustal shortening and thickening (Isacks et al. 1986; Sheffels 1990).

Isacks' model for crustal thickening in the Andes implies heating and weakening of the crust in a convergent regime, a contemporaneous change in plate velocity and direction, and subsequent compression by imbrication of the upper, continental crust since Miocene times. This results in rather sudden crustal thickening, and, by isostatic compensation, uplift of the Altiplano region (Benjamin et al. 1987). Immediately (i.e. within a few million years) following rapid uplift the Altiplano region was covered by vast dacitic to rhyodacitic ignimbrites. This ignimbrite "flare-up" (de Silva 1989) documents crustal heating and melting in response to the crustal thickening event.

3 Principal Components of Subduction Zone Magmas

We now briefly review the possible influences and summarize the possible components in arc magma genesis.

1. Melts and/or fluid phases derived from melting and/or dehydration of subducted oceanic crust represent the most important chemical, but not volumetric, components of all ocean/ocean and ocean/continent subduction-related magmas. These melts or fluids (slab-component) carry their distinctive geochemical fingerprints into the overlying mantle wedge, which represents the prime source for arc magmas.
2. Partial melts form from the asthenospheric mantle in the wedge between the downgoing slab and the overlying lithosphere. The fluid phase liberated from the slab may directly lower the melting temperature of mantle peridotite or form a hydrated mantle peridotite through metasomatic reaction. There are two ways in which such altered mantle could melt. (i) Metasomatized, low-density mantle may eventually rise and melt by decompression. The degree of melting would depend on the distance of diapiric uprise, and hence on the thickness of the asthenospheric wedge which, in turn, is inversely related to the thickness of overlying crust and lithosphere (Plank and Langmuir 1988). However, it is not clear to us why magmas should be unable to rise without further melting and why a diapir should stop melting when it reaches the base of the crust or mantle lithosphere. (ii) Alternatively, cooling of the mantle wedge by a cold downgoing slab could be compensated by counter-convection of hot asthenosphere into the wedge, thus superimposing hot molten asthenosphere above the cold slab.
3. Partial melts from the mantle lithosphere of the overlying plate may also contribute to arc magmas. Lithospheric sources, whilst well documented from other tectonic settings (e.g. intra-plate activity) as being chemically and isotopically enriched, are controversial as a source for an enriched component in subduction settings (Rogers and Hawkesworth 1989; Davidson et al. 1990).
4. Partial or bulk melts from lower and/or upper crust may form in response to magmatic heating during prolonged input of mantle magmas and burial during crustal interaction. Such crustal melts may form where ponding at the base of the crust due to density contrast, or storage at higher crustal levels, results in cooling and crystallization of primary mantle magmas. Heat transfer into the surrounding country rock produces partial crustal melts with characteristic

geochemical and isotopic signatures. Mixing between mantle magmas and crustal melts then yields hybrid magmas.

4 Results

4.1 Description of Volcanic Centres

Thirty-seven centres have been studied and sampled in varying detail: between 2 (a monolithological dome, Co. Porquesa) and 165 (Nevados de Payachata complex) samples were collected from each complex, some of which could be stratigraphically subdivided into different units using morphological criteria and contact relations. Details are given in Horn (1991), whilst a full description of Volcán Parinacota and the Nevados de Payachata complex may be found in Wörner et al. (1988) and Davidson et al. (1990).

Volcán Tacora (5980 m a.s.l., 17°43'S/69°47'W; 9 samples, 8 analysed) at the Peru-Chile border consists of steeply dipping, glacially eroded mafic andesite lava flows (with and without amphibole). One sample gave a K-Ar whole rock age of 0.49 Ma (unpublished data). Samples were taken from a young moraine-wash at around 4500 m on the lower southeastern flank. The limited sample set includes plagioclase/clinopyroxene and plagioclase/amphibole mafic andesites, some of juvenile bomb morphology. Nevertheless, the collection is representative because shape, erosional features and rock colour all indicate that the volcano is entirely composed of mafic and intermediate andesites (55.9-61.1% SiO_2), some of which contain olivine phenocrysts.

Volcán Taapaca (5775 m a.s.l., 18°15'S/ 69°30'W; 8 samples, 7 analysed) is a lithologically very uniform complex of phenocryst-rich silicic andesite domes (62.9 - 64.9% SiO_2). Morphology and degree of alteration distinguishes three eruptive phases, all of which produced non-welded block-and-ash flows around the western flank. The easternmost dome is related to the youngest phase, and the most recent activity is represented by a pyroclastic flow from dome collapse into the headwaters of the valley above the village of Putre. Samples cover all dome phases, their associated pyroclastics and abundant mafic amphibole-rich clots (54.0% SiO_2).

Nevados de Payachata (6438 m a.s.l., 18°10'S/ 69°10'W) comprises two coalescing stratovolcanoes: the older, Volcán Pomerape (c. 0.2 Ma), is formed by a cluster of andesitic domes and flows that have been deeply eroded by glaciers. The younger, Volcán Parinacota (c. 0.3 Ma to recent), is a cone of andesitic to dacitic flows and domes which formed upon a basement of amphibole andesites and clusters of rhyodacitic to rhyolitic domes. This older cone collapsed towards the west and produced a 110 km^2 debris avalanche deposit. The younger cone is composed mostly of amphibole-free mafic andesites with the most recent activity represented by flows and spatter cones issuing from an almost N-S oriented fissure at the southern base of the cone. Some 153 samples were collected from this complex, of which 132 were analysed for major and trace elements (52.3-75.2% SiO_2) as well as isotopic compositions (Wörner et al. 1988; Davidson et al. 1990). The sample collection also includes representative samples from older surrounding centres (Caquena, Chucullo), which belong to the same geochemical lineage. Detailed petrological and microprobe studies of Nevados de Payachata lavas are underway (cf. Entenmann et al. 1990).

Volcán Lauca and associated ignimbrite (5140 m a.s.l., 18°39'S/69°23'W; 51 samples, 37 analysed). Lauca volcano comprises a cone of rather monotonous amphibole andesite lava flows, and has collapsed at its centre during a major ignimbrite-forming eruption. Two types of andesite are present (altered plagioclase-rich andesite and younger, fresh fine-grained silicic andesite), both of which are older than the Lauca ignimbrite. These rocks span a range of 58.7 to 76.1% SiO_2. Wörner et al. (1988) report an age of 10.5±0.5 Ma for the Lauca andesite. The Lauca ignimbrite has its maximum thickness and degree of welding in outcrops within the northern breach of the caldera. A dome has formed at the northeastern rim of the caldera with fresh plagioclase vitrophyre at its base. This centre and the upper Miocene ignimbrites of the Lauca Basin have now been extensively sampled and the sample collection covers the petrographic range of lavas.

El Rojo Norte (18°28'S/69°12'W, 1 sample, 55.1% SiO_2) is a small isolated monogenetic cinder cone overlying the volcaniclastic sediments of the Lauca Basin. Erosion has formed a very subdued mound of reddish scoria of about 50 m height and 300 m diameter at the base. Fresh dense bombs of basaltic andesite with olivine phenocrysts have been sampled.

Volcán Guallatiri (6071 m a.s.l., 18°25'S/ 69°15'W; 19 samples, 7 analysed) forms a glaciated cone built on andesitic lavas. A large central mound of smooth morphology forms the summit area, with numerous solfaric fumaroles at the southwestern flank and near the summit. Its youthful morphology was interpreted from Landsat images (Francis and de Silva pers. commun.) as a recent dome. However, in comparison with typical domes on similar volcanoes in the area (e.g. V. Tacora) its size is too large and morphological smoothness too perfect, so that it may represent scree and erosion around the flanks of a deeply hydrothermally altered summit plug. We found large breadcrust bombs (up to 50 cm across) at the northern lower flanks, and these were probably transported there by glaciers. This suggests that small volume eruptions may have occurred intermittently with fumarolic activity. Samples were collected from around the northern, western and southwestern flanks to elevations up to 4800 m a.s.l.. Pyroxene andesites dominate (57.3-62.5% SiO_2), but one bomb of fresh rhyolitic obsidian (74.3% SiO_2) was found.

Arintica-Puquintica, Salar de Surire (5597-5780 m a.s.l., 18°40'S/69°00'W; 21 samples, 13 analysed) are two clustered cones made up of andesitic flows and domes of two eruptive phases from each centre. Landsat images (de Silva 1989 written commun., 1989) show postglacial lava flows in the summit area of Arintica. These, however, have not been sampled. Fresh columnar-jointed amphibole andesite bombs from the southern upper flanks were initially taken as evidence for rather recent activity. However, a K-Ar whole rock date of 0.64 Ma (Wörner et al. unpubl. data) appears far too old, and may be due to excess Ar in the quenched bomb. Highly altered rhyolitic breccias in the summit region of Puquintica indicate significant differentiation within this volcanic complex. The total range of SiO_2 measured is 54.1-73.8%.

Volcán Isluga (5501 m a.s.l., 19°09'S/58°50'W; 51 samples, 22 analysed) forms a cluster of flows and domes overlying a Miocene rhyolitic ignimbrite and older, strongly altered dacite centres (Co. Quimsachatas, 0.57 Ma, unpubl. date). The degree of alteration, morphology, K-Ar age dating (Wörner et al. unpubl.) and chemistry are distinct for three eruptive phases (Isluga I, II, III). Petrographic and chemical variability is rather large (58.3-64.8% SiO_2). The most mafic rocks are from the youngest phase erupted along a ridge towards the east (0.096 Ma, unpubl. date), overlying rocks of the older phases. Erosion into the summit plateau has exposed deeper sections of tephra deposits towards the north. A small phreatic crater, about 250 m in diameter and about 80 m deep, dissects the eastern margin of the summit area exposing phreatomagmatic deposits which are heavily altered by solfatoric activity. A sulphuric steam cloud rises up to 200 m above this crater.

Cerro Porquesa (ca. 4600 m a.s.l., 19°59'S/68°46'W; 3 samples, 2 analysed) is a large dome cluster comprising at least three phases of two to three lobes each. The morphology of the younger domes suggests minor, if any, glaciation and thus a rather young age. Aerial photography and limited sampling indicate rather uniform rhyodacite composition (ca. 68% SiO_2). A thin (2 m) rhyolitic ignimbrite (69.5% SiO_2), presumed to have been derived from the Porquesa domes, fills the valley to the south of Porquesa. This underscores the recent eruptive activity at Porquesa. Porquesa is unusual in representing a "failed ignimbrite", of which similar examples exist further south (e.g. Chaco dacite, de Silva et al. 1988).

Volcán Irrutupuncu (5165 m a.s.l., 20°44'S/68°34'W; 13 samples, 7 analysed) is a comparatively small two-phase stratovolcano nested into a southwest facing amphitheatre of an older glaciated and altered cone. The northern, older part of the cone shows a collapse scar and a small avalanche deposit towards the west. The younger cone is formed from steeply dipping, "hanging", plagioclase-andesite lava flows, some of which have lost their connection with the crater source by slumping. Others have formed block and ash flow deposits upon gravitational collapse. The resulting debris forms several avalanche deposits towards the southwest. Most prominent are two pristine flows and a "mega"-breadcrust bomb of 3 m in diameter at the crater rim. The crater is breached towards the south and shows strong fumarolic activity. Samples completely cover the small petrographic range of older and younger lava flows as well as the various pumice types. The younger lava flows overlie a fresh plinian pumice deposit which shows excellent petrographic evidence of andesite-dacite magma mixing. It is notable that, with the exception of the mafic mixing member of hybrid pumices (59.9% SiO_2), pumice and lava flows are of similar major element composition (60.6-65.6% SiO_2) and phenocryst assembly. However, phenocryst size, abundance, eruption style and matrix type are quite different. Possibly, injection of mafic magma triggered the premature eruption of an Irrutupuncu andesite magma which normally would have stagnated and crystallized for longer to produce the high viscosity, coarsely phyric andesite flows of V. Irrutupuncu. Just to the southwest of Irrutupuncu, a recent thin (1.5 m), subplinian ignimbrite rests upon Pleistocene salar sediments; V. Irrutupuncu is a likely source.

El Rojo Sur (20°50'S/68°37'W; 1 sample) is identical in size, shape and rock type (55.1% SiO_2) to El Rojo Norte about 260 km to the north.

Volcán Olca (5407 m a.s.l., 20°56'S/68°30'W; 8 samples, 6 analysed) is a dome-and-flow complex forming an E-W oriented 5-km-long ridge. Fumarolic activity is present near the centre of the ridge while the most recent lava flows issued from the westernmost sector. These are only mildly glaciated in the summit area (estimated) and overlie older, more glaciated flows on the lower flanks. Most recent activity is represented by phreatomagmatic and phreatic activity which left several nested explosion craters and typical surge deposits at the summit. Rock types are fairly uniform amphibole-bearing andesites (58.6-63.8% SiO_2). Alteration at the ridge is very intense and has led to mining activity at the summit.

El Miño (5611 m a.s.l., 21°11'S/68°35'W; 10 samples, 4 analysed) is a small stratocone with uniform, slightly glaciated slopes. Deep glacial scars are not developed, probably because of the smaller size and lower elevation of the cone. Moderately altered plagioclase-phyric andesites dominate. El Miño overlies an older, deeply glaciated centre to the east, which erupted more mafic, aphyric andesites. The total compositional range, however, is small (57.9-62.2% SiO_2).

Volcán Aucanquilcha (6176 m a.s.l., 21°11'S/68°35'W; 10 samples, 4 analysed) is a large complex of clustered stratovolcanoes forming an 8 km E-W chain. The central region of the edifice is strongly altered by fumarolic activity. Francis and Wells (1988) described a prominent collapse structure and resulting debris avalanche deposit towards the NNW. Deeply glaciated andesite flows are overlain by less eroded lavas at the base of the structure. Pristine morphology of some flows attests to rather recent (postglacial ?) eruptions within the past several tens of thousands of years, and thus extended eruptive activity at this complex, possibly over several Ma. Petrographic compositions range from andesite flows to dacite domes and dome complexes (62.8-65.7% SiO_2). Sampling includes basal flows as well as postglacial lavas from around the structure and thus covers the entire history and range in rock types.

Volcán Ollague (5863 m a.s.l., 21°37'S/68°11'W; 34 samples, 18 analysed) represents a typical complex, andesitic to rhyodacitic composite cone with evidence of several distinct eruptive periods. Old Ollague forms the base and eastern flank and is composed of altered andesite and dacite flows. The eastern summit region also comprises older rocks while the western flank consists of younger, mostly glaciated flows. Francis and Wells (1988) first described a debris avalanche deposit to the west and this probably formed a collapse scar within the older cone, later filled by younger Ollague flows. One flow overlying the avalanche has been dated at 0.8 ± 0.1 Ma by Francis and Rundle (1976). The present western summit feeds short postglacial andesite flows and hosts a vigorously active fumarole. A cluster of isolated rhyodacite domes, erupted at the northwest base of the cone, are not glaciated, possibly due to their lower elevation. Lava compositions are relatively diverse (52.9-67.4% SiO_2) from clinopyroxene - plagioclase andesites of the older flows to amphibole-bearing andesites and rhyodacites with mafic magma inclusions (clots) of basaltic andesite in the younger part of the cone. The variable presence of amphibole in andesites of similar bulk chemistry attests to a complex and variable magmatic plumbing system with time. Sampling covers all the southern, western and northwestern flanks and summit area of V. Ollague. The eastern part, on Bolivian territory, belongs to the older phase and is probably fully represented in our sample collection by material from the southern and northwestern part of the cone.

Cerro Porunita (21°19'S/68°34'W; 7 samples, 2 analysed) is a phreatomagmatic monogenetic tuff cone, about 100 m high and 600 m in diameter overlying the Ollague debris avalanche deposit. Lake deposits (carbonate algae mats) form lake-level marks of different elevation on debris avalanche hummocks. These are much thicker and more pronounced than on the tuff cone, which suggests that Cerro Porunita is significantly younger than avalanche deposition. The cone consists of pyroclastic layers with dense, glassy andesite clasts and abundant xenoliths (6 samples). The andesite is extremely fresh and contains olivine phenocrysts, despite its rather high SiO_2 content of 60.5 wt.%.

Cerro Chela (5644 m a.s.l., 21°24'S/68°30'W; 8 samples, 3 analysed) stratocone overlies a rhyolitic ignimbrite southwest of Salar de Carcote. Uniform slopes, radial ridges and widespread grey to reddish rock colours attest to a rather uniform mafic andesite composition of the glacially dissected stratocone (57.1-59.8% SiO_2). Cerro Chela is very similar to Cerro Palpana.

Cerro Palpana (6023 m a.s.l., 21°33'S/68°32'W; 9 samples, 5 analysed) is a dissected cone consisting mostly of mafic andesite lava flows, coarse-grained scoria layers and rare intercalated pumice tephra deposits. While these show regularly dipping strata around the cone, the summit area is rather flat and consists of horizontal lava flows and a subdued central dome-cluster. Erosion cuts deeply into the structure leaving regular radial ridges. An avalanche scar in the west is the source of a minor debris avalanche deposit at the lower flank. No postglacial rocks have been observed. However, morphological features suggest an age no older than about 1-2 Ma. Rock compositions are similar to Co. Chela and are rather uniform where sampled on the southern ridge (57.6-58.9% SiO_2). More mafic andesites contain rare phenocrysts of plagioclase, and some have olivine. Considering evidence for an overall uniform lithology, we believe that our limited sampling on the lower southern flank around 4200 to 4600 m a.s.l.. could well be representative for the entire volcano.

Cerro de las Cuevas (5294 m a.s.l., 21°35'S/ 68°29'W; 4 samples, 2 analysed) consists of a central andesite dome surrounded by (overlying?) extensive and thick clinopyroxene-olivine andesite lava flows which contain abundant quartz xenocrysts. Sampling was limited to the eastern side and may not be representative of this centre (52.9% SiO_2). A young spatter cone (1 sample) rests upon the northeastern flank with extremely fresh and glassy aphyric andesite (59.3% SiO_2). Quartz xenocrysts are also found here and appear to be a typical feature of the Cerro de las Cuevas centre.

Cerro Cebollar (5716 m a.s.l., 21°37'S/68°28'W; 6 samples, 2 analysed) is the oldest of the Cebollar/Las Cuevas/Palpana N-S volcanic chain. Judging from morphological preservation it may be several million years old. Irregular remnants of altered andesite/dacite flows (61.7-61.8% SiO_2) are overlain by a scattered distal rhyolite pumice. This probably correlates with andesitic air fall pumice from near the Salar de San Martin caldera further north.

Volcán Azufre (5846 m a.s.l., 21°47'S/68°14'W; 2 samples, both analysed) is a large, Mio/Pliocene stratovolcano with several eruptive phases producing andesite to dacite flows. We possess only limited samples of the older structure. A significantly younger (1.5±0.1 Ma, Baker and Francis 1978) dome cluster (Cerro Chanca, 4554 m a.s.l.; 21°46'S/ 68°18'W) at the western base of Azufre erupted dacites with variable mafic inclusions (4 samples, all analysed, 62.0-66.8% SiO_2).

Volcán Poruña (3400 m a.s.l., 21°53'S/68°30'W; 1 sample analysed) is a postglacial satellite scoria cone about 800 m in diameter, 100 m high and situated 10 km west of San Pedro-San Pablo. The latter has been extensively studied by O'Callaghan and Francis (1986). We sampled a levee of a small flow issuing from V. Poruña towards the east. The lava is fresh and has phenocrysts of pyroxene, plagioclase and olivine (SiO_2 = 60.7 wt %). There are small gabbroic inclusions and a clinopyroxene xenocryst. Volcán Poruña marks the southernmost extent of our sampling traverse.

This sampling is considered representative for most centres and analytical data are now available (Table 1) for the petrographic spectrum present at each volcano studied along this traverse of the CVZ volcanic front.

It is important to note that over a N-S distance of 150 km, between V. Isluga at 19°9'S and V. Irrutupuncu (20°44'S), there is only one Plio/ Pleistocene centre (Porquesa dome at 19°59'S), suggesting a volcanic gap in the Alto de Pica region ("Pica gap"). Within the same segment of the volcanic chain, the western Andean Pre-cordillera is unusual in lacking any tectonic horsts, which are typical for the entire CVZ further to the north and south. Instead, the Alto de Pica area is characterized by an undisturbed rhyolite-ignimbrite ramp extending down from the Altiplano towards the Longitudinal Valley of northern Chile.

5 Analytical Results

Representative geochemical analyses are listed in Table 1. Analytical methods include standard XRF techniques (available in the geochemical laboratories of Bochum and Mainz), loss on ignition (LOI in Mainz), H_2O determinations by Karl Fischer titration (in Bochum) and Fe^{2+}/Fe^{3+} determination on selected samples by potentiometric titration. Stable isotope analyses were performed at the UK NERC Isotope Geosciences Laboratory and radiogenic isotope ratios were determined at the Department of Earth Sciences at Oxford University. More details of the analytical techniques are given in Wörner et al. (1988).

In order to make all the data comparable we use only LOI and H_2O - free normalized analytical data, except for a very few literature data where a full bulk chemical analysis was not available. We have also reanalysed 25 rocks to identify systematic inter-laboratory differences. All the major and most trace elements compare very well between the two laboratories (Bochum and Mainz). The exception is Nb analyses, which show unsystematic scatter for values lower than about 20 ppm. Slight systematic interlaboratory variations (e.g. for Zr) have been accounted for by empirical correction of the Bochum data. A full list of geochemical data may be obtained from the authors upon request.

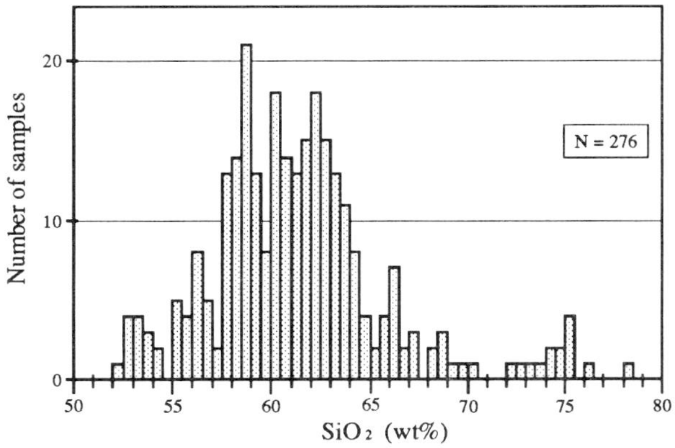

Fig. 3. SiO_2 frequency distribution for 276 individual samples from centres between 17.5° and 22 °S. Note the unusually mafic basalts and basaltic andesites which were found in this study

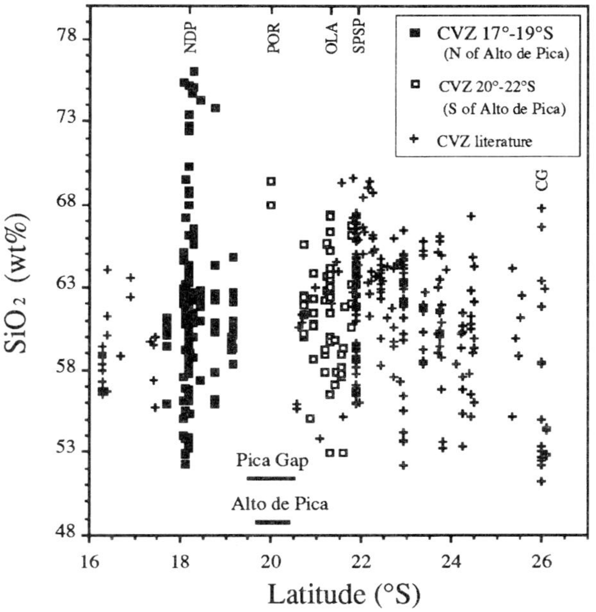

Fig. 4. SiO_2 latitudinal variations based on own and literature data. Data from Wörner et al. (1988), Wörner et al. (1990) and other literature sources referenced in these contributions. A full list of references may be obtained from the authors upon request. Note the location of the "Pica gap" where young volcanism (< 3 Ma) is absent except for the Co. Porquesa megadome. Individual centres: *NDP* Nevados de Payachata, *POR* Co. Porquesa, *OLA* Ollague, *SPSP* San Pedro San Pablo, *CG* Cerro Galan

5.1 Major and Trace Elements

Hildreth and Moorbath (1988) calculated regression lines in variation diagrams to identify "baseline" geochemical values for each volcanic centre. Petrographic and major element variability at many centres of our N-S traverse are not large enough to calculate regression lines with which to establish chemical trends in the same way. We therefore chose to simply apply a SiO_2-filter, using only samples with <62 wt.% SiO_2 (H_2O-free, normalized to 100%). There are a number of graded and stepwise geochemical variations along the N-S traverse. Here we give only a short description and brief interpretation. A full discussion of the data set will be given elsewhere. In summary we observe :

1. There is an overall range in SiO_2 of 52 to 78 wt.%. The majority of lavas fall between 58 and 65 wt.% SiO_2 (Fig. 3).

2. TiO_2, K_2O, Ba and Zr abundances, as well as abundance ratios such as TiO_2/Y, at below 62 wt.% SiO_2 are higher for centres north of the Pica gap. The Nevados de Payachata is simply the most extreme example of this "enriched" chemical signature (Figs. 5-9). This overall trend should reflect variations in source components and their relative contributions (composition, mineralogy and degree of partial melting in the crust and/or variable input from the subduction zone).

3. The data define two geochemically distinct sectors in the N-S traverse. The observed geochemical changes take place between 19°09'S and 20°44'S, in a region where young (i.e. < ca. 4 Ma) volcanism is almost absent (Pica gap, Fig. 2).

5.2 Isotopic Variation

Sr- and Pb-isotope ratios in mafic andesite lavas (< 62% SiO_2) also show systematic variations along the volcanic front (Figs. 10-12). The figures and discussion include data for the CVZ south of 22°S. $^{87}Sr/^{86}Sr$ ratios in the southern sector (20°44'-27°S) range from 0.7055 to 0.7090 (Figs. 10 and 11). In contrast, the northern sector (north of the Pica Gap) has a smaller Sr-isotopic range from 0.7065 to 0.7075. This compares with the smaller range in SiO_2 (except for Nevados de Payachata, 18°15'S) in the northern sector.

Pb isotopes show a number of notable features: (1) $^{207}Pb/^{204}Pb$, and $^{206}Pb/^{204}Pb$ ratios are significantly lower in the northern CVZ sector than south of the Pica gap (Fig. 12a and b). (2) In contrast, $^{208}Pb/^{204}Pb$

Table 1. Representative chemical and isotope analyses of 28 volcanic complexes of the CVZ volcanic front between 17.5 ° and 22 °S

Sample	TAC-06	CAQ-02	TAP-03	GUG 182	CHU 173	AJO 177	LAU 005	CHP 098	GUL-17	SUA-06	IS1-05	IS2-12	IS3-29	POR 2	IRU 1c
SiO_2	55.68	56.57	53.39	58.36	55.78	60.75	58.66	63.24	57.37	61.74	58.36	60.75	60.75	67.05	59.55
TiO_2	1.31	0.80	1.81	0.69	1.64	0.69	0.75	0.68	1.42	0.86	1.32	1.06	0.91	0.54	0.84
Al_2O_3	16.98	18.72	16.70	18.68	16.58	17.27	18.04	15.89	16.85	16.47	16.83	16.85	17.42	15.88	17.62
Fe_2O_3T	7.80	7.78	8.42	7.04	8.43	6.14	7.24	4.34	7.66	5.64	7.35	6.48	5.88	2.95	6.36
MnO	0.11	0.15	0.09	0.28	0.10	0.11	0.14	0.08	0.09	0.10	0.10	0.10	0.10	0.05	0.10
MgO	4.41	3.55	4.27	1.59	4.16	2.41	2.84	1.75	3.57	2.25	3.19	2.37	2.08	0.98	2.68
CaO	7.09	7.45	6.88	6.16	6.70	5.59	6.55	4.24	5.93	4.81	5.97	5.35	5.13	2.95	5.66
Na_2O	3.84	3.58	4.23	3.83	4.44	3.78	3.66	4.21	4.11	4.07	3.95	4.41	4.12	4.67	4.11
K_2O	2.00	1.91	2.26	2.26	2.32	2.44	1.81	3.05	2.57	2.90	2.61	2.91	2.91	3.30	2.17
P_2O_5	0.41	0.21	0.75	0.30	0.57	0.22	0.22	0.20	0.49	0.35	0.41	0.41	0.43	0.20	0.24
LOI				1.17		0.52		1.47						1.04	1.26
Total	99.63	100.74	98.80	100.38	100.71	99.91	99.93	99.14	100.05	99.18	100.09	100.68	99.74	99.61	100.59
V	182	161	186	101	187	105	124	74	166	94	160	126	86	48	121
Cr	10	7	77		103	13	9	27	60	16	9	5	1	9	14
Co	31	39	33	42	45	26	34	37	34	27	34	36	29	7	34
Ni	30	9	45	6	48	8	7	10	32	4	8	1	1	6	7
Cu	52	52	45	36	42	39	42	26	50	29	33	32	87	12	39
Zn	96	79	111	82	113	81	89	74	104	80	95	88	130	62	88
Rb	46	79	42	86	31	70	53	92	60	78	80	92	96	94	62
Sr	739	584	1213	681	1085	565	534	601	1007	815	712	731	779	664	685
Y	19	19	17	20	19	19	22	12	18	18	18	18	18	7	16
Zr	181	131	243	202	214	156	164	143	216	203	214	219	219	146	132
Nb	10	11	14	13	9	11	28	12	12	13	13	13	13	8	7
Ba	894	759	1286	913	1140	825	712	966	1292	1192	874	1081	1088	979	720
Pb		11		15	15	14	13								
$^{87}Sr/^{86}Sr$	0.70625	0.70584		0.70608	0.70667	0.70686	0.70665		0.70670	0.70644		0.70585	0.70589	0.70582	
$^{206}Pb/^{204}Pb$	18.116	18.228		18.185	18.046	18.283	18.203		18.093	18.132		17.888	18.243	18.534	
$^{207}Pb/^{204}Pb$	15.613	15.610		15.597	15.596	15.619	15.615		15.624	15.612		15.597	15.617	15.607	
$^{208}Pb/^{204}Pb$	38.113	38.420		38.431	38.198	38.556	38.518		38.392	38.394		37.946	38.410	38.520	
Age (Ma)						7.06		6.60		0.64					
Latitude (°S)	17.72	18.07	18.08	18.13	18.23	18.26	18.28	18.30	18.42	18.75	19.15	19.15	19.15	19.98	20.73

Sample	ELR 1	OLC 6	MIN 2	AUC 1	OLA 13	PORU 2	CHE 8	PUN 1	CAR 1	PAL 4	CUEV 4	CEB 5	CHAN 3	AZU 2	SP 01
SiO_2	54.90	59.10	61.78	64.58	62.15	60.52	57.08	56.54	58.06	58.07	59.74	61.57	65.24	60.05	56.42
TiO_2	1.69	1.04	0.64	0.62	0.81	1.05	0.88	0.98	1.01	0.98	0.96	0.68	0.52	0.85	0.87
Al_2O_3	15.86	16.99	16.89	16.10	16.69	17.28	16.97	18.80	17.18	17.81	16.92	17.36	16.07	16.19	16.53
Fe_2O_3T	8.61	7.06	5.25	3.90	5.47	6.38	7.38	6.78	7.22	6.94	6.57	5.57	3.82	6.16	7.40
MnO	0.10	0.09	0.08	0.07	0.08	0.08	0.10	0.08	0.09	0.09	0.09	0.09	0.06	0.08	0.10
MgO	4.29	3.73	2.78	1.87	2.37	3.33	4.67	2.73	3.82	2.99	3.83	2.20	1.60	3.81	5.69
CaO	6.76	5.97	4.98	3.83	4.75	5.77	6.96	6.90	6.34	6.19	5.90	5.00	3.57	6.17	6.98
Na_2O	4.27	4.11	4.05	4.37	4.12	4.30	3.80	4.18	4.04	4.25	4.41	4.19	3.98	3.26	3.64
K_2O	2.51	2.47	2.69	2.94	2.89	2.38	1.84	1.96	2.42	2.22	2.06	2.75	3.56	2.33	1.71
P_2O_5	0.73	0.27	0.20	0.18	0.24	0.25	0.24	0.26	0.24	0.27	0.26	0.19	0.15	0.21	0.22
LOI	0.30	0.19	1.06	1.56	0.49		0.10	0.81	0.17	0.39		0.87	1.13	1.58	0.61
Total	100.02	101.02	100.40	100.02	100.06	101.34	100.02	100.02	100.59	100.20	100.74	100.47	99.70	100.69	100.17
V	157	136	94	72	97	128	143	167	150	149	142	95	71	138	169
Cr	128	62	55	31	31	56	93	24	61	32	113	14	11	84	206
Co	27	18	12	20	9	16	22	16	20	18	20	13	9	24	25
Ni	81	18	18	10	8	20	37	24	21	23	37	10	7	22	60
Cu	45	51	44	36	10	30	82	59	34	60	54	22	25	28	33
Zn	141	94	74	69	82	113	90	79	100	98	101	67	62	86	93
Rb	44	71	62	83	93	49	38	50	65	50	42	86	138	71	37
Sr	1347	654	596	567	515	609	616	618	563	652	711	549	493	573	578
Y	26	21	13	11	18	17	16	21	19	18	15	17	15	19	19
Zr	265	189	140	158	179	180	133	166	171	171	158	145	162	146	124
Nb	20	10	7	8	10	11	5	7	10	8	7	8	9	9	9
Ba	1914	796	794	1030	858	961	634	681	767	730	728	777	903	701	593
Pb			11	13											
$^{87}Sr/^{86}Sr$	0.70654	0.70580	0.70550	0.70604	0.70691	0.70671	0.70562	0.70556	0.70624	0.70555	0.70592	0.70575	0.70608		0.70663
$^{206}Pb/^{204}Pb$	17.873	18.676	18.650	18.678	18.776	18.671	18.689	18.606	18.714	18.660	18.699	18.753	18.724		18.737
$^{207}Pb/^{204}Pb$	15.602	15.636	15.623	15.624	15.646	15.637	15.623	15.616	15.624	15.619	15.637	15.636	15.624		15.638
$^{208}Pb/^{204}Pb$	38.113	38.559	38.591	38.566	38.657	38.478	38.544	38.480	38.535	38.529	38.593	38.644	38.657		38.674
Age (Ma)															
Latitude (°S)	20.88	20.93	21.18	21.21	21.30	21.31	21.40	21.42	21.44	21.55	21.59	21.61	21.78	21.79	21.89

Notes: Fe_2O_3T = Total iron calculated as Fe_2O_3. Individual centres: TAC = Tacora, CAQ = Caquena, TAP = Taapaca, GUG = Guane Guane, CHU = Chucullo, AJO = Ajoya, LAU = Lauca, CHP = Choquelimpie, GUL = Guallatiri, SUA = Arintica/Puquintica, IS = Isluga, POR = Porquesa, IRU = Irruputuncu, ELR = El Rojo Sur, OLC = Olca, MIN = Miño, AUC = Aucanquilcha, OLA = Ollaque, PORU = Porunita, CHE = Chela, PUN = Puntilla, CAR = Carcote, PAL = Palpana, CUEV = Las Cuevas, CEB = Cebollar, CHAN = Chanca, AZU = Azufre, SP = San Pedro San Pablo. A full list of geochemical analyses may be obtained from the authors upon request.

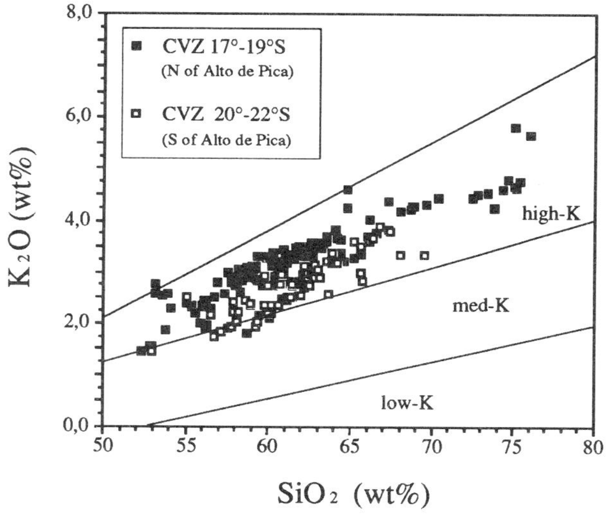

Fig. 5. SiO_2-K_2O variations. Note the significant differences for centres north and south of the Pica gap (Alto de Pica)

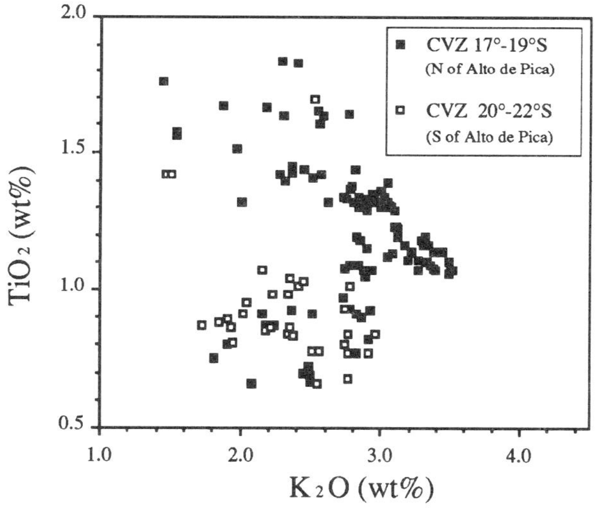

Fig. 6. K_2O-TiO_2 variations. There is a clear separation between northern and southern centres, although the northern population is dominated by samples from the Nevados de Payachata

ratios are similar in volcanics from the Peruvian CVZ and the southern sector in northern Chile, but the area around the Nevados de Payachata complex (18°-19°S) is unusual in having lower $^{208}Pb/^{204}Pb$ ratios down to 37.75 (Fig. 12c); and (3) Pb-isotope ratios change gradually with increasing $^{206}Pb/^{204}Pb$ and

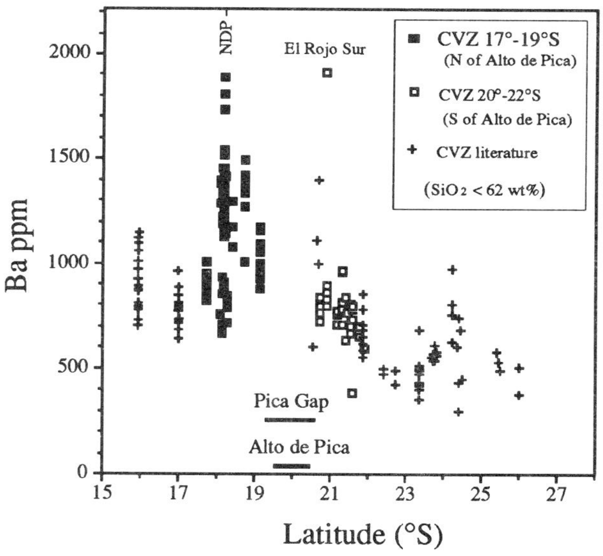

Fig. 7. Ba variations along centres of the active volcanic front (from North to South in latitude °S) for samples with SiO_2 contents < 62 wt.% along the volcanic chain between 17° and 22°S based on own and literature data. The bulk of the data forms a trend of increasing Ba from S to N with a peak at the Nevados de Payachata *NDP* - "anomaly". It is important to note that the monogenetic basaltic andesite of El Rojo Sur has Ba contents characteristic of northern centres, although it is located in the south. Individual centres and references as in Fig. 4

$^{207}Pb/^{204}Pb$ ratios from 26° to 16°S (Fig. 12a and b). The mafic monogenetic centre of El Rojo Sur is notable because it is located in the southern sector but has the relatively unradiogenic $^{206}Pb/^{204}Pb$-isotope ratios typical of the northern sector.

6 Discussion and Conclusions

Rogers and Hawkesworth (1989) explain the isotopic and trace element variations in Andean magmas through time by increasing contributions from old mantle lithosphere below the Brazilian Shield, because the focus of magma generation in the subduction zone is slowly migrating east towards the ancient crust and mantle lithosphere below South America.

Old, enriched mantle lithosphere is known to produce low-degree, incompatible element-enriched partial melts in an intra-plate setting. However, it will probably not do so above a subduction zone where the degree of melting is generally regarded as

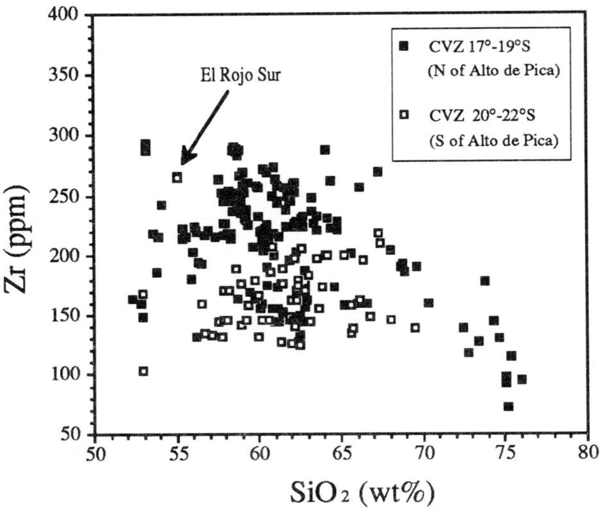

Fig. 8. Zr-SiO_2 variations. There is a tendency for centres from the north (*filled squares*) to be higher in Zr at a given SiO_2 content compared with centres from the south (*open squares*)

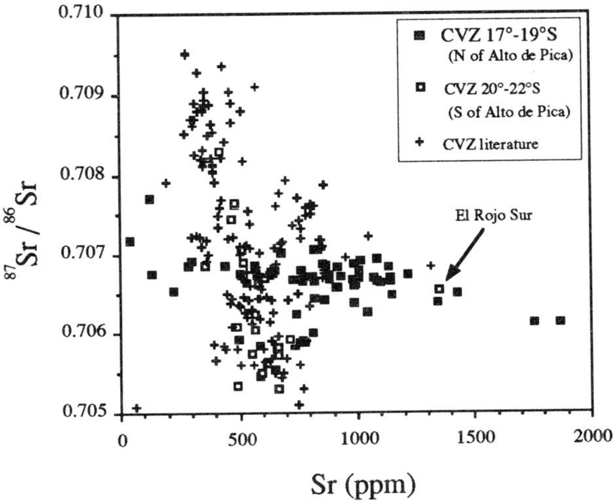

Fig. 10. $^{87}Sr/^{86}Sr$ versus Sr contents. There are clear differences in trends for the northern and southern centres: small isotopic variations in the north at variable Sr contrast with variable $^{87}Sr/^{86}Sr$ values in the south which are inversely correlated with Sr abundances. References as in Fig. 4

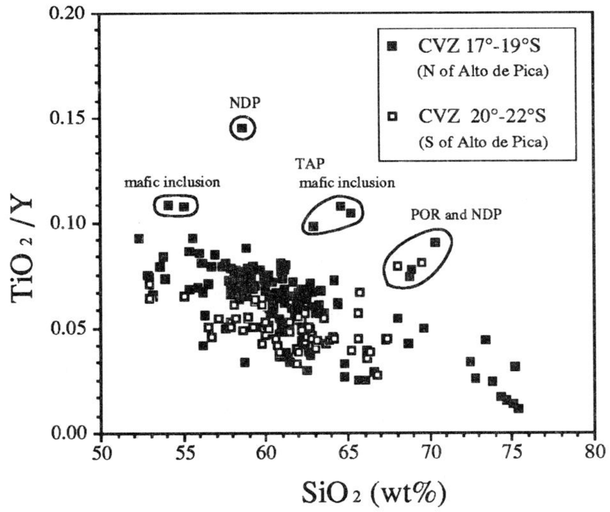

Fig. 9. TiO_2/Y-SiO_2 variations. There are three types of samples: Centres (1) north and (2) south of the Pica gap in the Alto de Pica region and (3) some unusual samples representing mafic inclusions in andesites, and the Porquesa dome. Individual centres as in Fig. 4 except for Volcán Taapaca (*TAP*)

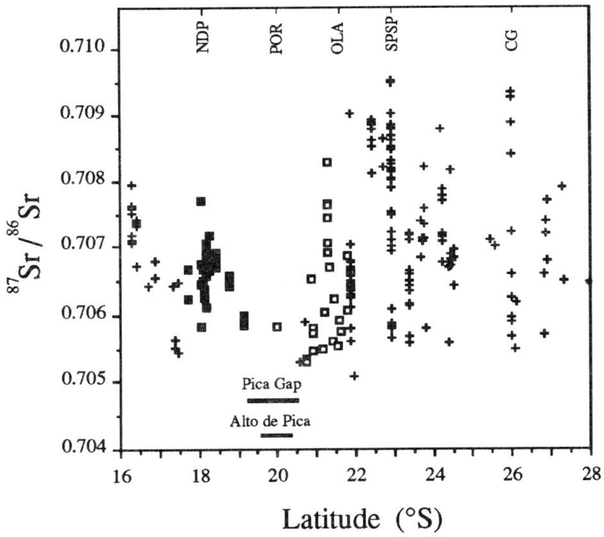

Fig. 11. $^{87}Sr/^{86}Sr$ versus latitude. For further description and discussion see text. Individual centres and references as in Fig. 4

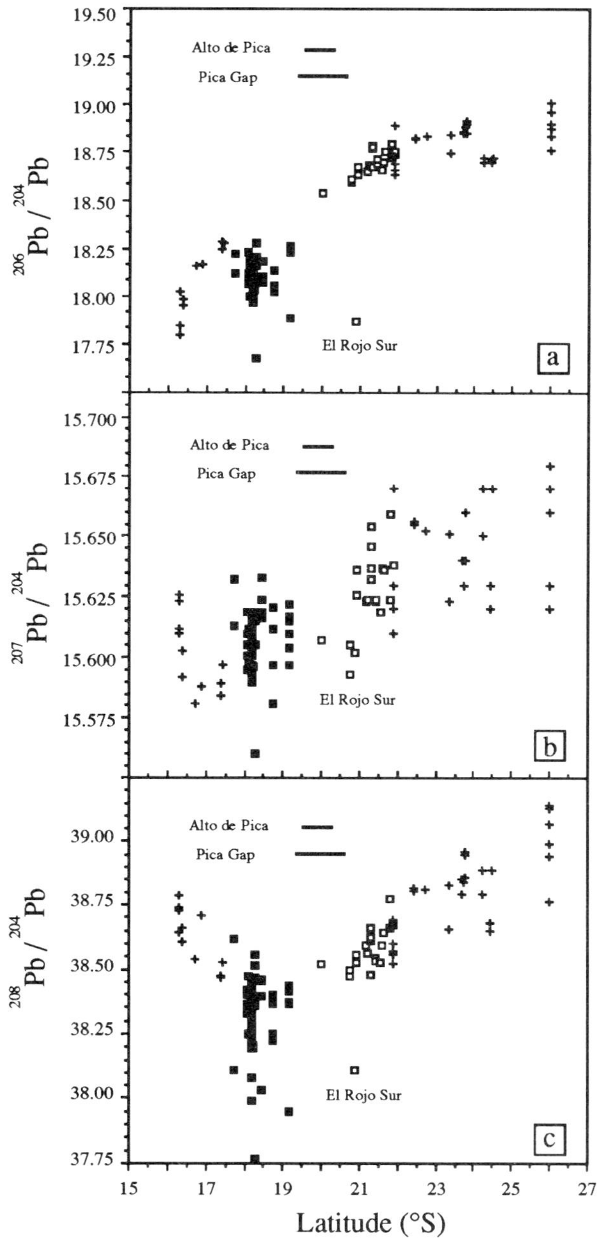

Fig. 12. Regional variation in Pb isotope compositions along the CVZ Andean arc. (a) Note the declin-ing values of $^{206}Pb/^{204}Pb$ isotopic ratios from N to S and the transition between the two sectors at the Pica gap in the Alto de Pica region. (b) Note the subtle decline in $^{207}Pb/^{204}Pb$ values from N to S. There is no steep transition between the two sectors in the Pica gap region. (c) Note the general decline in $^{208}Pb/^{204}Pb$ values from N to S. The region north of the Alto de Pica is characterized by unusually low $^{208}Pb/^{204}Pb$. It is important to note that the monoge-netic basaltic andesite of El Rojo Sur has isotopic cha-racteristics of northern centres although it is located in the south. Individual centres and references as in Fig. 4

large (Davidson et al. 1987; Nye and Reid 1987). In addition, the correlation between crustal thickening and temporal variations in isotopic ratios in the CVZ is striking. Moreover, during the past 200 Ma of magmatism in the CVZ, low-degree melting, enriched, lithospheric components will have become exhausted by continuing magma production and will become smaller, not larger, with time. For more arguments against a significant contribution by the upper plate lithospheric mantle see Davidson et al. (1990, 1991) and McMillan et al. (in prep.).

In contrast, there is compelling evidence for large-scale crustal involvement in Andean magmas throughout the history of the subduction zone. Hildreth and Moorbath (1988) have presented a geochemical experiment in which they showed in a sector of the southern Andean volcanic front between 33.5 and 37.5°S, that systematic chemical and isotopic variations correlate with crustal thickness (and hence with the probability of, or degree of, crustal contamination). Other factors could be convincingly excluded because additional possible controls on magma composition such as subduction geometry, magma production rate, type and amount of sediment in the trench, as well as age and composition of the subducted lithosphere and overlying continental crust, are essentially constant in that region.

Hildreth and Moorbath (1988) have shown that contamination produces geochemical characteristics that are typical and constant for one volcano but differ slightly for the next. This suggests that contamination processes act under similar conditions at the deep roots of Andean volcanic complexes. Such contamination processes must operate at constant rates and involves similar components during the typical life-span of Andean volcanoes.

We have documented a number of morphological, petrographic, geochemical and isotopic changes between 20°45'S and 19°09'S (V. Irrutupuncu to V. Isluga), which we correlate with two distinct crustal domains separated by a major crustal boundary. Our arguments for this interpretation are:

1. Subduction setting, distance from the trench, height above the Benioff zone, sediment type and occurence in the trench, and thickness of the crust all remain constant between 16° and 26°S. Our sample traverse covers a previously existing gap between 17° and 22°S. The only variable parameter within this segment of the volcanic front is the age and composition of the continental crust, which was sampled by MASH zone magmatism: Proterozoic unradiogenic basement (Arequipa and Belen) in the north (Tilton and Barriero 1980; Damm and Todt

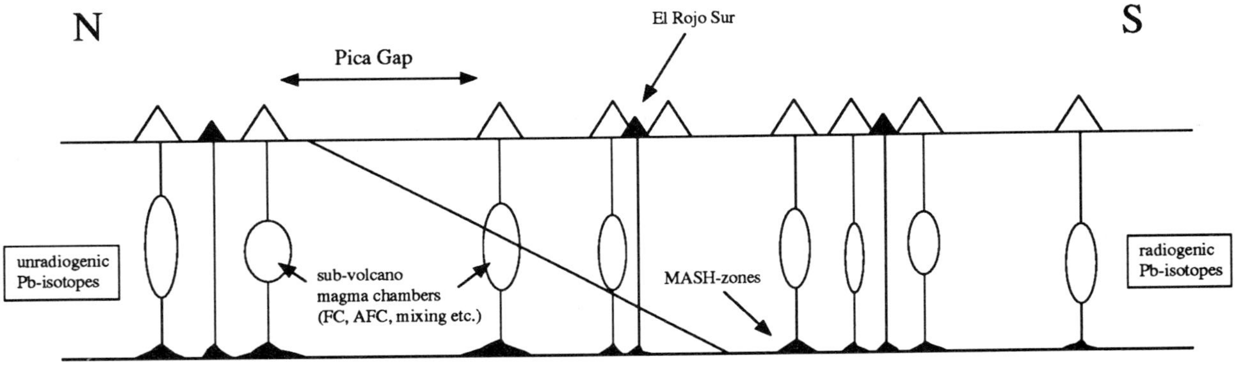

Fig. 13. Highly schematic N-S section showing the inclined crustal domain boundary and the location of MASH-interaction zones within the crust. For further discussion see text

1986) to Palaeozoic, more radiogenic old basement rocks in the south (Bahlburg et al. 1988). In this context, the observed changes in isotopic composition of the volcanic rocks are most probably related to crustal changes rather than to any other parameters. It seems superfluous to postulate major Pb-isotopic variations in the underlying mantle lithosphere, as Rogers and Hawkesworth (1989) did. The shift towards less radiogenic Pb isotopes in the magmas clearly correlates with relatively unradiogenic Pb in crustal basement rocks from the same region (Arequipa). A special case exists along the volcanic front in the Belen area where $^{208}Pb/^{204}Pb$ ratios in the lavas are relatively unradiogenic (Fig. 12c). Basement rocks in this region around Belen are also low in $^{208}Pb/^{204}Pb$ (Damm 1989 pers. commun.) suggesting a long-term Th depletion in addition to low U in basement rocks.

2. Sr isotope ratios are significantly elevated compared with typical mantle values. Sr isotopes are almost constant in northern centres (e.g. Nevados de Payachata) over the entire range of magma compositions from basalts to rhyolites. The more southerly centres show correlation between Sr-isotope ratios and parameters of differentiation (e.g. SiO_2 , Sr). This correlates with evolution of larger magma systems in the south as deduced from SiO_2 systematics. In addition, systematic changes in Sr isotopic variations are observed between the sectors north and south of the Pica gap, correlating with changes in Pb isotopes. Based on these arguments control by different mantle sources is unlikely.

3. Major and trace element variations are subtle (Figs. 5-9), but support the distinction into two sectors based mainly on isotopic composition. General chemical trends compare with the SVZ results of Hildreth and Moorbath (1988), who demonstrated that variable crustal contributions from different depths are a likely explanation. However, while these authors attributed geochemical variations in erupted magmas to variable crustal thickness, our observations indicate control of magma composition by crustal composition and age. Regional geochemical variations along our N-S traverse are thus interpreted as reflecting interaction between mantle-derived magmas and melts from two different crustal domains.

Within this context important evidence may be obtained from the most mafic lavas. One unusual mafic basaltic andesite from a small monogenetic cinder cone (El Rojo Sur) at 20°50'S is of particular interest. Although it is located in the southern sector, it has relatively unradiogenic $^{206}Pb/^{204}Pb$ and low Sr isotope ratios more typical of the northern sector. In addition, its trace element content, particularly in high Sr, Ba and most other incompatible elements, is similar to monogenetic centres found only in the north.

Assuming that crustal contributions control isotopic compositions in CVZ magmas, these observations imply that small undifferentiated magmatic systems still represent the "northern crustal domain", while more evolved, larger (and shallower ?) systems show increasing involvement of the "southern crustal domain". We propose that mafic magmas coming from the deepest MASH levels (e.g. El Rojo Sur) have an isotopic signature solely derived from the lowermost (Pb-unradiogenic) crust. Larger magmatic systems in this region, however, show an increasing contribution from the southern crustal domain. Apparently both domains are present in the vertical crustal column, with only El Rojo Sur "tapping" the northern type basement at a greater depth. We speculate that the domain boundary, which strikes out at the surface in the Pica area, is inclined towards the south, with "northern-type" basement rocks wedging out at the base of the crust (Fig. 13). Within this transition zone, more evolved rocks interact at deep and shallower crustal levels (larger MASH zone).

In the future we will investigate older (Cretaceous) intrusive rocks as well as Tertiary ignimbrites from this transition zone in order to determine chemical and isotopic variations in a N-S traverse through time, and therefore prior to, and during, crustal thickening.

6.1 Implications for Growth of the Continental Crust at Subduction Zones

It follows from our discussion that widespread andesitic magmatism in the Andes is the result of large-scale crustal involvement and recycling in conjunction with intra-crustal differentiation of mantle and hybrid magmas. The net input into the continental crust at subduction zones is regarded as broadly basaltic in composition. It is now generally accepted that melts from the mantle wedge are basaltic, whether derived from enriched mantle lithosphere or depleted asthenosphere. Clearly, significant growth of crust of andesitic bulk composition cannot occur in ocean-continent subduction zones such as the Central Andes by simply adding basaltic melts.

Arndt and Goldstein (1989) proposed that basaltic magma at the base of the crust may stagnate, cool, partially crystallize and thereby assimilate crustal rocks. Early crystallized, high density minerals (e.g. olivine, pyroxene) have similar physical properties to rocks of the Earth's mantle (peridotite, composed mainly of olivine, pyroxenes and an Al-bearing phase such as spinel, or garnet). These rocks form the lower parts of underplated complexes and may eventually delaminate, founder, and be recycled into the mantle lithosphere leaving behind, as an addition to the crust, a complementary, differentiated component. In this model of Andean magmatism some ultramafic material may founder from the base of MASH zones, be recycled to the lithospheric mantle and generate a more evolved "andesitic" contribution to crustal growth at Andean-type subduction zones.

Acknowledgements. We gratefully acknowledge help and hospitality from the officials of the town of Ollague and the CONAF personnel of Lauca, Surire, and Isluga National Parks who facilitated the fieldwork. Technical and analytical assistance has been provided by R. Goodwin and P. Taylor of Oxford University. Funding for this work was provided by NFS grants EAR 8319766 (J.P.D.) and EAR 8318916 (R.S.H.), and by DFG grant Wo 362/5-1 (G.W.). This work is a contribution to IGCP Project 249 "Andean Magmatism and its Tectonic Setting".

References

Arndt NT, Goldstein SL (1989) An open boundary between lower continental crust and mantle: its role in crust formation and crustal recycling. Tectonophysics 161: 201-212

Bahlburg H, Breitkreuz Ch, Giese P (eds) (1988) The southern Central Andes. Contributions to structure and evolution of an active continental margin. Lecture Notes in Earth Sciences 17. Springer, Berlin Heidelberg New York., 261pp

Baker MCW, Francis PW (1978) Upper Cenozoic volcanism in the Central Andes - ages and volumes. Earth Planet Sci Lett 41: 175-187

Benjamin MT, Johnson NM, Naeser CW (1987) Recent rapid uplift in the Bolivian Andes: evidence from fission-track dating. Geology 15: 680-683

Damm KW, Todt W (1986) Geochemie, Petrologie und Geochronologie der Plutonite und des metamorphen Grundgebirges in Nordchile. Berl Geowiss Abh (A) 66, I: 73-146.

Davidson JP, Dungan MA, Ferguson KM, Colucci MT (1987) Crust-magma interactions and the evolution of arc magmas: The San Pedro-Pellado complex, southern Chilean Andes. Geology 15: 443-446

Davidson JP, Ferguson, KM, Colucci MT, Dungan MA (1989) The origin and evolution of magmas from the San Pedro-Pellado volcanic complex, S. Chile: multicomponent sources and open system evolution. Contrib Mineral Petrol 100: 429-445

Davidson JP, McMillan NJM, Moorbath S, Wörner G, Harmon RS, Lopez-Escobar L (1990) The Nevados de Payachata volcanic region (18°S/69°W, N. Chile) II. Evidence for widespread crustal involvement in Andean magmatism. Contrib Mineral Petrol 105: 412-432

Davidson JP, Harmon RS, Wörner G (1991) The source of central Andean magmas: some considerations. Geol Soc Am Spec Publ (in press)

DePaolo DJ (1981) Trace element and isotopic effects of combined wall rock assimilation and fractional crystallization. Earth Planet Sci Lett 53: 189-202

de Silva SL (1989) Altiplano-Puna complex of the Central Andes. Geology 17: 1102-1106

de Silva SL, Francis PW (1990) Potentially active volcanoes of Peru - observations using Landsat thematic mapper and Space Shuttle imagery. Bull Volcanol 52: 286-301

de Silva SL, Self S, Francis PW (1988) The Chaco dacite revisited. EOS 69 (44): 1487

Entenmann J, Wörner G (1990) Constraints on the origin of geochemical variations of Central Andean mafic andesites. Abstr IAVCEI Conf, September 1990, Mainz, FRG

Francis PW, Rundle CC (1976) Rates of production of the main magma types in the Central Andes. Geol Soc Am Bull 87: 474-480

Francis PW, Wells GL (1988) Landsat thematic mapper observations of debris avalanche deposits in the Central Andes. Bull Volcanol Geotherm Res 50: 258-278

Francis PW, Sparks RSJ, Hawkesworth CJ, Thorpe RS, Pyle DM, Tait SR, Mantovani MS, McDermott F (1989) Petrology and geochemistry of the Cerro Galan caldera, northwest Argentina. Geol Mag 126: 515-547

Harmon RS, Barreiro BA, Moorbath S, Hoefs J, Francis PW, Thorpe RS, Deruelle B, McHugh J, Viglino JA (1984) Regional O-, Sr-, and Pb-isotope relationships in Late Cenozoic calc-alkaline lavas of the Andean Cordillera. J Geol Soc London 141: 803-822

Hildreth WE, Moorbath S (1988) Crustal contributions to arc magmatism in the Andes of central Chile. Contrib Mineral Petrol 98: 455-499

Horn S (1991) Intra- und intervulkanische Variationen entlang der Zentralen Vulkanzone in Nordchile (17,5°S - 22°S): Petrographische und geochemische Implikationen. Diplom-Thesis, Universität Mainz, 120 pp (unpubl)

Isacks BL (1988) Uplift of the Central Andean plateau and bending of the Bolivian orocline. J Geophys Res 93: 3211-3231

Isacks BL, Kay SM, Fielding EJ, Jordan T (1986) Andean volcanism: icing on the cake. EOS Abstr 67: 1073.

James DE (1971a) Plate tectonic model for the evolution of the Central Andes. Geol Soc Am Bull 82: 3325-3346

James DE (1971b) Andean crustal and upper mantle structure. J Geophys Res 76: 3246-3271

McMillan N, Davidson JP, Harmon RS, Lopez-Escobar L, Moorbath S, Wörner G Mechanism of trace element enrichment related to crustal thickening: the Nevados de Payachata region, northern Chile. in prep

Nye CJ, Reid MR (1986) Geochemistry of primary and least fractionated lavas from Okmok volcano, Central Aleutians: Implications for arc magmatism. J Geophys Res 91 (B10): 10271-10287

O'Callaghan LJ, Francis PW (1986) Volcanological and petrological evolution of San Pedro volcano, Provincia El Loa, North Chile. J Geol Soc Lond 143: 275-286

Pilger RH (1981) Plate reconstructions, aseismic ridges, and low angle subduction beneath the Andes. Geol Soc Am Bull 92: 448-456

Plank T, Langmuir CH (1988) An evaluation of the global variations in the major element chemistry of arc basalts. Earth Planet Sci Lett 91: 171-185

Rogers G, Hawkesworth CJ (1989) A geochemical traverse across the north Chilean Andes: evidence for crust generation from mantle wedge . Earth Planet Sci Lett 91: 271-285

Schwarz G, Martinez E, Bannister J (1986) Untersuchungen zur elektrischen Leitfähigkeit in den zentralen Anden. In : Giese P (ed) Forschungsberichte aus den Zentralen Anden (21° - 25° S). Berl Geowiss Abh (A) 66: 49-72

Sheffels BM (1990) Lower bound on the amount of crustal shortening in the central Bolivian Andes. Geology 18: 812-815

Thornburg TM, Kulm LD (1987) Sedimentation in the Chile trench: depositional morphologies, lithofacies, and stratigraphy. Geol Soc Am Bull 98: 33-52

Tilton GR, Barreiro BA (1980) Origin of lead in Andean calc-alkaline lavas, southern Peru. Science 210: 1245-1247

Wigger P (1986) Krustenseismische Untersuchungen in Nord-Chile und Süd-Bolivien. In : Giese P (ed) Forschungsberichte aus den Zentralen Anden (21° - 25° S). Berl Geowiss Abh (A) 66: 31-48

Wörner G, Harmon RS, Davidson JP, Moorbath S, Turner DL, McMillan NJ, Nye C, Lopez-Escobar L, Moreno H (1988) The Nevados de Payachata volcanic region (18°S/69°W, N.Chile) I. Geological, geochemical, and isotopic observations. Bull Volcanol 50: 287-303

Wörner G, Moorbath S, Harmon RS (1990) Isotopic variations in Central Andean lavas. Abstr IAVCEI Conf, September 1990, Mainz, FRG

Partial Melting in the Lower Crust: New Constraints on Crustal Contamination Processes in the Central Andes

CHRIS HAWKESWORTH and CHRIS CLARKE

Abstract. The granitoids of the Idaho Batholith were generated in the late Cretaceous when the crust in that area was 60-70 km thick, and they therefore constrain models for crustal melting and crustal contamination in areas of unusually thickened crust, such as the Central Andes. The volumetrically dominant granodiorites and granites exhibit unfractionated Rb/Ba and Sr/Nd, relatively low P, Ti, Y and HREE abundances, and Sr, Nd and Pb isotopes consistent with partial melting of crustal source rocks with no significant contribution from mantle derived magmas. It is argued that they were generated by partial melting of tonalitic source rocks, in the presence of residual garnet, at depth in the continental crust, and that they may therefore be used to evaluate lower crustal contamination processes in the Central Andes.For the Central Andes there are now a significant number of Nd isotope analyses which offer tight constraints on the Nd isotope composition of any crustal endmember. Mass balance considerations indicate that if crustal contamination was responsible for the Nd and Sr isotope ratios of selected andesites in the Central Andes, the amounts of contamination are ~35-70%. Moreover, the uncontaminated magmas must themselves have had relatively high Nd (and Sr) abundances, whether that reflects fractional crystallization and/or source or partial melting processes. It would appear that lower crustal contamination models for the Central Andes andesites either imply that the isotope ratios were changed with little affect on the observed trace element patterns, or that the added contaminant is relatively low in SiO_2 and has none of the characteristic features of the lower crustal melts from the Idaho Batholith.

1 Introduction

Despite a number of detailed studies the role of the lower continental crust in the generation of magmas in the Central Andes remains deeply controversial. The recent volcanic rocks of this area are characterized by higher average SiO_2 and $^{87}Sr/^{86}Sr$ ratios than those in magmatic rocks from elsewhere along the Andean chain (Hickey et al. 1984; Hildreth and Moorbath 1988; Davidson et al. 1990). In general, Sr isotope ratios >0.707 are accompanied by higher SiO_2 and lower Sr contents, consistent with upper crustal melting and contamination of mantle derived magmas with upper crustal material. However, many recent volcanic suites have $^{87}Sr/^{86}Sr$ ratios of 0.705-0.706, and these have been variously attributed to old, trace element enriched source regions in the upper mantle, and to contamination in the lower crust.

Correspondence to: Chris Hawkesworth, Department of Earth Sciences, The Open University, Walton Hall, Milton Keynes, MK7 6AA, UK

Studies which compared the isotope and trace element ratios of Recent volcanic rocks *along* the Andean chain were struck by the observation that $^{87}Sr/^{86}Sr$ ratios, for example, are higher in those magmas erupted in areas where the continental crust is thicker, and so they tended to invoke contamination processes within the thickened segments of crust (e.g. Hildreth and Moorbath 1988; Davidson et al. 1990). In contrast, Rogers and Hawkesworth (1989) investigated the geochemical and isotope variations in 185-0 Ma magmatic rocks on a section at right angles to the strike of the Andes at 22°S. They also recognized a progressive increase in $^{87}Sr/^{86}Sr$ from 0.704-0.706, particularly in the lower silica andesites, as magmatism migrated eastwards with time. Since the main crustal thickening event in this segment of the Andes started in the late Miocene, it was argued that crustal thickening was responsible for the increased amount of upper crustal melting in the 15-0 Ma rocks, but that the shift from 0.704 to 0.706 in the 185-0 Ma rocks primarily reflected variations in the upper mantle source regions as magmatism migrated eastwards.

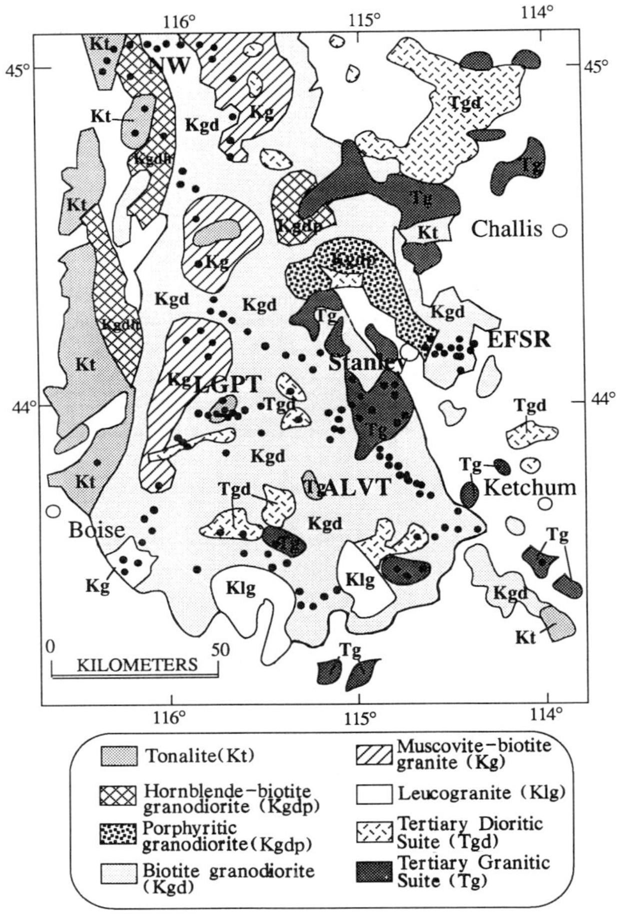

Fig. 1. Simplified geological map of the Atlanta lobe of the Idaho Batholith showing the general distribution of the Cretaceous and Tertiary granitoids and the sample localities (black circles) of this study

Any analysis of likely contamination processes in the lower continental crust is hampered by the shortage of samples from the lower crust, and by the paucity of information on the geochemistry of partial melts generated under lower crustal conditions. Conditions in the Central Andes are further complicated by the unusual thickness of crust in that area. Thus, in order to evaluate the composition of partial melts generated at depth in an unusually thickened segment of continental crust we have undertaken a detailed study of the Atlanta Lobe of the Idaho Batholith. These rocks were generated in the late Cretaceous when the crust had been thickened to a minimum of 60-65 km during the Sevier compressional events in the western USA. The compositions of the resultant lower crustal melts are compared with the upper crustal melts in the Andes, and mass balance calculations are presented which constrain both the amount and nature of the contaminant required if the isotope and trace element features of basaltic andesites in the Central Andes are attributed to crustal contamination processes. Throughout this discussion upper and lower crustal melts refer to melts which are inferred to have been generated, rather than emplaced, in the middle to upper, and the lower continental crust respectively.

2 Background Geology

Different magmatic phases in the Idaho Batholith were emplaced between 90 and 75 Ma (Lewis et al. 1987), and the batholith is widely regarded has having been formed in response to crustal thickening in the Sevier/Laramide orogenes (Patiño-Douce et al. 1990). The associated thrust faults are truncated by the batholith, and the best estimates of the amount of crustal shortening are ~100 km (Cooney and Harms 1984). An independent constraint on the timing of the crustal thickening event is also available in the Aptian palaeontological age obtained on the earliest synorogenic conglomerates (Heller et al. 1986). Seismic data indicate that the present thickness of the continental crust in the Idaho region is 35-40 km (Mabey and Webring, 1985), and estimates of the depth of crystallization are available from studies of the magmatic and country rocks. Zen (1988) used the presence of late magmatic epidote in the tonalites and granodiorites of the Idaho Batholith to suggest pressures of crystallization of ~8 kb, and the country rocks were metamorphosed to upper amphibolite facies at the time of intrusion. Thus, rocks at the present erosion surface were probably covered by 25-30 km of overburden, consistent with the existence of 60-70 km of crust in the late Cretaceous.

3 Geochemistry

Some 150 samples of the Cretaceous and Tertiary rocks of the Idaho Batholith (Fig. 1) have been analysed for major and trace elements, and Sr, Nd and Pb isotopes have been determined on a representative subset (Clarke 1990). The Cretaceous rocks range from tonalites through to leucogranites, with the majority being weakly peraluminous biotite granodiorites (Fig. 2 and Table 1). In the area studied there is no simple petrogenetic link between the tonalites and granodiorites with less than 68-69% SiO_2 and the volumetrically dominant higher SiO_2 granodiorites and muscovite-biotite granites, and in

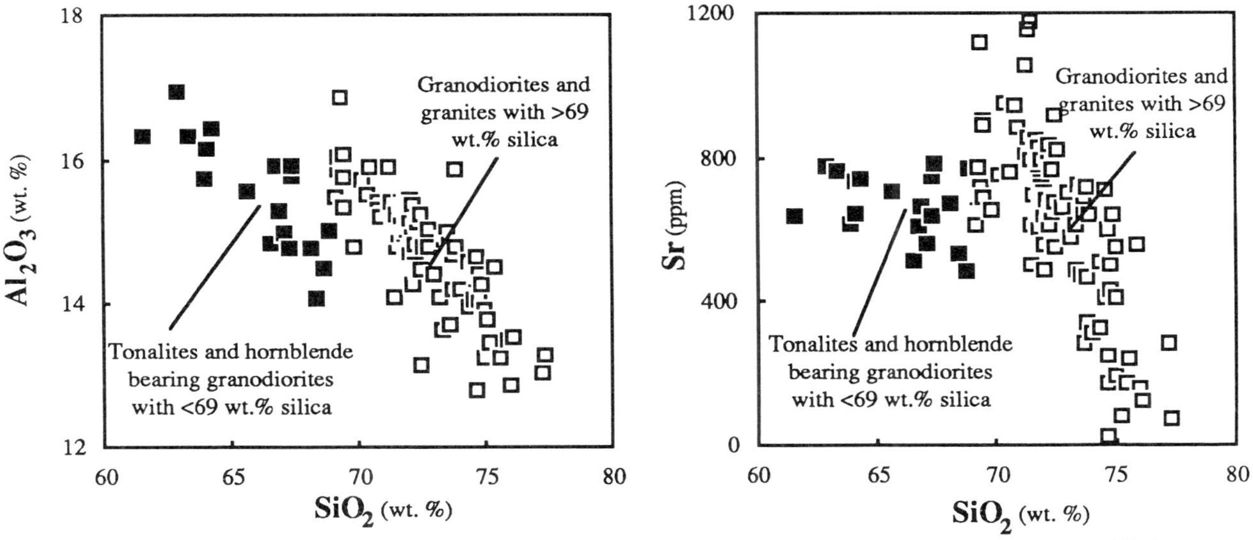

Fig. 2. Variation of Al_2O_3 and Sr versus SiO_2 for the Cretaceous granitoids of the Atlanta lobe. The *filled squares* represent the lower silica (<69 wt % SiO_2) tonalites and granodiorites which commonly contain hornblende, and which appear not to be simply related to the higher silica (>69 wt % SiO_2) granodiorites and granites represented by the *open squares*

Table 1. Average compositions of granitoids of the Idaho batholith and of Central Andean ignimbrites

	Idaho batholith	Central Andes
	Biotite Granodiorites (n=56)	Ignimbrites (n=117)
SiO_2	71.60	68.10
TiO_2	0.27	0.54
Al_2O_3	15.02	15.59
Fe_2O_3	1.81	3.49
MnO	0.05	0.07
MgO	0.49	1.39
CaO	2.08	3.58
Na_2O	4.20	3.35
K_2O	3.45	3.64
P_2O_5	0.08	0.14
Ba	1352	673
Rb	93.0	162
Sr	758	310
Zr	171	151
Nb	20.7	14.1
Y	12.0	25.5

this contribution we are primarily concerned with the higher SiO_2 rocks. In most of the Cretaceous rocks, ε_{Sr} = 30 to 75 and ε_{Nd} = -5 to -12 (Fig. 3), and there is little systematic change in isotope ratios with increasing SiO_2. $^{206}Pb/^{204}Pb$ ratios on whole rocks and separated feldspars range from 18.6 to 20.4, and the linear arrays on Pb isotope diagrams are consistent with source ages of ~1.5 Ga (Clarke 1990).

Feldspar and garnet are common aluminous minerals in the continental crust; garnet tends to replace feldspar at higher pressures, and each has a distinctive trace element signature. Figure 4 summarizes the variations in Sr and Y with SiO_2 in the Cretaceous rocks with >69% SiO_2. Y and Sr both decrease as SiO_2 increases from 70 to 73%, but at higher SiO_2 Y changes little and Sr continues to fall. Although 73% SiO_2 is an arbitary boundary the data in Fig. 4 indicate that it marks a change in the controlling mineral assemblages.

The rocks with >73% SiO_2 exhibit a greater range in LIL element abundances, and these have been successfully modelled by late stage fractional crystallization of a granitic assemblage (Clarke 1990). The striking feature of the 70-73% SiO_2 rocks is the depletion of HREE and Y with increasing SiO_2: this might be attributed to residual accessory phases, but they are not consistent with other trace element trends, and so it appears that garnet was a controlling phase. Experimental work indicates that

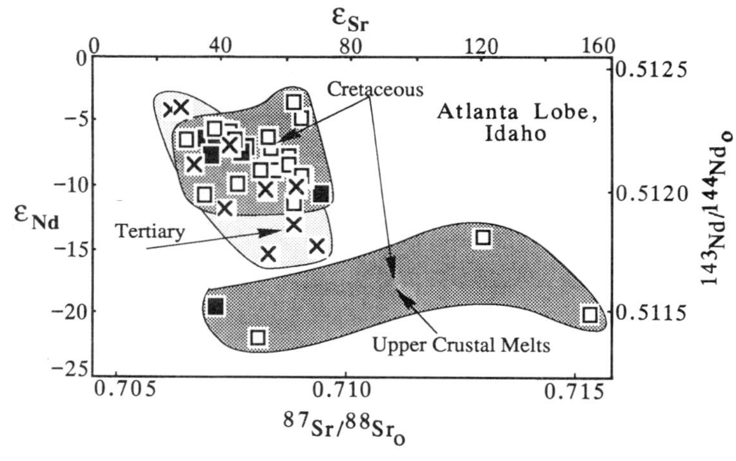

Fig. 3. ε_{Nd} versus ε_{Sr} for the Cretaceous and Tertiary granitoids of the Atlanta lobe. The Cretaceous granitoids show no systematic variation in isotope ratios with silica, but the Tertiary granitoids define a trend of increasing ε_{Sr} with decreasing ε_{Nd} thought to be due to mixing between mantle- and crustally derived material. A subset of Cretaceous granitoids, mainly leucogranites from the southern Atlanta lobe, have anomalously high ε_{Sr} and anomalously low ε_{Nd} and are interpreted to be largely derived from upper crustal material. *X* Tertiary granitoids; *open squares* Cretaceous granitoids >69% SiO_2; *filled squares* Cretaceous granitoids <69% SiO_2

garnet only occurs on the liquidus of granitic magmas if they are highly peraluminous and at relatively shallow pressures (Thompson 1988). Thus garnet is unlikely to have been a liquidus phase (which might have participated in fractional crystallization processes) in these Idaho Batholith magmas since they are only weakly peraluminous and they crystallized at depths >25 km. Rather, it is argued that garnet was a residual phase, and that the geochemical variations observed in the 70-73% SiO_2 rocks primarily reflect partial melting processes.

A series of partial melting models was explored, and the preferred model is illustrated in Fig. 5. It was inferred that partial melting took place in the deep crust under fluid absent conditions with the degree of melting controlled by reactions involving the breakdown of micas and amphiboles up to ~1000 °C (Clarke 1990). In addition the parent/daughter element ratios were constrained by the inferred source age of ~1.5 Ga, and the initial Nd and Sr isotope ratios. The source rocks were calculated to be tonalitic in composition, consistent with the general I-type nature of the Idaho Batholith rocks, the degree of melting for the biotite granodiorites was ~30-20%, and the residual mineralogy is given in Fig. 5. The shaded field in Fig. 5 indicates the likely range of source compositions calculated for different partition coefficients appropriate for intermediate to rhyolitic compositions (after Henderson 1982). Moreover, that range in source compositions is greater than the difference in source compositions calculated for different samples with 69-70% SiO_2.

The minor and trace element patterns in the calculated source rocks have several distinctive features (Fig. 5):

(1) Their high Ba/Nb and Sr/Nd ratios are typical of magmas from subduction related settings (Pearce 1983), suggesting that the source rocks in the region of the Idaho Batholith were themselves generated in a subduction environment, albeit 1.5 Ga ago. There is no isotope or trace element evidence that the source rocks had previously been depleted significantly by, for example, granulite facies metamorphism.

(2) The abundances of Sr and the LREE are high, both in comparison with Rb, Ba, K and Th, and with the Sr and LREE contents of other subduction related crustal magmas. For Sr this is a result of little residual plagioclase, and consequently Rb/Sr in the biotite granodiorite magmas is only slightly higher than that calculated for their source rocks, Rb/Sr = 0.11 and 0.07 respectively.

(3) The residual garnet bearing assemblage has fractionated Sm/Nd by ~25%, in addition to Y and the HREE.

In summary, the Idaho granodiorites and granites are I-type granitoids generated in equilibrium with residual garnet by melting reactions involving the breakdown of plagioclase feldspar and amphibole. No significant contribution from asthenospheric mantle derived melts has been recognized, and specifically the model Sr, Nd and Pb ages are all consistent with partial melting Proterozoic crustal source rocks. It is inferred that the source rocks were probably tonalitic in composition, and that they melted to leave a residual garnet-bearing mafic granulite typical of high grade metamorphic conditions in the lower crust. Partial melting took place at depths greater than the estimated emplacement depth of 25-30 km, and the crust was probably 60-70 km thick at the time of magmatism. Thus the Idaho Batholith rocks are taken as examples of the crustal melts generated by partial melting at depth within a segment of

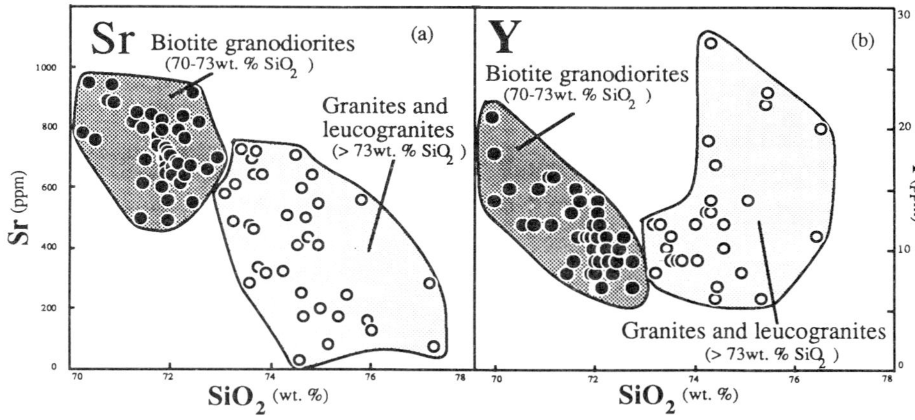

Fig. 4. Variation of Sr and Y versus SiO_2 for the granitoids of the Atlanta lobe with greater than 70 wt % SiO_2. This sample set has been split into two subsets where the *filled circles* represent the granitoids with 70-73 wt % SiO_2, and are generally biotite granodiorites in which both Sr and Y decrease with increasing silica. The *open circles* are generally biotite- and muscovite-biotite-bearing granites and leucogranites with >73 wt % SiO_2 in which Sr continues to fall, but Y no longer decreases. The arbitrary boundary at 73 wt % SiO_2 is interpreted to represent a change from the samples whose compositions are primarily controlled by partial melting in the presence of residual garnet, to those with >73% SiO_2 whose compositions also reflect subsequent fractional crystallization, primarily of plagioclase and alkali feldspars

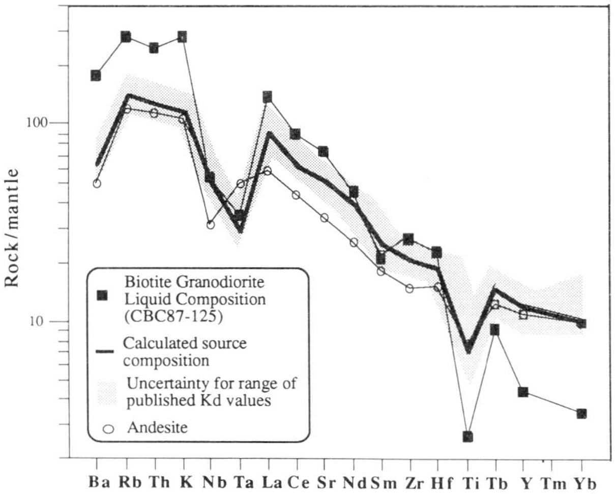

Fig. 5. Calculated trace element compositon of the biotite granodiorite source rock assuming 30% batch equilibrium modal partial melting and modelling back from a given biotite granodiorite parental liquid (CBC87-125) where the preferred residue was a mafic granulite containing 40% clinopyroxene, 28% orthopyroxene, 10% garnet, 20% plagioclase and 2% minor and opaque phases. The Kd values taken to calculate the bulk distribution coefficient of this assemblage were taken from Henderson (1982), and references therein, and from Pearce and Norry (1979). The estimated bulk D values of this assemblage are D_{Ba} = 0.1, D_{Rb} = 0.7, D_{Th} = 0.07, D_K = 0.04, D_{Nb} = 0.76, D_{Ta} = 0.23, D_{La} = 0.53, D_{Ce} = 0.60, D_{Sr} = 0.66, D_{Nd} = 1.1, D_{Sm} = 1.52, D_{Zr} = 0.7, D_{Hf} = 0.75, D_{Ti} = 2.92, D_{Tb} = 2.07, D_Y = 2.67, D_{Yb} = 3.94 and the uncertainties for the range of possible Kd values are approximately represented by the *shaded region*. The model andesite compositon is given for comparison after Taylor and McLennan (1985) and the primordial mantle normalizing factor is after Thompson et al. (1983)

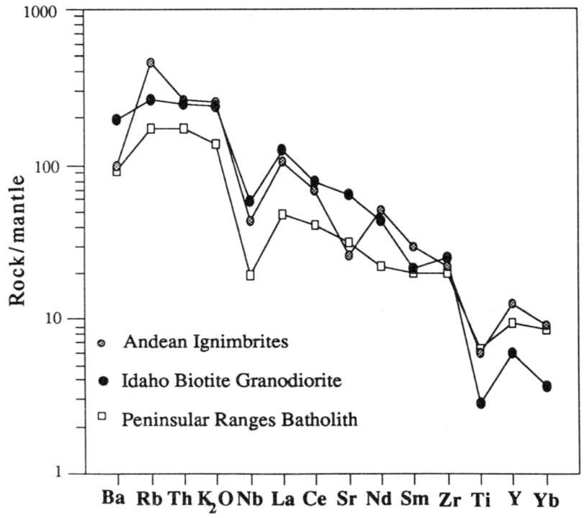

Fig. 6. Mantle normalized trace element variation diagram comparing the average composition of the biotite granodiorites from the Atlanta lobe of the Idaho Batholith (see Table 1) with that of ignimbrites from the Central Andes (largely from De Silva 1987), and the Peninsula Ranges Batholith (Gromet and Silver 1987)

unusually thickened continental crust, and as such they indicate the likely composition of partial melts generated in the lower crust of the Central Andes over the last 10 Ma.

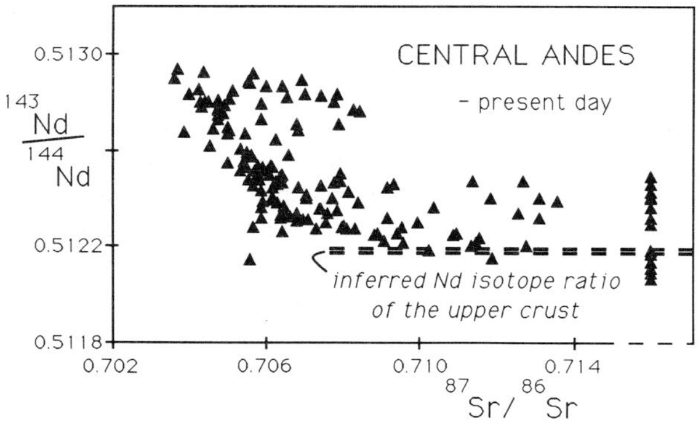

Fig. 7. Present day Nd and Sr isotope ratios in extrusive and intrusive igneous rocks, and a few sediments, from the Central Andes (data from the literature, e.g. Rogers and Hawkesworth 1989; Miller and Harris 1989; and references therein)

3.1 Comparison of Upper and Lower Crustal Melts

Having argued that the biotite granodiorite magmas in the Idaho Batholith are examples of lower crustal melts in an area of thickened continental crust, they may now be compared with typical middle to upper crustal melts from, for example, the Central Andes (Table 1). Their trace element patterns are illustrated in Fig. 6, where they are also compared with an average composition for the Peninsular Range Batholith (Gromet and Silver 1987) which exhibits some intermediate characteristics. The striking features are that the upper crustal melts from the Andes have low Sr abundances and Sr/Nd ratios, distinctive high Rb/Ba, and relatively unfractionated HREE and Y. In contrast the lower crustal melts from Idaho have unfractionated Rb/Ba and Sr/Nd, and significantly depleted HREE and Y, suggesting that there is a clear shift from dominant feldspar control for the upper crustal melts to garnet for the lower crustal melts. All three average compositions exhibit a significant negative Ti anomaly consistent with a residual Ti-bearing phase(s) in the source of both the upper and the lower crustal melts.

4 Discussion: The Effects of Lower Crustal Contamination on the Magmas of the Central Andes

A critical step in the evaluation of any crustal contamination model is the provision of realistic constraints on the isotope composition of the crustal material. Rb/Sr is very variable in crustal rocks, and so with time the continental crust exhibits a wide range in $^{87}Sr/^{86}Sr$. In contrast, Sm/Nd is relatively restricted in crustal rocks, and so it is much easier to estimate the Nd isotope ratio of likely crustal endmembers. There have been several Nd isotope studies on Andean rocks with the result that some 250 analyses are available to constrain the Nd isotope composition of the crust at the time of magmatism (see Fig. 7). A number of the rocks analysed are sediments and upper crustal melts, and these tend to have the lower $^{143}Nd/^{144}Nd$ ratios. Thus from the array of data in Fig. 7 a reasonable estimate of $^{143}Nd/^{144}Nd$ in the upper crust of the Central Andes is ~0.51220. Moreover, since most estimates of Sm/Nd, and hence with time $^{143}Nd/^{144}Nd$, in the lower crust suggest that it is slightly higher than that in the upper crust (De Paolo 1988), the value 0.51220 is regarded as a minimum for $^{143}Nd/^{144}Nd$ in the lower crust of this area. This figure is used in

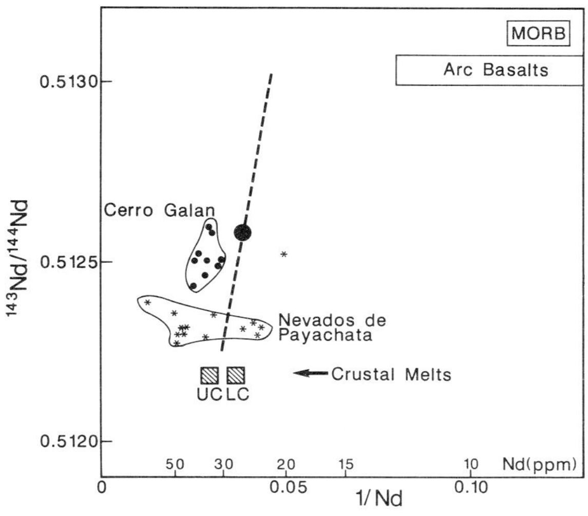

Fig. 8. $^{143}Nd/^{144}Nd$ versus 1/Nd illustrating the Nd isotope ratios and Nd abundances of MORB, typical island arc tholeiites, and andesites from Cerro Galan and Nevados de Payachata in the Central Andes (Francis et al. 1989; Davidson et al. 1990). Also plotted are upper (*UC*) and lower (*LC*) crustal melts, with the Nd abundances from Fig. 6 and the Nd isotope ratio of Andean crust estimated from Fig. 7. The large *filled circle* represents the calculated parental magma composition for the Cerro Galan andesites (Mantovani and Hawkesworth 1990). The *dashed line* illustrates the possible mixing relations involved in crustal contamination models for the Andean andesites

the subsequent mass balance calculations, but it is noted that to adopt a higher, and possibly more realistic, Nd isotope ratio would increase the calculated amounts of crustal contamination in the samples considered.

Figure 8 summarizes the variation in Nd isotope and element abundances in selected island arc basalts, Central Andean andesites, and typical upper and lower crustal melts. The Nd abundances of the crustal melts are from Fig. 6, and the Nd isotope ratios are those inferred from Fig. 7. The data from just two Andean suites are illustrated: those from Cerro Galan (Francis et al. 1989) have been the subject of a detailed assimilation-fractional crystallization inversion study (Mantovani and Hawkesworth 1990) and so the calculated parental magma is also plotted. In contrast, the rocks from the Nevados de Payachata region have even lower Nd isotope ratios, and they have been modelled in terms of lower crustal contamination of "normal arc magmas" generated in a depleted mantle wedge (Davidson et al. 1990). A key feature of such Central Andean andesites is that they have Nd abundances which are similar to those in crustal melts, and this offers useful constraints for any mixing models. One interpretation of the Andean andesite data in Fig. 8 is that they reflect derivation from mantle source regions which were both old and relatively enriched in minor and trace elements. However, Fig. 8 may also be used to evaluate the effects of crustal contamination models.

In detail, contamination processes are likely to be complex, but they must be consistent with the mass balance implications from Fig. 8. The most common crustal contamination models suggest that the parental magmas have Sr and Nd isotope ratios similar to those of island arc tholeiites. Mixing results in straight lines in Fig. 8, so that mixing between an arc tholeiite magma and a crustal endmember requires the latter either to have improbably high Nd contents (>50 ppm), or higher $^{143}Nd/^{144}Nd$ ratios (i.e. >0.51220). However, if higher $^{143}Nd/^{144}Nd$ ratios are invoked the calculated amount of contamination in any andesite suite increases sharply.

More realistic contamination models may be developed if the mantle-derived magmas have higher Nd contents (20-30 ppm). These could reflect magma differentiation processes, perhaps associated with crustal contamination, smaller degrees of partial melting, and/or higher trace element abundances in the source of the Central Andean magmas. Moreover, there are undoubtedly magmas from areas such as the southern Andes and the Aleutians which have both high Nd contents and high $^{143}Nd/^{144}Nd$ ratios. Accepting uncontaminated magmas with 21 ppm Nd and $^{143}Nd/^{144}Nd$ = 0.51305, and a crustal contaminant with 35 ppm Nd and $^{143}Nd/^{144}Nd$ = 0.51220 (i.e. along the dashed line in Fig. 8), yields estimates of 35% for the amount of crustal material in the calculated parental magma at Cerro Galan, and ~70% in a typical Nevados de Payachata andesite with $^{143}Nd/^{144}Nd$ = 0.51232. Such figures should be viewed with considerable caution, but they do highlight the large amounts of crustal contamination that are required by models which invoke normal island arc magmas in the generation of the Central Andean andesite suites. Moreover, the calculated amounts of crustal material may be erroneously low because it was assumed that the uncontaminated magmas had 21 ppm Nd, rather than 10-12 ppm which might be more typical of basalts from depleted source regions in the mantle wedge.

Figure 9 illustrates the trace element distribution patterns of the calculated parental magma at Cerro Galan, a basaltic andesite from Nevados de Payachata, and an average island arc tholeiite. The two Andean magmas have very similar trace element patterns, except for Ba and Nb, and the relatively high Nb and Ta abundances in the Cerro Galan rocks have generally been attributed to a larger within plate component (e.g. Francis et al. 1989), which may in turn be linked to the position of Cerro Galan inland of the main Cordillera. The striking feature is that the component responsible for the difference in the trace element patterns of a typical island arc tholeiite and those observed in the Central Andean suites, must have a relatively smooth trace element pattern (apart for Ba and Nb) and thus be significantly different from that observed in the Idaho granitoids. The mass balance calculations using Nd isotopes indicated that 35 and 70% contamination were required (Fig. 8) and yet crustal melts are characterised by significant fractionation of P/Sm, Ti/Y, Ti/Nb and by high SiO_2 contents (Figs. 6 and 8), none of which are a feature of the Central Andes andesites.

More complex models are hampered by the lack of knowledge on whether the increase in Nd abundances from those typical of island arc rocks (Figs. 8 and 9) to the values >20 ppm inferred from the mixing models, was due to fractional crystallization and/or source or partial melting processes. The latter are more likely to be dominant, and if they are responsible for Nd abundances in the uncontaminated magmas being >20 ppm, the trace element patterns of those magmas will be very similar to those of the Andean andesites plotted in Fig. 9. Thus, the mass balance considerations illustrated on Fig. 8 further indicate that if crustal contamination is invoked to explain the Nd and Sr isotope ratios of the Andean andesites, it may not have been primarily responsible for the observed trace element patterns, in which case the isotope and trace element signatures are in effect decoupled.

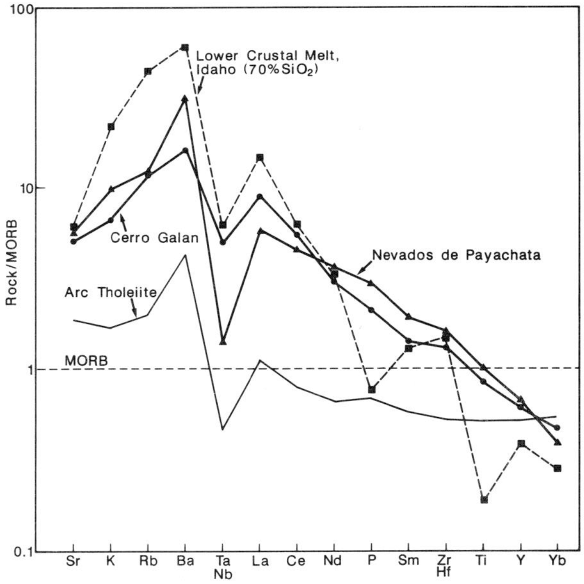

Fig. 9. MORB nomalized trace element distribution patterns in an average arc tholeiite, two selected andesites from the Central Andes (Francis et al. 1989; Davidson et al. 1990), and a lower crustal melt from the Idaho Batholith (Clarke 1990). If the Central Andes andesites are to be generated by the addition of material to an arc tholeiite magma, the minor and trace element pattern of that material bears little resemblance to the Idaho lower crustal melt

5 Summary

The granitoids of the Idaho Batholith were generated at a time when the continental crust was 60-70 km thick in that area. The results of a detailed study (Clarke 1990) suggest that the biotite granodiorites and granites were generated by partial melting of tonalitic source rocks, in the presence of residual garnet, at depth in the continental crust. The resultant magmas had high SiO_2 contents, unfractionated Rb/Ba and Sr/Nd, and relatively low P, Ti, Y and HREE abundances. It is argued that these may be representative of magmas generated at depth within thickened continental crust, and that they may therefore be used to evaluate lower crustal contamination processes in the Central Andes (Hildreth and Moorbath 1988; Davidson et al. 1990).

There are now a significant number of Nd and Sr isotope analyses on igneous and sedimentary rocks from the Central Andes, and these offer good constraints on the Nd isotope composition of the crust in this area. Mass balance considerations indicate that if crustal contamination was responsible for the Nd and Sr isotope ratios of selected andesites in the Central Andes, the amounts of contamination are ~35-70%. Moreover, the uncontaminated magmas must themselves have had relatively high Nd (and Sr) abundances, whether that reflects fractional crystallization and/or source or partial melting processes. It would appear that crustal contamination models for the Central Andes andesites either have to argue that the isotope ratios were changed with relatively little effect on the observed trace element patterns, or that the added contaminant is relatively low in SiO_2 and has none of the characteristic features of the lower crustal melts from the Idaho Batholith.

References

Clarke CB (1990) The geochemistry of the Idaho Batholith in the western United States Cordillera. PhD Thesis, Open University, Milton Keynes, 395 pp

Cooney PJ, Harms TA (1984) Cordilleran metamorphic complexes: Cenozoic extensional relics of Mesozoic compression. Geology 12:550-554

Davidson JP, McMillan NJ, Moorbath S, Wörner G, Harmon RS, Lopez-Escobar L (1990) The Nevados de Payachata volcanic region (18°S/69°W, N. Chile). II. Evidence for widespread crustal involvement in Andean magmatism. Contrib Mineral Petrol, 105: 412-432

DePaolo DJ (1988) Neodymium isotope geochemistry: an introduction. Springer, Berlin Heidelberg New York, 187 pp

De Silva SL (1987) Large volume explosive silicic volcanism in the Central Andes of N. Chile. PhD thesis, Open University, Milton Keynes, 410 pp

Francis PW, Sparks RSJ, Hawkesworth CJ, Thorpe RS, Pyle DM, Tait SR, Mantovani MS, McDermott F (1989) Petrology and geochemistry of the Cerro Galan caldera, northwest Argentina. Geol Mag 126:515-547

Gromet LP,Silver LT (1987) REE in the Peninsular Ranges batholith. J Pet 28:75-125

Grove TL, Baker MB (1984) Phase equilibrium controls on the tholeiitic versus cala-alkaline differentiation trends. J Geophys Res 89 B5:3253-3274

Heller PL, Chambers HP, Coogan JC, Hagen ES, Shuster MW, Winslow NS, Lawton TF (1986) Time of initial thrusting in the Sevier orogenic belt, Idaho, Wyoming and Utah. Geology 14:388-391

Henderson P (1982) General geochemical proporties and abundances of the REE. In: Henderson P (ed) Rare earth element geochemistry. Elsevier, Amsterdam, pp12-35

Hickey RL, Gerlach DC, Frey FA (1984) Geochemical variations in volcanic rocks from central-south Chile (33°-42°S): implications for their petrogenesis. In: Harmon RS, Barreiro BA (eds) Andean Magmatism, chemical and isotopic constraints. Shiva, Nantwich, pp72-95

Hildreth W, Moorbath S (1988) Crustal contributions to arc magmatism in the Andes of central Chile. Contrib Mineral Petrol 98:455-489

Lewis RS, Kiilsgaard TH, Bennett EH, Hall W (1987) Lithological and chemical characteristics of the central and southeastern part of the southern lobe of the Idaho batholith. US Geol Surv Prof Pap 1436:171-196

Mabey DR, Webring MW (1985) Regional geophysical studies in the Challis quadrangle. In: McIntyre DH (ed) Symposium on the geology and mineral deposits of the Challis 1° x 2° quadrangle, Idaho. US Geol Surv Bull 1658:69-79

Mantovani MS, Hawkesworth CJ (1990) An inversion approach to assimilation and fractional crystallization processes. Contrib Mineral Petrol 105:289-302

Miller JF, Harris NBW (1989) Evolution of continental crust in the Central Andes; constraints from Nd isotope systematics. Geology 17:615-617

Patiño-Douce AE, Humphreys ED, Johnston AD (1990) Anatexis and metamorphism in tectonically thickened continental crust exemplified by the Sevier hinterland, western North America. Earth Planet Sci. Lett 97:290-315

Pearce JA (1983) The role of sub-continental lithosphere in magma genesis at active continental margins. In: Hawkesworth CJ, Norry MJ (eds) Continental basalts and mantle xenoliths. Shiva, Nantwich, pp 236-249

Pearce JA, Norry MJ (1979) Petrogenetic implications of Ti, Zr, Y and Nb variations in volcanic rocks. Contrib Mineral Petrol 69:33-47

Rogers G, Hawkesworth CJ (1989) A geochemical traverse across the north Chilean Andes: evidence for crust generation from the mantle wedge. Earth Planet Sci Lett 9: 271-285

Taylor SR, McLennan SM (1985) The continental crust: its composition and evolution. Blackwell, Oxford, 312 pp

Thompson AB (1988) Dehydration melting of crustal rocks. Rend Soc Ital Mineral Pet 43:41-60

Thompson RN, Morrison MA, Dickin AP, Hendry GL (1983) Continental flood basalts.... arachnids rule OK? In: Hawkesworth CJ, Norry MJ (eds) Continental basalts and mantle xenoliths. Shiva, Nantwich, 211 pp

Thorpe RS, Potts PJ, Francis PW (1976) Rare earth data and petrogenesis of andesite from the north Chilean Andes Contrib Mineral Petrol 54:65-78

Zen E-an (1988) Tectonic significance of high pressure plutonic rocks in the Western Cordillera of North America. In: Ernst WG (ed) Metamorphism and crustal evolution of the western United States. Rubey, vol 7, Prentice-Hall, Englewood Cliffs, 7:42-63

State of In Situ Stress in Northern Chile and in Northwestern Argentina

K.H.SCHÄFER and M.DANNAPFEL

Abstract. In situ stresses have been determined at rock surfaces by the overcoring method at 45 locations in northern Chile and in northwestern Argentina between 22° and 26° S. From the Pacific coast to the Subandean Ranges we recorded the actual state of stress of all prominent morphotectonic units. There are four distinct regional stress fields: (1) The west Chilean stress field comprises the complete Coastal Cordillera between the Pacific coast and the western margin of the Longitudinal Valley. While horizontal tensile stresses oriented E-W prevail on the Mejillones Peninsula, along the coast north of Antofagasta and in the area of Taltal, horizontal compressive stresses (σHmax = σ1) directed NE-SW predominate in the entire Coastal Cordillera. σHmax reveals mean values of 9 MP, a whereas σHmin (= σ2) oriented NW-SE yields -1 MPa on average. In situ stresses of the west Chilean stress field signify an actual uplift of the Coastal Cordillera bordered on the west (Mejillones Peninsula) and on the east (Atacama Fault) by active normal faults striking N-S. (2) The Central Chilean stress field includes the Longitudinal Valley and the western part of the Chilean Precordillera. From west to east compressive σHmax directed NW-SE to N-S decreases from 40 to 7 MPa, whereas tensile σHmin oriented NE-SW to E-W increases from 0 to -7 MPa. Generally σHmax trending NW-SE corresponds to σ2 in the west and to σ1 in the east of the Central Chilean stress field, indicating actual normal faulting in the west (Atacama Fault, Longitudinal Valley) and strike-slip or thrust faulting in the east (Calama). (3) The Central Andean stress field extends from the Preandean Cordillera Domeyko across the Preandean Depression, Western Cordillera, and Puna, and includes the western area of the Eastern Cordillera. Maximum horizontal in situ stresses are generally oriented E-W yielding average values of 14 MPa (N-S, σHmin = 0.7 MPa) in the western part of the Central Andean stress field (Cordillera Domeyko, Preandean Depression, Western Cordillera), whereas in the Puna lower E-W compressive stresses only occasionally exceed 10 MPa. Horizontal tensile stresses oriented NW-SE and of -5 to -15 MPa along the Olacapato - El Toro lineament indicate its probable recent strike-slip movement. (4) The east Andean stress field is defined by the eastern part of the Eastern Cordillera or may extend to the east into the Subandean Ranges. Maximum horizontal stresses are compressive in the NE-SW direction and vary between 2 and 17 MPa. In situ horizontal stresses across the uppermost crust of the Andes between 22° and 26° S are dependent on the thickness of the rigid Andean crust rather than on topographic elevation.

1 Introduction

The converging Nazca and South American plates induce directional stresses to their rigid subducted and obducted compartments. The investigated E-W traverse (65° - 70° W, 22° - 26° S) is part of an area studied for focal mechanisms of earthquakes which occurred during 1962-1970 (Stauder 1973).

There is evidence that at depths of 30 - 60 km within the underthrusting Nazca plate, maximum compressive stress and plate motion correspond in the azimuth of N 85° E and in the dip of 20° E. Focal mechanisms of intermediate deep earthquakes, revealing tensional stress directed N 65° E and within the Nazca plate which dips beneath the Central Andes 30° E, and of deep (>500 km) earthquakes and stresses of the Nazca plate under the eastern Andes do not induce stresses on the South American plate by ductile decoupling. As there are shallow hypocentres of earthquakes under the Coastal Cordillera, so these exist beneath the Subandean Ranges of the eastern Andes too (Chinn and Isacks 1983). Their focal mechanisms indicate the presence of maximum horizontal compressive stresses oriented E-W to ENE-WSW. There is a seismic gap in the upper plate between the Subandean Ranges and the Coastal Cordillera.

Correspondence to: Karlheinz Schäfer, Abt. Geologie der Universität, Universitätsstr. 30, D-8580 Bayreuth

Seismic activity (Wigger 1988), recent plutonism (Schwarz et al. 1986) and volcanism, neotectonics such as thrusting and folding (Schwab 1985, Reutter et al. 1988), actual normal and wrench faulting (Okada 1971, Salfity 1985, Scheuber et al. 1986, Strecker et al. 1989), current uplift of marine terraces (Ratusny & Radtke 1988) and present basin subsidence (Abele 1988) all establish the contemporaneous building of the Andes.
An assumed in situ stress field with an E-W directed maximum horizontal compressive stress component deriving from the E-W convergence at the western active margin of the South American plate and from the E-W divergence at its eastern passive margin, would not explain the N-S alignment of alternating uplifted and subsided crustal segments that dominate the Andean morphology. An Andean transect from west to east shows the uplifted Coastal Cordillera bordered by normal faults, the downthrown Longitudinal Valley followed by the uplifted Chilean Precordillera, which is limited to the east either by the subsided Preandean Depression or by the broad uparched Central Andes.

We investigated in situ rock stresses of all these structural units in order to determine whether amount and orientation of actual stresses relate to the tectonic setting.

2 Applied Method

To determine in situ stresses the overcoring method has been applied in mines, in the open pit of Chuquicamata or, because of the lack of sufficient quarries, at any horizontal fresh rock surface at the bottom of fluvial eroded canyons or at raised hard rock beaches. We avoided sites where the rock quality, in terms of texture and structure, did not promise good results.

Local topography-induced gravitational stresses could be corrected by the calculation of horizontal stresses occurring at the foot of quarry walls due to the weight of the vertical rock masses according to $\sigma hor = \sigma vert \cdot \nu / (1-\nu)$.

At the base of a 20-m-high quarry wall composed of rocks of a density of 2.5 and with a Poisson ratio (ν) of 0.2, gravitational stress is 0.125 MPa. At a distance of more than 5 m from the quarry wall, gravitational stress is less than 0.1 MPa, which is the lowest stress value considered in this study.

Gravitational stress induced by regional topography has been determined by finite elements and produced values that could be ignored. Thermally induced stresses and rock anisotropies exert a major influence on our data. Diurnal and seasonal temperature fluctuations cause thermal stresses to a depth of about 0.5 m (daily) or to more than 10 m (annually) (Hooker and Duvall 1971).

The anisotropies, e. g. of the coefficient of thermal expansion, of Young's moduli and of Poisson's ratio, have been determined at each overcored specimen in the horizontal directions N-S (0°), NW-SE (135°) and E-W (270°) by means of the same strain rosette in the laboratory as was earlier measured by in situ strain release in the field.

Rocks at depths of 10 to 20 m below the surface are subjected to an annual mean temperature and are not thermally stressed. Thus, thermally induced stresses have been eliminated by a correction factor derived from the temperature difference between annual mean temperature and measured rock temperature at a particular site, and the specific coefficient of thermal deformation, given by $\varepsilon T0 = \varepsilon T1 - \alpha(T1-T0)$. Table 1 shows the procedure for the elimination of thermally induced stresses and the application of anisotropic Young's moduli and Poisson's ratio.

All the rocks examined proved to be thermally and elastically anisotropic. The anisotropies significantly affected the direction and amount of in situ stress and therefore could not be disregarded. They were determined by means of a strain gauge attached to the specimen that was not modified after overcoring.

3 Sites and Data of In Situ Stress

Results for in situ stress have been obtained at 45 sites in northern Chile and northwestern Argentina between 22° and 26° S. Table 2 shows the location, lithology and age of the rocks investigated, the orientation of maximum horizontal stress ($\Theta\sigma Hmax$) and the amount of maximum and minimum horizontal stress.

The directions of $\sigma Hmax$, which can be either a maximum compressive stress or a minimum tensile stress, are shown in Fig. 1, superimposed on the prominent morphotectonic units of the Andes.

Ten locations are in the Coastal Cordillera, on the Mejillones Peninsula and in the coastal areas north and south of Antofagasta (C1-C5, C15-C17, C19, C23). We recorded E-W directed maximum tensile stress on the Mejillones Peninsula and at the raised coastal beaches north of Antofagasta. Also, tensile stress perpendicular to the compressive $\sigma Hmax$ appears at the coast south of Paposo (C16) and north

Table 1. The procedure for elimination of thermally induced stress and the application of anisotropic Young's modulus and Poisson's ratio. Thermal and elastic anisotropies significantly influence the direction and amount of in situ stress

Sample No:	C18/4		
Site:	Andes Mines, Sierra Gorda, N. Chile		
Lithology, stratigraphy:	Rhyolite, Jurassic		
	a = 0°	b = 135°	c = 270°
Measured strain at site (ε) [μm/m]:	εa	εb	εc
	236	260	226
Azimuth of maximum horizontal deformation [$\Theta\varepsilon$Hmax]:		140°	
Annual mean temperature (T0):		18.0 °C	
Temperature of overcored specimen (T1):		21.0 °C	
Coefficient of thermal expansion (α) [10^{-6}/K] :	αa	αb	αc
	9.3	9.2	9.8
Strain at annual mean temperature (removed from thermal strain) $\varepsilon T0 = \varepsilon T1 - \alpha(T1-T0)$ [μm/m]:	εaT0	εbT0	εcT0
	208	232	197
Young's modulus (E) [MPa 10^5]:	Ea	Eb	Ec
	0.276	0.230	0.184
Poisson's ratio (ν):	νa	νb	νc
	0.38	0.34	0.19
Calculation of tectonic stress: $\sigma = \varepsilon T0\ E/(1-\nu)$ [MPa]	σa	σb	σc
	9.26	8.08	5.13
Amount of maximum horizontal stress (σHmax):		9.4 MPa	
Amount of minimum horizontal stress (σHmin):		4.9 MPa	
Azimuth of maximum horizontal stress ($\Theta\sigma$Hmax):		168°	

of Taltal (C17) (Fig.1). In the copper mine of Michilla (C15), at Coloso (C19), and at Punta de Lobos (C23) horizontal stress is compressive in all directions and is highest in the NE-SW orientation with an average value of 14.7 MPa. The mean value of σHmax at the Pacific coast and on the Mejillones Peninsula is 5.7 MPa and σHmin amounts to -4.8 MPa, while in situ stresses rise to mean values of 14.7 MPa (σHmax) and 6.4 MPa (σHmin) in the Coastal Cordillera.

In the Longitudinal Valley in situ stress was determined at five locations. Three locations are east to northeast of Antofagasta (C6, C7, C10), revealing a high compressive σHmax directed N-S- to NW-SE and with an average value of 33.8 MPa. σHmin oriented E-W- to NE-SW indicates a low mean stress (0.1 MPa) which is tensile north of Carmen Alto (C7) and tensile in general because of the high compressive stresses perpendicular to it. Two locations are in the north (21° S) of the area investigated at the transition from the Longitudinal Valley to the Chilean Precordillera (C24, C25). Although the two locations are close to each other (3 km), the orientation and amount of σHmax differs considerably between them. One location (C24) shows high compressive stress directed NE-SW while the other location (C25) shows a high tensile stress in the same direction. It is assumed that a prominent fault running N-S marking the transition between the Longitudinal Valley and the Chilean Precordillera is tectonically active in the

Table 2. Numbering of sites corresponds to that in Fig. 1

No.	Location	Stratigraphy/lithology	No. sites/ stress determ.	ΘσHmax (ø) [°]	σHmax [MPa]	σHmin [MPa]
C1	Juan Lopez, Mejillones	Palaeozoic magmatite	4/3	176	4.3	-8.7
C2	5 km N Antofagasta	Jurassic fanglomerate	3/6	68	2.1	-15.9
C3	Mejillones, facing Santa Maria island	Palaeozoic magmatite	3/3	9	3.2	0.8
C4	Mejillones, 5 km N C3	Palaeozoic magmatite	1/3	169	9.0	-1.5
C5	Mejillones, 8 km N C3	Palaeozoic magmatite	2/-	73	-	-
C6	6 km E Baquedano	Mesozoic andesite	3/9	2	34.6	3.6
C7	21 km N Carmen Alto	Mesozoic andesite	4/9	112	26.7	-5.9
C8	18 km E San Pedro de Atacama	Pleistocene ignimbrite	5/12	56	19.7	-9.3
C9	2 km SE Calama	Cretaceous andesite	3/6	132	6.8	3.8
C10	23 km SSE Mantos Blancos	Devonian metamorphite	4/3	108	40.1	2.6
C11	44 km E San Pedro de Atacama	Pleistocene andesite	1/3	108	16.6	-20.2
C12	Cordillera Domeyko, 39 km W San Pedro de Atacama	Cretaceous pelite	4/9	85	15.2	-1.4
C13	2 km SE Toconao	Pleistocene ignimbrite	4/12	113	9.0	-0.3
C14	Cordillera de la Sal, 29 NE San Pedro de Atacama	Pliocene ignimbrite	2/3	46	11.8	1.1
C15	Michilla	Jurassic volcanite	6/15	84	19.4	10.8
C16	Sto. Domingo, 10 km S Paposo	Jurassic volcanite	2/3	32	0.4	-1.9
C17	14 km N Taltal	Jurassic volcanite	3/9	110	15.1	-1.6
C18	Andes Mines, 3 km W Sierra Gorda	Cretaceous andesite	6/15	150	11.8	-7.0
C19	Coloso, 19 km S Antofagasta	Palaeozoic gabbro	3/6	47	7.2	1.3
C20	Chuquicamata, Mina Sur	Palaeozoic granodiorite	4/12	112	16.9	0.0
C21	10 km S Toconao	Pleistocene ignimbrite	1/3	125	9.8	5.7
C22	Pukara, 3 km NNW San Pedro de A.	Miocene ignimbrite	3/6	94	14.9	4.0
C23	3 km N Punta de Lobos, 87 km S Iquique	Jurassic diorite	2/6	48	17.6	7.1
C24	Quebrada Blanca	Cretaceous andesite	1/3	53	84.0	7.7
C25	Quebrada Blanca	Cretaceous andesite	2/6	130	-6.2	-43.0
C26	Laguna Aguas Calientes	Pliocene ignimbrite	3/9	118	10.0	1.0
C27	20 km ESE C26	Pliocene ignimbrite	4/12	32	29.0	8.0
A28	Mina Norma	Quaternary playa carbonate	2/6	95	10.0	-16.1
A29	Casa Zorro	Quaternary playa carbonate	3/9	110	-0.3	-3.6
A1	73 km N Cafayate	Cretaceous sandstone	4/12	75	4.5	0.5
A2	15 km NE Cafayate	Tertiary sandstone	5/-	57	-	-
A3	Anfiteatro, 45 km N Cafayate	Cretaceous sandstone	4/12	169	15.0	10.0
A4	15 km N Salta	Ordovician quartzite	1/3	133	-3.4	-14.7
A5	5 km S Huacalera	Ordovician quartzite	3/3	57	1.7	-1.6
A6	El Carmen	Ordovician quartzite	1/3	54	13.5	-1.0
A7	20 km SW San Antonio de los Cobres	Tertiary andesite	2/6	114	-5.0	-10.0
A8	18 km SW San Antonio de los Cobres	Tertiary andesite	4/12	86	9.0	3.0
A9	12 km E San Antonio de los Cobres	Precambrian greywacke	2/6	158	-10.0	-30.0
A10	San Antonio de los Cobres	Precambrian greywacke	3/6	71	14.7	-2.7
A11	2 km SW Sta. Rosa de Tastil	Precambrian quartzite	2/-	117	-	-
A12	4.7 km NW Alfacito	Precambrian granite	1/-	141	-	-
A13	1 km N Sta. Rosa de Tastil	Precambrian granite	2/6	94	13.2	5.3
A14	La Merced	Cretaceous limestone	2/6	6	8.0	-1.5
A15	15 km W Chicoana	Cretaceous limestone	2/3	61	2.4	1.8

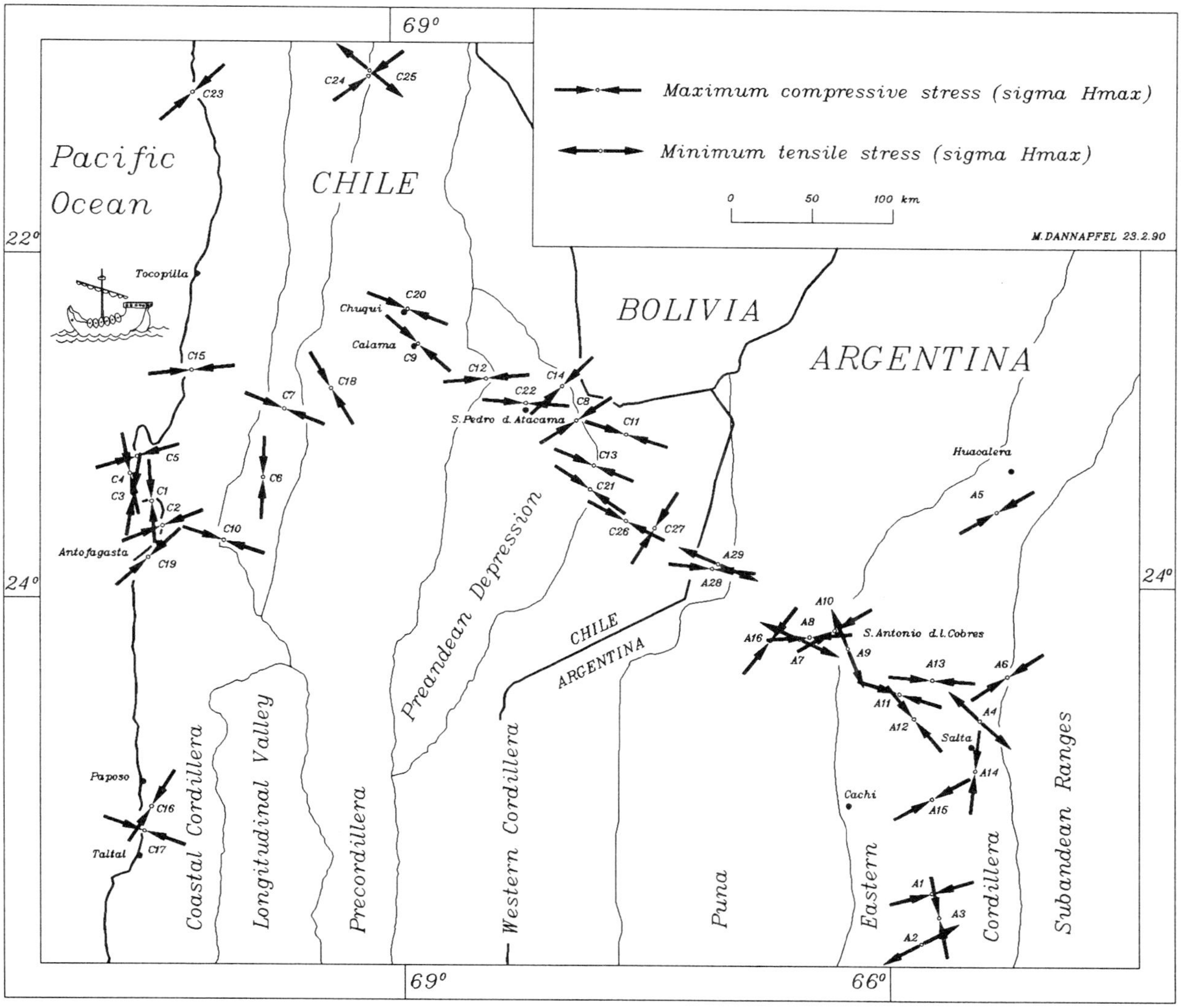

Fig. 1. Numbers of the locations refer to those shown on Table 2, where site characteristics, and the direction and amount of σHmax and σHmin are also indicated

area under study.

In situ stress has been determined at four locations in the Chilean Precordillera (C9, C12, C18, C20). At three locations (C9, C18, C20) around and west of Calama compressive σHmax is directed NW-SE. One location (C12) shows a compressive σHmax oriented E-W. The mean values of σHmax and σHmin are 12.7 and -1.2 MPa respectively. From west to east across the Longitudinal Valley and the Chilean Precordillera there is a consistency in the orientation of σHmax, while compressive σHmax values decrease and tensile σHmin increases.

In situ stress has been determined at five locations in the Preandean Depression. All the rocks studied are ignimbrites. At two locations (C14, C22) the ignimbrites are part of the Cordillera de la Sal, which has been folded and thrust since the Miocene (Wilkes and Görler 1990, this Vol.). Compressive σHmax (11.8 MPa) of the Pliocene ignimbrite (C14) is directed NE-SW and strikes almost parallel to the axis of the Cordillera de la Sal at its northeastern end; σHmax (14.9 MPa) of the Miocene ignimbrite (C22) is oriented E-W and runs perpendicular to the fold axis and thrust faults of the Cordillera de la Sal.

Three locations (C8, C13, C21) are at the eastern margin of the Preandean Depression where ± 1-million-year-old Pleistocene ignimbrites, derived from the young volcanoes of the Western Cordillera, dip under the Holocene deposits of the Salar de Atacama. Orientation of σHmax is NE-SW (C8) or NW-SE (C13, C21) and the average compressive stress is 12.8 MPa. The average σHmin is tensile (-1.3 MPa).

In the Western Cordillera of the Andes in situ stress was studied in five locations. While three locations (C11, C26, C27) are in Pleistocene andesite and Pliocene ignimbrites, two (A28, A29) are in quarries of Quaternary playa carbonates. Except at C27, where the direction of σHmax is 32°, we recorded a general WNW-ESE trend of maximum horizontal compressive stress. The average stress of σHmax is 13.1 MPa, while that of σHmin generally oriented NNE-SSW diminishes to -6.2 MPa.

In situ stresses have been determined at five locations in the Puna, with four of them in Tertiary andesites and Precambrian greywackes at San Antonio de los Cobres and about 10-20 km west and east of it (Fig. 1 and Table 2). The other location is at Olacapato where the in situ strain of a Tertiary sandstone has been measured. The direction of σHmax varies between ENE-WSW and NW-SE around San Antonio de los Cobres, while at Olacapato a NE-SW orientation was recorded. Average σHmax values drop to 3.3 MPa, the lowest of the entire Andes transect. Also, σHmin is at its lowest at -7.7 MPa.

We recorded in situ stress in 11 locations of the Eastern Cordillera of the Andes. Sediments, metamorphic and plutonic rocks of Tertiary to Precambrian age were available for the stress studies. The E-W to NW-SE trend of σHmax still continues west and south of Salta but in the eastern part of the Eastern Cordillera, maximum horizontal compressive stresses are predominantly oriented NE-SW.

Stress values of σHmax (6.9 MPa) and σHmin (-0.2 MPa) have risen in comparison to the stresses recorded for the Puna but are still lower than all other mean stresses recorded in the different morphological units to the west.

4 Discussion and Conclusions

In situ horizontal stresses in northern Chile and northwestern Argentina are probably a function of the thickness of the rigid Andean crust. In the west of the area investigated, where the rigid crust is thickest, we recorded highest σHmax values in the Coastal Cordillera and Longitudinal Valley (Fig. 2). At the Pacific coast and on the Mejillones Peninsula σHmax and σHmin are considerably lower than in the adjacent Coastal Cordillera and this may be due to tectonic erosion and crustal thinning from the trench to the coast. The Coastal Cordillera may not yet undergo tectonic erosion at its rigid crustal base, rather the convergent Nazca and South America plates induce an uplift of the Coastal Cordillera. Elevated marine terraces confirm uplift of the Coastal Cordillera at 1.7 mm/a over the past 3000 years (Ratusny and Radtke 1988). The Longitudinal Valley and the Chilean Precordillera are part of one westward-tilted crustal block with erosion at the uplifted Precordillera side and deposition of the eroded material on the subsided Longitudinal Valley side. Since the Coastal Cordillera is uplifted faster than the downward incision of its valleys advances, it operates as a dam preventing sediment transport from the Longitudinal Valley to the coast (Abele 1988).

Magma rise beneath the Western Cordillera and the Preandean Depression leads to the upwelling of material of low electrical resistivity to about 10 km below the surface (Schwarz et al. 1986) and this may be responsible for the tilting of the Longitudinal Valley-Precordillera crustal block by buoying it on its eastern side (Fig. 2).

North of 22° S the Longitudinal Valley-Precordillera block breaks into two parts, separating the Longitudinal Valley from the Precordillera as a graben. In situ stresses change from higher compressive σHmax in the Longitudinal Valley to lower compressive σHmax in the Chilean Precordillera, and from lower tensile σHmin in the Longitudinal Valley to higher tensile σHmin in the Precordillera.

We have no explanation for the high compressive stresses in the Longitudinal Valley. We recorded lower compressive σHmax and high tensile σHmin in the Quebrada Huatacondo farther north (21° S), where the Longitudinal Valley is a graben bordered by normal faults. In the Rhinegraben or close to its bordering master faults compressive σHmax is also lower than in some distance to the Rhinegraben.

In contrast to the expected E-W oriented tensile σHmin resulting from the N-S alignment of young and active volcanoes in the Western Cordillera, we recorded a compressive σHmax generally directed E-W and with an average value of 13.1 MPa. σHmin is oriented N-S revealing besides the Puna the highest tensile stresses that average to -6.2 MPa. Although magma rises from the subducted Nazca plate N-S-canalized to depths of about 10 km beneath the

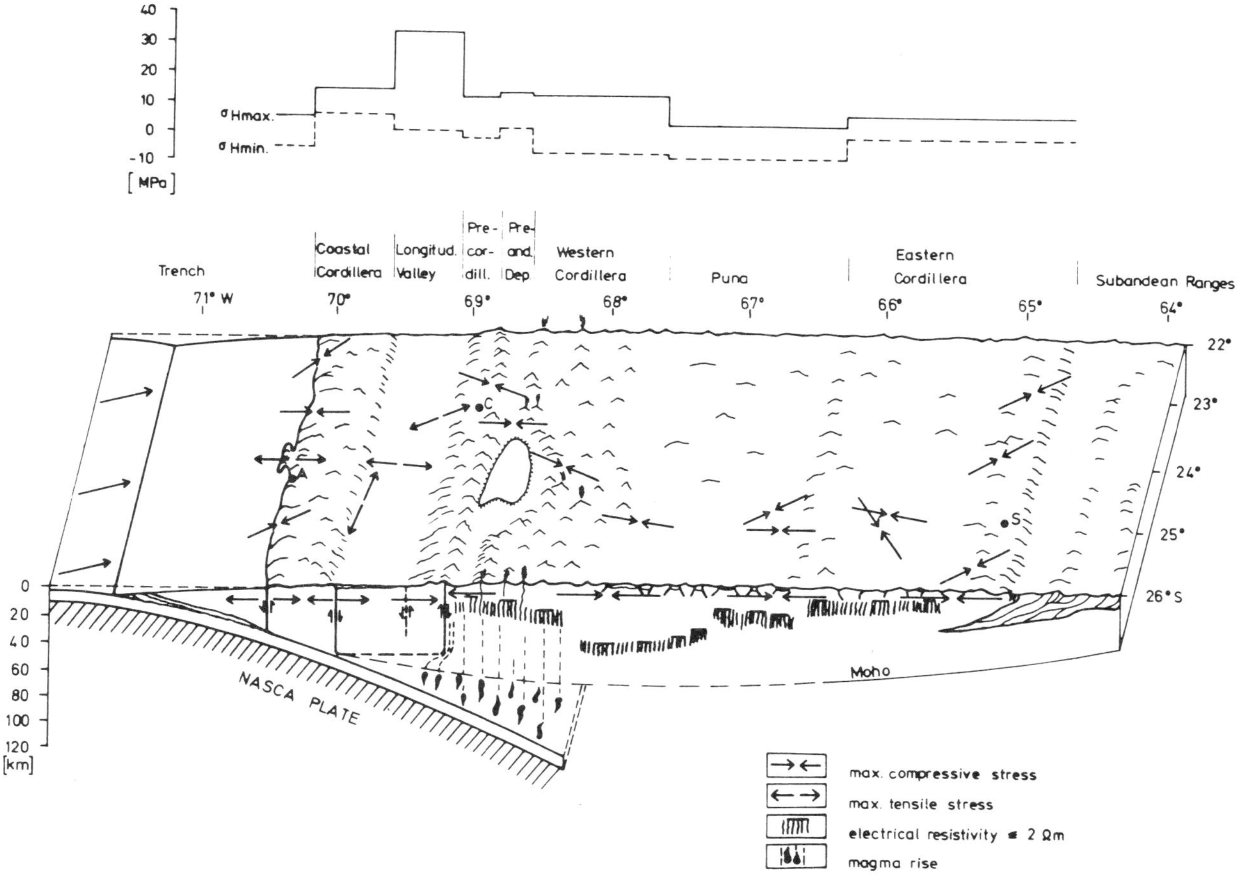

Fig. 2. Block diagram of the Andes in northern Chile and northwestern Argentina using geological and geophysical data of a cross section at 21°25'S (from Reutter et al. 1988). The high mean value of sHmax in the Longitudinal Valley is directed NW-SE to N-S and is, therefore, in contrast to the generally E-W oriented sHmax of all other morphological units. The variation in mean values of sHmax and sHmin as seen in the *upper part* of the figure, is probably a function of the thickness of the rigid Andean crust. A Antofagasta, C Calama, S Salta

surface, splitting and magma injection into the rigid upper crust is directed E-W due to tensile σHmin oriented N-S. In the area investigated there exist many examples of E-W necking of the Western Cordillera crust by volcanoes such as the presently active Lascar volcano and five neighbouring Holocene craters extending E-W to the Aquas Calientes volcano. Other Quaternary pairs of volcanoes oriented E-W are Taco Purico-Santa Barbara, Licancabur-Juriques, Colorado-Curiquinca and Torta-Tocorpuri.

E-W alignment of Cenozoic volcanic effusives in the Puna corresponding to tensile in situ stresses oriented N-S which average -7.7 MPa, reveals E-W in situ fissuring of the rigid part of the crust even though it is thicker than in the adjacent Western and Eastern Cordilleras.

Acknowledgements. We thank the German Research Foundation (DFG) for supporting this study, and colleagues of the Berlin research group, of the Universidad de Norte at Antofagasta, and of the Universidad Nacional de Salta for helpful discussions.

References

Abele G (1988) Geomorphological west - east - section through the north Chilean Andes near Antofagasta. In: Bahlburg H, Breitkreuz C, Giese P (eds) Lecture Notes in Earth Sciences 17. Springer, Berlin Heidelberg New York, p 153-168

Chinn DS, Isacks BL (1983) Accurate source depths and focal mechanisms of shallow earthquakes in western South America and in the New Hebrides island arc. Tectonics 2:529-563

Hooker VE, Duvall WI (1971) In situ rock temperature. Stress investigation in rock quarries. Bureau of Mines, Rep of Investigations, Denver Colorado, 7589,12 p.

Okada A (1971) On the neotectonics of the Atacama fault zone region - preliminary notes on late Cenozoic faulting and geomorphic development of the coast range of northern Chile. Bull Dep Geogr Univ Tokyo 3:47-65

Ratusny A, Radtke U (1988) Jüngere Ergebnisse küstenmorphologischer Untersuchungen im 'Großen Norden' Chiles. Hamb Geogr Stud 44:31-46

Reutter K-J, Giese P, Götze H-J, Scheuber E, Schwab K, Schwarz G, Wigger P (1988) Structures and crustal development of the Central Andes between 21° and 25° S. In: Bahlburg H, Breitkreuz C, Giese P (eds) Lecture Notes in Earth Sciences 17. Springer, Berlin Heidelberg New York, pp 231-261

Salfity JA (1985) Lineamentes Transversales al Rumbo Andino en el Noroeste Argentino. In: IV Congr Geol Chileno, Antofagasta, T I: 2-119 - 2-137

Scheuber E, Rössling R, Reutter K-J (1986) Strukturen der chilenischen Küstenkordillere zwischen Paposo und Antofagsta. Berl Geowiss Abh A66:209-224

Schwab K (1985) Basin formation in a thickening crust - the intramontane basins in the Puna and the Eastern Cordillera of NW-Argentina (Central Andes). In: IV Congr Geol Chileno, Antofagasta, T I: 2-138 - 2-158

Schwarz G, Martinez E, Bannister J (1986) Untersuchungen zur elektrischen Leitfähigkeit in den zentralen Anden. Berll Geowiss Abh A66:49-71

Stauder W (1973) Mechanisms and spatial distribution of Chilean earthquakes with relation to subduction of the oceanic plate. J Geophys Res 78:5033-5061

Strecker MR, Cerveny P, Bloom AL, Malizia D (1989) Late Cenozoic tectonism and landscape. Development in the foreland of the Andes: Northern Sierras Pampeanas (26° - 28° S), Argentina. Tectonics 8:517-534

Wigger PJ (1988) Seismicity and crustal structure of the Central Andes. In: Bahlburg H, Breitkreuz C, Giese P (eds) Lecture Notes in Earth Sciences 17. Springer, Berlin Heidelberg New York, p 209-229

Wilkes E, Görler K (1990) Evolution of the Cordillera del la Sal, Northern Chile. In: Structure and evolution of the Central Andes in northern Chile, southern Bolivia and northwestern Argentina. Final workshop, May 23-25,1990, Abstr vol, FU TU Berlin, p 102-103

Large Events, Seismic Gaps, and Stress Diffusion in Central Chile

SERGIO E. BARRIENTOS

Abstract. The time and space distribution of rupture segments along south-central Chile suggest the high probability of of a large magnitude earthquake occurring in the proposed seismic gap (34.3°-37.2°S) located between the 1960 and 1985 rupture regions. Three lines of evidence support the occurrence of a magnitude 8+ event within the next couple of decades. First the repeat time of large magnitude earthquakes is 90±6 years; the last such shock was in 1928. The second line of evidence is related to a possible cause-effect relationship between events within the gap. The distribution of events exhibits a coupling of earthquakes from north to south. The average inter-occurrence time is 16±6 years and the last quake in the northern part was in 1985. The last line of evidence arises from an analysis of all the data in this century. A southward migration of ca. 7 km/year is apparent in the sequence indicating that the possible "stress front" will arrive at the seismic gap within the first decade of the next century. The observed velocities and recurrence periods are consistent with a stress diffusion model. This analysis by no means excludes the possibility of earlier activity, such as occurred in 1971 prior to the 1985 mainshock.

1 Introduction

Long term earthquake forecasting is based on the seismic gap concept (Fedotov 1965; Mogi 1969; Kelleher 1972; McCann et al. 1978). Those segments not ruptured during the past few decades are considered as probable sites of future activity. McCann et al. (1978) introduced the concept of seismic potential to categorize the seismic gaps as a function of time elapsed since the previous event. Nishenko and McCann (1981) and Nishenko (1985) incorporated the idea of a recurrence period for any given segment in order to reflect the imminence of a large event. In particular, Nishenko (1985) applied these ideas to the Chilean and southern Peruvian margins.

In this century Chile has been struck on average by a magnitude ±8 earthquake every decade, one of them being the 1960 earthquake which was the largest event recorded since the beginning of instrumental seismology. The historical record in south-central Chile is the most complete record of large earthquakes in this region because it was the first area to be colonized. It began in 1570 with the first accounts provided by Spanish settlers. In this chapter the 400-year record, combined with the seismic gap concept and a stress diffusion model, is used to assess future activity in the region between the 1960 and 1985 rupture zones.

Correspondence to: S. E. Barrientos Departamento de Geología y Geofísica, Universidad de Chile, Casilla 2777, Santiago-Chile

2 Data

Based on the work by Montessus de Ballore (1911), Lomnitz (1971), Kelleher (1972), Nishenko (1985), and Ramírez (1988) 17 shocks of magnitudes ±8 are considered here. A brief description of each event is given; magnitudes and rupture lengths are taken from Lomnitz (1971) and Ramírez (1988) respectively.

The sequence began in February, 1570, with what is known as the first Concepción damaging earthquake. The majority of houses were destroyed and several cracks appeared in the ground. A large tsunami was associated with the mainshock and aftershocks were still being felt 5 months later. The magnitude was estimated at between 8 and 8½ with a rupture length of approximately 200 km. In 1575, the first Valdivia damaging earthquake took place, it completely destroyed five villages and the tsunami generated was comparable with that associated with the 1960 event. The estimated magnitude was 8½ and the rupture length was somewhat less than that observed for the 1960 event. In 1647, the great Santiago earthquake occurred and 20% of the population was killed. The Quillota valley, to the northwest, was completely devastated. Lomnitz (1983) locates the epicentre offshore Valparaíso and estimates the magnitude at ca. 8½. The rupture length was about 400 km. The second Concepción damaging earthquake magnitude 8 took place in 1657, destroying the majority of houses. Most of the people died because of the tsunami that flooded the area. Montessus de Ballore (1911) estimated a rupture length of 300 km. The largest event in the Valparaíso region took place in 1730. Kelleher (1972) estimates a rupture length of between 350 and 450 km, overlapping the 1906 and 1943 rupture regions. In 1737 the second damaging earthquake in Valdivia occurred; it devastated the city of Valdivia and several cities on Chiloé Island, about 200 km south; the estimated magnitude was about 8. The third large Concepción earthquake magnitude 8½ took place in 1751; it produced the largest tsunami ever observed in the Concepción area and brought ruin to several cities 200 km away from Concepción to the north. In 1822, Valparaíso was struck by an earthquake of magnitude 8½. From the extension of coastal uplift the rupture length must have been about 200 km.

The fourth devastating Concepción earthquake in 1835 is one of the most thorougly documented. Santa María Island, just north of the Arauco Peninsula, was uplifted by about 3 m, Quiriquina Island by 2.5 m and Talcahuano, in the bay of Concepción, by about 1.5 m. About half an hour later a large tsunami with waves up to 8-10 m hight flooded the bay. The estimated magnitude ranges between 8 and 8¼ and the rupture length was about 400 km (Nishenko 1985). Two years later, in 1837, the third Valdivia earthquake took place. This event was comparable with the 1575 and 1737 earthquakes and was somewhat smaller than the 1960 earthquake. Crack openings were observed in Chiloé Island where the tsunami effects were noticeable. Lemus Island (45.2°S, 74.5°W) was uplifted by about 2.5 m. The estimated magnitude was at least 8 and the rupture length was about 600 km.

In this century, the first large earthquake was the 1906 Valparaíso event. Gutenberg and Richter (1954) estimated the magnitude at 8.6. The coast was uplifted by between 40 and 50 cm in the Zapallar-Quintero (32.5°S) and Pichilemu-Llico (34.5°S) areas. A small tsunami, with maximum amplitude reaching about 1 m above high tide, accompanied the event. The rupture length was about 360 km (Comte et al. 1986). Sixteen years later in 1928 the Talca earthquake occurred. It devastated the cities of Talca and Constitución and caused damage from Valparaíso to Concepción. A maximum amplitude of 1.5 m above high tide was generated by the tsunami; in the coastal locality of Putú (35.3°S) the beach was uplifted and the sea receded by about 200 m. Richter (1958) estimated the magnitude as 8.4.

The 1939 event completely destroyed the city of Chillán. The area of major destruction was confined to the Central Valley between Linares (35.8°S) and Los Angeles (37.5°S). The absence of tsunami and coastal elevation changes,and a reported focal depth of 80-90 km (ISS) indicate that this event was probably not a thrust event. Preliminary body wave modelling of the 1939 event (Campos and Kausel 1990) suggests that it is associated with a normal fault at about 90 km depth, implying that it is not a thrust-type earthquake and that it is similar, in both depth and source mechanism, to the December 9, 1950, northern Chile earthquake of magnitude 8.0 (Kausel and Campos 1989). Therefore, this event is not considered in the statistics of thrust events.

The two most extensively studied earthquakes in Chile, the 1960 and 1985 events (Fig. 1), bound the region wich forms the subject of this study. The 1960 event ruptured more than 900 km, from the Arauco Peninsula (37.3°S) to the Taitao (46.8°S) Peninsula. Maximum fault displacement of about 40 m, located mainly offshore, contributed to a seismic moment of 1 to 2 x 10^{31} dyn-cm (Kanamori 1977; Linde and Silver 1989; Barrientos and Ward 1990). Nearly all the important cities from Concepción to Puerto Montt suffered severe damage. Remarkable land level changes were observed over an area 200 km wide by 1000 km long. The city of Valdivia subsided by about 2 m and two islands, Guafo and Guamblin, were uplifted by more than 4 m (Plafker and Savage 1970). The earthquake induced a tsunami that spread over the Pacific Ocean and caused hundreds of deaths as far away as Japan.

The 1985 earthquake in Central Chile, of magnitude 7.8 is probably the best-documented earthquake in Chile. Body and surface wave data (Christensen and Ruff 1986; Monfret and Romanowicz 1986; Choy and Dewey 1988) in combination with geodetic estimates (Barrientos 1988) revealed the source process and slip distribution along its rupture length. It corresponds to an approximately 160-km-long fault with an average slip of about 2 m. It was preceded in 1971 by an event of surface wave magnitude 7.5 in its northern extension (Korrat and Madariaga 1986). As will be shown in the following analysis, rupture zones of both the 1960 and 1985 earthquakes bound a region of high earthquake potential (seismic gap).

The rupture lengths of the large magnitude events are plotted against time in Figs 2 and 3. Prior to the 1800s the available descriptions of damage cannot assure a precise length of rupture. This situation improves in the 1800s with Darwin's first accounts of

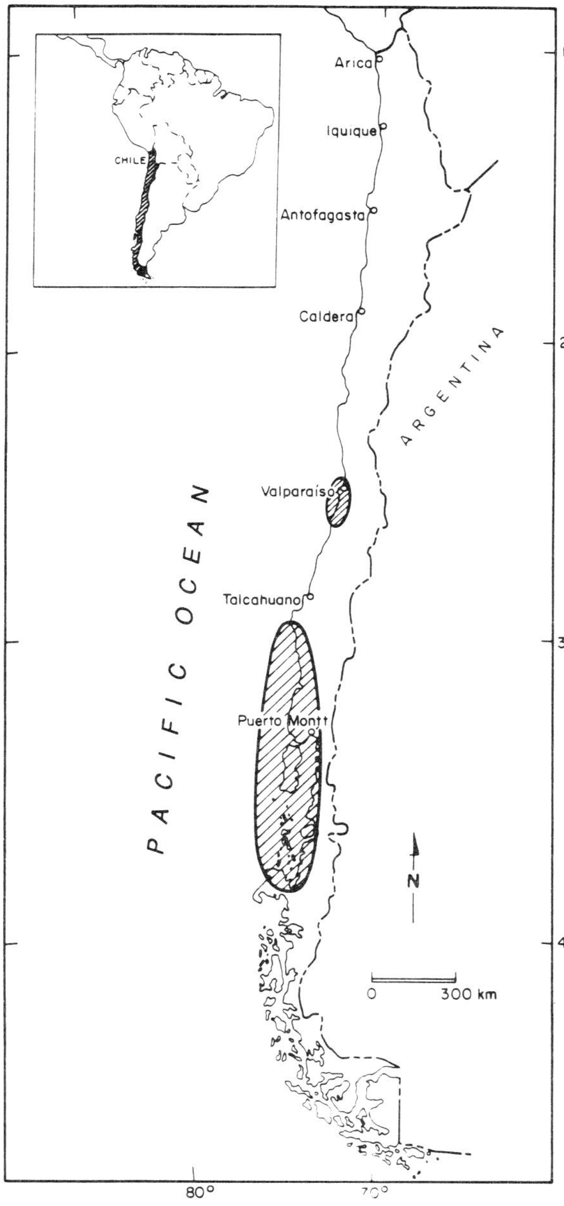

Fig. 1. The most recent two large events in central Chile. The 1960, May 21-22, M_W=9.4 and the 1985 Valparaíso, M_W=8.0, events bound the present seismic gap located between 34.3° and 37.2°S

the Concepción earthquakes (Darwin 1851). Because the analysis is based mostly on the time of earthquake occurrence, large errors associated with some estimated rupture lengths will not effect the following discussion.

3 Stochastic Analysis

The proposed seismic gap lies (Fig. 2) between the 1960 (about 900 km) and the 1985 (160 km) rupture regions, within the horizontal dashed lines. This region was last struck by earthquakes in 1928 (Talca) and 1939 (Chillán) with events of magnitude 8+.

The recurrence time of large magnitude earthquakes, as evidenced by the 1570, 1657, 1751, 1835, and 1928 events, is 90±5 years. This value is close to that observed for the events in the Valparaíso region immediately to the north, where considerations of six large earthquakes gives a value of 82±6 years. As in this region, the regularity of repeat earthquakes is not consistent with a time predictable model (Comte et al. 1986). Assuming that the time intervals between large shocks are described by a Weibull two-parameter distribution (Nishenko 1985), the probabilities of occurrence of the next event given that the prior mainshock took place in 1928, are shown in Fig. 4 (solid lines). Poisson distributions are not considered in this analysis because they give estimates of the future occurrence of large earthquakes which are independent of the time elapsed since the previous earthquake and are, therefore, unrealistic.

In 1990 was a 63% chance of a large event taking place in the next 30 years. The probability decreases to 10% when considering the next 20 years and to 0.8% for the next 10 years. All probabilities increase accordingly as time progresses and by 1995 it is almost certain that a large shock should take place within 30 years. Since the database is the same, these values reproduce the previous findings of Nishenko (1985).

Another noteworthy aspect of the sequence is related to a possible cause-effect relationship between events within the gap. From Fig. 2 and Fig. 3 it is possible to observe that there is a coupling of earthquakes from north to south. This is clear for the sequences 1570-1575, 1647-1657, 1730-1751, 1822-1835, and 1906-1928, which give intervals of 5, 10, 21, 13, and 22 years respectivly. Since the average inter-occurrence time is 14±6 years, and the last quake was in 1985, the next one to the south is expected to occur in around the year 2000. Kelleher (1972) has pointed out that, with the exception of the 1751 event, all magnitude ±8 earthquakes between 32° and 46°S occurred from 2 to 22 years after an earlier earthquake to the north. The conditional probabilities of occurrence given that the last event took place in 1985, and assuming a two-parameter Weibull distribution, are shown in Fig. 4 (dashed

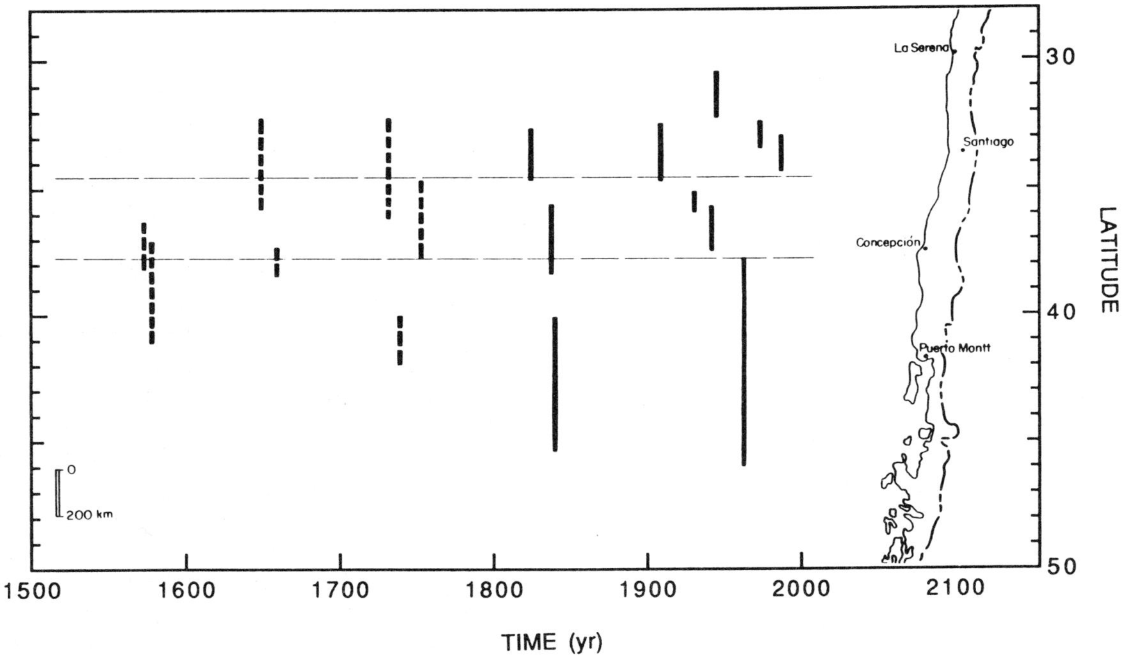

Fig. 2. Rupture lengths in south-central Chile. *Solid lines* (after 1800) represent rupture zones which are better constrained than the pre-1800 zones (*dashed lines*). Inter-occurrence periods of 90±6 years are determined for the region within the *horizontal dashed lines*. Within the gap a rupture zone in the north is followed, 14 ±6 years later, by a large event in the south

lines). When compared with the previous statistics, the next large event should occur earlier. In 1990 there was a 41% chance of it occurring within the next 10 years; this is 50 times greater than the previous estimate.

4 Stress Diffusion Model

The last line of evidence arises from an analysis of the whole sequence in the 1900s. The pattern 1906-1928-1939(?)-1960 (Fig. 3) suggests a long rupture in several stages, even though the 1939 event was probably not of the thrust type. The southward migration velocity of about 7 km/year is, depicted by the horizontal dashed line in Fig. 3. This southward migration is consistent with the two known rupture propagation directions determined for the 1960 (Press et al. 1961) and 1985 (Christensen and Ruff 1986; Monfret and Romanowicz 1986) events. Kelleher (1972) concluded that the region ruptures every century on a north to south progression of several earthquakes. Prior observations of the migration of seismic activity were noted by Mogi (1969) and particularly in the North Anatolian Fault by Toksöz et al. 1979). Furthermore, a propagating deformation front has been postulated to explain space-dependent seismic activity and water-well changes prior to the 1975 Haicheng earthquake (Scholz 1977).

If stress is applied to an elastic body, the stress front moves through the body at the velocity of elastic waves. A different medium is required to decrease the propagation velocity. The theoretical model proposed below is based on that published by Bott and Dean (1973) in which a horizontal two-dimensional elastic plate overlies a viscous asthenosphere (Fig. 5). A similar model was later proposed by Anderson (1975) to calculate the recurrence interval of great earthquakes at convergent boundaries and the separation of decoupling and lithospheric earthquakes. The rigidity modulus μ and thickness d characterize the elastic plate and the asthenosphere is characterized by the viscosity η and thickness b. It can be shown that considering the horizontal forces acting on a small length of the elastic plate, neglecting inertial forces, the

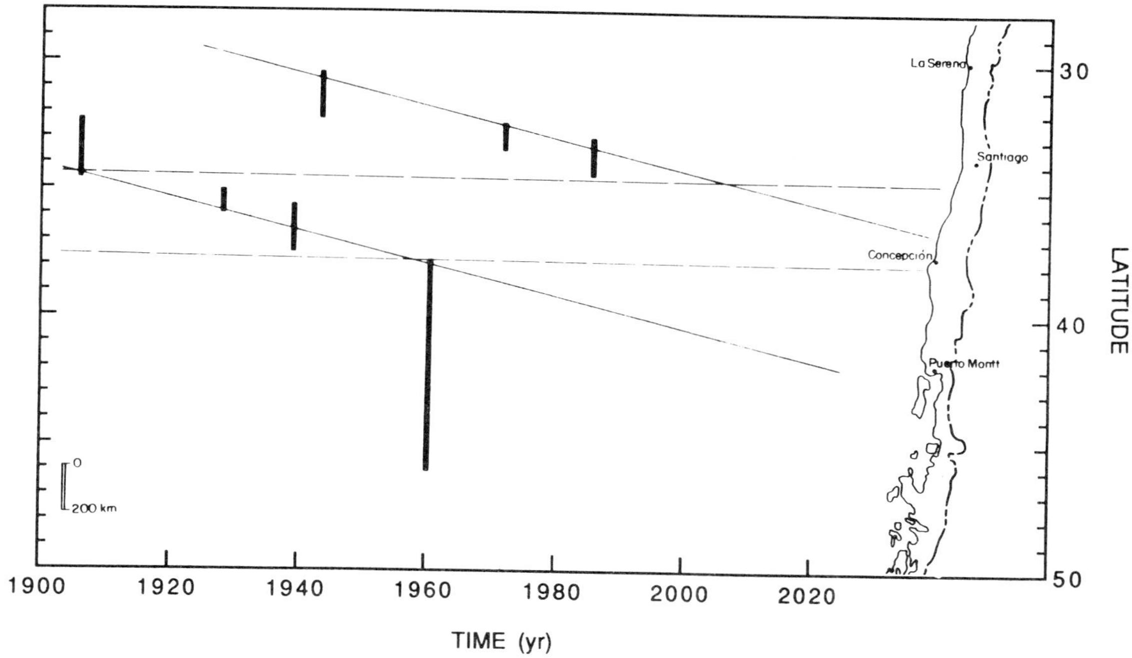

Fig. 3. Time-expanded version of Fig. 2. Two earthquake sequences are seen when only the post-1900 events are plotted. Two "rupture fronts" seem to activate all events in both sequences. The stress fronts, separated by about 100 years, propagate from north to south with a velocity of 7 km/year

displacement $u(x,t)$ is governed by

$$\frac{\partial^2 u}{\partial x^2} = \sigma \frac{\partial u}{\partial t}, \qquad (1)$$

in which $\sigma = \eta / bd\mu$ for an assumed linear velocity-depth distribution (Bott and Dean 1973).

Imposing a periodic perturbation on one end, i.e. $S(t) = S_0 \sin(\omega t)$, the appropriate solution of Eq. (1) is:

$$u(x,t) = -\frac{S_0}{\sqrt{2k\mu}} e^{-kx} \cos\left(\omega t - kx + \frac{\pi}{4}\right)$$

and

$$S(x,t) = S_0 e^{-kx} \sin(\omega t - kx),$$

in which k can be written as $\sqrt{\dfrac{\omega\eta}{(2bd\mu)}}$

and velocity v as $2\sqrt{\dfrac{\pi\mu\, bd}{\eta T}}$.

Table 1 summarizes the propagation velocity as a function of the period, assuming $d = 100$ km and $\mu = 5 \times 10^{10}$ Pa for the elastic lithosphere, and $b = 250$ km and $\eta = 10^{21}$ Pa for the viscous asthenosphere.

For this particular case, the recurrence period (horizontal distance between the two oblique lines in Fig. 3) is about 100 years (90 years from recurrence of large events) and the slope of each oblique line, which represents the velocity of the propagation pulse, is about 7 km/year, totally consistent with the values in Table 1. The oblique lines are restricted to intersect the only two known sites of rupture initiation, those of the 1960 and 1985 events.

Table 1. Parameters of stress diffusion

T (yr)	v (km/yr
1	70.4
10^2	7.04
10^4	0.704
10^6	0.0704

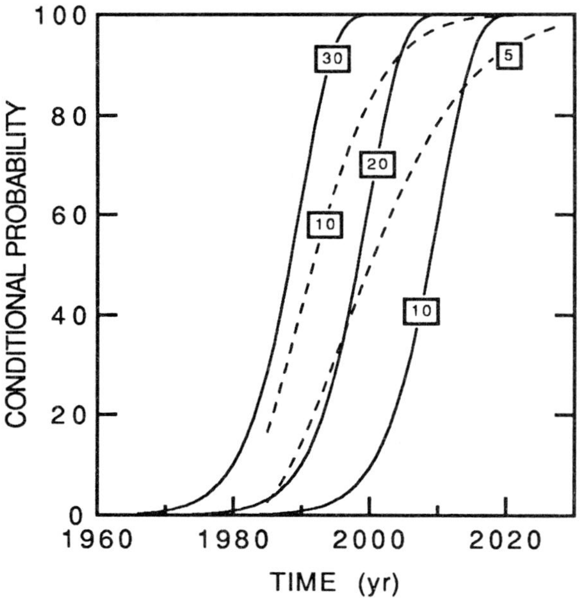

Fig. 4. Weibull conditional probabilities of earthquake occurrence as a function of time. *Solid lines* Intervals of 10, 20, and 30 years given that the last thrust event took place in 1928; *dashed lines* intervals of 5 and 10 years given that the 1985 event was the last one that occurred in the northern region

The propagating pulse now approaching the seismic gap region (upper oblique line) will arrive in around the year 2005 (±10).

5 Conclusions

The region between 34.3° and 37.2°S in central Chile, bounded by the 1960 and 1985 rupture zones, has regularly been exposed to large earthquakes. A recurrence period of 90 ± 5 years is deduced from the historical record that covers the last 400 years. Conditional probabilities of occurrence, assuming a two-parameter Weibull distribution, reach 63% for the next 30 years.

An independent criterion to estimate the time of occurrence of the next event in the region, based on the inter-occurrence time interval between shocks in the north and south within the seismic gap, indicates a 41% probability level for the next 10 years. This by no means excludes the possibility of earlier activity such as occurred in 1971 prior to the 1985 mainshock.

A stress diffusion model is used to explain the southward migration and rupture propagation direction of large events in central Chile. The velocity of the "stress pulse" is about 7 km/year. It will arrive at the seismic gap region in around the year 2005 (±10).

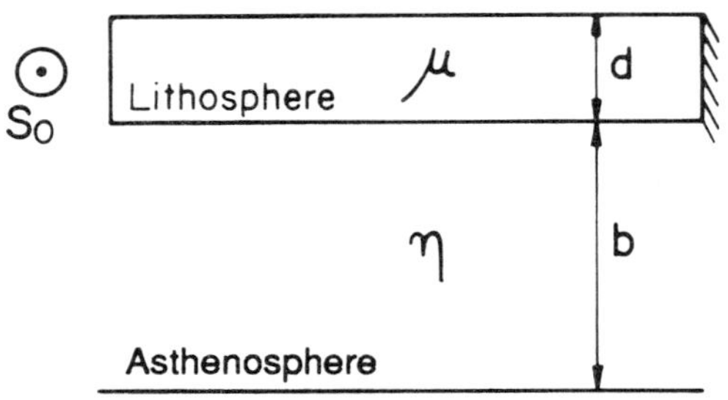

Fig. 5. The elastic lithosphere, of thickness d and rigidity μ, rests on the viscous asthenosphere of thickness b and viscosity η. One end of the plate is subjected to a shear periodic pulse, $S_0 \sin(\omega t)$, which propagates to the right at a velocity depending on the period of the pulse (Table 1)

The seismic gap extends for 320 km. If this region ruptures in two segments the associated probable magnitudes range from 8.0 to 8.2. But if only one large event ruptures the region, its moment would reach 7 x 10^{28} dyn-cm (M_W=8.4) according to the moment-rupture length relationship for Chilean earthquakes, proposed by Ramírez (1988).

Acknowledgements. This work was partially supported by the Fondo Nacional de Ciencia y Tecnología (FONDECYT) and by the American Association for the Advancement of Science under a grant from the MacArthur Foundation to support cooperation between US and Chilean scientists.

References

Anderson D (1975) Accelerated plate tectonics. Science 187:1077-1079

Barrientos S (1988) Slip distribution of the 1985 central Chile earthquake. Tectonophysics 145:225-241

Barrientos S, Ward SN (1990) The 1960 Chile earthquake: slip distribution from surface deformation. Geophys J Int 103:589-598

Bott MHP, Dean DS (1973) Stress diffusion from plate boundaries. Nature 243:339-341

Campos J, Kausel E (1990) The large 1939 intraplate earthquake of southern Chile. Seismol Res Lett 61:43

Choy G, Dewey J (1988) Rupture process of an extended sequence: teleseismic analysis of the Chilean earthquake of March 3, 1985. J Geophys Res 93:1103-1118

Christensen DG, Ruff LJ (1986) Rupture process of the March 3, 1985 Chilean earthquake. Geophys Res Lett 13:721-724

Comte D, Eisenberg A, Lorca E, Pardo M, Ponce L, Saragoni R, Singh S.K, Suárez G (1986) The 1985 central Chile earthquake: a repeat of previous great earthquake in the region? Science 233:449-453

Darwin C (1851) Geological observations on coral reefs, volcanic islands and on South America: being the geologist of the voyage of the Beagle, under the command of Captain Fitz Roy RN, during the years 1832-1836. London, 768 pp

Fedotov SA (1965) Regularities of the distribution of strong earthquakes in Kamchatka, the Kuril Islands, and northeast Japan. Tr Inst Phys Earth Acad Sci USSR 36:66-93

Gutenberg B, Richter CF (1954) Seismicity of the earth and associated phenomena, 2nd edn., Princeton University Press, Princeton, 310 pp

Kanamori H (1977) The energy release in great earthquakes. J Geophys Res 82:2981-2987

Kausel E, Campos J (1989) The Ms = 8 tensional earthquake of December 9, 1950, in northern Chile and its relation to the seismic potential of the region. IASPEI Abstr p 235

Kelleher J (1972) Rupture zones of large South American earthquakes and some predictions. J Geophys Res 77:2087-2103

Korrat I, Madariaga R (1986) Rupture of the Valparaíso (Chile) gap from 1971 to 1985. In: Das S, Boatwright J, Scholz CH (eds) Earthquake source mechanics, AGU, Washington D C, pp 247-258

Linde AT, Silver PG (1989) Elevation changes and the great 1960 Chilean earthquake: support for a seismic slip. Geophys Res Lett 16:1305-1308

Lomnitz C (1971) Major earthquakes and tsunamis in Chile during the period 1535 to 1953. Geol Rundsch 59:938-960

Lomnitz C (1983) On the epicenter of the great Santiago earthquake of 1647. Bull Seismol Soc Am 73:885-886

McCann WR, Nishenko SP, Sykes LR, Krause J (1978) Seismic gaps and plate tectonics: seismic potential for major plate boundaries. In Proceedings of Conference VI. Methodology for identifying seismic gaps and soon-to-break gaps. US Geol Surv Open-File Rep 78-943: 441-584

Mogi K (1965) Sequential occurrences of recent earthquakes. J Phys Earth 16:30-36

Mogi K (1969) Some features of recent seismic activity in and near Japan. Bull Earthq Res Inst Tokyo Univ 46:1225-1236

Monfret T, Romanowicz B (1986) Importance of on scale observation of first arriving Rayleigh wave trains for source studies: example of the Chilean event of March 3, 1985, observed in the GEOSCOPE and IDA networks. Geophys Res Lett 13:1015-1018

Montessus de Ballore F (1911) Historia sísmica de los Andes Meridionales al Sur del Paralelo XVI. Litografía Barcelona, Santiago, vol 5, 407 pp

Nishenko SP (1985) Seismic potential for large and great interplate earthquakes along the Chilean and southern Peruvian margins of South America: quantitative reappraisal. J Geophys Res 90:3589-3615

Nishenko SP, McCann WR. (1981) Seismic potential for the world's major plate boundaries. In: Simpson DW, Richards PG (eds) Earthquake prediction: an international review. Maurice Ewing Ser, vol 4, AGU, Washington D C, pp 20-28

Plafker G, Savage J C (1970) Mechanism of the Chilean earthquakes of May 21 and 22, 1960. Geol Soc Am Bull 81:1001-1030

Press F, Ben-Menahem A, Töksoz M N (1961) Experimental determination of earthquake fault length and rupture velocity. J Geophys Res 66:3741-3485

Ramírez D (1988) Estimación de algunos parámetros focales de grandes terremotos históricos chilenos. Tesis para optar al grado de Magister en Geofísica, Universidad de Chile, Santiago

Richter CF (1958) Elementary seismology. Freeman, San Francisco, 768 pp

Scholz CH (1977) A physical interpretation of the Haicheng earthquake prediction. Nature 267:121-124

Toksöz MN, Shakal AF, Michael AJ (1979) Space-time migration of earthquakes along the North Anatolian fault zone and seismic gaps. Pure Appl Geophys 117:1258-1270

GEOLOGICAL EVOLUTION

Tectonic Development of the North Chilean Andes in Relation to Plate Convergence and Magmatism Since the Jurassic

EKKEHARD SCHEUBER, TOMISLAV BOGDANIC, ARTURO JENSEN and KLAUS-J. REUTTER

Abstract. Since the early Jurassic the magmatic arc of the north Chilean Andes has been displaced from the Coastal Cordillera to the Western Cordillera. This eastward migration happened stepwise and four successive, ± stationary arc systems can be distinguished. The deformation history of the arc systems in relation to plate convergence and igneous activity shows that the magmatic arc, a zone of relative crustal weakness, reacted very sensitively to changing conditions of plate convergence. Both long-term continuous deformations and short-term tectonic events are recognized. They reflect periods of more or less steady state conditions and relatively sudden changes of subduction parameters respectively. Two major periods of different deformational styles, related to differing plate configurations and accompanying convergence obliqueness, can be distinguished: (1) 200-90 Ma (sinistral convergence obliqueness >45°) with general (trans-)tension during tectonic phases and interphases, and (2) since 90 Ma (dextral convergence obliqueness <45°) with transpression during phases and slight extension in interphases.

1 Introduction

The structural evolution of the Central Andes has largely been governed by the convergent motion between the South American upper plate and the subducting oceanic plate system of the Pacific. There is no doubt that the strain pattern within the edge of the upper plate is controlled by the parameters of plate convergence and, hence, if these parameters change with time, the strain pattern will change too. Therefore, it should be possible to ascribe the varying stages of geological evolution of the Andes to special and changing conditions of plate convergence.

Jarrard (1986) grouped the strain regimes within the overriding plates of modern subduction zones into seven strain classes ranging from the most extensional class 1 to the most compressional class 7 (Table 1). He showed that the best prediction of the strain regime occurs with a combination of the following three independent variables: (1) (trench-normal component of) convergence rate, (2) intermediate slab dip (from the trench to the 100-km depth), and (3) either slab age or absolute motion of the overriding plate with respect to the trench. All these variables operate in directions consistent with the hypothesis that the dependent variable coupling between the plates at their common boundary has the dominating influence on strain regime. As these strain regimes are considered a continuum, it should be possible to attribute a strain regime number between 1 and 7 to

Table 1. Strain classification of modern arc systems (after Jarrard 1986)

Strain Class	Description	Examples in modern subduction zones
1	active backarc spreading	Marianas,Tonga
2	incipient or very slow backarc spreading; high heat flow, thinned continental crust and thick sediment fill with growth-faulted grabens	Ryukyu, Izu-Bonin
3	mildly tensional, arc volcanism within an actively subsiding region, graben formation	Middle America New Zealand
4a	neutral, little evidence of either compression or extension	Lesser Antilles Cascades
4b	gradient: arc-forearc compression, backarc extension	Aleutians Alaska Peninsula
5	mildly compressional, gentle folds and thrusts	SW Japan, Java Sumatra, S Chile
6	moderately compressional, moderate folds, reverse faults	Colombia, Ecuador Peru, Alaska
7	very strong compressional, strong folding, imbricate thrusts	North Chile Central Chile

Correspondence to: E. Scheuber, Fachrichtung Geologie, Freie Universität Berlin, Malteserstr. 74-100, D-1000 Berlin 46

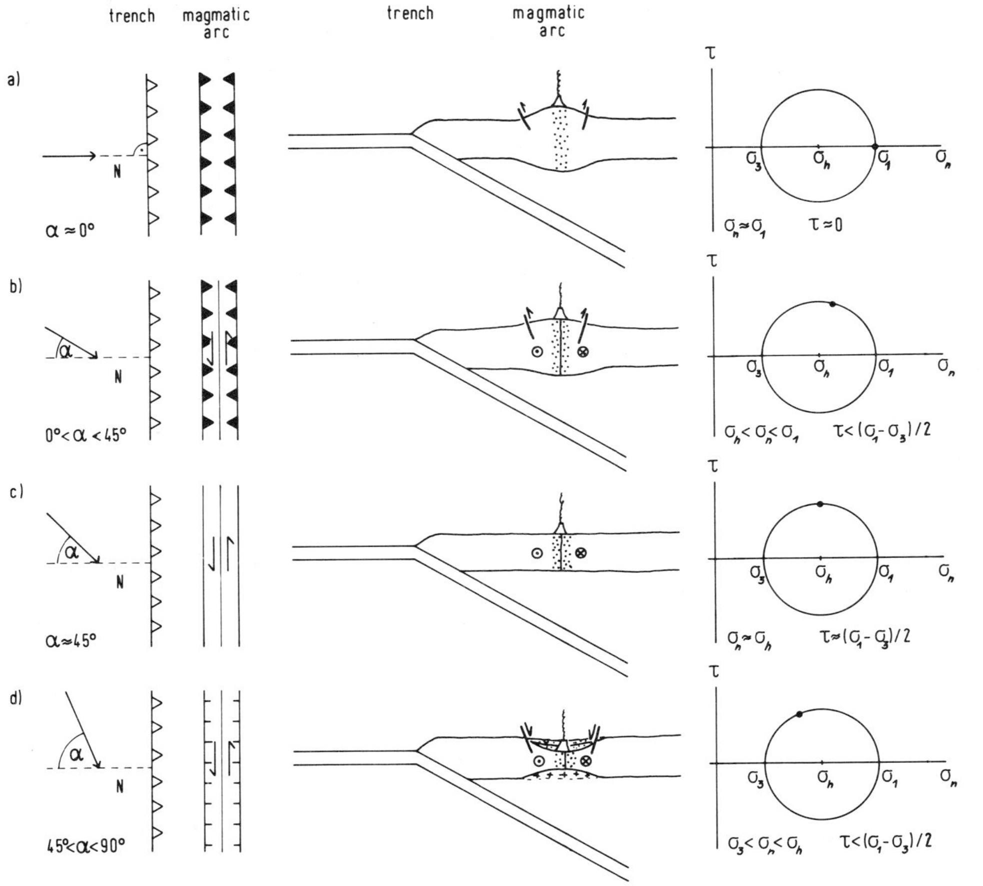

Fig. 1. The effect of subduction obliqueness on the deformational regime in the crust of the upper plate (prerequisites: convergence rate and plate coupling are sufficiently high). *Left column* map view; *middle column* cross section; *right column* Mohr circles showing the normal and shear stresses acting on the magmatic arc, which is reduced to a vertical plane oriented parallel to the plate boundary. In addition to arc-parallel shearing, obliqueness also leads to arc-normal movements: shortening occurs if obliqueness is <45° and extension if obliqueness is >45°. See text for explanation of **a-d**

all the different and transient palaeogeological situations in the evolution of the Central Andes and to draw conclusions about the effective parameters of convergence.

Another very important factor for the deformational regime operating within the upper plate is the angle between the plates' movement vector and the trench normal (convergence obliqueness) which, in the upper plate, may cause strike-slip movements parallel to the plate boundary (Fitch 1972; trench-linked strike-slip faults, Woodcock 1986). Such trench-linked strike-slip faults are indicative of the degree of coupling at the plate boundary, because weak coupling would cause lateral movements in the subduction zone (Beck 1983), while strong coupling produces strike-slip faulting within the forearc and/or arc of the upper plate. Such movements have occurred several times in the Central Andean history (Reutter and Scheuber 1988).

However, oblique convergence also causes movements normal to the plate boundary. Scheuber and Reutter (1992) have proposed a model according to which the angle of convergence obliqueness also determines compressive or tensional stress regimes in the upper plate (Fig. 1). If obliqueness is small ($\approx$ 0°, Fig. 1a), the normal stress acting on the magmatic arc which represents a relatively weak vertical zone oriented parallel to the plate boundary, is about equal to σ_1 (parallel to the plate's motion vector); this results in orogen-normal shortening and crustal thickening. In the case of obliqueness between 0 and 45° (Fig. 1b), a shear stress is set up along the magmatic arc with τ_{max} at $\alpha = 45°$, while the normal stress is smaller than σ_1 but exceeds the hydrostatic stress. This setting should result in shortening plus strike-slip (transpression). The special case of $\alpha = 45°$ should produce pure trancurrence (Fig. 1c); obliqueness of $\alpha > 45°$ should lead to a normal stress that is smaller than the hydrostatic stress and also to a shear stress component so that extension plus strike-slip (transtension) are expected (Fig. 1d). The effects of oblique convergence, proposed in this model, probably add to the strain produced by the other parameters of plate convergence mentioned above.

In this chapter, based on field studies in northern Chile within the segment between 21° and 25°S, we

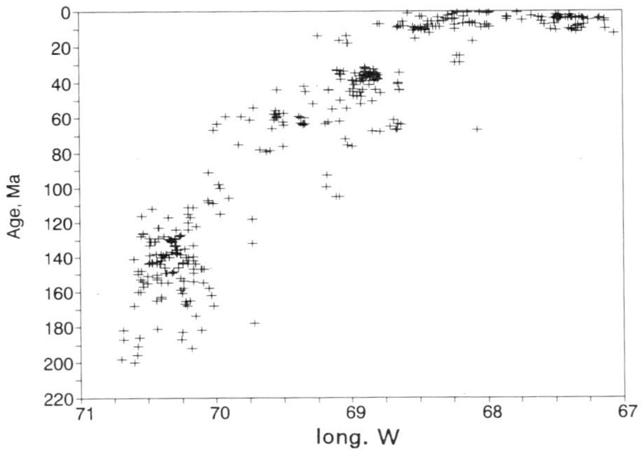

Fig. 2. Compilation of available isotope age data from north Chile between 21° and 26°S (406 age values included). The diagram shows the eastward migration of igneous activity since the early Cretaceous. Data from: compilation by Maksaev (1990): 297 age values (from various authors, 1965-1989); datings by Maksaev (1990), 58 age values; datings by Döbel et al. (1992), 18 age values; datings by Scheuber and Hammerschmidt (1991), 23 age values; datings by Pichowiak (this Vol.), 3 age values; datings by Andriessen and Reutter (this Vol.), 7 age values

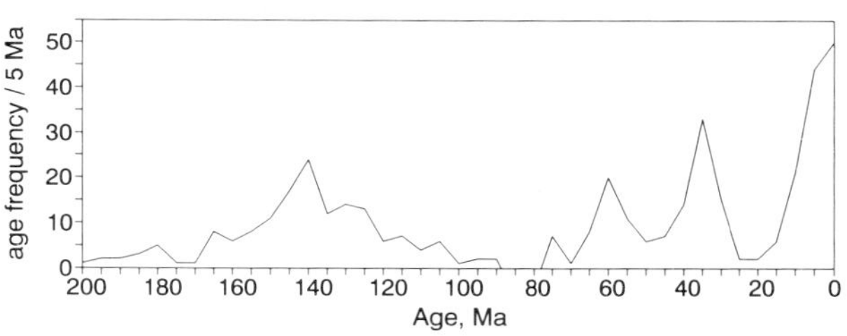

Fig. 3. The frequency of isotope age data in classes of 5 Ma (data base same as in Fig. 2)

consider only the deformations which have affected the active continental border since the early Jurassic (Andean Cycle, Coira et al. 1982). From that time to the Holocene, plate convergence was probably continuous, although its parameters have been subject to considerable variations. These variations influenced the strain regime and, hence, the tectonic setting of the magmatic arc and its respective forearc and backarc areas. In the study area, the wellknown (e.g. Coira et al. 1982) and well documented shift of the axis of the magmatic arc from west to east during the Andean Cycle is perhaps the best example of changing conditions at the active continental margin.

The compilation of all available isotope age data for northern Chile (Fig. 2) suggests that this migration was not characterized by jumps, but was a gradual process. Nevertheless, times of accelerated migration of igneous activity and separations of high and low activity periods (Fig. 3) allow the distinction of at least four arc systems (Fig. 4): (1) a Jurassic-early Cretaceous arc in the Coastal Cordillera, (2) a mid-Cretaceous arc in the Longitudinal Valley, (3) a late Cretaceous-Palaeogene arc in the Chilean Precordillera, and (4) the modern arc in the Western Cordillera.

These arc systems and their adjacent forearc and backarc areas had a specific and gradually developing tectonic evolution, which was sometimes interrupted by short-term events either during the lifetime of an arc system or between one arc system and the next one. Figure 5 gives a synopsis of the structural evolution of the north Chilean magmatic arc system since the Jurassic. In the following sections, this tectonic history will be reviewed and the attempt will be made to characterize the prevailing strain regimes and draw conclusions about the changing conditions of plate convergence.

2 The Jurassic-Early Cretaceous Arc System

The centre of igneous activity of the Jurassic-early Cretaceous arc system was situated in the Coastal Cordillera. Isotope ages range between 200 and 90 Ma, indicating that there was some areal overlapping with the 110 to 90 Ma old mid-Cretaceous arc system, whose activity was centred farther to the east in the Longitudinal Valley. To the east the Jurassic-early Cretaceous magmatic arc was bounded by a backarc basin which was installed upon an older continental crust. Subsidence started in the late Triassic and its deposits were marine until the

Table 2. Amount of rock units in the Coastal Cordillera between 20°15'-25°S

rock unit	km2	per cent
volcanic rocks	9436	37
plutonic rocks	10222	40
total igneous rocks	*19658*	*77*
pre Andean units	4171	16
younger units	1652	7
total	25481	100

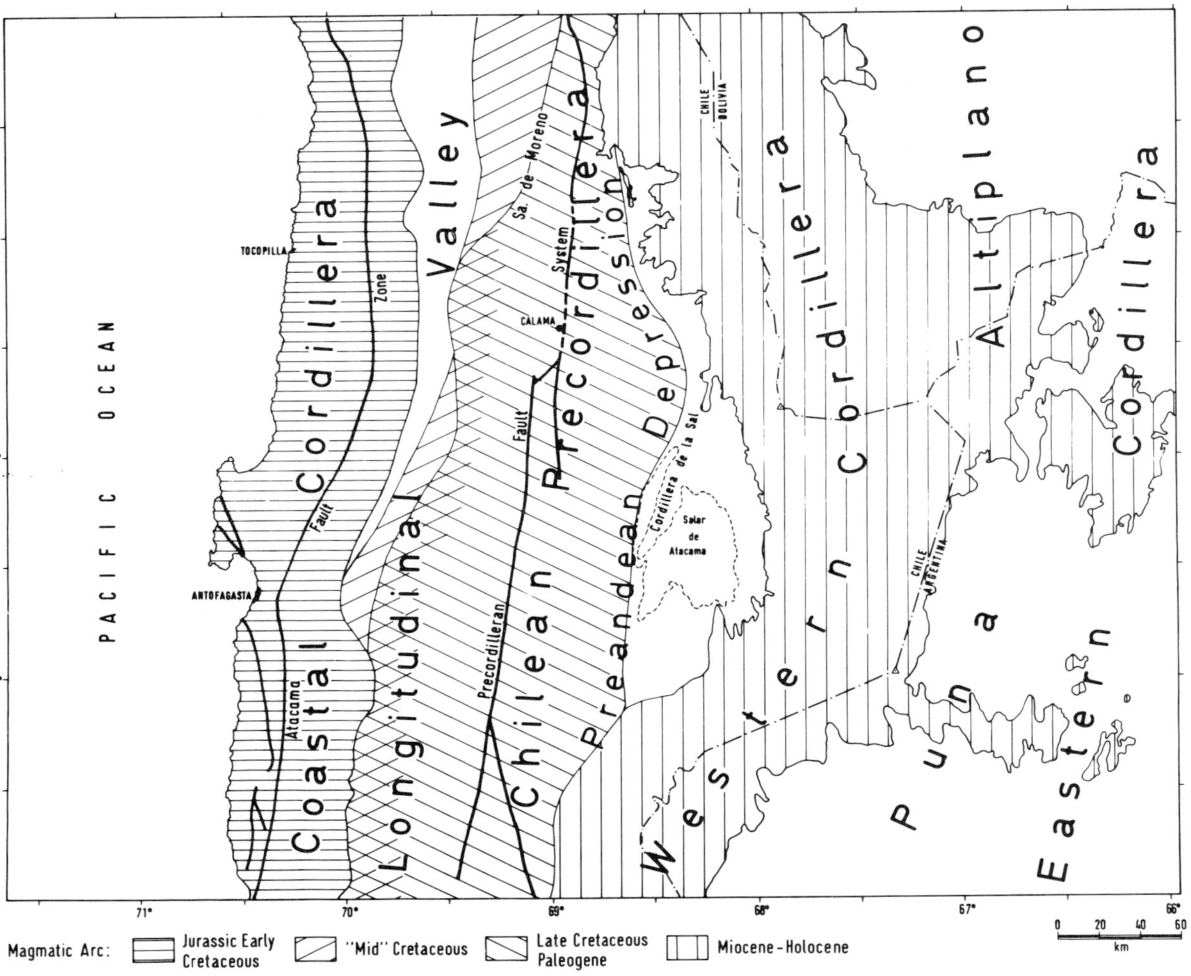

Fig. 4. The distribution of the four magmatic arcs that developed in the southern Central Andes (20°-26°S) since the early Jurassic (Andean Cycle, Coira et al. 1982)

Kimmeridgian when they gradually became continental. The forearc lies offshore and is possibly made up of Palaeozoic accretionary complexes.

The Jurassic-early Cretaceous magmatic arc is composed of large quantities of basic to intermediate igneous rocks consisting of lavas (La Negra Formation), large and small plutons and of numerous andesitic to dacitic subvolcanic stocks and dykes. The Jurassic-early Cretaceous volcanic and plutonic rocks cover some 77% of the area of the Coastal Cordillera (37% volcanics, 40% plutonic rocks, Table 2), although dykes and smaller sheet-like intrusions are not considered. The volcanics are calc-alkaline basalts to andesites. Their average thickness is about 3800-5000 m (Boric et al. 1990), however, in some places, e.g. near Antofagasta, it may exceed 10 km (Buchelt and Tellez 1988). There are several facts that indicate a deposition of the volcanics more or less at sea level: Early Sinemurian and Bajocian marine intercalations are found within the volcanics of the Coastal Cordillera north of 21°S and south of 25°S (Davidson et al. 1976; Naranjo and Puig 1984). Near Arica (18°30'S) the volcanics are conformably overlain by marine upper Oxfordian to lower Kimmeridgian sediments (García 1967). The deposition of the volcanics was thus coupled to strong subsidence of at least parts of the arc crust, which may be attributed to graben structures and pull-apart basins. The constant lateral thickness of the single lava flows of some 10 m, and the fact that volcanic breccias have not been described, points to an extrusion of the volcanic products by fissure eruptions

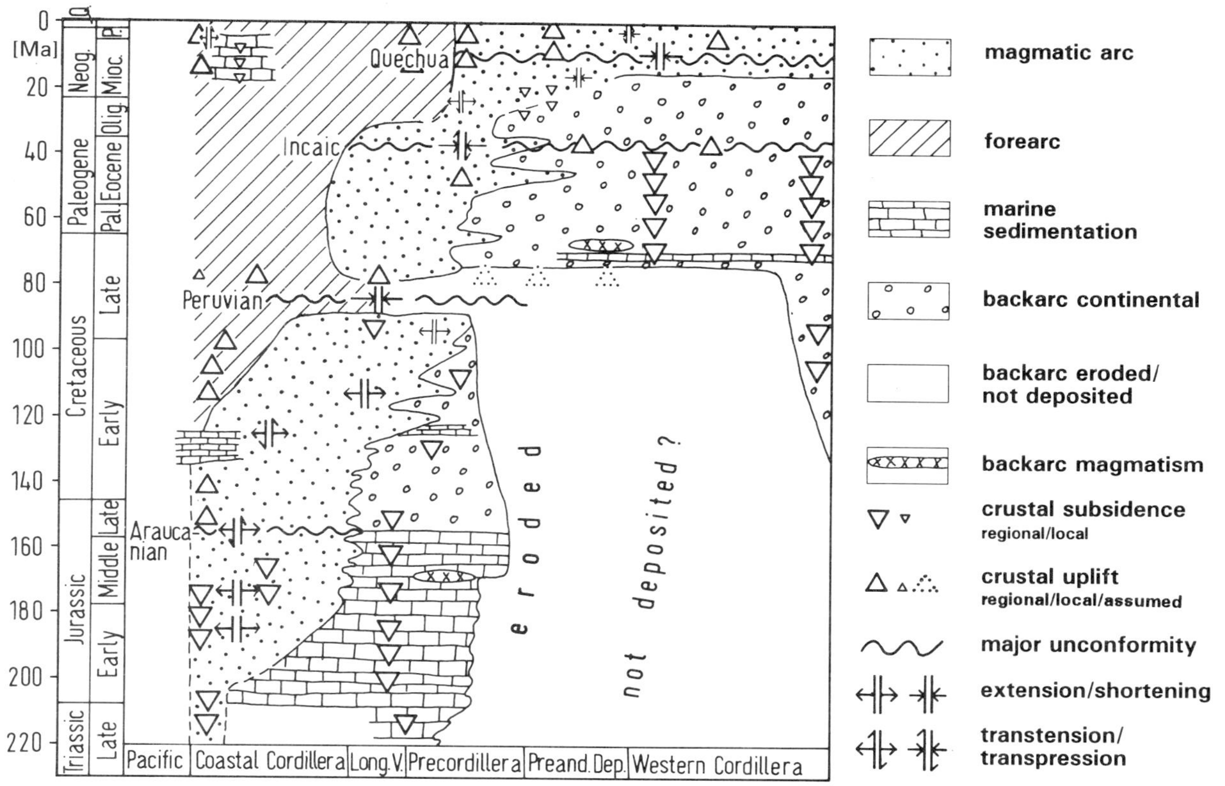

Fig. 5. Major aspects of the geological and structural development of the north Chilean Andes since the early Jurassic

rather than by single feeders. Volcanism was contemporaneous with the intrusion of huge batholiths. Geochemical data show that the plutonics are deep-level equivalents of the volcanics (Pichowiak et al. 1990). Throughout the Jurassic-early Cretaceous igneous rocks are mantle derivates without or with extremely little contamination by continental crust (e.g. Sr_i of about 0.703, Pichowiak this Vol.).

2.1 Deformation history of the Jurassic-early Cretaceous magmatic arc

The magmatic and tectonic activity were contemporaneous in this arc system. In the arc itself, extension normal to the arc and strike-slip movements parallel to the arc can be detected, to such a degree that a transtensional stress and deformation pattern becomes evident.

Arc-normal extension does not directly manifest itself in tectonic structures, but can be deduced from the following features:

(1) In some places numerous mafic to felsic dykes and subvolcanic stocks cover >40% of the area between the main branches of the Atacama Fault Zone (Fig. 6). Most of the dykes are oriented parallel to the trace of this fault zone (NNE-SSW).

(2) Frequent N-S linearity of plutonic intrusives (Rössling 1988: see geological map) points to an extensional regime at the time of their emplacement.

(3) The presence of mantle-derived gabbroic to dioritic intrusions at rather shallow levels is indicative of crustal thinning. The large Coloso Gabbro south of Antofagasta shows cumulate layering. Phanerozoic layered gabbros are normally related to rift zones or spreading centres (cf. Hyndman 1985), where they represent the pulses of opening of the magma chamber.

(4) Outcrops of Preandean rocks which show that the Coastal Cordillera was built up by continental material before the Jurassic make up only some 15% of the area of the Coastal Cordillera (Table 1) whereas Jurassic-early Cretaceous igneous rocks constitute some 77%. As these igneous rocks do not

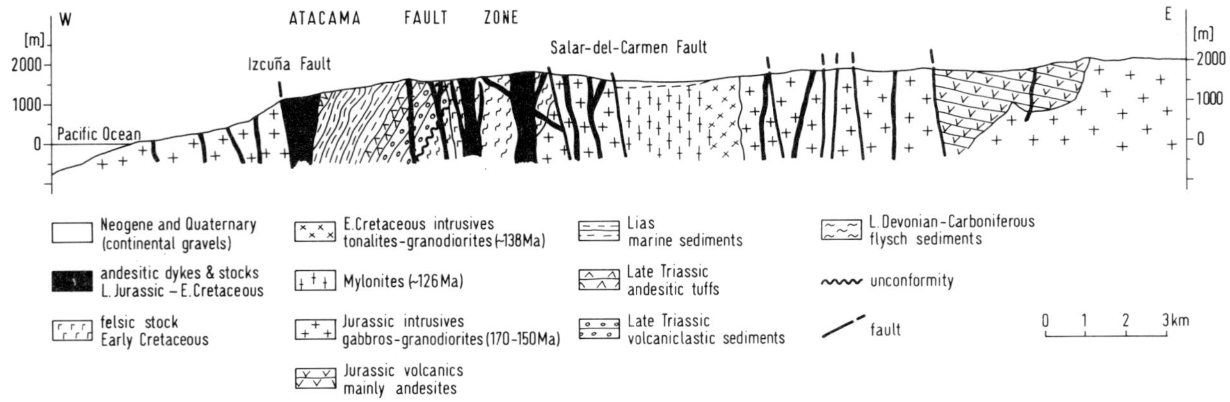

Fig. 6. Cross section through the Coastal Cordillera at 24°53'S showing the concentration of dykes and smaller sheet-like intrusions to the area of the Atacama Fault Zone

show a significant contamination by continental material it can be concluded that the originally existing continental crust has been replaced by this mantle derived material. Even where deep levels are exposed (e.g. south of Antofagasta) no remnants of the basement are found. High P-wave velocities of about 6.8 km/s in the Coastal Cordillera extending to depths of about 30 km (Wigger 1988) also exclude a normal continental basement beneath the Jurassic-early Cretaceous igneous units.

(5) In the western part of the Coastal Cordillera Palaeozoic-early Jurassic strata are homoclinal and dip steeply to the west or east (Fig. 6), but they are completely devoid of tight or even isoclinal folds and/or repetitions of strata successions that would be necessary to explain the steep dips by shortening. Thus the homoclinal dip of the beds can only be interpreted by block rotation due to crustal extension.

(6) Strong crustal subsidence affected not only the arc but also the backarc, implying that crustal thinning as a consequence of an extensional stress regime affected an area at least 200 km wide.

Arc-parallel strike slip movements generated a belt of foliated rocks that is linked to the Atacama Fault Zone (AFZ). The AFZ is considered to be the major and most continuous structure of the Coastal Cordillera (Arabasz 1971) and can be traced over more than 1000 km from ≈20° (Iquique) to ≈30°S (La Serena). Naranjo et al. (1984) reported the existence of mylonite zones from the AFZ north of Chañaral with early Cretaceous K-Ar ages (hornblende: 126 ± 10 Ma). Scheuber et al. (1986) mentioned mylonite zones which contain S-shaped vertical folds indicating sinistral strike-slip movements to have occurred along the AFZ. Hervé (1987) mapped plutons which show a sinistral displacement along some branches of the AFZ (K-Ar whole rock age of a mylonite: 139 ± 5 Ma). In a study of high to low grade mylonites from the AFZ Scheuber (1987), and Scheuber and Andriessen (1990) showed that microstructural features such as S-C fabrics, S-bands, asymmetric porphyroclast systems, and quartz-c axes preferred orientation reveal a uniformly sinistral sense of shear. South of Antofagasta two sets of ductile shear zones are found along the AFZ, a Jurassic one deformed under amphibolite facies conditions and an early Cretaceous one formed in the greenschist facies. For both sets of shear the age of deformation has been determined using the Rb-Sr and the $^{40}Ar/^{39}Ar$ methods (Scheuber and Hammerschmidt 1991). For the late Jurassic shear zones two deformation steps could be determined, one before 152 ± 1 Ma (hornblende $^{40}Ar/^{39}Ar$, biotite Rb-Sr) and one at 143 ± 0.3 Ma (biotite Rb-Sr and $^{40}Ar/^{39}Ar$). The deformation age of the early Cretaceous greenschist facies shear zone is 125.3 ± 0.3 Ma (biotite Rb-Sr and $^{40}Ar/^{39}Ar$). The ages are contemporaneous to the period of major intrusive activity in the Coastal Cordillera (Fig. 3) illustrating the combined action of magmatism and tectonism. The close temporal relationship between intrusion and deformations can also be inferred from transitions of magmatic flow structures to structures of plastic deformation in late Jurassic shear zones as described by Gonzalez (1990).

The deformation ages of the late Jurassic shear zones (Araucanian tectonic event, Riccardi 1990, and references therein) correspond to a major angular unconformity and the beginning of coarse conglom-

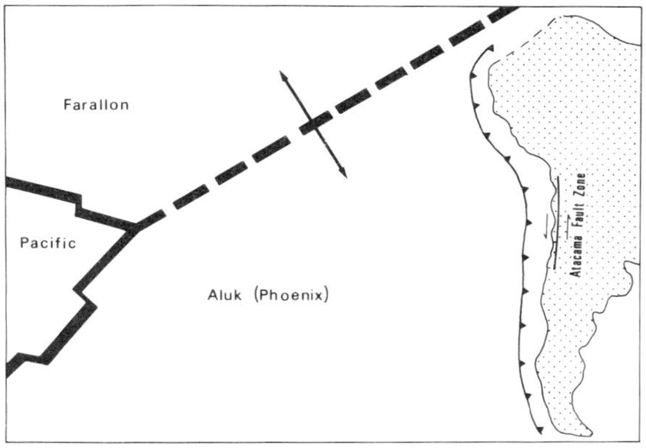

Fig. 7. Reconstruction of the plate configuration in the SE Pacific at ≈150 Ma. (After Larson and Pitman III 1972; Zonenshayn et al. 1984) Sinistral subduction obliqueness exeeded 45° resulting in a transtensional regime (cf. Fig. 1 d)

eratic sedimentation in the Coastal Cordillera (Fm. Caleta Coloso, Tithonian-Valanginian) north of 24°. The 125 Ma deformation of the greenschist facies shear zone occurred at the beginning of strong and rapid uplift of the Coastal Cordillera starting between 130 and 120 Ma ago (Maksaev 1990; Scheuber and Andriessen 1990; Andriessen and Reutter this Vol.).

In contrast to the magmatic arc, the narrow marine backarc basin was characterized by tectonic quiescence to some very slight backarc rifting which is indicated by minor occurrences of middle Jurassic basalts in the Chilean Precordillera. Early Cretaceous alkaline igneous rocks from northwest Argentina also point to foreland rifting (Galliski and Viramonte 1988). The thickness distribution of Jurassic backarc deposits indicates two periods of greater subsidence rates, one in the Sinemurian-Toarcian corresponding to the installation of the magmatic arc, and one in the Kimmeridgian during which a gradual change from marine to continental deposition took place (Prinz et al. this Vol.). In the magmatic arc this change corresponds to the culmination of intrusive and tectonic activity at ≈152 Ma. For the strong 125 Ma strike-slip movements no corresponding tectonic features have been described from the backarc basin. In summary, during the Jurassic-early Cretaceous, tectonic activity was largely restricted to the area of the magmatic arc which accommodated most of the imposed regional strain rate, due to a strongly reduced strength as a consequence of heating and intrusion of liquids. The backarc area only reflected, by subsidence, the crustal extension of the arc.

The deformations in the Jurassic-early Cretaceous arc system are consistent with available data of plate configurations of that time (Larson and Pitman III 1972, Zonenshayn et al. 1984). The Aluk (Phoenix) plate moved with a very high angle of obliqueness (≈60°) against South America (Fig. 7). From this, according to the model outlined above, the transtensional regime (normal stress < hydrostatic stress) can be deduced, and this is in agreement with the observed structures. The high angle of obliqueness also corresponds to the Jurassic tectonics of Peru (Jaillard et al. 1990), north of the Bolivian orocline. Here a subduction-related volcanism is only locally developed and sinistral strike-slip movements and extensions normal to the plate boundary are the prevailing deformations. Jaillard et al. (op.cit.) suggest that these tectonics are consistent with a sinistral transform plate boundary.

3 The Mid-Cretaceous Magmatic Arc System

During the early Cretaceous, the centre of igneous activity shifted eastward to a position in the previous backarc basin and the present Longitudinal Valley, although it extends into the adjacent parts of the Coastal Cordillera and Chilean Precordillera. A sequence of andesitic lavas about 2000 m thick (Empexa-Fm., Galli 1957; Estratos del Río Seco, Quebrada-Mala Fm., Charrier and Muñoz this Vol.), was deposited conformably upon a sequence of clastic sediments up to 3000 m thick with marine and some volcanic intercalations (Kimmeridgian-Barremian: Western Sequence, Bogdanic 1990). The migration of the arc from the Coastal Cordillera towards the Longitudinal Valley was a gradual process. Towards the south, where both arcs overlap, there is no unconformity between the extrusive products of the two arcs, and thus, the distinction from the former arc becomes somewhat arbitrary. The same is true for the character of magmatism which shows great similarities with the Jurassic-early Cretaceous one (Pichowiak this Vol.).

The lavas frequently alternate with sediments that are partly marine to the south of the segment considered here (Formación Aeropuerto, Naranjo and Puig 1984). The lower age limit of the mid-Cretaceous volcanics is given by underlying partly marine sediments of Hauterivian-Barremian age, while the upper limit is constrained by the 76 to 78-Ma San Cristobal intrusive complex (Maksaev 1990, Pichowiak this Vol.) and unconformably overlying volcanics of the late Cretaceous-Palaeogene arc (Fm. Chile-Alemania,

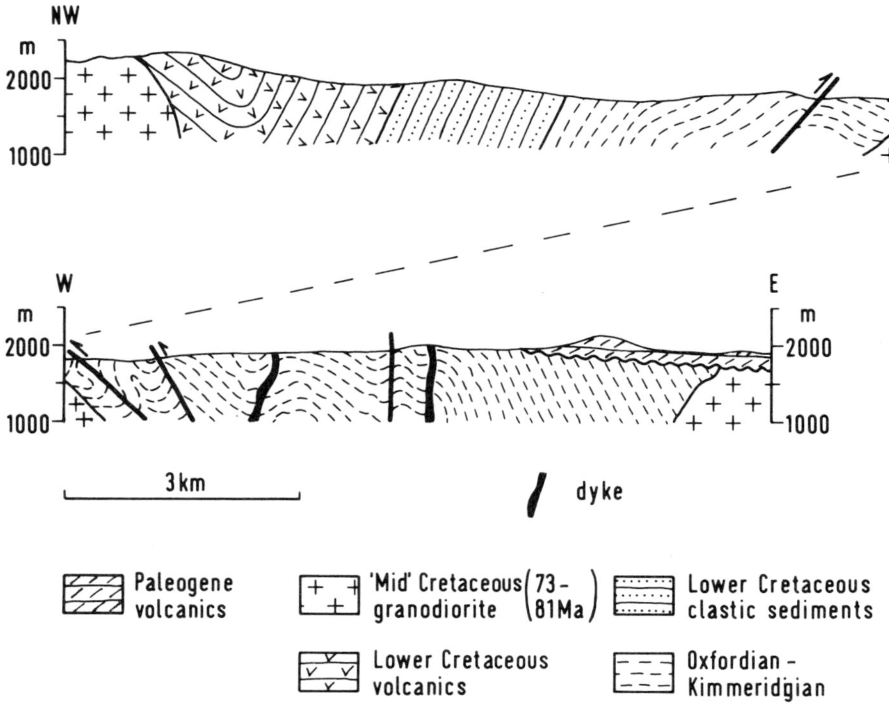

Fig. 8. Cross sections of the Longitudinal Valley east of Antofagasta at 23°45'S, 69°30'W showing the tectonics of the mid-Cretaceous magmatic arc. Lower Cretaceous volcanics and sediments were deposited conformably upon Jurassic-Lower Cretaceous sediments. The whole sequence was folded during the Peruvian phase (between 90 and 80 Ma)

oldest isotope age: ≈72 Ma, Naranjo and Puig 1984; Herrmann and Zeil 1989). This stratigraphic position corresponds to geochronological data of ≈115-90 Ma for the igneous rocks (Ulriksen 1979; Marinovic and Lahsen 1984; Rogers 1985; Döbel 1989; Andriessen and Reutter this Vol.). The mid-Cretaceous magmatic arc was followed by a gap in igneous activity that lasted some 10 Ma. For the time between 90 and 80 Ma, no isotope ages are reported from either the study area or the neighbouring areas in north Chile, northwest Argentina or southwest Bolivia. The gap in igneous activity may be due to the passing by of the Aluk-Farallon spreading centre (see below).

Information about the tectonics related to this arc system is rather scarce because most parts of the mid-Cretaceous arc are covered by younger formations of the Chilean Longitudinal Valley. As the mid-Cretaceous magmatic arc was installed within a subsiding basin without any marked angular unconformity between the sedimentary substrate and the lavas (Fig. 8), it may be concluded that extensional tectonics of the arc area (and former backarc area) continued up to the mid-Cretaceous. An internal angular unconformity within the volcanic-sedimentary sequences (between Fm. Quebrada Mala and Estratos del Rio Seco: Charrier and Muñoz this Vol.) can be attributed to the extensional tectonics. Large scale extensional tectonics of probably Aptian to Cenomanian age (124.5-90.4 Ma, Harland et al. 1990) have also been described by Mpodozis and Allmendinger (1991) for the Chilean Precordillera east of Copiapó (≈27°S).

In contrast to the preceding arc system, no directly adjoining backarc basin was developed. It is probable that, to the east, the arc bordered on hilly lowlands which did not receive any sedimentation during that time. However, about 400 km to the east, in the Bolivian Altiplano and the Puna of northwestern Argentina, a sedimentary basin developed, where mostly continental and marine (Cenomanian, Bolivia) sediments were deposited (Marquillas and Salfity 1988). The extensional nature of these depocentres is documented by mid-Cretaceous basaltic extrusions (120-90 Ma, Bossi and Wampler 1969, Valencio et al. 1976).

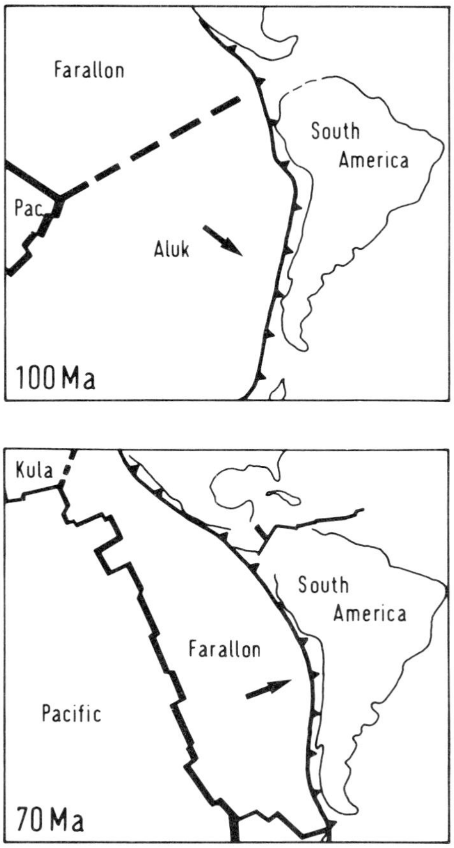

Fig. 9. The change in the SE Pacific plate configuration in the late Cretaceous. (After Zonenshayn et al. 1984)

The extensional strain regime was replaced by a compressional one with a tectonic event during the late Cretaceous. Deformation affected an area about 100 km wide corresponding more or less to that of the magmatic arc whose volcanic activity was terminated. The whole Jurassic-early Cretaceous sequence was subject to strong orogen-normal shortening which led to intense folding and thrusting, partly affecting also the pre-Jurassic basement (Fig. 8, Jensen 1985). In some places, e.g. east of Antofagasta, mid-Cretaceous volcanic rocks developed a foliation and axial plane cleavage is observed in upright folded Jurassic sediments. Strike-slip deformations within the arc have not yet been reported either in relation to the mid-Cretaceous extensional tectonics or in relation to the late Cretaceous compressional tectonics, but this may be due to the poor knowledge of this arc area. However, along major structures strike-slip movements may have occurred within the arc.

Similar to the late Jurassic events, the backarc area of the mid-Cretaceous system was not affected by the compressional tectonics, in either its elevated or its basinal parts. This is evidenced by the lack of definite angular unconformities with overlying late Cretaceous to Palaeogene sediments and volcanics to the east of the Chilean Precordillera. However, it is probable that the widespread stratigraphic gaps between these younger formations and the underlying Triassic to Palaeozoic rocks are related to mid-Cretaceous crustal uplift and erosion (cf. Fig. 5).

The deformational age is constrained by the upper age limit of the deformed rocks (≈90 Ma) and by the postdeformational emplacement of granitic-monzonitic plutonic bodies (78-76 Ma, Maksaev et al. 1988a; Pichowiak, this Vol.), as well as by an angular unconformity between the deformed sequence and the overlying volcanic sequence of the subsequent late Cretaceous-Palaeogene arc system (starting at ≈72 Ma, Naranjo and Puig 1984, Herrmann and Zeil 1989). All these events are correlated and are comparable with the effects of the Peruvian tectonic phase which was first described for Perú and western Bolivia as a compressive phase (Steinmann 1929); its age was described as Santonian (86.6-83 Ma, Harland et al. 1990) by Mégard (1987).

In contrast to the preceding and the following arc systems, deformations affected the mid-Cretaceous arc system only after its magmatic activity had ceased, i.e. during the 90 to 80-Ma magmatic quiescence. However, as deformation concentrated on the terminating magmatic arc, it can be concluded that the arc was a hot, still weak zone in the crust of the upper plate. The change in the deformational regime at the end of the mid-Cretaceous arc reflects the major change in the plate configuration in the southeast Pacific between 110 and 70 Ma (Fig. 9). During this time span the spreading centre between the Aluk and the Farallon plates (possibly a continuation of the Tethyan rift, Jaillard et al. 1990) migrated towards the south, and, as a result, the Aluk-South America convergence was replaced by one between the Farallon and the South American plates. While the Aluk-South America convergence was at a very high angle of sinistral convergence obliqueness, the convergent plate motion between Farallon and South America had a dextral component at a lower angle of obliqueness. Thus, a completely different stress and strain regime can be inferred for the time following the change in plate movements. During the Aluk-South America convergence,

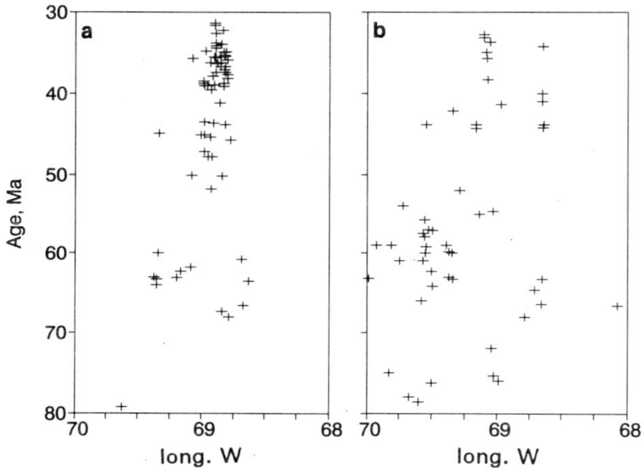

Fig. 10. Distribution of isotope age values in the Chilean Precordillera, *a* north of the line Calama-Antofagasta; *b* south of the line Calama-Antofagasta (Source of data same as in Fig. 2)

orogen-normal extension occurred, whereas orogen-normal shortening is assumed to have occurred during the Farallon-South America convergence. The period of igneous quiescence between 90 and 80 Ma (Figs. 2 and 3b), which coincides with the reorganization of Pacific plates during the mid-Cretaceous, may be a result of subduction of the southward migrating spreading axis. A modern example of a gap in igneous activity due to the subduction of an oceanic ridge is the subduction of the Chile ridge underneath the southern Andes at 50°S; this also produces a gap in the volcanic chain (Herron et al. 1981; Ramos et al. 1991).

4 The Late Cretaceous-Palaeogene Arc System

This magmatic arc system, which was again emplaced to the east of the preceding magmatic arc system, was active from the late Cretaceous to the Oligocene. Isotope age determinations of volcanic and plutonic rocks fall between 80 and 30 Ma. Figures 3 and 10 show two maxima of igneous activity, one at 75-55 Ma, more distinct in the Longitudinal Valley south of the line Antofagasta-Calama, and the other one at 48-35 Ma centred on the Chilean Precordillera north of Calama. As both parts of this magmatic arc show differences in their tectonic setting they will be dealt with separately.

4.1 Maastrichtian-early Eocene

During this time the magmatic arc was more than 100 km wide (Fig. 10a). In contrast to the preceding arc systems of the Andean Cycle, the volcanics of this time (basaltic and rhyolitic lavas, acid tuffs and ignimbrites; Chile Alemania Fm.) were deposited in the western parts of the arc, directly upon Palaeozoic to Cretaceous rocks, with a marked angular unconformity at the base. The substrate of the arc, which had been subject to shortening after the extinction of the preceding arc system, was located above sea level. At its eastern side the arc was bordered by a broad backarc basin which extended to the east as far as the Eastern Cordillera of southwest Bolivia and northwest Argentina. Here, crustal extension can be inferred from Maastrichtian to lower Palaeocene marine sedimentation (Salta Group, Marquillas and Salfity 1988), and strongly alkaline basic volcanics of 78-76 Ma (Reyes et al. 1976; Valencio et al. 1976) and 65-60 Ma (Omarini et al. 1988). In the north Chilean part of the backarc basin mainly continental red sandstones and conglomerates up to 2 km thick were deposited (north of Calama: Eastern Sequence (Bogdanic 1990); east and southeast of Calama: Tonel Fm. of the Purilactis Group (Charrier and Reutter this Vol.)) indicating the

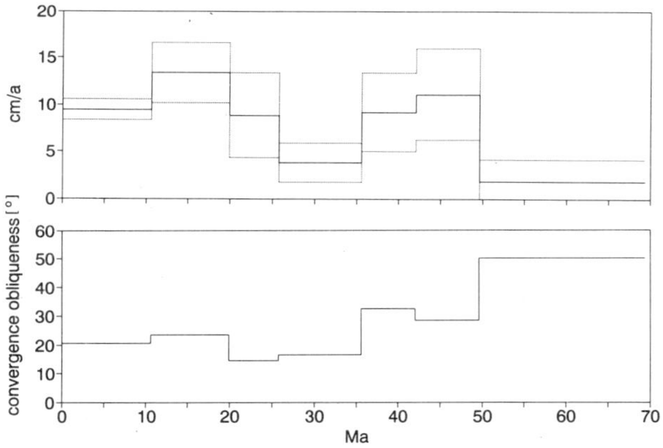

Fig. 11. *Above* convergence rate (cm/a) and *below* convergence obliqueness between the Farallon/Nazca and the South American plates since the uppermost Cretaceous. (After Pardo-Casas and Molnar 1987.) For the convergence rate the error range is also shown (*light lines*), for the convergence obliqueness only the roughly determined mean value has been used (cf. Pardo-Casas and Molnar 1987: Figs. 4 and 5). Data have been interpolated for 25°S between the values for 20° and 30°

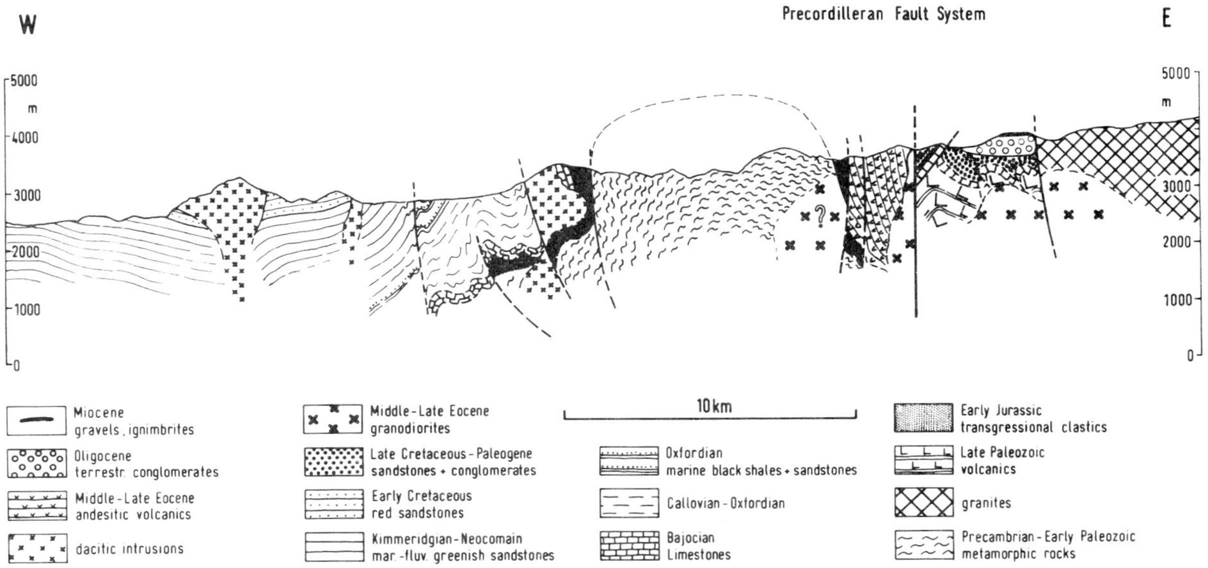

Fig. 12. Cross section through the Chilean Precordillera at 21°S (Quebrada Choja). The section shows the incorporation of the Preandean basement (Precambrian-late Palaeozoic) into an anticlinal structure which originated during the Incaic phase (≈38 Ma). The structure is intruded by granodioritic to dacitic stocks of Eocene age. The Precordilleran Fault System is also shown

transition between the elevated arc and the backarc area. Extensional stress conditions for this backarc area are recorded by a Maastrichtian marine ingression, and by alkali-basaltic flows of ≈68 Ma (Döbel et al. 1992).

During the Maastrichtian-early Eocene the deformational regime was characterized by relative tectonic quiescence to slightly extensional conditions in the arc and, in the backarc/foreland, by the described strong crustal extension. In the magmatic arc some extension can be inferred from the presence of early Eocene calderas in the Longitudinal Valley at 24°30'S (Herrmann and Zeil 1989) and in the Chilean Precordillera east of Copiapó (27°45'S, Rivera and Mpodozis 1991). Large scale arc-parallel strike-slip faults have not been described from this arc.

The time interval Maastrichtian-early Eocene was characterized by a low convergence rate of <5 cm per year (Fig. 11). A rather low degree of coupling between the plates is expected which, in turn, should produce an extensional strain regime in the upper plate. The relatively high degree of convergence obliqueness may have increased the extensional regime (cf. Fig. 1).

4.2 Late Eocene-early Oligocene

During this time the magmatic arc became narrower and the centre was situated in the Chilean Precordillera north of Calama (Fig. 10b) where calc-alkaline volcanics up to 2000 m thick were deposited. Volcanic activity extended to the south along the western border of the Maastrichtian-Eocene backarc basin where the lavas interfinger with sediments. Volcanism was accompanied by the emplacement of mainly acid plutons. The onset of igneous activity was linked with uplift of the arc's crust as is documented in the deposition of >3000 m of conglomerates in the backarc basin (Purilactis Fm. s. str. of the Purilactis Group, Charrier and Reutter this Vol.). In contrast to the Maastrichtian-early Eocene, there are no indications of backarc/foreland extension.

In late Eocene times the arc was affected by the strong deformations known as the Incaic tectonic phase (Steinmann 1929; Noble et al. 1979). The deformation history is initially governed by a transpressional regime of arc-normal shortening and arc-parallel dextral strike-slip faults. Shortening is well illustrated in the Chilean Precordillera (Fig. 12) which is built up by long more or less upright anticlines (Chong and Reutter 1985: 25% of shortening) with Preandean basement cores consisting of Precambrian to Palaeozoic sedimentary, metamorphic and plutonic rocks. Smaller fold structures are developed in the sedimentary and volcanic cover rocks. Dőbel et al. (1992) could establish the ages of folding during the Incaic phase by $^{40}Ar/^{39}Ar$ datings of volcanic rocks located immediately below and above a 50° angular unconformity. As both ages (38.45 ± 0.61 and 38.54 ± 0.87 Ma respectively) are identical, it can be concluded that folding took place during a period of <1.5 Ma.

During the Incaic phase, arc-parallel strike-slip movements started along the Precordilleran Fault System (Reutter et al. 1991). A contemporaneous activity of shortening and transcurrent movements can be inferred from the observation that some of the Precordilleran folds and thrusts are oriented en echelon, obliquely (NW-SE) to the prevailing N-S trend of the orogen, and normally to the direction of shortening in the dextral system. On the other hand, strike-slip movements lasted until at least 35 Ma. This has been shown by Maksaev et al. (1988b) who reports a 34 Ma age in mylonites from shear zones cutting through a 35 Ma old pluton. The longer duration of transcurrent movements compared with folding can also be inferred from the observation that second order vertical folds were generated in stratified rocks, where strike-slip faults cut through vertically dipping flanks of first order horizontal folds (Reutter et al. 1991).

In contrast to the preceding arc systems deformation at the end of the Eocene (Incaic phase) was not restricted to the magmatic arc. The backarc region of the late Cretaceous-Palaeogene arc was also subject to strong shortening so that the backarc sediments and parts of their substrate were involved in east-vergent folds and reverse faults. Crustal shortening that extended to the Eastern Cordillera of Bolivia and Argentina (Reutter et al. 1988) must have contributed to an important crustal thickening in the arc area.

Reconstructions of plate movements during the Eocene are consistent with the observed structures (Fig. 11). Around 48 Ma there was a sharp increase in the Farallon-South America convergence rate and this coincides with the onset of igneous activity after the early to mid-Eocene. This increase must have resulted in an intensified coupling between the plates. Together with the 30° dextral obliqueness the plate configuration corresponds to the transpressional regime observed in the arc.

5 The Miocene-Holocene Arc System

Around the Oligocene-Miocene boundary, the modern magmatic arc (Central Volcanic Zone of the Andes) was installed in the Western Cordillera, east of the preceding extinct arc system. In the segment considered the arc is very wide as it extends as far as the western reaches of the Eastern Cordillera. Although isotope ages indicate igneous activity since 28 Ma (Figs. 2 and 3), ages older than 17 Ma are very rare, and there is a further remarkable increase in the number of data since 10 Ma which corresponds to the Quechua tectonic phase (Steinmann 1929; Mégard 1984).

The lavas and ignimbrites of the new arc overlie unconformably different Palaeozoic to Oligocene rocks of the shortened backarc area of the late Cretaceous-Eocene arc system. It can be supposed that where the new arc was installed the crust was relatively thick due to the late Eocene shortening. However, slowing convergence rates during the Oligocene (Fig. 11, Pardo-Casas and Molnar 1987) and the resulting decrease in the compressive stress in the upper plate led to some crustal collapse and crustal thinning in the course of isostatic adjustments. The formation of intramontaneous basins (e.g. Salar de Atacama (probably along preexisting structures), Upper Loa Valley, and the probably still active depressions within the Puna), and the rapid uplift of Eocene-Oligocene plutons and older rocks during the Oligocene (Maksaev 1990; Andriessen and Reutter this Vol.) give evidence of the tectonics of this interarc stage. Geologically this period is well marked by the formation of a broad peneplain over the whole segment of the Andes, and by a pronounced angular conformity beneath a thick cover of unconsolidated continental gravels (Gravas de Atacama) of late Oligocene to early Miocene age, which interfinger with the volcanic products near the Western Cordillera.

The deformations in the modern arc system are largely governed by arc-normal shortening. Miocene and younger structures of intense crustal shortening can be observed in the Western Cordillera and its border regions. In the Western Cordillera deformations occurred at ≈23, ≈10 Ma and ≈4-5 Ma (Lahsen 1982). According to Kussmaul et al. (1975), in the Altiplano of southwest Bolivia shortening took place during the early Miocene (23-24 Ma), the late Miocene (13-14 Ma) and the late Pliocene (≈3 Ma). In the Puna of northwestern Argentina, west and east vergent conjugate reverse fault systems developed in the early Miocene, in the middle Miocene, in the Pliocene and in the Quaternary (Schwab 1970, 1985). In the southeastern border of the Puna plateau Allmendinger (1986) has documented the youngest deformations. He could define a late Pliocene deformation (after ≈2.35 Ma ago, before early Quaternary) with NW-SE shortening (vertical extension), and a Quaternary deformation dominated by strike-slip movements (E-W shortening, N-S extension). Allmendinger attributed this Quaternary kinematic change to a change in either the geometry of the subducted Nazca plate or plate convergence. To the west of the Western Cordillera, on the border of the Salar de Atacama depression (Cordillera de la Sal, Cordon de Lila), after Oligocene extension and collapsing, reverse faults (Megafalla Tucucaro, Niemeyer 1984) and anticlines (Cordillera de la Sal, Wilkes 1991; Wilkes and Görler this Vol.) developed from the Miocene to the present. To summarize, the structures of the modern magmatic arc indicate more or less continued deformations (mainly arc-normal shortening) from the Oligocene-Miocene boundary until the present. However, deformations were strongest during the Quechua (≈10 Ma) and the Diaguita (5-3 Ma) tectonic phases.

The most important deformation related to the modern arc system is the development of the fold-and-thrust belt of the Subandean Ranges of Bolivia and northwestern Argentina since the late Miocene. Strong crustal shortening of up to 140 km (Kley et al. 1991) leads to the underthrusting of the Andean foreland (corresponding to the Brazilian Shield) beneath the Andes. Generally, the backarc deformations started later than those of the arc (Jordan and Gardeweg 1989; Mégard 1989). This can be explained by Isacks' (1988) model which proposes two stages for the Neogene compressive deformations: (1) a pervasive and widespread horizontal shortening, and (2) a concentration of deformation along the eastern side of the orogen (Eastern Cordillera, Subandean Ranges) triggered by an increased convergence rate in the late Miocene. The later phase resulted in the strong uplift of the present arc to a height of some 4000 m.

In contrast to the shortening in the arc and backarc regions the western part of the forearc is characterized by steep fractures. For the Coastal Cordillera Armijo and Thiele (1990) ascribed these faults to an E-W extension caused by subduction-related underthrusting and the existence of east-dipping ramps.

In the modern magmatic arc system orogen-parallel strike-slip displacements seem to be of minor importance. This is due to the low angle of convergence obliqueness (<20°). Nevertheless, on the eastern scarp of the Salar de Atacama depression right-stepping Riedel shears indicate local left lateral strike-slip movements with a throw unlikely to exceed 100 m. Orogen-parallel sinistral strike-slip movements have also been reported by Armijo and Thiele (1990) from the forearc region where a reactivation of the Atacama Fault occurred in the Coastal Cordillera. These authors interpret the sinistral movements, which are in contradiction with the present minor right lateral component of Nazca convergence, as due to the clockwise rotation of the Central Andes south of the Arica bend.

The more or less continuous deformations in the modern magmatic arc system correspond to special conditions of plate convergence since the early Miocene: (1) convergence rates are very high (Fig. 11), (2) the subducting Nazca plate is relatively young (≈40 Ma at the trench, Herron 1972), resulting in a low slab-dip angle of <30°, (3) the motion of the South American plate in a direction opposite to that of the Nazca plate overcompensates subduction rollback, and (4) a convergence obliqueness of <20°. All these factors lead to strong compressional stresses at the interface of the plates, a good coupling between the plates and, thus, to an intense stress transmission to the upper plate. As a consequence, the modern Central Andes have the world's strongest compressional strain regime which resulted in a crustal thickness of some 70 km and a very broad zone of crustal shortening.

6 Discussion

6.1 Continuous Deformation and Tectonic Events

The tectonics of the active continental margin of the Central Andes is an expression of the changing conditions of plate convergence. Both long-term continuous deformations, and short-term tectonic events can be recognized, which should respectively reflect periods of more or less steady state conditions and relatively sudden changes of subduction parameters (e.g. Fig. 11). Some of the events can be correlated with global tectonic phases whereas other deformations are only of regional or local significance.

Examples of long-term continuous deformation, which may pertain to the stress pattern induced by a uniform plate movement, are: (1) the Jurassic and early Cretaceous crustal subsidence under growing lava formations and a probable contemporaneous crustal uplift of deeply intruded plutons in the arc area; (2) the Maastrichtian-early Eocene tectonic quiescence to mild extension in the arc, coupled with long-term subsidence and extension in the backarc; and (3) superimposed on the Quechua and Diaguita tectonic phases, the continued Miocene to recent shortening and uplift in the modern magmatic arc.

Short-term tectonic events are better known because of their marked structural expression, and they can be better appreciated in terms of plate tectonics and global events than can the long-term deformations.

The tectonic event that affected the Jurassic arc during the Kimmeridgian (Araucanian phase) corresponds to a worldwide tectonic phase which is probably linked with major changes in the plate configurations. For the Peruvian and Colombian segment of the Andes, Jaillard et al. (1990) have attributed these late Jurassic events to a geodynamic revolution which is marked by the end of the Tethyan breakup, and by the beginning of the Atlantic rifting. The Araucanian deformations of northern Chile are also contemporaneous with the $\approx 155 \pm 3$ Ma Nevadian phase in western North America as dated by Schweickert et al. (1984). The early Cretaceous (126 Ma) movements along the Atacama Fault Zone have no correspondence with other areas of the arc system. On the other hand, these deformations mark the beginning of uplift and termination of the arc, and the beginning of the eastward displacement of igneous activity. The 126 Ma movements may, thus, be related to the dynamic evolution of the arc rather than to the result of an externally imposed tectonic phase.

The late Cretaceous tectonic event (Peruvian phase, between 90 and 80 Ma), induced by the major plate reorganization in the southeast Pacific, as shown in Fig. 9, is prominent in the many parts of the Andean domain (Vicente et al. 1973; Coira et al. 1982; Mégard 1987), although in some segments it is not present (e.g. Godoy 1991) In northern Chile, this phase not only terminates the generally extensional tectonics of the previous arc systems but also initiates the slightly extensional to compressional conditions governing the active continental margin since the very late Cretaceous (see below).

The late Eocene change of plate-motion directions and velocities (well documented by the bend in the Hawaii-Emperor ridge) is recorded in the Andes as the Incaic phase. In the segment under study it produced strong transpressional deformations, consisting of folding in arc and backarc, and dextral strike slip restricted to the magmatic arc area. The Incaic compressional event was replaced by an Oligocene period of minor compressional stress and consequent isostatic relaxations with the formation of intramontaneous basins perhaps as a consequence of a decreased convergence rate and/or less effective plate coupling. For the Oligocene deformations no corresponding structures have been reported in other parts of the Andes; it may thus be interpreted as a local event.

Finally, the late Miocene Quechua and the Pliocene Diaguita phases, which again coincide with tectonic events in other parts of the world, interfere in the activity of the Miocene-Holocene arc, but they accelerate this activity rather than intercept it. They only cause an increase in the compressional strain regime (again up to the Jarrard strain class 7) and initiate the uparching of the Central Andes to their present height, as well as initiating the apparently continuously ongoing development of the Subandean foreland fold-and-thrust belt.

6.2 Development of the strain regime

Due to the changing conditions in plate convergence the structural development of the Central Andes described in the previous sections shows a wide range of different strain regimes. In order to obtain a more quantitative idea of the deformation history of the Central Andean segment, an attempt has been made to group the deformations into Jarrard's (1986) strain classes (Table 1). The strain regime of tectonic phases and of interphases is shown in Table 3 and Fig. 13. In order to show the relation between deformation and igneous activity. Fig. 13 also shows the frequency of isotope ages. Obviously, from the observed structures the strain regime for the tectonic phases is far better known than the one for the interphases. The classification of the interphases is thus somewhat uncertain. According to their strain classification it is possible to distinguish two main periods with contrasting types of tectonics: (1) between 200 and 90 Ma a general extensional to transtensional strain regime during tectonic phases and interphases, and (2) since 90 Ma a period of strong transpression/compression during tectonic phases and slight extension to compression in interphases.

The differences in tectonic style between the periods can be attributed to different plate configurations (Fig. 9). For period (1) the convergence obliqueness was always $>45°$, whereas for period (2) the angle was $<45°$.

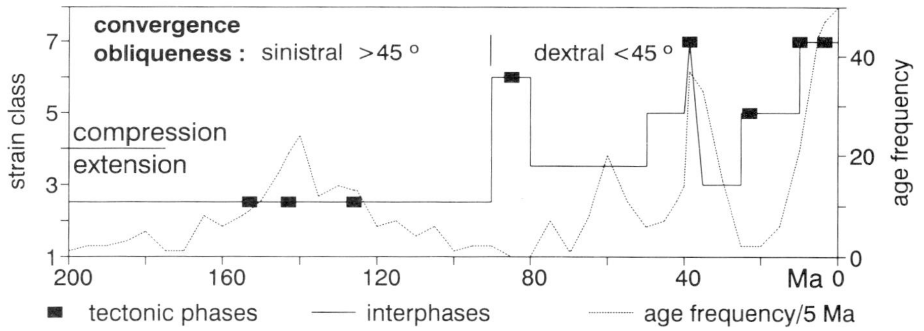

Fig. 13. The tectonic evolution of the north Chilean Andes in terms of Jarrard's (1986) strain classes and the frequency of isotope ages (5 Ma classes; data source same as in Fig. 2). Two periods with contrasting tectonics can be distiguished: (1) before 90 Ma with a sinistral convergence obliqueness of >45° and (2) after 90 Ma with a dextral convergence obliqueness of <45° to nearly trench-normal convergence. During period (1) deformations are (trans-)tensional in phases and interphases, while period (2) is characterized by (trans-)pression in phases and tectonic quiescence to extension in interphases

For period (1) arc-normal extension is expected for: (i) high convergence rates and/or a high degree of plate coupling, and also for (ii) low convergence rates and/or low plate coupling. In case (i), which should correspond to tectonic phases, extension should be produced according to the model outlined in Fig. 1 (normal stress is less than the hydrostatic stress at obliqueness >45°, Fig. 1). Case (i) can be applied to the tectonic phases of the Jurassic and early Cretaceous. Here, the existence of trench-linked strike-slip faults (e.g. the Atacama Fault Zone) implies that there was good coupling between the plates, and this, in turn, implies that extension was not the result of suduction rollback, very steep slab dip, and/or movement of the upper plate in the same direction as the subducting plate. Case (ii), which can be applied to the interphases, also led to crustal extension. The reason is that a decreasing convergence rate leads to a decoupling of the plates, which also operates in the direction of extensional strain. However, low convergence rates may be the reason for the lower tectonic activity during interphases.

For the period (2) the development of plate convergence rates and obliqueness is rather well constrained (Fig. 11). Tectonic phases can be correlated with times of increased plate convergence rate (Incaic, Quechua). It can be concluded that a high convergence rate leads to an increased coupling between the plates and, thus, to the observed structures of shortening. During the interphases the convergence rate, together with the horizontal stress, decreases, leading to a decoupling to some extent and thus to crustal extension. This effect has been described by Daly (1989) for Colombia where low angle faults in the forearc operated as reverse faults during periods of high convergence rates and as normal faults when convergence rates decreased.

Figure 13 also shows that there is no simple relationship between igneous activity and deformations in the magmatic arc. Some maxima in igneous activity can be correlated with deformational phases, while other phases are related to igneous quiescence; one maximum in igneous activity occurred during tectonic quiescence. Clearly related to maxima in isotope age frequency, and thus to increased convergence rates and plate coupling, are the late Jurassic-early Cretaceous phases, the Incaic phase and the Quechua events. In contrast to these phases, the mid-Cretaceous (Peruvian phase, 90-80 Ma) and late Oligocene (≈23 Ma) deformations are more related to periods of magmatic quiescence, although they can be correlated to reorganizations of the plate system and/or a displacement of the magmatic arc. During both phases the centre of igneous activity had been offset to the east, during the Peruvian phase a complete plate reorganization took place in the southeast Pacific, and during the late Oligocene the Farallon-South America convergence rate started to increase and may thus have influenced the deformational regime. The 60 Ma maximum in igneous activity occurred during tectonic quiescence and this is consistent with the low convergence rate at that time.

Table 3. The development of strain classes in the Central Andes since the early Jurassic

Time	strain class	indication	strike-slip
Sinemurian - Oxfordian (200-153 Ma)	2-3	arc volcanism in subsiding region, arc-normal extension (dykes,layered gabbro), backarc basalts	?
Late Jurassic events(≈153 and ≈143 Ma): Araucanian Phase	2-3	arc-normal extension (dykes), graben formation	strong sinistral (Atacama Fault Zone)
Kimmeridgian - Barremian (153-126 Ma)	2-3	arc-normal extenison (e.g. dykes)	?
Early Cretaceous event (≈126 Ma)	2-3	arc - normal extenison (e.g. dykes), beginning of alkaline backarc/foreland magmatism	strong sinistral (Atacama Fault Zone)
Aptian - Cenomanian (126-90 Ma)	2-3	arc volcanism in subsiding region, strong alkaline backarc/foreland magmatism	?
Peruvian Phase (90-80 Ma)	6	strong folding, thrusts	?
Late Campanian - early Eocene (80-48 Ma)	3-4	slight extension in the arc (calderas), partly marine backarc/foreland basin with alkaline magmatism	not present
Middle - Late Eocene (48-39 Ma)	4	increased arc uplift, continental sedimentation in backarc, no alkaline backarc magmatism	?
Incaic Phase (39-38 Ma)	7	thrusting and folding in arc and backarc	str. dextral (Precordilleran Fault S.)
E.- M. Oligocene (35-25 Ma)	3	basin formation (collaps structures)	(sinistral)
L. Oligocene - L. Miocene (25-10 Ma) (deformational event at ≈23 Ma)	5	continued but gentle folding, reverse faults	(sinistral)
Quechua Phase (≈10 Ma)	7	onset of very strong shortening in the backarc (fold and thrust belt), continued shortening in the arc	weak sinistral
Late Miocene - present (10-0 Ma), including Diaguita Phase (≈4.5 Ma)	7	continued shortening in arc and backarc, some	weak sinistral

6.3 Magmatic arc tectonics

The geological record of the Andean segment between 21° and 25°S shows that especially the magmatic arc areas within the active continental margin were very sensitive to the prevailing strain regime, which was apparently expressed in the quantity and composition of the magmas produced as well as in the tectonic development. In comparison with the adjacent forearc and backarc areas, the strength of the magmatic arc is necessarily greatly reduced by the much higher heat flow and temperature level, and by the presence of liquid magmatic bodies at different levels in the crust, so that a concentration of the deformation can be expected here. Indeed, arc-parallel shear movements caused by oblique subduction ran longitudinally through the respective arcs at least during the Jurassic, early Cretaceous and Eocene-Oligocene. Special extensional and compressional tectonics within the arc area have also been described (Reutter and Scheuber 1988; Scheuber and Reutter 1992). It can now be added that the concentration of important tectonic events to the magmatic arc proves again that in this area the contemporaneous strain and its variations are well registered.

Acknowledgements. This research is part of the project "Mobility of Active Continental Margins" supported by the DFG (German Research Foundation) and by the Freie Universität Berlin. Support was also given by the Universidad Catolica del Norte, Antofagasta.

References

Allmendinger RW (1986) Tectonic development, southeastern border of the Puna Plateau, northwestern Argentine Andes. Geol Soc Am Bull 97: 1070-1082

Arabasz WJ (1971) Geological and geophysical studies of the Atacama Fault Zone in northern Chile. Ph D, Calif Inst Tech, Pasadena, 275 pp (unpubl)

Armijo R, Thiele R(1990) Active faulting in northern Chile: ramp stacking and lateral decoupling along a subduction plate boundary. Earth Planet Sci Lett 98: 40-61

Beck ME (1983) On the mechanism of tectonic transport in zones of oblique subduction. Tectonophysics 93: 1-11

Bogdanic T (1990) Kontinentale Sedimentation der Kreide und des Alttertiärs im Umfeld des subduktionsbedingten Magmatismus in der chilenischen Präkordillere. Berl Geowiss Abh A123: 1-117

Boric R, Diáz F, Maksaev V (1990) Geología y yacimientos metaliferos de la región de Antofagasta. Servicio Nacional de Geología y Mineria - Chile, Boletin 40: p 246

Bossi GE, Wampler M (1969) Edad del Complejo Alto de Las Salinas y Formación El Cadillal segun el metodo K-Ar. Acta Geol Lilloana 10: 141-160

Buchelt M, Tellez C (1988) The Jurassic La Negra Formation in the area of Antofagasta, Northern Chile (lithology, petrography, geochemistry). In Bahlburg H, Breitkreuz C, Giese P (eds) The southern Central Andes, Lecture Notes in Earth Sciences 17. Springer, Berlin, Heidelberg, New York, pp 171-182

Chong G, Reutter KJ (1985) Fenomenos de tectonica compresiva en las Sierras de Varas y de Argomedo, Precordillera Chilena, en el ambito del paralelo 25° sur. IV Congr Geol Chil Actas 2: 2-219 - 2-238

Coira B, Davidson J, Mpodozis C, Ramos V (1982) Tectonic and magmatic evolution of the Andes of northern Argentina and Chile. Earth Sci Rev 18: 303-332

Daly MC (1989) Correlations between Nazca/Farallon plate kinematics and forearc basin evolution in Ecuador. Tectonics 8: 769-790

Davidson J, Godoy E, Covacevich V (1976) El Bajociano marino de Sierra Minillas (70°30' long. W - 26° lat. S) y Sierra Fraga (69°50' long. W - 27° lat S), Provincia de Atacama, Chile: Edad y marco geotectónico de la Formación La Negra en esa latitud. I Congr Geol Chil Actas: 255-272

Döbel R (1989) Geochemie und Geochronologie alttertiärer Vulkanite aus der Präkordillere Nordchiles zwischen 21° und 23°30'S. Ph D, Berlin, 152 pp (unpubl)

Döbel R, Friedrichsen H, Hammerschmidt K (1992) Implication of 40Ar/39Ar dating of early Tertiary volcanic rocks from the north Chilean Precordillera. Tectonophysics 202 55-81

Fitch TJ (1972) Plate convergence, transcurrent faults, and internal deformation adjacent to Southeast Asia and western Pacific. J Geophys Res 77: 4432-4460

Galli C (1957) Las formaciones geológicas en el borde occidental de la Puna de Atacama, sector de Pica, Tarapacá. Minerales 12: 14-26

Galliski MA, Viramonte JG (1988) The Cretaceous paleorift in north-western Argentina: a petrologic approach. J S Am Earth Sci 1: 329-342

García F (1967) Geología del Norte Grande de Chile. Simposium sobre el Geosinclinal Andino. Soc Geol Chile 3: 1-138

Godoy E (1991) El corrimiento del fierro a la dicordancia intrasenoniana en el Rio Cachapoal, Chile Central. VI Congr Geol Chil Resumenes Expandidos: 635-639

Gonzalez G (1990) Patrones estructurales, modelo de ascenso, emplazamiento y deformación del Pluton de Cerro Cristales, Cordillera de la Costa al sur de Antofagasta, Chile. Memoria de Titulo Universidad Catolica del Norte Antofagasta 135 pp

Harland WB, Armstrong RL, Craig LE, Smith AG, Smith DG (1990) A geologic time scale 1989. Cambridge University Press

Herrmann R, Zeil W (1989) Tectonics and volcanism in the north Chilean Longitudinal Valley (24°30' - 25°15' S). Zentralbl Geol Paläontol I: 1065-1073

Herron EM (1972) Sea-floor spreading and the Cenozoic history of the east-central Pacific. Geol Soc Am Bull 83: 1671-1692.

Herron EM, Cande SC, Hall BR (1981) An active center collides with a subduction zone: a geophysical survey of the Chilean margin triple junction. Geol Soc Am Mem 154: 683-702

Hervé M (1987): Movimiento sinistral en el Cretacico Inferior de la Zona de Falla Atacama al Norte de Paposo (24°S), Chile. Revista Geológica de Chile 31: 37-42

Hyndman DW (1985) Petrology of igneous and metamorphic rocks 2nd edn. McGraw-Hill, New York, 786 pp

Isacks B (1988) Uplift of the Central Andean plateau and bending of the Bolivian orocline. J Geophys Res 93 B4: 3211-3231

Jaillard E, Soler P, Carlier G, Mourier T (1990) Geodynamic evolution of the northern and Central Andes during early to middle Mesozoic times: a Tethyan model. J Geol Soc Lond 147: 1009-1022

Jarrard RD (1986) Relations among subduction parameters. Rev Geophys 24: 217-284

Jensen A (1985) El Sobreescurrimiento de Cerro Laberinto. IV Congr Geol Chil Actas A2: 84-103

Jordan TE, Gardeweg M (1989) Tectonic evolution of the late Cenozoic Central Andes (20°-33°S). in: Ben-Avraham Z (ed) The evolution of the Pacific Ocean margins. Monographs on Geol and Geophys, Oxford University Press: pp 193-207

Kley J, Reutter KJ, Scheuber E (1991) Die zentralen Anden - Geologische Strukturen eines aktiven Kontinentalrandes. Geogr Rundsch 3/1991: 134-142

Kussmaul S, Jordan L, Ploskonka E (1975) Isotopic ages of Tertiary volcanic rocks of southwest Bolivia. Geol Jahrb B14: 111-120

Lahsen A (1982) Upper Cenozoic volcanism and tectonism in the Andes of northern Chile. Earth Sci Rev 18: 285-302

Larson RL, Pitman III WC (1972) World-wide correlation of Mesozoic magnetic anomalies, and its implications. Geol Soc Am Bull 83: 3645-3662

Maksaev V (1990) Metallogeny, geological evolution, and thermochronology of the Chilean Andes between latitudes 21° and 26° South, and the origin of major porphyry copper deposits. PhD Thesis Dalhousie University Halifax Canada 554 pp

Maksaev V, Boric R, Zentilli M, Reynolds PH (1988a) Metallogenetic implications of K-Ar, ^{40}Ar-^{39}Ar, and fission track dates of mineralized areas in the Andes of northern Chile V Congr Geol Chil Actas 1: B65-B86

Maksaev V, Zentilli M, Reynolds PH (1988b) ^{40}Ar-^{39}Ar geochronology of porphyry copper deposits of northern Chilean Andes. V Congr Geol Chil Actas 1: B109-B133

Marinovic S, Lahsen A (1984) Hoja Calama. Carta Geol de Chile 58. Serv Nac Geol Min, Santiago, 140 pp

Marquillas R, Salfity JA (1988) Tectonic framework and correlations of the Cretaceous-Eocene Salta Group; Argentina. In: Bahlburg H, Breitkreuz C, Giese P (eds) The southern Central Andes, Lecture Notes in Earth Sciences 17. Springer, Berlin, Heidelberg, New York, pp 119-136

Mégard F (1984) The Andean orogenic period and its major structures in central and northern Peru. J Geol Soc Lond 141: 893-900

Mégard F (1987) Cordilleran Andes and marginal Andes: a review of Andean geology north of the Arica elbow (18°S). In: Monger JW, Francheteau J (eds) Circum - Pacific orogenic belts and evolution of the Pacific Ocean Basin, Am Geophys Union Geodyn Ser 18: 71-95

Mégard F (1989) The evolution of the Pacific Ocean margin in South America north of Arica elbow (18°S). In: Ben-Avraham Z (ed) The evolution of the Pacific Ocean margins. Monographs on Geol and Geophys. Oxford University Press, pp 208-230

Mpodozis C, Allmendinger R (1991) Extensión Cretacica a gran escala y napas extensionales en la Precordillera de Copiapó: la región de Puquios - Sierra Fraga, Región de Atacama, Chile. VI Congr Geol Chil Resumenes Expandidos: 208-212

Naranjo JA, Puig A (1984) Hojas Taltal y Chañaral. Carta Geol de Chile 62-63, Serv Nac Geol Min. Santiago 140 pp

Naranjo JU, Hervé F, Prieto X, Munizaga F (1984) Actividad Cretacica de la Falla Atacama al Este de Chañaral: Milonización y Plutonismo. Comunicaciones 34: 57-66

Niemeyer H (1984) La Megafalla Tucucaro en el extremo sur del Salar de Atacama: una antigua zona de cizalle reactivada en el Cenozoico. Comunicaciones 34: 37-45

Noble DC, McKee EH, Mégard F (1979) Early Tertiary "Incaic" tectonism in the Andes of central Peru. Geol Soc Am Bull 90: 903-907

Omarini RH, Salfity JA, Linares E, Viramonte JG, Gorustovich S (1988) Petrología, Geoquimica y Edad de un filón Lamproítico en el Subgrupo Pirgua (Alemanía - Salta). Universidad Nacional de Jujuy Argentina: in press

Pardo-Casas F, Molnar P (1987) Relative motion of the Nazca (Farallon) and South American plates since late Cretaceous time. Tectonics 6: 233-248

Pichowiak S, Buchelt M, Damm KW (1990) Magmatic activity and tectonic setting of the early stages of the Andean cacle in northern Chile. Geol Soc Am Special Paper 241: 127-144

Ramos VA, Kay SM, Márquez M (1991) La dacita Cerro Pampa (Mioceno - Provincia de Santa Cruz, Argentina): Evidencias de la colision de una dorsal oceanica. VI Congr Geol Chil Resumenes Expandidos: 747-751

Reutter KJ, Scheuber E (1988) Relation between tectonics and magmatism in the Andes of northern Chile and adjacent areas between 21° and 25° S. V Congr Geol Chil Actas 1: A345-A363

Reutter KJ, Scheuber E, Helmcke D (1991) Structural evidence of orogen-parallel strike slip displacements in the Precordillera of northern Chile. Geol Rundsch 80: 135-153

Reutter KJ, Giese P, Götze HJ, Scheuber E, Schwab K, Schwarz G, Wigger P (1988) Structures and Crustal Development of the Central Andes between 21° and 25° S. In: Bahlburg H, Breitkreuz C, Giese P (eds) The southern Central Andes, Lecture Notes in Earth Sciences 17. Springer, Berlin, Heidelberg, New York, pp 231-261

Reyes FC, Salfity JA, Viramonte JG, Gutierrez W (1976) Consideraciones sobre el vulcanismo del Subbgrupo Pirgua (Cretácico en el norte argentino). VI Congr Geol Argent Actas I: 205-223

Riccardi AC (1988) The Cretaceous System of southern South America.- Geol Soc Am Memoir 168: 168 pp

Rivera O, Mpodozis C (1991) Volcanismo explosivo del Terciario inferior en la Precordillera de Copiapó, Región de Atacama. VI Congr Geol Chileno Resumenes Expandidos: 213-216

Rogers G (1985) A geochemical traverse across the north Chilean Andes. PhD Thesis, The Open University Milton Keynes, 333 pp (unpubl)

Rössling R (1988) Petrologie in einem tiefen Stockwerk des jurassischen magmatischen Bogens in der nordchilenischen Küstenkordillere südlich von Antofagasta. Berl geowiss Abh A 112: 73 pp

Saleeby JB, Geary EE, Paterson SR, Tobisch OT (1989) Isotope systematics of Pb/U (zircon) and $^{40}Ar/^{39}Ar$ (biotite-hornblende) from rocks of the Central foothills terrane, Sierra Nevada, California. Geol Soc Am Bull 101: 1481-1492

Scheuber E (1987): Geologie der nordchilenischen Küstenkordillere zwischen 24°30' und 25°S - unter besonderer Berücksichtigung duktiler Scherzonen im Bereich des Atacama-Störungssystems. Thesis, FU Berlin, 157 pp (Unpubl)

Scheuber E, Andriessen PAM (1990) The kinematic and geodynamic significance of the Atacama Fault Zone, northern Chile. J Struct Geol 12: 243-257

Scheuber E, Hammerschmidt K (1991) $^{40}Ar/^{39}Ar$ and Rb-Sr data from ductile shear zones from the Atacama Fault Zone (AFZ), northern Chile, an attempt to determine the age of deformation. Terra Abstr 3: p 364

Scheuber E, Reutter KJ (1992) Relation between tectonics and magmatism in the Andes of northern Chile and adjacent areas between 21° and 25°S. Tectonophysics 205: 127-140

Scheuber E, Rössling R, Reutter KJ (1986) Strukturen der chilenischen Küstenkordillere zwischen Paposo und Antofagasta. Berl geowiss Abh A 66: 209-224

Schwab K (1970) Ein Beitrag zur jungen Bruchtektonik der argentinischen Puna und ihr Verhältnis zu den angrenzenden Andenabschnitten. Geol Rundsch 59: 1064-1087

Schwab K (1985) Basin formation in a thickening crust - the intermontane basins in the Puna and the Eastern Cordillera of NW-Argentina (Central Andes). IV Congr Geol Chil Actas 2: 138-158

Schweickert RA, Bogen NL, Girty GH, Hanson RE, Merguerian C (1984) Timing and structural expression of the Nevadan orogeny, Sierra Nevada, California. Geol Soc Am Bull 95: 967-979

Steinmann G (1929) Geologie von Peru. Winter, Heidelberg, 448 pp

Tobisch OT, Paterson SR, Saleeby JB, Geary EE (1989) Nature and timing of deformation in the foothills terrane, central Sierra Nevada, California: Its bearings on orogenesis. Geol Soc Am Bull 101: 401-413

Ulriksen C (1979) Regional geology, geochronology and metallogeny of the Coastal Cordillera of Chile between 25°30'and 26° south. M Sc Thesis, Dalhousie University Canada, 221 pp

Valencio DA, Giudice A, Mendia JE, Oliver GJ (1976) Paleomagnetismo y edades K/Ar del Subgrupo Pirgua, provincia de Salta, República Argentina. Séptimo Congr Geol Argent Actas I: 527-542

Vicente JC, Charrier R, Davidson J, Mpodozis C, Rivano S (1973) La Orogenesis Subhercinica: fase mayor de la evolución paleogeográfica y estructural de los Andes agentino-chileno centrales. V Congr Geol Argentino Actas V: 81-98

Wigger P (1988) Seismicity and crustal structure of the Central Andes. In Bahlburg H, Breitkreuz C, Giese P (eds) The southern Central Andes, Lecture Notes in Earth Sciences 17. Springer, Berlin, Heidelberg, New York, pp 209-229

Wilkes E (1991) Die Geologie der Cordillera de la Sal, Nordchile. Berl geowiss Abh A 128: 73 pp

Woodcock NH (1986) The role of strike-slip fault systems at plate boundaries. Philos Trans R Soc Lond A 317: 13-29

Zonenshayn LP, Savostin LA, Sedov AP (1984) Global paleogeodynamic reconstructions for the last 160 million years. Geotectonics 18: 181-195

K-Ar and Fission Track Mineral Age Determination of Igneous Rocks Related to Multiple Magmatic Arc Systems Along the 23°S Latitude of Chile and NW Argentina

PAUL A.M. ANDRIESSEN and KLAUS-J. REUTTER

Abstract. K-Ar and fission track mineral age determinations and fission track length measurements were performed on several plutons which together form a cross section along 23°S lat. running from the Coastal Cordillera of northern Chile to the Cordillera Oriental of northwestern Argentina. The spectrum of apparent mineral ages of the different plutons, ranging from about 530 to 30 Ma, support the present general ideas about the magmatic and tectonic evolution of this area. A strong relation is seen between the mineral ages and variations of the subduction regime. The plutons intruded at different depths; some of them belong to the Jurassic-early Cretaceous magmatic arc system 10 to 12 km deep, while others of the mid-Cretaceous magmatic arc system and especially those related to the late Cretaceous-Palaeogene magmatic arc systems were emplaced at shallow depths. These plutons were brought to the surface in the final emplacement of the arc systems by tectonic uplift. The mineral ages of Palaeozoic basement rocks, situated farther to the east, show a more complex history with final uplift during Eocene-Oligocene tectonism.

1 Introduction

The study of the geochronology of cooling and uplift histories of intrusive bodies is important for understanding the evolution of areas characterized by strong magmatic activity, areas such as the Andes. Of special interest is the subduction-controlled development of the magmatic arc systems of the Central Andes which, due changes in plate configuration and variations in the parameters of plate convergence, have been reorganized several times during the Phanerozoic. For this purpose, nine plutons, forming a more or less W-E cross section along 23°S lat. from the coast of northern Chile near Antofagasta to the Puna of northwestern Argentina (Fig. 1), have been the subject of an extensive mineral dating programme. 32 K-Ar and fission track age determinations and nine length measurements of spontaneous fission tracks in apatite were made. A minimum of three and maximum of five different minerals were analysed in each igneous rock sample.

The mineral dating systems can be used as geochrono-thermometers to establish cooling and uplift histories when several coexisting minerals are analysed. Each mineral has its own characteristic blocking temperature or temperature zone, above which the particular elements of the decay system can move freely in and out of the lattice structure of the crystal. Below the blocking temperature (or temperature zone), the free pathway of the elements is blocked and the radiogenic products are captured within the system. A similar behaviour is seen for fission tracks, because tracks only accumulate when the temperature drops below the so-called partial annealing zone. The temperature intervals at which the K-Ar mineral systems are closed lies above approximately 300 °C and those at which fission tracks in the minerals apatite, zircon and sphene are stable lie below 300 °C. Therefore, fission track dating along with other mineral dating systems is a very powerful tool for reconstructing emplacement and cooling histories of igneous rocks.

The interpretation and, hence, the geological significance of the mineral ages obtained depends strongly on the open or closed behaviour of the particular dating system. Several coexisting minerals may yield a pattern which permits a straightforward interpretation. Concordant mineral ages, for example, indicate fast cooling, while a pattern of decreasing apparent ages, according to the closure temperatures, suggests a slow and continuous cooling. Of course, coexisting minerals from one rock may also yield widely discordant ages, and this requires a careful interpretation before any conclusion about geological

Correspondence to: Paul A. M. Andriessen, Laboratorium voor Isotopengeologie z.w.o., De Boelelaan 1085, 1081 HV Amsterdam, The Netherlands

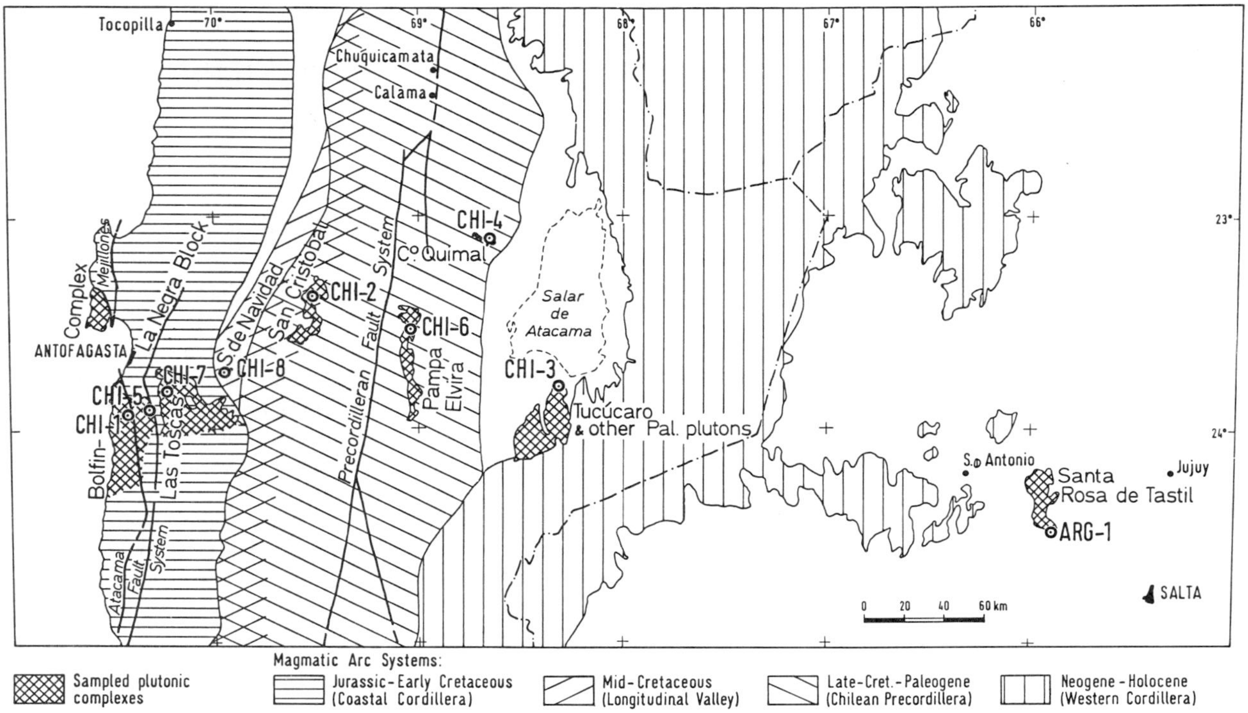

Fig. 1. Sketch map between 22° and 25°S lat. of northern Chile and northwestern Argentina. The succeeding magmatic arc systems and the sampled plutonic complexes are indicated

processes can be made. It is also possible that such discordant apparent ages have no real geological meaning because they are the result of disturbed systems.

In the geological evolution of the Central Andes, a Palaeozoic Preandean Orogenic Cycle and a Meso-Cenozoic Andean Cycle can be distinguished (Coira et al. 1982). The Preandean development is characterized by continental accretion and migration of the magmatic, tectonic and sedimentary processes from east to west, i.e. from the area of the Eastern Cordillera in Argentina to that of the Coastal Cordillera in Chile. Orogenic events occurred in late Precambrian-early Cambrian times, in the Ordovician, and during the Carboniferous and Permian (Coira et al. 1982; Miller 1984; Damm et al. 1990, this Vol.). The samples ARG-1, CHI-3, and CHI-6 were taken from plutons which can be attributed to these orogenies respectively.

The Palaeozoic basement rocks and structures of the Central Andes were tectonically, thermally and magmatically overprinted by the Mesozoic-Cenozoic Andean Cycle. In contrast to the Palaeozoic Preandean Orogenic Cycle, the Andean development is characterized by migration of the arc-related magmatic and tectonic processes towards the east and a consequent retreat of the continental margin (Coira et al. 1982). In the segment studied, four magmatic arc systems successively developed since the Jurassic: (1) a Jurassic-early Cretaceous arc in the Coastal Cordillera, (2) a mid-Cretaceous arc in the Longitudinal Valley, (3) a late Cretaceous-Palaeogene arc in the Precordillera (Sierra de Moreno, Cordillera Domeyko), and (4) a Miocene-Holocene arc in the Western Cordillera (Scheuber and Reutter 1992; Scheuber et al. this Vol.). The samples CHI-1, CHI-5 and CHI-7 came from (1), CHI-8 and CHI-2 from (2) and CHI-4 from (3). No samples from rocks belonging to the Miocene-Holocene arc were analysed in this study.

2 Analytical Procedures and Interpretation of the Mineral Dates

Standard laboratory techniques were applied for separation of hornblende, biotite, sphene, zircon and apatite. K contents were determined by flame

Table 1. Fission track analytiv data

Sample	Mineral	ρ_s* (x 10^6 t/cm^2)	ρ_i* (x 10^6 t/cm^2)	t ± 2σ (x 10^6 yr)	Number of grains	ρ_{glass}* (x 10^5 t/cm^2)	Zeta	Var(%)	Chi^2 Probability %
CHI-1	Apatite	1.4 (2343)	0.72 (1212)	119 ± 13	100	0.201 (3318)	308.1	4.6/4.71	
	Zircon	7.75 (567)	4.18 (153)	119 ± 22	6	2.098 (3238)	308.1		97
	Sphene	5.01 (554)	11.26 (623)	136 ± 14	5	0.300 (1837)	11134		85
CHI-5	Apatite	0.41 (616)	0.29 (437)	85 ± 13	50/51	0.193 (2129)	308.1	10/8.57	
	Zircon	9.74 (731)	6.10 (229)	102 ± 17	6	2.098 (3238)	308.1		55
CHI-7	Apatite	1.6 (595)	1.2 (456)	78.7 ± 11.6	50/51	0.193 (2129)	308.1	8.86/9.93	
	Zircon	19.61 (1435)	9.38 (343)	134 ± 19	6	2.098 (3238)	308.1		78
	Sphene	12.79 (672)	26.15 (687)	150 ± 15	7	0.300 (1837)	11134		68
CHI-8	Apatite	1.09 (445)	9.38 (1911)	69.2 ± 7	11	0.536 (3368)	11134		
	Zircon	7.77 (656)	6.02 (254)	73.9 ± 11.9	6	1.868 (2883)	308.1		30
	Sphene	1.71 (496)	6.97 (1010)	72.4 ± 7.7	7	0.266 (1632)	11134		92
CHI-2	Apatite	1.0 (869)	1.0 (903)	59.2 ± 7.5	51/51	0.201 (3318)	308.1	6.72/5.52	
	Zircon	25.51 (1436)	25.72 (724)	64.3 ± 9.8	6	2.098 (3238)	308.1		<1
CHI-6	Apatite	0.30 (444)	0.35 (518)	50.8 ± 7.6	50/50	0.193 (2129)	308.1	7.39/9.70	
	Zircon	18.24 (1266)	5.45 (189)	213 ± 34	6	2.098 (3238)	308.1		15
CHI-4	Apatite	0.62 (1049)	0.56 (972)	64.5 ± 7.9	101/102	0.193 (2129)	308.1	4.69/4.43	
	Zircon	7.73 (1175)	8.71 (662)	57.1 ± 7.3	6	2.098 (3238)	308.1		44
	Sphene	1.90 (551)	8.76 (1269)	66.3 ± 9.5	7	0.266 (1632)	11134		<1
CHI-3	Apatite	0.12 (356)	0.19 (569)	38.6 ± 5.6	100/100	0.201 (3318)	308.1	6.72/5.52	
	Zircon	16.51 (1208)	4.32 (158)	243 ± 41	6	2.098 (3238)	308.1		73
ARG-1	Apatite	1.1 (1801)	2.2 (3672)	30.3 ± 3.0	100/100	0.201 (3318)	308.1	3.92/4.55	
	Zircon	27.86 (980)	5.29 (93)	373 ± 110	6	2.098 (3238)	308.1		4

* In parenthesis are the actual counted tracks

photometry with a lithium internal standard and caesium chloride-aluminium nitrate buffer. Ar was extracted in a resistor-heated all-metal vacuum furnace and analysed by isotope dilution techniques using an ^{38}Ar spike; the measurements were made by the static method with a modified MAT III mass-spectrometer. Fission track dating was performed according to the technique described by Andriessen and Bos (1986), using the population method for apatite and the external detector method for sphene, zircon and apatite CHI-8. A zeta calibration factor of 308.1 in combination with NBS glass 962 was used for the age calculation of apatite and zircon; for sphene a zeta calibration factor of 11134 in combination with NBS glass 963 was used. Zeta calibration is based on the repeated analysis of Fish Canyon apatite and zircon, FTBM zircon, and Mount Dromedary sphene (Hurford 1990). The constants are those recommended by Steiger and Jäger (1977). Confined track length measurements were made on horizontal tracks using a dry 80x objective and a digitizing tablet plus projection tube arrangement. Only fully etched tracks in grains whose polished surfaces were approximately parallel to the c-axis were measured.

The analytical accuracy of the K concentrations is within 1% and the uncertainty of the argon-spike calibration within 2%. These estimated limits of relative error are the sum of the known sources of possible systematic error and the precision of the total analytical procedures. Where the pooled data were significant in the Chi-square test at 5%, the errors for fission track age were calculated using the "conventional method" (Green 1981), otherwise a mean age of the individual grains is calculated. The analytical results are given in Tables 1 (fission track data), 2 (K-Ar data), and 3 (compilation of the mineral ages and length measurements of confined spontaneous fission tracks in apatite).

Mineral dating systems can be used both as geochronometers and geothermometers, because every mineral has its own characteristic closure temperature (zone). In this study the following closure temperatures have been used: 500±50 °C for K-Ar hornblende, 325±50 °C for K-Ar biotite, 275±50 °C for fission track sphene, 225±25 °C for fission track zircon, and 100±25 °C for fission track apatite (Wagner 1968; Purdy and Jäger 1976; Gleadow and Lovering 1978; Naeser 1979; Harrison and McDougall 1980; Harrison 1981; Zaun and Wagner 1985). The time-temperature relation of each pluton, using the closure temperature concept, is shown in Fig. 2.

The mean length and distribution of the measured confined spontaneous fission tracks in apatite will provide additional information on the lower temperature (below 125 °C) history of the rock

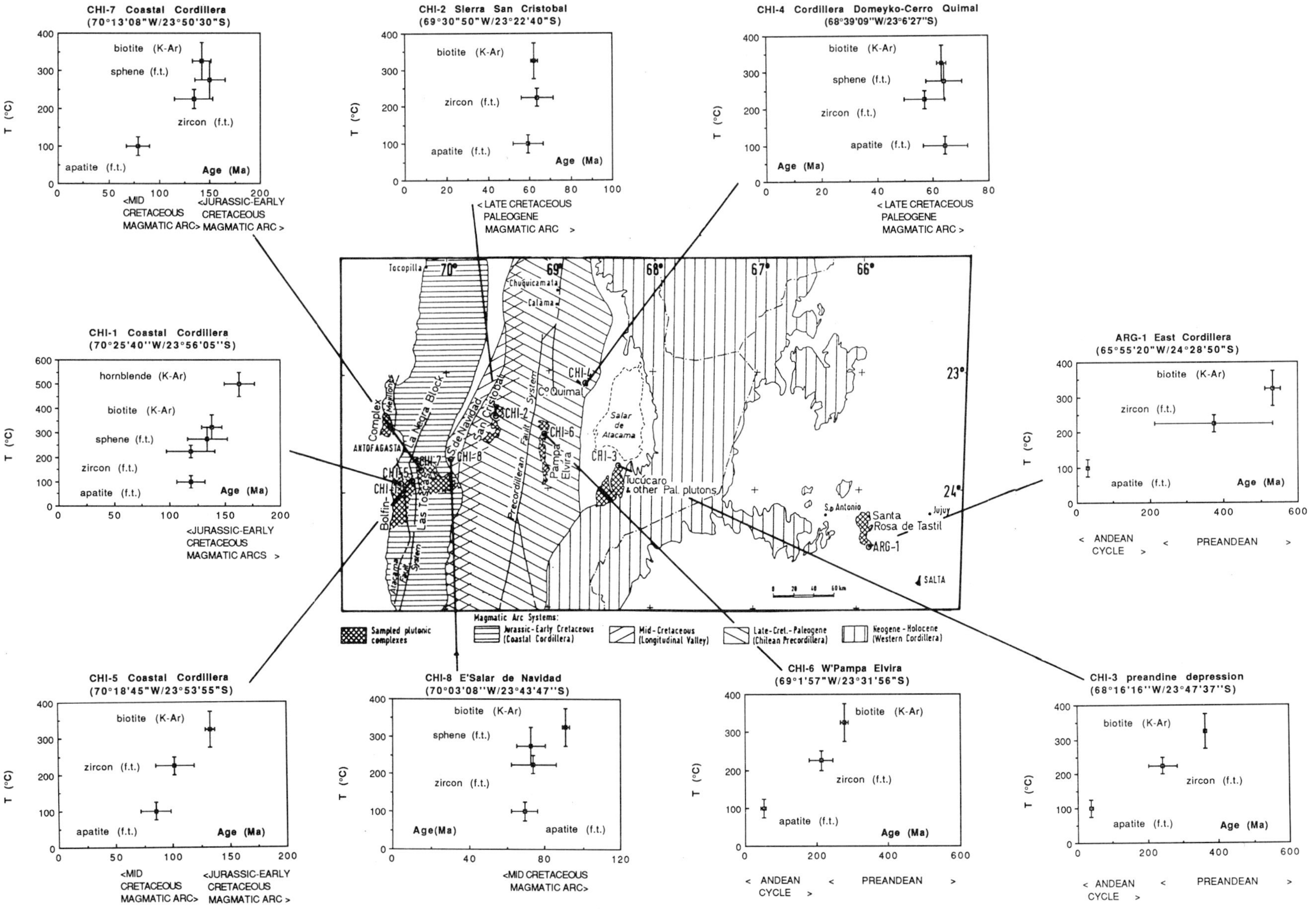

Fig. 2. Time-temperature relations for the individual plutons based upon the closure temperature concept of the various geochronometers

Table 2. K-Ar analyses of minerals of plutonic rocks from Chile and Argentina

Sample	Mineral	K (%wt)	Rad Ar (ppb wt)	$^{40}Ar_{rad}$ / $^{40}Ar_{tot}$ (%)	Age (Ma)
CHI-1	Hornblende	0.462	5.906	56.4	
		0.462	5.522	35.3	163 ± 14
			4.935	35.3	
	Biotite	7.21	75.61	4.6	
		7.19	72.39	44.7	138 ± 9
			66.35	35.1	
CHI-5	Biotite	5.54	52.68	57.0	
		5.53	52.99	40.3	133 ± 4
CHI-7	Biotite	6.74	68.04	39.5	
		6.73	68.90	18.0	
			75.22	31.1	
			64.29	32.2	142 ± 9
CHI-2	Biotite	7.34	32.4	53	
		7.35	32.1	54	62.3 ± 1.7
CHI-4	Biotite	7.74	34.2	56	
		7.75	35.0	49	63.3 ± 1.7
CHI-8	Biotite	6.15	39.3	48	
		6.13	39.8	33	91 ± 2
CHI-6	Biotite	5.50	122	24	
		5.47	114	31	
			110	32	
			115	24	279 ± 11
CHI-3	Biotite	7.07	219	24	
		7.06	203	18	
			196	37	
			200	7	364 ± 5
ARG-1	Biotite	7.52	338	11	
		7.51	321	8	
			309	9	532 ± 21

Table 3. Age results of northern Chile and northwestern Argentina

Sample	Mineral	K-Ar age (Ma)	Fission track age (Ma)	Mean length (μm)
CHI-1	Hornblende	163 ± 14		
70°25'40"W/	Biotite	138 ± 9		
23°56'05"S	Sphene		134 ± 18	
	Zircon		119 ± 22	
	Apatite		119 ± 13	14.75 ± 1.3
				16.68 ± 0.77*
CHI-5	Biotite	133 ± 4		
70°18'45"W/	Zircon		102 ± 17	
23°53'55"S	Apatite		85 ± 13	11.83 ± 1.86
CHI-7	Biotite	142 ± 9		
70°13'20"W/	Sphene		150 ± 15	
23°50'30"S	Zircon		134 ± 19	
	Apatite		78.7 ± 11.6	12.24 ± 1.52
CHI-8	Biotite	91 ± 2		
70°03'08"W	Sphene		72.4 ± 7.7	
23°43'47"S	Zircon		73.9 ± 11.9	
	Apatite		69.2 ± 7.0	13.06 ± 1.65
CHI-2	Biotite	62.3 ± 1.7		
69°30'50"W/	Zircon		63.8 ± 7.8	
23°22'40"S	Apatite		59.2 ± 7.5	15.36 ± 1.3
CHI-6	Biotite	279 ± 11		
69°1'57"W/	Zircon		213 ± 34	
23°31'56"S	Apatite		50.8 ± 7.6	12.74 ± 2.76
CHI-4	Biotite	63.3 ± 1.7		
68°39'09"W/	Sphene		64.1 ± 6.5	
23°6'27"S	Zircon		57.1 ± 7.3	
	Apatite		64.5 ± 7.9	14.66 ± 1.3
CHI-3	Biotite	364 ± 5		
68°16'16"W/	Zircon		243 ± 41	
23°47'37"S	Apatite		38.6 ± 5.6	13.36 ± 1.2
ARG-1	Biotite	532 ± 21		
65°55'20"W/	Zircon		373 ± 160	
24°28'50"S	Apatite		30.3 ± 3	12.58 ± 2.06
				16.47 ± 0.66*

* Mean length of induced tracks

(Gleadow et al. 1983, 1986). A narrow distribution and a mean length of between 14 and 15 μm indicates rapid cooling, and is commonly observed in volcanic rocks or shallow intrusions. Reduced lengths indicate that the host rock spent some time in the temperature zone where tracks are partially annealed. The mean length is shorter, the distribution becomes broader and shows a negative skewness to shorter track lengths. A more complex thermal history as the result of later heating, e.g. magmatic activity and/or burial history, may cause partial annealing of the tracks accumulated before the reheating, and a so-called mixed age is the result. The contribution of small tracks in the fission track length measurement will increase and bimodal distributions can even be obtained (Gleadow et al. 1986; Green 1986). The distribution of the lengths of the spontaneous fission tracks in apatite of all the analysed samples is shown in Fig. 3, together with the length distribution of induced tracks of CHI-1 for comparison.

3 Regional significance of the mineral ages

In Fig. 4, the mineral dates and some Rb-Sr whole rock analyses from Pichowiak et al. (1990) are plotted in an E-W cross section along 23°S lat. running from the Chilean coast near Antofagasta to the Argentine Puna. The spectrum of mineral dates of the different plutons, ranging from 530 to 30 Ma (Table 3), agrees with the present general ideas about the tectonic and magmatic evolution of the Central Andes of northern Chile and northwestern Argentina (Coira et al. 1982; Berg and Baumann 1985; Ramos et al. 1986; Scheuber and Reutter 1992; Scheuber et al. this Vol.). Two groups of mineral dates are distinguished and these show different trends:

1. Mineral dates of Palaeozoic igneous rocks, belonging to the Preandean Cycle, become younger from east to west.

2. Mineral dates of intrusives pertaining to the Meso-Cenozoic Andean Cycle show an opposite trend as they become younger from west to east.

The older 'trend', based on only three Palaeozoic igneous basement rocks, agrees with the ideas about the evolution of the Palaeozoic Andes (Cordani et al. 1973; Miller 1984; Ramos et al. 1986; Damm et al. 1990; Damm et al. this Vol.). All three Palaeozoic basement rocks have in common discordant mineral dates, suggesting a complex history. The significance of this will be discussed later.

Igneous rocks younger than 200 Ma were emplaced as a consequence of magmatic arc activity

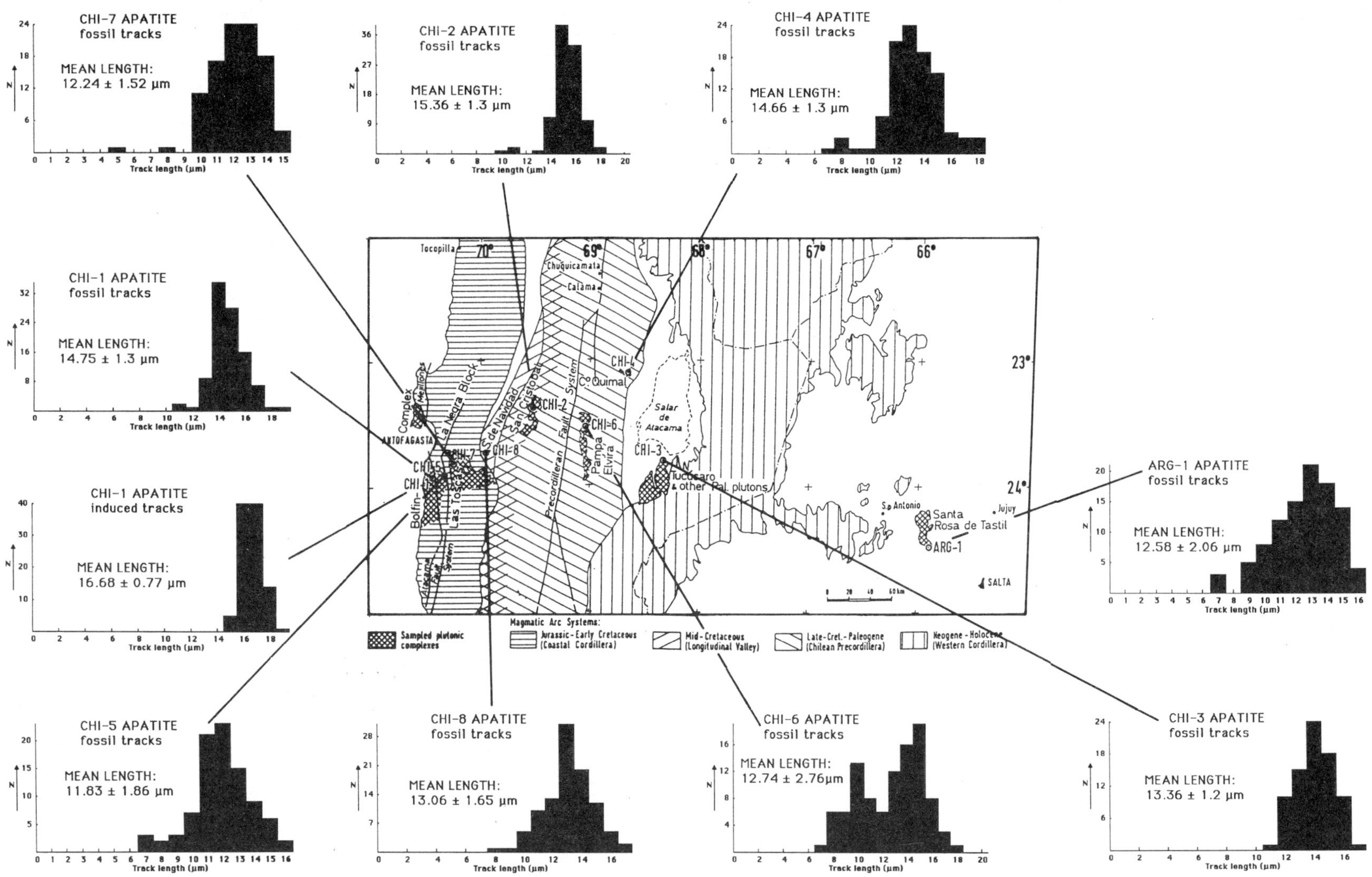

Fig. 3. Distribution of spontaneous fission track lengths of each pluton. For comparison the length distribution of the induced fission tracks for CHI-1 is also shown

during the Andean Cycle. During this time, the magmatic arc activity shifted to the east by ≈200 km (Coira et al. 1982). This occurred, however, not in a continuous way. In discrete geographic positions parallel to the trench, four magmatic arcs successively developed and had, for some tens of million years, a history of their own in a relatively fixed position (Scheuber and Reutter 1992). After the extinction of an earlier arc system, a new one was emplaced with its axis situated 50-100 km farther to the east. The final stages of the three fossil arc systems of the Andean Cycle, i.e the Jurassic-early Cretaceous arc, the mid-Cretaceous arc, and the late Cretaceous-Palaeogene arc, were, in each case, determined by tectonic events. When interpreting the individual mineral dates, it must be borne in mind that each arc system was separated from the next one by a relatively short distance and partly overlapped each other. Thus, the magmatic, tectonic and thermal effects of one pluton will have affected the adjacent one so that clearly differentiated steps of arc magmatism cannot be expected (Scheuber et al. this Vol.: Fig. 2).

In Fig. 4, the boxed areas enclose points from the four magmatic arc stages in time and space. Some of the mineral age data of the igneous rocks fit well into one of the boxes and may be considered as exponents of a particular magmatic arc system. Other mineral dates, however, are too young or too old and do not fit into the boxes. This indicates that these mineral systems may represent the early exponents of a subsequent event or late exponents of a preceding event. In spite of these problems, the data clearly confirm the eastward migration of magmatic arc systems by a total of some 200 km since the Jurassic, corresponding to a long term average migration rate of about 1 km/Ma. The eastward migration of the magmatic arc and consequent retreat of the Chilean Trench towards the continent are most probably related and, as pointed out by von Huene and Scholl (1991), the data can be used to estimate the long term rates and amounts of material eroded from the Andean continental margin of northern Chile.

In addition to the above mentioned trend, the data allow another division of the plutons according to their cooling histories. Plutons showing more or less concordant mineral ages were subject to fast cooling (e.g. CHI-4, CHI-8), whereas those having strongly discordant but regularly decreasing mineral ages reveal a steady post-emplacement cooling (e.g. CHI-1, CHI-5). A third group shows irregularly decreasing mineral ages as a consequence of a complex cooling history (e.g. CHI-2, CHI-7, and all Palaeozoic plutons).

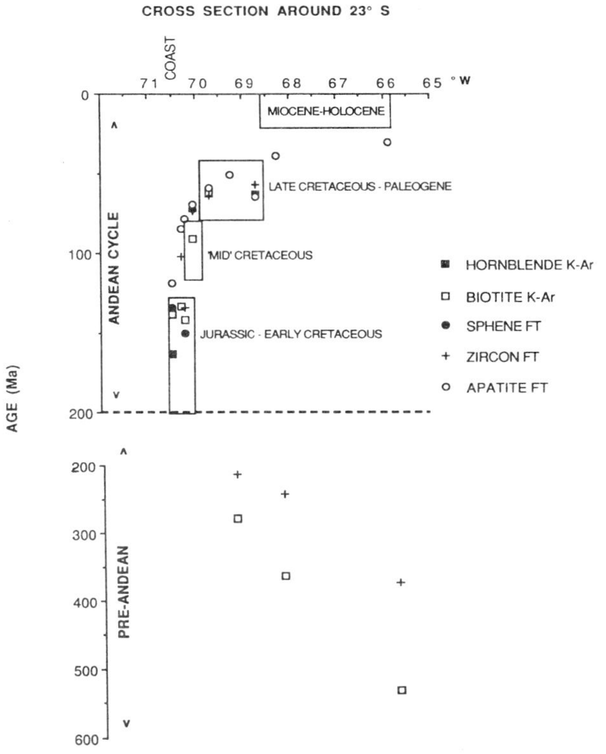

Fig. 4. Cross section around 23°S lat. of northern Chile and northwestern Argentina with the obtained mineral ages. Also indicated are the different magmatic arc systems in time and space

4 The Jurassic-Early Cretaceous Arc

The samples CHI-1, CHI-5, and CHI-7 were collected in the Coastal Cordillera from intrusive rocks of the Jurassic-early Cretaceous magmatic arc (Fig.1; for more geological details see Scheuber et al. this Vol.). In this order they represent a W-E section from the deepest, probably central, part of that arc towards its eastern margin.

CHI-1 was taken from a sheared metadiorite situated west of the western main branch of the Atacama fault zone. The diorite belongs to the so-called Bolfin complex, which consists of mantle-derived basic to intermediate intrusive and metamorphic rocks and is considered to represent a middle or lower crustal level of the Jurassic-early Cretaceous magmatic arc (Scheuber and Reutter, 1992). According to Rössling (1989) the diorite intruded at a depth of some 12-15 km.

CHI-5 is a bio-hbl-granite and is situated between the two main branches of the Atacama fault zone within a block of mainly granodioritic rocks. Intrusion depth is about 7-11 km (Rössling 1989: 2-3 Kb) and, hence, is more shallow than that of CHI-1. The U-Pb zircon age of the sampled rock (207±29 Ma) points to an early Jurassic age according to Damm et al. (1986), who also mention a nearby intrusion of hbl granodiorite giving a U-Pb zircon age of 152±10 Ma. Towards the north, along the strike, this block borders, at a transverse normal fault, on another block whose andesitic lava flows (total thickness ≈10 km, La Negra Formation) represent the volcanic surface level of the arc of that time.

CHI-7 belongs to the Cerro Las Toscas granite massif and is situated east of the Atacama fault zone in an area corresponding to the eastern side of the Jurassic-early Cretaceous magmatic arc. The U-Pb analyses of zircons revealed an age of 174±16 Ma (Damm et al. 1986). Near the sample point the granite is pervaded by N-S directed gabbroic dykes, probably also of Jurassic age, indicating not only an intense heating but also extensional tectonics at that time. The top of the pluton intruded into lower Palaeozoic metasediments, which in this region are overlain by a late Palaeozoic to early Liassic sedimentary sequence and the Jurassic arc lavas. As the thickness of the Jurassic lava sequence outside the central zone of the arc was probably much less than the 10 km mentioned before, the thickness of the roof of the pluton at the time of its intrusion can be very roughly estimated at 5 to 8 km.

As the three sampled plutons of the Coastal Cordillera (CHI-1, CHI-5, CHI-7) have almost concordant biotite K-Ar ages of ≈138 Ma, despite their different intrusion levels, the conclusion can be drawn that these ages reflect cooling due to tectonic uplift. With regard to the fission track ages, the subsequent history of the three parts of the Coastal Cordillera develops differently. In the westernmost part, early Cretaceous tectonism manifests itself in sinistral shear movements along the Atacama fault zone (Scheuber and Andriessen 1990) and rapid uplift revealed by unconformable clastics on top of the La Negra Fm. (Rössling 1989). The fission track length measurements of CHI-1 apatite show a mean length of 14.7±1.3 μm and this is a strong indication that this western pluton came to the surface relatively fast around 120 Ma ago. Continuous and conformable sedimentation in the backarc area during that time gives evidence that the major tectonic movements were confined to the magmatic arc and particularly to its central part.

The fact that several Rb-Sr whole rock, K-Ar and Ar-Ar mineral and U-Pb zircon age determinations of igneous rocks of the western part of the Coastal Cordillera provided early Cretaceous ages (Halpern 1978; Berg and Baumann 1985; Damm et al. 1986; Maksaev et al. 1988; Hervé and Marinovic 1989; Maksaev 1990) and that apatites yielded similar fission track ages and fission track length distributions as sample CHI-1 (Damm et al. 1990; Pichowiak et al. 1990, Maksaev 1990, Damm et al. this Vol.) strongly suggests that, in this whole segment, the tectonically most active time of the arc system in the Coastal Cordillera was towards the end of its intrusive history. Tectonic events of the magmatic arc system at 152, 143 and 126 Ma were revealed by detailed geochronological investigations of mylonitic rocks along the Atacama fault zone with both Rb-Sr whole rock and ^{40}Ar-^{39}Ar mineral dating systems (Scheuber et al. 1990; Scheuber et al. this Vol.).

In the Coastal Cordillera tectonic activity continued into the late Cretaceous as shown especially by the 85±13 Ma apatite age of CHI-5. The mean lengths of the confined spontaneous fission tracks of CHI-5 and CHI-7, at 11.8±1.9 and 12.2±1.5 μm, respectively, and the length distributions show that the host rocks rose up more slowly than CHI-1, spending more time at temperatures where partial annealing took place. The apparent apatite ages of approximately 80 Ma for both samples coincide more or less with the age of the intrusion of the Sierra San Cristobal pluton (CHI-2) in the mid-Cretaceous arc system, and these ages are slightly younger than the Peruvian event of tectonic shortening that affected this latter arc. Thus, the two plutons of CHI-5 and CHI-7 of the Coastal Cordillera were finally uplifted to the surface during the late Cretaceous when they had become part of the forearc.

5 The Mid-Cretaceous Magmatic Arc System

Andesitic lavas with ages of approximately 100 Ma (Rogers 1985) and overlying marine Jurassic backarc sediments and Lower Cretaceous continental sandstones, and plutons part of which have a slightly younger age, indicate the existence of a magmatic arc system during mid-Cretaceous times in the area of the Longitudinal Valley and western parts of the Chilean Precordillera (Sierra de Moreno and Cordillera de Domeyko). The samples CHI-8 and CHI-2 represent igneous rocks of mainly granodioritic composition which can be ascribed to the mid-Cretaceous arc system.

CHI-8 was taken from a granitic stock which intruded Palaeozoic sediments at a depth of probably not more than 5 km, considering the thickness of the Palaeozoic and Mesozoic sequences. A time gap of some 20 Ma between the K-Ar biotite age of 91 Ma and the fission track ages of sphene, zircon, and apatite can be ascribed to uplift that occurred during the transition from the mid-Cretaceous arc system to the late Cretaceous-Palaeogene arc system. Within the large errors limits, the fission track ages are concordant at about 72 Ma, suggesting a tectonic event that finally brought the pluton to the surface. The mean length of the confined fission tracks in apatite at 13.1 ± 1.7 μm and the negative skewness of the distribution are, however, indicative of a steady cooling history (Gleadow et al. 1986).

CHI-2 comes from the large granodioritic-tonalitic Sierra San Cristobal pluton emplaced at a depth of only 1 to 3 km as indicated by its intrusion in a folded sequence of Jurassic-lower Cretaceous sediments and overlying mid-Cretaceous andesitic lavas (Muñoz et al. 1989). The sediments were deposited in the backarc region of the previous Jurassic-early Cretaceous arc system, while the lavas correspond to the effusive activity of the Mid-Cretaceous arc system. The San Cristobal pluton is dated at 78-76 Ma (Maksaev 1990: ^{40}Ar-^{39}Ar; Pichowiak this Vol.: Rb-Sr whole rock). However, all the analysed mineral systems yielded concordant early Tertiary ages of around 60 Ma. The fission track lengths of apatite, with a mean value of 15.4 ± 1.3 μm and a narrow distribution, strongly suggest rapid cooling from temperatures above 125 °C to surface temperature. Similar results for this pluton have been obtained by Maksaev (1990).

The age of folding of the sedimentary sequence mentioned above is limited to the time interval between the intrusion of the Sierra San Cristobal pluton and a presumed age of the mid-Cretaceous andesitic lavas of about 100 Ma. The deformation thus corresponds to the Peruvian phase of Santonian age (Mégard 1987). The intrusion itself may be regarded as post-tectonic and the final magmatic activity of the mid-Cretaceous arc system. The K-Ar biotite system of CHI-2 indicates temperatures in the Sierra San Cristobal pluton at the sample location of above some 300 to 325 °C during the earliest Tertiary, before cooling to surface temperature occurred. As the post-intrusive tectonic activity in this region was rather modest, early Tertiary cooling must be explained by strong degradation of the land surface. This was perhaps a consequence of rapid upwarping related to the emplacement of the subsequent late Cretaceous and Palaeogene magmatic arc system in the Chilean Precordillera. Lavas and ignimbrites of this age overlie unconformably the folded older rocks in the region of Sierra San Cristobal. Another possible explanation is reheating or continued heating of the area by shallow subvolcanic intrusions associated with initial phases of the late Cretaceous-Palaeogene arc system, which spatially overlapped the previous one.

If we look at the very regular curve of progressively younger apatite fission track ages from CHI-5 in the Coastal Cordillera to CHI-2 in the Longitudial Valley and samples at even more easterly located sites (Fig. 4), then it becomes evident that the cooling of the crust advanced continuously eastwards, in contrast to the stepwise migration of the magmatic arc systems.

6 The Late Cretaceous-Palaeogene Magmatic Arc System

About 72 Ma ago the late Cretaceous-Palaeogene arc system was emplaced in the north Chilean Precordillera and in the adjacent parts of the Chilean Central Valley, where it overlapped the area of the previous mid-Cretaceous arc system. In this western part, the rather irregularly distributed late Cretaceous-Palaeogene lavas and pyroclastics were deposited upon an erosional surface and covered, with an angular unconformity, the structures of the preceding Peruvian phase shortening and also the San Cristobal pluton of CHI-2. As pointed out above, the CHI-2 ages of cooling from more than 300 °C to surface temperatures during the Palaeocene are due to processes related with the late Cretaceous-Palaeogene arc system. In the eastern part of this arc and outside the area of Peruvian phase deformation, the extrusive products covered disconformably Mesozoic and late Palaeozoic sedimentary and volcanic sequences as well as late Palaeozoic intrusive complexes (e.g. that of sample CHI-6). Still farther to the east and near the location of sample CHI-3 (Cerro Quimal pluton), the elevated magmatic arc area bordered, probably along normal faults, on a contemporaneously subsiding sedimentary basin which is considered to have been an epicontinental backarc basin (Charrier and Reutter this Vol.). It extended eastwards as far as the Eastern Cordillera of Argentina. Samples CHI-3 and ARG-1 were taken from the crystalline basement which had been buried under a cover more than 2 km thick of mostly continental backarc sediments.

CHI-6 was gathered from a late Palaeozoic granite,

cropping out in the core of a basement anticline in the eastern part of the late Cretaceous-Palaeogene magmatic arc area. The biotite K-Ar system yielded an early Permian age of 279±11 Ma, similar to the Rb-Sr whole rock age of 285±32 Ma (Baeza and Pichowiak 1988). The volcanic and plutonic rocks of late Palaeozoic age are interpreted as magmatic arc products of a late Carboniferous to early Triassic subduction regime (Coira et al. 1982; Breitkreuz 1990). The zircon fission track age of 213±34 Ma may be an expression of uplift and erosion prior to the deposition of Jurassic marine and early Cretaceous continental sediments (this is the backarc area of the Jurassic-early Cretaceous arc system; Prinz et al. this Vol.), and most probably represents a mixed age.

In the Palaeozoic granite of CHI-6, only the apatite fission track age of 50 Ma is related to the late Cretaceous-Palaeogene arc system. The bimodal length distribution, with a mean length of 12.7±2.8 μm, is typical for partial annealing of existing tracks, indicating a period of temperature increase and subsequent cooling to temperatures where annealing hardly takes place. The period of heating could correspond to an episode of burial beneath Mesozoic sediments and/or late Cretaceous-Palaeogene arc volcanics. The cover must have been thick enough to assure sufficient heating for partial annealing of fission tracks in apatite. Uplift of the basement anticline and consequent erosion is considered to have caused the slow cooling of the late Palaeozoic granite in the anticlinal core during the Eocene.

The Cerro Quimal pluton, from which CHI-4 was collected, is situated near the border between the elevated area of the late Cretaceous-Palaeogene arc system to the west and its backarc to the east, where sediments up to 3-4 km thick were deposited (Purilactis Group; Charrier and Reutter this Vol.). Previous K-Ar determinations from Cerro Quimal gave an age of 66.4±1.4 Ma and 64.6±1.1 Ma, similar to the ones obtained in this study (Maksaev et al. 1988). The pluton shows rapid cooling below 325 °C, because biotite, sphene, zircon and apatite ages coincide with a mean age of 62±6 Ma. The mean track length of apatite at 14.7±1.3 μm and the rather narrow distribution strongly favour a rapid ascent to the surface. Crustal uplift occurred to the west of the postulated normal fault system along which the backarc basin subsided (Charrier and Reutter this Vol.: Fig. 5), and erosion of the possibly 2-4-km-thick sequence of late Palaeozoic volcanics and Jurassic sediments at the roof of the pluton was facilitated by the presumably elevated topography of the arc area at the border to the extensional backarc area. Cerro Quimal and surroundings served as a source area for the clastic backarc sequences until the end of the Eocene.

The Chilean Precordillera was subject to a thermal and tectonic paroxysm in the late Eocene and early Oligocene (Incaic phase; see Scheuber et al. this Vol.). During that time shortening reached a maximum and, in combination with important dextral orogen-parallel strike slip movements (e.g. West Fissure of the Chuquicamata copper mine; Reutter et al. 1991), reveals the influence of strong dextral transpression. As our samples do not include plutonic material of that time, we report here published data from igneous bodies of Chuquicamata, situated in the Precordillera approximately 100 km to the north of Cerro Quimal: U-Pb 43±10 and 42±10 Ma, Rb-Sr whole rock 34.2±4 Ma, K-Ar 50-34 Ma, ^{40}Ar-^{39}Ar 39-32 Ma, and apatite fission track 38-31 Ma (Maksaev et al. 1988; Maksaev 1990). The age data show rapid cooling after intrusion.

7 Eastern Palaeozoic Basement Intrusives

CHI-3 and ARG-1 represent the basement of the backarc region of the late Cretaceous-Palaeogene arc. It is characterized by the widespread deposition of mostly continental sediments which, at least locally, attained a thickness of more than 2 km. The modern (Miocene to present) magmatic arc system of the Western Cordillera was emplaced in this former backarc area.

CHI-3 is from the Ordovician Tucúcaro pluton whose U-Pb zircon age is 450±11 Ma (Damm et al. 1986, 1990, this Vol.), and whose biotite K-Ar and zircon fission track ages are late Devonian and late Permian respectively. The geological interpretation of these dates is probably complex because of the influence of post-intrusive tectonic processes, clearly demonstrated by the apatite fission track cooling ages of 38.6±5.6 Ma. Thus, relatively late in its history, the pluton was situated at a depth of at least 3-4 km and the final uplift occurred in the Tertiary. The estimation of such a burial depth follows from the length measurement of the confined spontaneous fission tracks in apatite, which show no tracks smaller than 10 μm. This points to total annealing of the pre-existing tracks and is contrary to CHI-6 from the Palaeozoic core of a basement anticline in the Precordillera and whose apatite tracks were only partially annealed. The mean length of 13.4±1.2 μm and the negative skewness of the distribution are typical of a steady cooling history, from the total annealing zone (>125 °C) to the partial annealing zone (125 to 75 °C), and to the surface temperature conditions (Gleadow et al. 1986; Green et al. 1989). The Tucúcaro pluton was uplifted during the Incaic phase in the late Eocene, when strong shortening tectonics affected not only the magmatic arc in the Precordillera but also the respective backarc area.

Sample ARG-1 from the Santa Rosa de Tastil pluton in the Eastern Cordillera in Argentina is located some 300 km to the east of CHI-3. The granite intruded folded sediments are of late Precambrian to early Cambrian (?) age (Puncoviscana Group), and are unconformably overlain by the mid- to late Cambrian Meson Group. There are several radiometric ages which are more or less concordant with our K-Ar biotite age of 532 ± 21 Ma (Bachmann et al. 1987: U-Pb ≈ 533 Ma; Damm et al. 1990: ^{207}Pb-^{206}Pb zircon ≈ 546 Ma). The K-Ar biotite age is therefore interpreted as an expression of the early cooling history related to uplift and erosion preceding the sedimentation of the Meson Group. The fission track zircon age of 373 ± 160 Ma is most certainly a mixed age and no geological significance is given to it. The apatite fission track system, however, registers the Andean history. The broad distribution of the measured confined fission track lengths, the negative skewness and the mean length of 12.6 ± 2.1 μm all point to a continuous cooling during the Tertiary. The length distribution with relatively many tracks < 12 μm indicates that movement from the total annealing zone through the partial annealing zone and into the non-annealing zone must have been slow. The mid-Oligocene apatite fission track age (30.3 ± 3 Ma) must be viewed in connection with the late Eocene Incaic tectonic phase. Tectonic shortening commenced in the late Cretaceous-Palaeogene arc at about 38.5 Ma (Scheuber et al. this Vol.) and affected successively, for an unknown time and width, the backarc converting it to a complex foreland fold-and-thrust belt. As the sample must have spent a relatively long time in the partial annealing zone, it is proposed that the final uplift was caused by isostatic adjustment of the crust, which had been thickened by tectonic and/or sedimentary processes, as a consequence of a drop of the compressive stresses during the Oligocene.

8 Conclusions

1. The obtained mineral ages for the igneous rocks across Chile and northwestern Argentina indicate a succession of magmatic arcs and arc-related tectonic events. For the Palaeozoic Preandean Cycle, an east to west trend of decreasing age is obtained. In the Meso-Cenozoic Andean Cycle, the opposite trend is observed and ages decrease from west to east. This trend clearly demonstrates the eastward shift of the magmatic arc from the beginning of the Jurassic to recent times. The magmatic arc shifted stepwise to the east so that four arc systems of different age and location can be recognized. The data show that, with respect to the apatite fission track ages, the eastward retreat of the forearc-facing side of the arc systems was gradual rather than stepwise. The inland migration of some 200 km for the axis of arc magmatism in northern Chile since the early Jurassic, some 180-200 of Ma ago, indicates a long term average rate of 1 km/Ma.

2. Between the emplacement age of the igneous rocks belonging to the Andean Cycle and the mineral ages, a time-gap is generally observed. The high value of the mean length of the confined spontaneous fission tracks in apatite at 14-15.5 μm, together with the fact that several mineral age systems give concordant ages, indicate rapid uplift to the (near)surface amounting to 3 to >10 km. These quantities are supported by field observations. The cooling history of the plutons shows that during the evolution of this region, tectonic events accompanied the final activity of each magmatic arc system and/or the initial activity the succeeding one.

3. For the Palaeozoic basement rocks analysed in the Precordillera and eastwards, the apatite fission track ages and length measurements show a burial depth of some 3-4 km for the late Cretaceous-Palaeogene arc and backarc regions. At the end of the late Cretaceous-Palaeogene arc system, tectonic shortening of the Incaic phase affected not only the area of the magmatic arc but also the backarc area, where sort of a foreland fold-and-thrust belt developed. Cooling of the basement occurred by tectonic upheaval during compression or subsequent isostatic adjustment of the thickened crust during the Oligocene.

In general, the resulting thermo-tectonic histories of igneous rocks, which intruded in magmatic arcs related to different episodes of active subduction of the oceanic crust beneath the continental South American plate, indicate time-temperature pathways with periods of fast cooling and rapid tectonic uplifts.

It seems that such phenomena are common features not only in zones of long ocean-continent interaction, e.g. in other parts of the Andes (Kohn et al. 1984; Benjamin et al. 1987), southern Alps of New Zealand (Kamp et al. 1989) and coastal mountains of British Columbia (Parrish 1985), but also in continental collision zones as the Alps (Hurford 1986, 1991) and Tibet-Himalaya (Zeitler 1985; Lewis 1990).

Acknowledgements. This chapter would not have been possible without the efforts of the entire staff of the NWO Laboratory of Isotope Geology, Amsterdam. The authors are particularly indebted to E.H. Hebeda, I.S. Oen, E.A.Th. Verdurmen, R.H. Verschure, J.N. Vogel-Eissens, J.R. Wijbrans and L. Ijlst. We thank Prof. von Huene and Prof. Wagner for their constructive review and comments on an earlier version. Mrs. R. Jong Loy kindly typed the manuscript. This work forms part of the research programme of the 'Stichting voor Isotopen-Geologisch Onderzoek', and is financially supported by the Netherlands Organisation for the Advancement of Pure Research (NWO). Field work was carried out within the scope of the geoscientific research programme "Mobilität aktiver Kontinentalränder" of Freie and Technische Universität Berlin in cooperation with the Universidad del Norte, Antofagasta, and was financially supported by the Deutsche Forschungsgemeinschaft.

References

Andriessen PAM, Bos A (1986) Post Caledonian thermal evolution and crustal uplift in the Eidfjord area, western Norway. Nor Geol Tidskr 66: 243-250

Bachmann G, Grauert B, Kramm U, Lork A, Miller H (1987) El magmatismo del Cámbrico medio-Cámbrico superior en el noroeste Argentino: Investigaciones isotópicas y geocronológicas sobre los granitóides de los complejos intrusivos de Santa Rosa de Tastil y Cañani. 10 Congr Geol Argent Tucumán Actas 4: 125-127

Baeza L, Pichowiak S (1988) Ancient crystalline basement provinces in the north Chilean Central Andes: relics of continental crust development since the Mid-Proterozoic. In: Bahlburg H, Breitkreuz C, Giese P (eds.): The Southern Central Andes, Lecture Notes in Earth Sciences 17, pp 3-24

Benjamin MT, Johnson NM, Naeser CW (1987) Recent rapid uplift in the Bolivian Andes: evidence from fission track dating. Geology 15: 680-683

Berg K, Baumann A (1985) Plutonic and metasedimentary rocks from the Coastal Range of northern Chile: Rb-Sr and U-Pb isotopic systematics. Earth Planet Sci Lett 75: 101-115

Breitkreuz C (1990) Late Carboniferous to Triassic magmatism in the Central and Southern Andes: the change from accretionary to an erosive plate margin mirrors the Pangea history. Symp Int Géodyn Andine Grenoble, Orstom, Paris, pp 359-362

Coira B, Davidson C, Ramos V (1982) Tectonic and magmatic evolution of the Andes of northern Argentina and Chile. Earth Sci Rev 18: 303-332

Cordani UG, Amaral G, Kawashita K (1973) The Precambrian evolution of South America. Geol Rundsch 62: 309-317

Damm KW, Pichowiak S, Todt W (1986) Geochemie, Petrologie and Geochronologie der Plutonite und des metamorphen Grundgebirges in Nordchile. Berl Geowiss Abh 66, pp 73-146

Damm KW, Pichowiak S, Harmon RS, Todt W, Omarini R, Niemeyer H (1990) Pre-Mesozoic evolution of the Central Andes: the basement revisited. Geol Soc Am Spec Paper 241: 101-126

Gleadow AJW, Lovering JF (1978) Thermal history of granitic rocks from Western Victoria: a fission track dating study. J Geol Soc Aust 25: 323-340

Gleadow AJW, Duddy IR, Lovering JF (1983) Fission track analysis: a new tool for the evolution of thermal histories and hydrocarbon potential. Aust Petrol Explor Assoc J 23: 93-102

Gleadow AJW, Duddy IR, Green PF, Lovering JF (1986) Confined fission track lengths in apatite: a diagnostic tool for thermal history analysis. Contrib Mineral Petrol 94: 405-415

Green PF (1981) A new look at statistics in fission track dating. Nucl Tracks 5: 77-86

Green PF (1986) On the thermo-tectonic evolution of northern England: evidence from fission track analysis. Geol Mag 123: 493-506

Green PF, Duddy IR, Laslett GM, Hegarty KA, Gleadow AJW, Lovering JF (1989) Thermal annealing of fission tracks in apatite 4: quantitative modelling techniques and extension to geological timescales. Chem Geol (Isotope Geosci Sect) 79: 155-182

Halpern M (1978) Geological significance of Rb-Sr isotopic data of northern Chile crystalline rocks of the Andean orogene between latitudes 23° and 27° South. Geol Soc Amer Bull 89: 522-532

Harrison TM (1981) Diffusion of ^{40}Ar in hornblende. Contrib Mineral Petrol 78: 324-331

Harrison TM, McDougall I (1980) Investigations of an intrusive contact, northwest Nelson, New Zealand. 1. Thermal, chronological and isotopic constraints. Geochim Cosmochim Acta 44: 1985-2003

Hervé M, Marinovic N (1989) Geocronología y evolución del Batolito Vicuña Mackenna, Cordillera de la Costa, sur de Antofagasta (24-25°S). Rev Geol Chile 16: 31-49

Huene R von, Scholl DW (1991) Observations at convergent margins concerning sediment subduction, subduction erosion, and the growth of continental crust. Rev Geophysics 29: 279-316

Hurford AJ (1986) Cooling and uplift patterns in the Lepontine Alps, south central Switzerland and the age of vertical movements of the Insubric fault line. Contrib Mineral Petrol 92: 413-427

Hurford AJ (1990) Standardization of fission track calibration: recommendation by the Fission Track Working Group of the IUGS, subcommission on geochronology. Chem Geol (Isotope Geosci Sect) 80: 171-178

Hurford AJ (1991) Uplift and cooling pathways derived from fission track analysis and mica dating: a review. Geol Rundsch 80: 349-368

Kamp PJJ, Green PF, White SH (1989) Fission track analysis reveals character of collisional tectonics in New Zealand. Tectonics 8: 169-195

Kohn P, Shagan R, Banks PO, Burkley LA (1984) Mesozoic-Pleistocene fission track ages on rocks of the Venezuela Andes and their tectonic implications. Geol Soc Amer Mem 162: 365-384

Lewis CLE (1990) Thermal history of the Kunlum Batholith, N. Tibet, and implications for uplift of the Tibetan plateau. Nucl. tracks 17: 301-309

Maksaev V (1990) Metallogeny, geological evolution, and thermochronology of the Chilean Andes between latitudes 21° and 26° south and the origin of major porphyry copper deposits, Ph D thesis, Dalhousie University, Canada, 554 pp

Maksaev V, Boric R, Zentilli M, Reynolds PH (1988) Metallogenic implications of K-Ar, ^{40}Ar-^{39}Ar and fission track dates of mineralized areas in the Andes of northern Chile. 5 Cong Geol Chil Santiago Actas: B65-B86

Mégard F (1987) Cordilleran Andes and marginal Andes: a review of Andean geology north of the Arica Elbow (18°S). In: Monger JW, Francheteau J (eds) Circum-Pacific orogenic belts and evolution of the Pacific Ocean Basin, Am Geophys Union Geodyn Ser 18: 71-95

Miller H (1984) Orogenic development of the Argentinian/Chilean Andes during the Palaeozoic. J Geol Soc Lond 141: 885-892

Muñoz N, Charrier R, Pichowiak S (1989) Cretácico superior volcánico-sedimentario (Formacion Quebrada Mala) en la región de Antofagasta, Chile, y su significado geotectónico. Contrib. Simposios Cretacico de America Latina, Buenos Aires, A: Eventos y registro sedimentario, pp 113-148

Naeser CW (1979) Fission track dating and geological annealing of fission tracks. In: Jäger E, Hunziker JC (eds) Lectures in Isotope Geology. Springer, Berlin, Heidelberg, New York, pp 154-169

Parrish RP (1985) Cenozoic thermal evolution and tectonics of the coast mountains of British Columbia: fission track dating, apparent uplift rates and patterns of uplift. Tectonics 6: 601-631

Pichowiak S, Buchelt M, Damm KW (1990) Magmatic activity and tectonic setting of the early stages of the Andean cycle in northern Chile. Geol Soc Amer Spec Paper 21: 127-144

Purdy JW, Jäger E (1976) K-Ar ages on rock-forming minerals from the Central Alps. Mem Inst Geol Univ Padova 30: 1-31

Ramos VA, Jordan TE, Allmendinger RW, Mpodozis C, Kay SM, Cortes JM Palma M (1986) Palaeozoic terranes of the central Argentina-Chilean Andes. Tectonics 5: 855-880

Reutter KJ, Scheuber E, Helmcke D (1991) Structural evidence of orogen-parallel strike slip displacements in the Precordillera of Northern Chile. Geol Rundsch 80: 135-153

Rogers G (1985) A geochemical traverse across the North Chilean Andes. Ph D thesis, Dep. Earth Sci, Open University, Milton Keynes, 333 pp

Rössling R. (1989): Petrologie in einem tiefen Stockwerk des jurassischen magmatischen Bogens in der nordchilenischen Küstenkordillere südlich von Antofagasta. Berl geowiss Abh A 112, 73 pp

Scheuber E, Andriessen PAM (1990) The kinematic and geodynamic significance of the Atacama fault zone. J Struct Geol 12: 243-257

Scheuber E, Hammerschmidt K, Teufel S, Friedrichsen H (1990) R-Sr isotopic data from mylonites from the Atacama fault zone: preliminary results. Final Workshop Structure and Evolution of the Central Andes in northern Chile, southern Bolivia and northwestern Argentina, Berlin May 23-25, pp 97 (Abstr)

Scheuber E, Reutter KJ (1992) Relation between tectonics and magmatism in the Andes of northern Chile and adjacent areas between 21° and 25°S. Tectonophysics 205: 127-140

Steiger RH, Jäger E (1977) Subcommission on geochronology: convention on the use of decay constants in geo- and cosmochronology. Earth Planet Sci Lett 36: 359-362

Wagner GA (1968) Fission track dating of apatites. Earth Planet Sci Lett 4: 411-415

Zaun PE, Wagner GA (1985) Fission track stability in zircons under geological conditions, Nucl. Tracks 10: 303-307

Zeitler PK (1985) Cooling history of the NW Himalaya, Pakistan, Tectonics 4: 127-151

Geothermal and Tectonic Evolution of the Eastern Cordillera and the Subandean Ranges of Southern Bolivia

JONAS KLEY and MARTIN REINHARDT

Abstract. Combined tectonic and geothermal investigations along a profile from Villamontes to San Vicente provide new ideas about the evolution of the Eastern Cordillera and Subandean Ranges of southern Bolivia. Sediment maturity was determined using reflectance and infrared spectroscopy measurements on organic matter as well as the crystallinity of illites and ranges from diagenetic in the Subandean Ranges to weak metamorphic in the Eastern Cordillera. The tectonic evolution is characterized by Hercynian movements producing a "Protocordillera" and the Andean orogeny with substantial crustal shortening which is compensated at a shallower level mainly in the Subandean Ranges while the Eastern Cordillera remains relatively stable.

1 Introduction

East of the present magmatic arc, the geological situation of Bolivia is characterized by four major tectonic and morphological units. These are, from west to east: (1) the Altiplano highplains, the mountain chains of (2) the Eastern Cordillera and (3) the Subandean Ranges and (4) the Chaco lowlands.

This study investigates a cross section through the Eastern Cordillera and Subandean Ranges in southern Bolivia (Fig. 1), where the young (Andean) and old (Preandean) evolution of the eastern Andes can be studied. Within the cross section three parts can be distinguished: (1) in the east the fold-and-thrust belt of the Subandean Ranges represents an area of very young Andean movements, (2) in the west the Eastern Cordillera exhibits a quite different tectonic style which is not yet fully understood; between these main features (3) a transition zone exists.

The aim of the project was to distinguish between Andean and Preandean movements and to characterize their tectonic styles. The methods used included tectonic field studies and profile balancing, while investigations of organic matter and clay minerals, with subsequent modelling of maturity evolution, where used to reconstruct the geothermal and tectonic history.

Correspondence to: J. Kley, Fachrichtung Geologie, Freie Universität Berlin, Malteserstr. 74-100, D-1000 Berlin 46

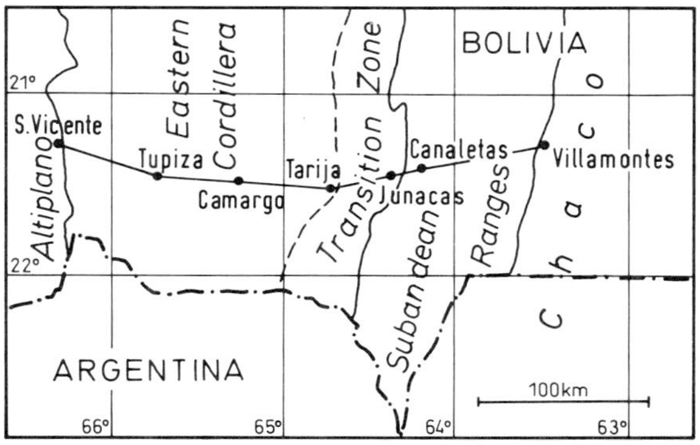

Fig. 1. Location of the study area

2 Stratigraphy

The pre-Quaternary sedimentary succession exposed in southern Bolivia comprises Precambrian to Neogene strata. It is almost exclusively composed of siliciclastic rocks which are of marine origin up to the Upper Devonian and are predominantly continental from the Pennsylvanian to the late Tertiary. The stratigraphy will be described in order from the oldest to the youngest formations. Where differences in the stratigraphic columns make it necessary, the stratigraphic successions will be treated separately for the single tectonic units introduced above (Fig. 2).

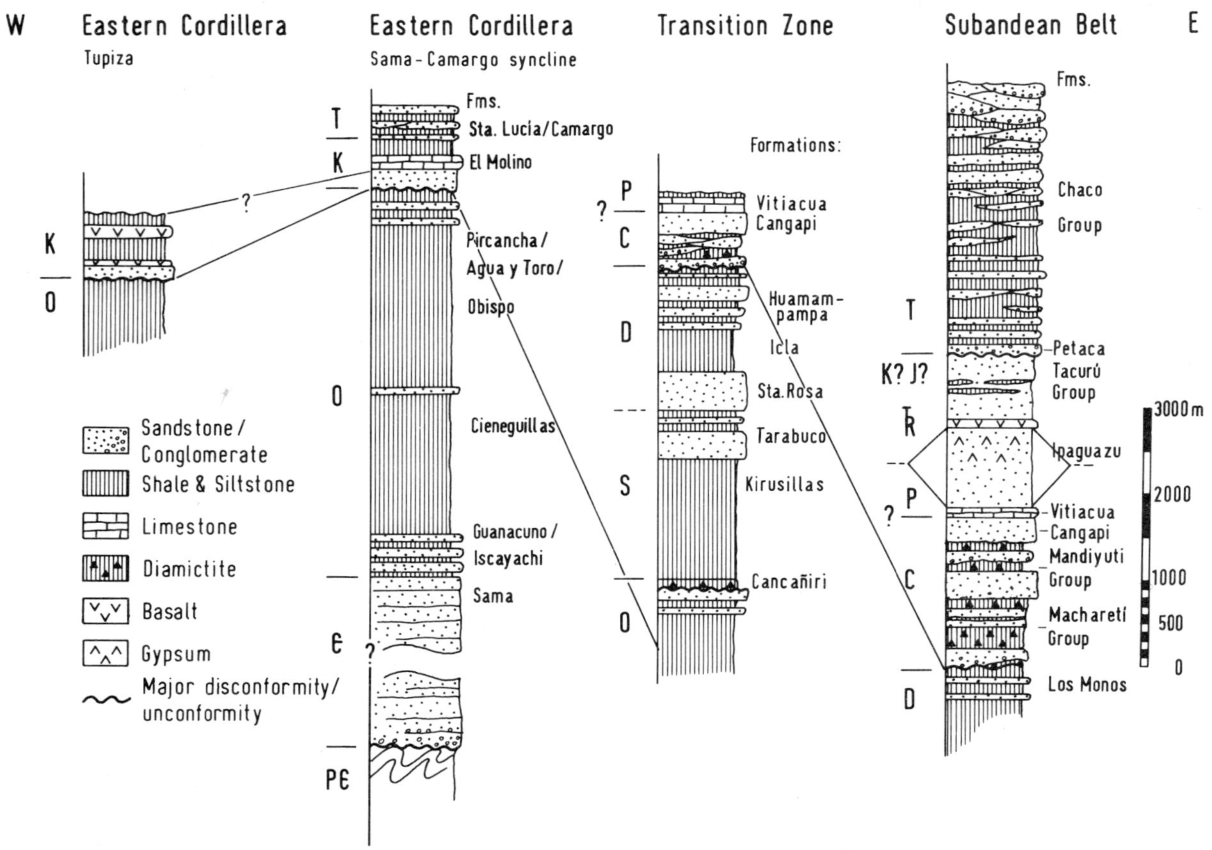

Fig. 2. Stratigraphic columns of Subandean Ranges, Transition Zone and Eastern Cordillera. Abbreviations: *T* Tertiary; *K* Cretaceous; *J* Jurassic; *Tr* Triassic; *P* Permian; *C* Carboniferous; *D* Devonian; *S* Silurian; *O* Ordovician; *ε* Cambrian; *Pε* Precambrian

2.1 Precambrian

Precambrian outcrops occur only in the southernmost part of Bolivia near the Argentine border. They consist of bluish-green phyllites and fine grained quartzites showing folds and slaty cleavage. These rocks are considered to be equivalent to the Puncoviscana Fm. of Argentina (Castaños and Rodrigo 1978).

2.2 Lower Palaeozoic

The Precambrian rocks are overlain with a marked angular unconformity by thick cross-bedded quartzites with conglomerates at their base. Their overall thickness is at least 1500 m and possibly as much as 3000 m. This sequence has been subdivided into three formations (Castaños and Rodrigo 1978) which are, from the base to the top: Condado, Torohuayco and Sama. The age of the Sama quartzites is assumed to be Upper Cambrian.

The Sama quartzites have a transitional upper limit to finer grained quartzitic sandstones with siltstone intercalations (Iscayachi Fm.) containing layers of poorly preserved shell debris. The overlying Guanacuno Fm., consisting of sand and siltstones, has a rich fauna mainly of trilobites which indicate a Lower Tremadocian age. These rocks grade upwards into a thick series of shales with some intercalated turbidites in the higher stratigraphic levels (Cieneguillas, Filoma, Obispo, Agua y Toro and Pircancha Fms). Graptolites give ages from Tremadocian at the base to middle Arenigian at the top of the sequence. The cumulative thickness of the Ordovician sediments is about 4500 m. The formation names cited above were introduced by Rivas et al. (1969), who studied the section along the road between the syncline of Camargo and the Sama Pass. It should be noted, however, that these lithostratigraphic divisions cannot easily be applied to neighbouring areas.

In the Transition Zone, the Ordovician shales described above grade rapidly upwards into a thin (ca. 150 m) sequence of quartzites showing wave ripples and desiccation cracks as well as many *Cruziana* traces, thus indicating a very shallow marine origin. Graptolites from shales directly underlying the quartzites are of early-middle Arenigian age.

The Arenigian quartzites are disconformably covered by the Ashgillian to Lower Llandoverian (Sempéré et al. 1988) Cancañiri Fm., a dark grey diamictite with a sandy matrix. It is only 100 m thick in the area studied.

This formation is overlain by dark shales with thin intercalations of fine grained sandstones (Kirusillas Fm./ Pampa Fm.). These grade upwards into a cross-bedded sandstone sequence with thin mudstone intercalations (Tarabuco Fm.) and finally into a cross-bedded medium- to coarse grained sandstone unit (Sta. Rosa Fm.). While a Silurian age is ascribed to the first two formations (Ludlovian; Mehl 1982), the exact position of the Silurian/Devonian boundary is controversial. Most authors now assume the Sta. Rosa Fm. to be Devonian in age.

There is a rapid transition to the fossiliferous Icla (Gamoneda) Fm., consisting mainly of bioturbate siltstones with brachiopods, tentaculites, conularians, trilobites and other fossils. Its age is late Siegenian to mid-Emsian (Isaacson 1977).

The Icla Fm. is overlain by the mid-Emsian to early Eifelian (Isaacson 1977) Huamampampa Fm., made up of several coarsening and thickening upward cycles ranging from siltstones to medium grained sandstones and even conglomerates, although the latter are rare.

In the Subandean Ranges, the oldest rocks exposed belong to the Givetian to Frasnian (Baby et al. 1989) Los Monos Fm. consisting mainly of dark siltstones and fine grained sandstones. In the area studied, the only outcrop of this formation is situated in the easternmost anticline of the Subandean Ranges near Villamontes. There, its exposed thickness is several hundred metres and the base is not visible. A probable equivalent of the Los Monos Formation, although less thick (0 - about 350 m) overlies the Huamampampa Fm. in some parts of the Transition Zone.

2.3 Upper Palaeozoic

The uppermost Devonian and at least part of the Mississippian are missing in the area studied (Helwig 1972, Rodrigo 1973). Carboniferous rocks are disconformably overlying different Devonian formations (Los Monos or Huamampampa). This sedimentary gap is linked to a major change from exclusively marine to predominantly continental sedimentation.

The Upper Mississippian and Pennsylvanian sediments show marked variations in facies and thickness from east to west. While in the east there is a thick succession (ca. 1500 m) of sandstones and diamictites showing good lateral continuity (Macharetí and Mandiyuti Group), in the west (Transition Zone) sandy and conglomeratic facies prevail. Here, the thickness is reduced to a few hundred metres and changes rapidly along with facies, indicating that the margin of the basin was situated close to the actual western outcrop limits of the Carboniferous. Tills(?) and tilloids containing faceted and striated pebbles indicate glacial influence throughout the Pennsylvanian (Helwig 1972).

2.4 Mesozoic and Cenozoic

2.4.1 Subandean Ranges and Transition Zone

In some places the following sandstones of the Cangapi Fm. cover sediments of the Mandiyuti Group with a marked disconformity. They have traditionally been combined with the conspicuous Vitiacua siliceous limestones and the Ipaguazú gypsiferous sandstones to form the Cuevo Group, which was presumed to be Triassic in age (Pareja et al. 1978). Palaeontological evidence for a Norian age of the shallow marine Vitiacua Fm., based on a pelecypod, is cited by Beltan et al. (1987). In contrast, Schlatter and Nederlof (1966) and Sempéré (1990b) assign to the Vitiacua Fm. an Upper Permian and probably Lower Triassic age based on palynological data. Sempéré considers the Cangapi Fm. to be of Pennsylvanian to Lower Permian age.

In the area studied, thicknesses of the above mentioned formations are: Cangapi about 400 m, Vitiacua 30-220 m, Ipaguazú (only to be found regionally) 0-1500(?) m.

There is an ongoing debate about the age of the overlaying Tacurú sandstones of fluvial and aeolian origin. Radiometric dating of a basalt which underlies these sediments in a restricted area of the Subandean Ranges yielded no unequivocal results. Oller and Sempéré (1990) assume sedimentary continuity between the Ipaguazú Fm. and the Tacurú Group. They suggest a Kimmeridgian age for the upper limit

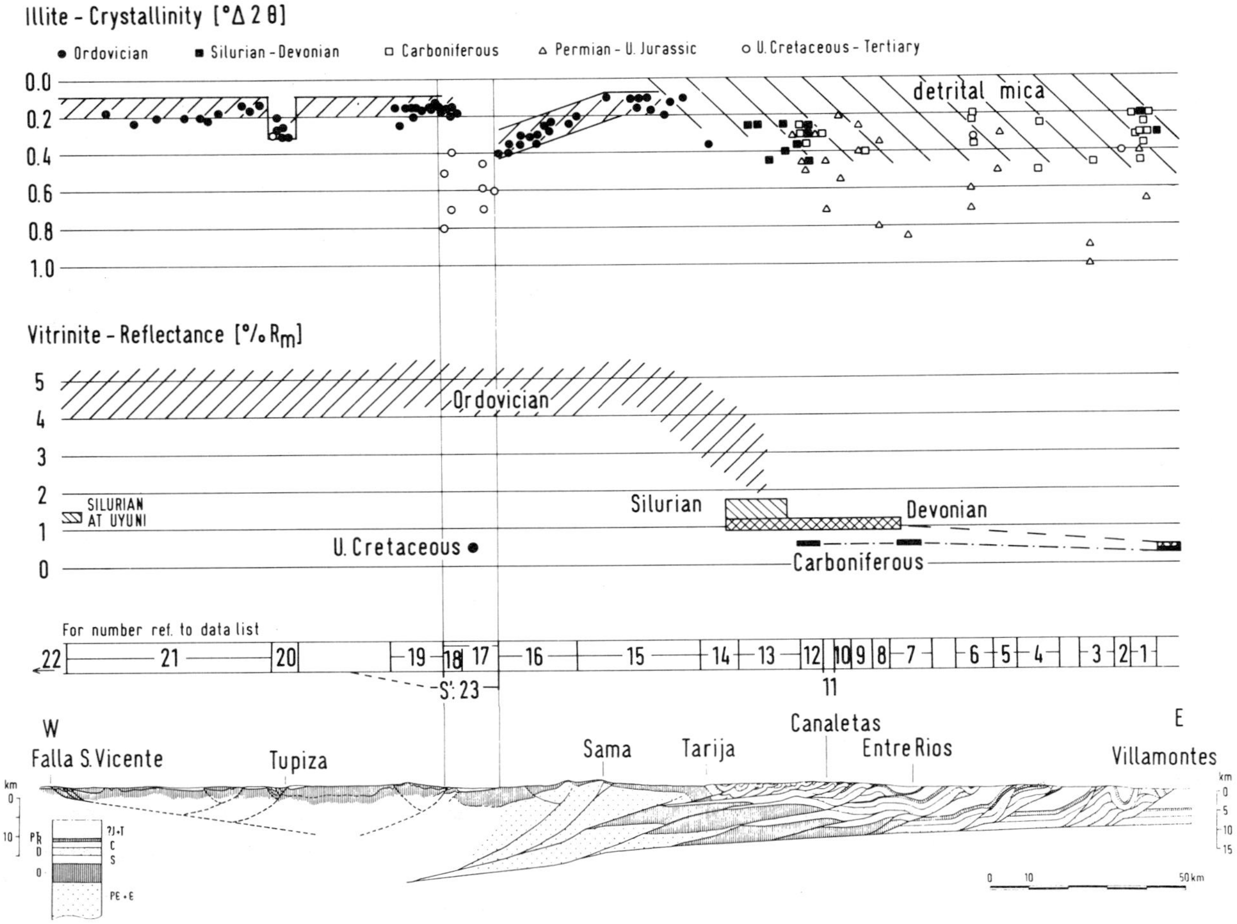

Fig. 3. E-W cross section with illite crystallinity and vitrinite reflectance data. Structures are drawn according to YPFB data (Perez 1971) and our own observations. The eastern part of the section is balanced up to the Sama anticline (15). Abbreviations: *Pε+ε* Precambrian and Cambrian; *O* Ordovician; *S* Silurian; *D* Devonian; *C* Carboniferous; *PTr* Permian/Triassic; *?J+T* ?Jurassic and Tertiary

of the Tacurú Group, ascribing phenomena of extensional tectonics documented in these sediments to the Upper Jurassic Araucan phase. Other authors (e.g. Pareja et al. 1978) attribute to the continental Tacurú sandstones an Upper Cretaceous age and correlate them with red sandstones of the Puca Group present in the Eastern Cordillera (see below). The Tacurú Group is 400 - 900 m thick in the area studied.

No pre-Quaternary sediments younger than the Vitiacua Fm. are preserved in the Transition Zone.

In the Subandean Ranges, after a long interval of non-deposition and erosion, sedimentation is resumed in the late Oligocene with the sandy to conglomeratic Petaca Fm. followed by a succession of fluvial sandstones and mudstones several kilometres thick called the Chaco Group. While the Petaca Fm. is well dated due to its vertebrate fossils, it is assumed that the Chaco Group reaches as far as the late Miocene and/or Pliocene (Pareja et al. 1978).

2.4.2 Eastern Cordillera

In the sector of the Eastern Cordillera considered here, the Ordovician is directly overlain by Cretaceous sediments with an angular unconformity. The Mesozoic cover is only preserved in the synclinal structures of Camargo and Tupiza where sediments and, in the latter case, also volcanic rocks of the Puca Group are exposed.

In the southern part of the Camargo syncline, red and white continental sandstones containing root casts are followed by the very shallow marine limestones and marls of the Campanian to Maastrichtian El Molino Fm. which, in turn, is overlain by the red mudstones and sandstones of the latest Cretaceous to Tertiary Santa Lucía and Camargo Fms..

At Tupiza, a light-coloured quartzitic sandstone is overlain by several hundred metres of red sandstones and mudstones with various intercalations of basalts. Sempéré et al. (1988) suggest that these rocks correlate with the basal sandstones of the southern Camargo syncline and are of a Turonian to Santonian age.

Conglomeratic to sandy Tertiary sediments (Oligocene-Pliocene) overlie the Ordovician in several basins near Tupiza and further west.

3 Tectonics

As outlined above, the section from Villamontes to San Vicente can be subdivided into three segments based on structural style and the stratigraphic level exposed. These segments are, from east to west: The Subandean belt between Villamontes and Entre Rios/Canaletas, a transition zone from Canaletas to Tarija; and the Eastern Cordillera proper, comprising the sector between Tarija and San Vicente. The cross section shown in Fig. 3 is based on unpublished data from Yacimientos Petrolíferos Fiscales Bolivianos (Perez 1971) and on our own observations.

3.1 Subandean Ranges

The Subandean belt is characterized by about six major east verging thrusts at 10 to 15-km intervals which are quite continuous along the strike (N-S). In its eastern part, antiformal stacks have developed (Dellapé and Patuel 1987) and no faults with great stratigraphic separation are exposed at the surface. Thrusts in the western Subandean Ranges show westward-dipping hanging-wall panels and compound ramp anticlines with eroded frontal parts. The widest stratigraphic separation observed on emerging thrusts is Mississippian to Neogene (corresponding to a thickness of more than 4000 m). Throughout the Subandean Ranges, a thick succession of predominantly sandy units (Mississippian to Neogene) is underlain by Devonian and older Palaeozoic strata, mainly shales and siltstones with some competent sandstone units. The exact position of the basal decollement within the lower shaly formations cannot be determined from surface data alone, as the deepest unit exposed is the Upper Devonian Los Monos Fm.. According to reflexion seismic and well data, the deepest detachment level in the Subandean Ranges is situated in the Silurian Kirusillas shales (Baby et al. 1989). The age of the deformation is late Miocene or younger, as the Neogene Chaco Group is involved in the structures.

Between Entre Rios and Canaletas a rapid rise in the topographic level occurs along with a change of tectonic style. The widely spaced thrusts and folds are replaced by thrust-bounded, immediately adjacent anticlines which become progressively tighter in a westerly direction (Fig.3).

3.2 Transition Zone

From Canaletas to Tarija, most outcrops are of Devonian and Silurian strata, with Carboniferous and Permo-Triassic rocks occuring only in tight synclines. No younger rocks are present, except for unconformingly overlying Pliocene to recent sediments. This area is called the Transition Zone (TZ) in this chapter and is equivalent to the "conjunto Tarija-Azurduy-Teoponte" of Sempéré et al. (1988). Folds in the TZ are smaller than in the Subandean Ranges and thrusts have reduced stratigraphic separation (about 2000 m), often bringing the Emsian in contact with the Permian. Moreover, most thrusts are not continuous along the strike but pass into folds over short distances, thus limiting the displacement that can be attributed to single faults. Vergence is eastward over most of the TZ with some westward thrusts being present in the westernmost parts near Tarija. After their initial formation, the structures underwent continuous or renewed shortening producing characteristic large scale kinks in fold limbs, hinge collapses in synclines with more competent units being squeezed out and, in a few cases, overturned thrusts (Fig.4). In the TZ, the depth to the nearest detachment can be determined from tight anticlines. We infer the basal decollement to rise from the Kirusillas shales (Silurian) in the west near Tarija to the Icla shales and siltstones (Devonian) in the east.

One might suspect that the differences in structural style between the Subandean belt and the TZ are due to lateral lithostratigraphic changes, and especially to the lack of a thick competent Mississippian-Pennsylvanian succession and a Mesozoic to Tertiary cover in the TZ. The reduced thickness of Carboniferous strata in the west certainly affects the size and shape of the structures. However, to attain

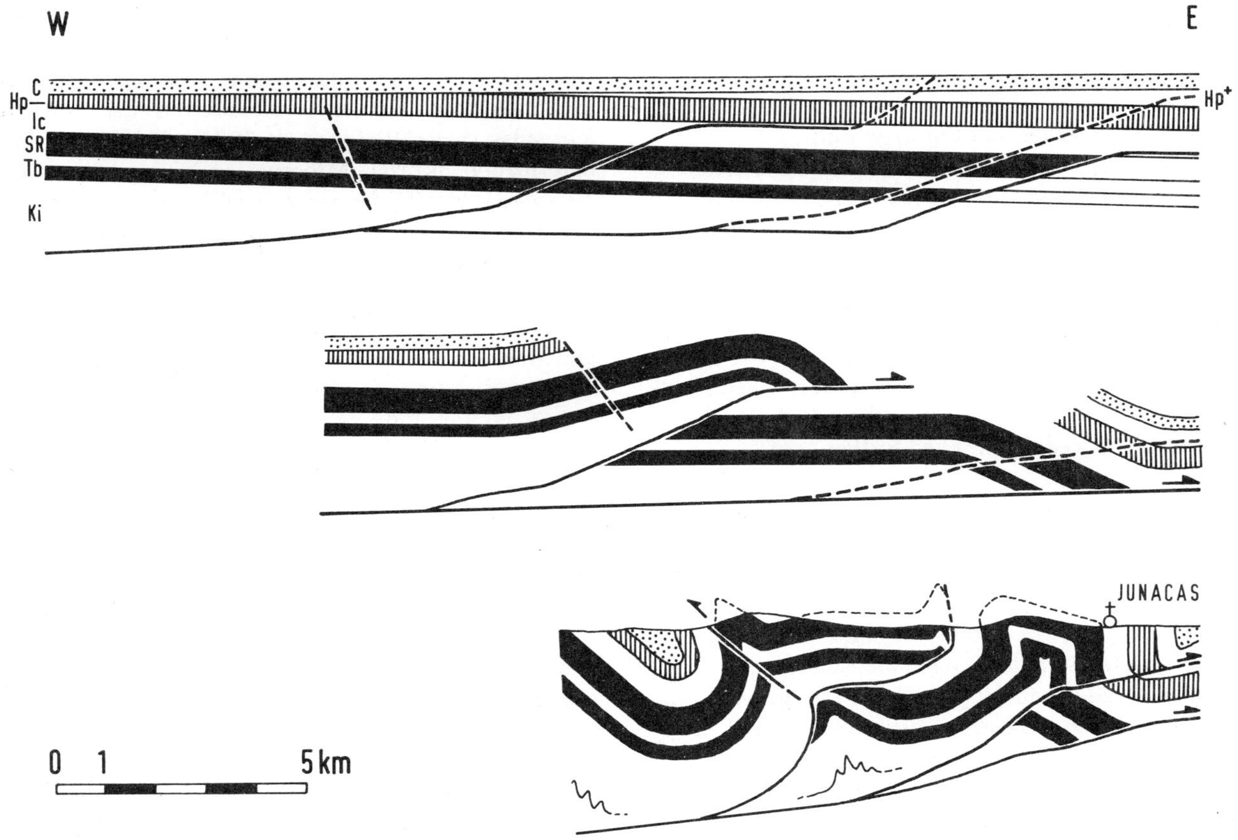

Fig. 4. Model for the development of structures in the Transition Zone. *C* Carboniferous; *Hp*$^{+}$ shaly facies overlying Hp; *Hp* Huamampampa sandstones; *Ic* Icla shales and siltstones; *SR* Santa Rosa sandstones; *Tb* Tarabuco sandstones with shaly intercalations; *Ki* Kirusillas shales. Note westward tapering of Hp^{+}, possibly due to slight tilt caused by Hercynian movements

the vitrinite reflectance values measured in Carboniferous sediments from the TZ, some additional cover of sediments (2-3 km) younger than Permo-Triassic is required. Considering the very tight structures which can be observed today, we deduce that most of this cover was probably eroded before the final structuration of the area.

West of Tarija, there is another spectacular rise in the topographic level from 2000 m to more than 4000 m, with Ordovician and Cambrian rocks surfacing in a huge anticline, the core of which includes the Precambrian sedimentary basement (Fig. 3.). This structure marks the eastern border of the Eastern Cordillera proper. It is best explained as being a large ramp anticline, with the TZ representing a wide and internally deformed hanging-wall "flat" segment of the same thrust sheet. (As pointed out above, the basal detachment of the TZ is not strictly parallel to bedding but rises at a shallow angle from west to east.) According to this theory, west verging thrusts in the western part of the TZ are passive roof back-thrusts.

Underlying the TZ, and a cause of its uplift in relation to the Subandean belt, there is probably another large anticlinal structure involving the basement. This can be inferred from a projection of structures farther south, near the Argentine border, which dip gently northwards (see Pareja et al. 1978).

A balanced cross section, based on the observations and assumptions outlined above produced a shortening value of about 140 km for the Subandean belt and the TZ (Fig.3).

It should be noted that the limits of the tectonic units introduced above for convenience do not correspond to single emergent thrusts with particularly large displacement.

3.3 Eastern Cordillera

The Eastern Cordillera is made up almost exclusively of early Ordovician sediments, mostly shales (Fig.3). The structural level is thus quite uniform, even if the great thickness of the Ordovician sediments (about 4500m) is taken into account. The border of the Eastern Cordillera with the Altiplano in the west is formed by a system of westward verging thrusts (Baby and Sempéré 1989). One important element of this thrust system is the emergent San Vicente fault. Younger rocks (Cretaceous and Tertiary) are only present in the Camargo syncline (see below) and around Tupiza. The Ordovician has a N-S trending, more or less vertical slaty cleavage which does not affect the Cretaceous rocks and provides proof of a compressional phase of Preandean age. The Cretaceous and Tertiary are involved in Andean folding and thrusting together with the underlying Palaeozoic rocks. According to Baby et al. (1990), deformation started in the late Oligocene which is markedly earlier than the onset of Andean tectonics in the Subandean zone.

The role of Preandean (mainly pre-Cretaceous) versus Andean deformation cannot be easily assessed. The very low angle unconformity between the Ordovician and the Cretaceous, present in many parts of the Camargo syncline, seems to indicate only minor folding before Cretaceous times. However, in many regions without a cover of younger rocks, the Ordovician shows non-parallel folds of up to several kilometres in size. The slaty cleavage is consistent with the more-or-less vertical axial planes of these folds, suggesting that they are Preandean structures which have not suffered any strong Andean overprint. This apparent contradiction could be solved by assuming that the Andean deformation was strongest in those sectors which had been little affected by earlier folding. It seems reasonable that the already intensely folded segments might have behaved like relatively rigid blocks during Andean compression.

The pinpointing of the exact age of the pre-Cretaceous deformation cannot be limited by direct geological observations from the Eastern Cordillera itself. Arguments for a late Devonian-earliest Carboniferous event, based on geothermal and stratigraphic data, will be discussed in the next section.

There is evidence that the Andean age structures of the Eastern Cordillera are mostly thick-skinned. In the core of an internal anticline within the Camargo syncline, Tremadocian quartzites are exposed, ruling out a detachment horizon in the mid-Ordovician shales for this structure at least. (No deeper potential detachment level is present in the stratigraphic column), The steep upthrust bounding the Camargo syncline in the west has an important dextral strike-slip component, for which north-plunging axes of drag folds are evidence. Major strike-slip movements are also assigned to faults of the Tupiza region (Hérail et al. 1990), thus excluding a purely thin-skinned style of deformation.

Taking this into account, the contribution of Eastern Cordilleran structures to overall Andean age shortening and crustal thickening seems to be minor in the transect considered here.

4 Geothermometry

The rocks exposed along the section are almost exclusively clastic sediments. Measurements of the reflectance of organic matter and determinations of illite crystallinity were carried out to investigate the diagenetic and metamorphic state of these rocks. The data obtained are listed in an appendix. The limits between diagenesis and metamorphism are given in Fig. 5.

4.1 Methods

4.1.1 Reflectance of Vitrinite and other Organic Matter

The mean reflectance (Rm) of organic matter (OM) was measured on polished surfaces of crushed rock specimens using a 50x lens and oil immersion. An advantage of this method is that it enabled us to distinguish between different components of the OM optically and thus avoid the measuring of reworked or oxidized particles. Along the section many sediments show a high degree of weathering and only a few formations are suitable for reflectance measurements.

Vitrinite s.str. originates from higher land plants and therefore occurs only in Devonian or younger sediments. In older strata the reflectance of other types of OM was measured and then normalized to vitrinite for comparison. In the Silurian, remains of primitive land plants are found and these have an optical behaviour similar to that of vitrinite, liptinite or bituminite. The reflectance values of 1-2% for all three types of OM are very similar in this range; e.g. bituminite reflectances in the range 1-2% R_m are very close to those of vitrinite (Jacob 1989, Fig.11). In Silurian and Ordovician strata, bituminite-like OM

method / metamorphic stage	Vitrinite-reflectance [% Rm]	Illite-crystallinity [mm]	[°Δ2Θ]
diagenesis			
boundary	2.7 - 3.4[1]	5.5[2]	.42[3] — .38[4]
very-low-grade (anchizone)			
boundary	~5[1]	3.5[2]	.25[3] — .21[4]
low-grade (epizone)			

1) after FREY et al. 1980, p. 199
2) after THOREZ 1976, p. 15
3) after KÜBLER 1984 in Frey 1987, p. 17
4) after KISCH 1980, p. 275

Fig. 5. Limits between diagenesis, very-low-grade metamorphism and low-grade metamorphism for illite crystallinity and vitrinite reflectance.

often takes the form of filaments, bands or hollow, round structures. In Ordovician rocks the maximum reflectance (R_{max}) of graptolites was measured on whole rock specimens, cut parallel or perpendicular to the bedding plane, and on crushed rock specimens. The correlation with vitrinite proposed by Goodarzi and Norford (1989) is valid for values from parallel cuts which (at least in our samples) are about 1.4 times higher than those from perpendicular cuts.

4.1.2 Infrared Spectroscopy (IRS) of Organic Matter

The preparation and measuring procedures were carried out according to Ganz and Kalkreuth (1991). In contrast to reflectance measurements, the whole OM is used and different organic components cannot be distinguished between. This leads to the problem of an unknown amount of reworked material which can shift the measured maturity to higher values. On the other hand, this method does not depend on a certain type of maceral (e.g. vitrinite) and allows investigations into different types of kerogen. The correlation with vitrinite has been elaborated on coals of a rank up to approximately 4% R_m (Ganz and Kalkreuth 1991). It has been shown that the minimum wave number (W_{min}) correlates well with the evolution of the vitrinite reflectance and is not influenced by the effects of weathering. In this study IRS values from weakly metamorphic Ordovician sediments show lower maturities than graptolite reflectances and illite crystallinities.

4.1.3 Illite Crystallinity (IC)

This method provides the best results in the range from strong diagenesis to low-grade metamorphism (epizone). The clay fraction of $<2\mu m$ is used (methods after Dunoyer de Segonzac 1969 and Thorez 1975, 1976, modified) and the half-height peak width of the 10Å peak of illite (in mm and °Delta 2Θ) is determined on the X-ray diffractogram. One problem concerning the interpretation of the data is that reworked material cannot be identified (as with IRS). Samples from the Subandean Ranges show a predominant amount of reworked low-grade metamorphic illites concealing the true (diagenetic) state of the sediments.

4.1.4 Fluid Inclusions

J. Mullis (Basel) determined p-T conditions for seven samples of quartz veins occuring in Lower Ordovician sediments. Two samples contain only saline water and give minimal homogenization temperatures (Th) of ≈270 °C; in four samples Th ranges between ca. 220 and 270°C and one sample gives Th values between 195 and 270 °C. The last five samples are undersaturated with a small amount of methane and, thus, the measured Th values are too low. According to J. Mullis (pers. commun.) the true inclusion temperatures probably range between about 240 and 270 °C.

4.2 Results of Geothermometric Investigations

The state of the sediments along the transect varies between very weakly diagenetic in the Subandean Ranges to low-grade metamorphic in the Eastern Cordillera. Details of the geothermal history will now be discussed.

4.2.1 Geothermal Modelling of the External Subandean Ranges

In the external Subandean Ranges the maturity of the Upper Devonian Los Monos Formation is very low, reaching 0.45-0.5% R_m at the surface as well as at a depth of 4200 m. Temperature and maturation history have been modelled on the basis of the Arrhenius equation, calculating, first, time-temperature indices (TTI) and then corresponding vitrinite reflectances (Wood 1988) (Fig. 6). Input data are burial history, activation energy (218 kJ/mol), frequency factor (5.45×10^{26}/Ma), surface temperature (10 °C) and geothermal gradient. During modelling the burial history and (restrictedly) geothermal gradient are

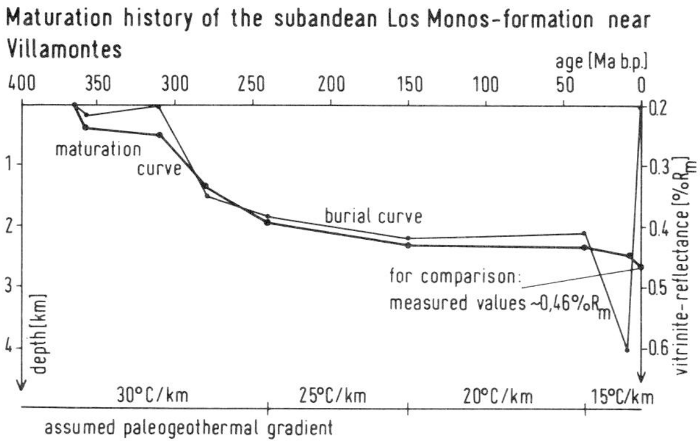

Fig. 6. Burial history and calculated maturation history for the Los Monos Formation (Upper Devonian) in the anticline of Aguarague near Villamontes

adjusted within reasonable limits to make modelled and measured data tally. The burial history is characterized by long periods of non-deposition or erosion (Lower Carboniferous, Cretaceous and Lower Tertiary are absent) and several formations cannot be given an exact date (see above), but lithology and facies suggest that at the most a few hundred metres of sediment are missing from the Mesozoic and Cenozoic record.

Thus, the assumed burial history (Fig.6) shows a period of subsidence during the Upper Carboniferous followed by a long interval of constant burial depth and a very short period of deep subsidence with subsequent uplift. The remaining variable factor for modelling is the evolution of the geothermal gradient. During much of its history the region considered in this study was relatively stable and, so, no unusually high or low gradient should be expected; but during Tertiary times the geothermal gradient dropped to values of 15-20 °C/km; such values can be measured today in boreholes (Robles 1987). The most effective model uses a geothermal gradient of $\leq$30 °C/km up to 150 Ma ago, which then decreases to 20 °C/km up to 35 Ma ago and subsequently to 15 °C/km to the present (Fig.6).

The maturity data of older parts of the Devonian could be determined for borehole samples made available by YPFB. The measured values can be simulated using the assumptions and data given above. Maximum temperatures of about 80 °C in the Los Monos Formation were reached within the last 5 Ma at a depth of 4 km. Checking this result with other methods, estimations of the recent maturity give values of approx. 0.43% R_m (after Bostick et al. 1979; T_{eff}=3 Ma) and 0.56% R_m (after Barker and Pawlewicz 1986). The latter value seems to be too high, probably because of the very short heating time which is not taken into account by the algorithm used: ($\ln[R_m]=0.0078\ T_{max} - 1.2$). The maturity of the OM corresponds to conditions at the lowest limit of the oil window.

4.2.2 Lateral Maturity Trends From the Subandean Ranges to the Transition Zone

Upper Carboniferous strata show a slight E-W increase in maturity ranging from about 0.38% R_m in the Subandean Ranges to 0.50% R_m in the Transition Zone. This can be related to a corresponding lateral increase in the geothermal gradient (Fig. 3). In underlying Devonian rocks, however, the increase is much more significant, with reflectance values rising from about 0.45% R_m in the Subandean Ranges to about 1.0% R_m in the western part. The resulting sudden decrease in rank from Devonian to Carboniferous strata in the Transition Zone requires an Upper Devonian?/Lower Carboniferous (Hercynian) heating event (see below).

4.2.3 What Happened in the Transition Zone During the Late Palaeozoic?

East of Tarija, the Devonian and Silurian strata underwent a different burial history from that of their equivalents further east in the Subandean Ranges (Fig. 7 and see above).

As previously mentioned, a sudden decrease in the degree of coalification (from about 1% R_m in the Devonian to about 0.5% R_m in the Upper Carboniferous) indicates that the coalification of the Devonian strata had been terminated before the Upper Carboniferous. Extrapolating the Upper Devonian value of 1% R_m to a surface value of about 0.25% R_m, using the vertical R_m gradient of 0.10-0.14 log%R_m/km of Devonian and Silurian strata, requires at least 4-6 km of additional sedimentary cover (Fig. 8). This layer of sediment must have been eroded before the deposition of the Upper Carboniferous began.

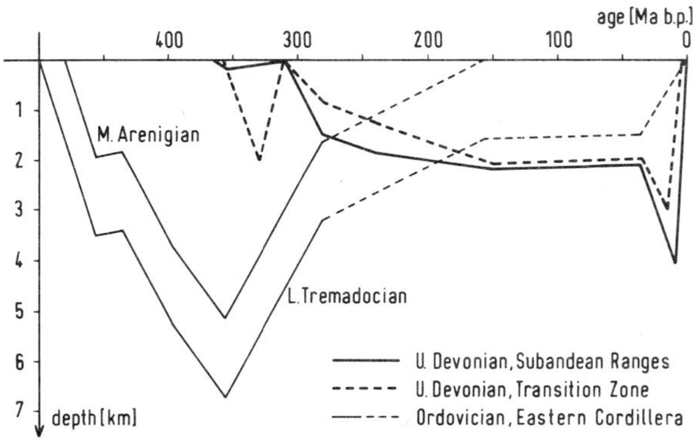

Fig. 7. Burial histories of Subandean Ranges, Transition Zone and Eastern Cordillera. During the Hercynian and the Andean orogeny, material eroded from the rising "Protocordillera" or Eastern Cordillera, respectively, was transported into the foredeep to the east (Transition Zone or Subandean Ranges, respectively)

4.2.4 The Eastern Cordillera: Age and Degree of Metamorphism

In the Eastern Cordillera illite crystallinity values for Ordovician rocks reach the zone of very-low-grade and low-grade metamorphism. In the eastern part (Sama anticline) at least, the degree of metamorphism is controlled by the stratigraphic position: The Ordovician sequence west of Tarija which was originally about 3.5 km thick suffered low-grade metamorphism at its base (Tremadocian) and very-low-grade conditions at its top (Arenigian).

These results are confirmed by graptolite reflectance values which also indicate weak metamorphism and show a similar stratigraphic dependence. An upper limit for the temperature can be set because no newly formed graphite occurs. However, a few samples reach a transitional stage between "vitrinite" and graphite indicating temperatures of 260-330 °C (according to Diessel et al. 1978). This temperature range is confirmed by fluid inclusion data from two samples giving homogenization temperatures of at least 270 °C (see Sect. 4.1.4).

Preandean tectonics left their mark in the form of a N-S trending, vertical slaty cleavage of Ordovician rocks, but the influence of Hercynian (Lower Carboniferous) and/or Ocloyic (Upper Ordovician) movements is still under discussion. Proof for Hercynian movements in Peru and northern Bolivia was described by Martinez (1980) who assumed that their influence continued farther south to the Eastern Cordillera of southern Bolivia and Argentina. In contrast, an Ocloyic phase affected the Argentinian Puna and the southern Altiplano (Coira et al. 1982). Data obtained during this study introduce new arguments for the dating of thermal events in the Eastern Cordillera.

Geothermometric data indicate temperatures of nearly 300 °C at the base of the thick Ordovician sequence which grades upwards into a litoral facies overlain by Silurian strata with no unconformity (north of Tarija). If one tentatively relates these temperatures to the Ocloyic orogeny and infers an elevated geothermal gradient of about 45 °C/km, a burial depth of about 6 km results for the lowest Tremadocian. However, only about 4 km of Ordovician sediments are present today and these were probably thickened considerably by tectonic compression. Thus, at least 1.5-3 km of additional Ordovician sediment are missing; this would have been deposited and eroded during Llanvirnian-Ashgillian times (Ocloyic orogeny).

Alternatively, the main heating may be of Hercynian age: at the end of the Devonian lowest Ordovician rocks were covered by about 6-7 km of Ordovician, Silurian and Devonian sediments. A surplus of 4-6 km of Uppermost Devonian and/or Lower Carboniferous would have to be added (see Sect. 4.2.3) due to the sudden decrease of coalification from the Devonian to the Carboniferous. The resulting sediment layer of 10-13 km would not require an elevated geothermal gradient to explain the high temperatures at the base of the Ordovician. In addition, facies distribution of Carboniferous rocks indicates the presence of an elevated area west of Tarija which has been lifted during the Lower Carboniferous. Thus, a Hercynian age of metamorphism seems more probable than an Ocloyic one.

5 The Camargo Structure

This important structure provides a good example of how results from the different methods used were combined to unravel the tectonic and thermal history of an area.

The Camargo syncline is about 150 km long and up to 10 km wide. It is one of the few localities in the Eastern Cordillera where Cretaceous and Tertiary sediments in a diagenetic state cover weakly metamorphic Ordovician rocks with a generally very slight angular unconformity (Fig. 9).

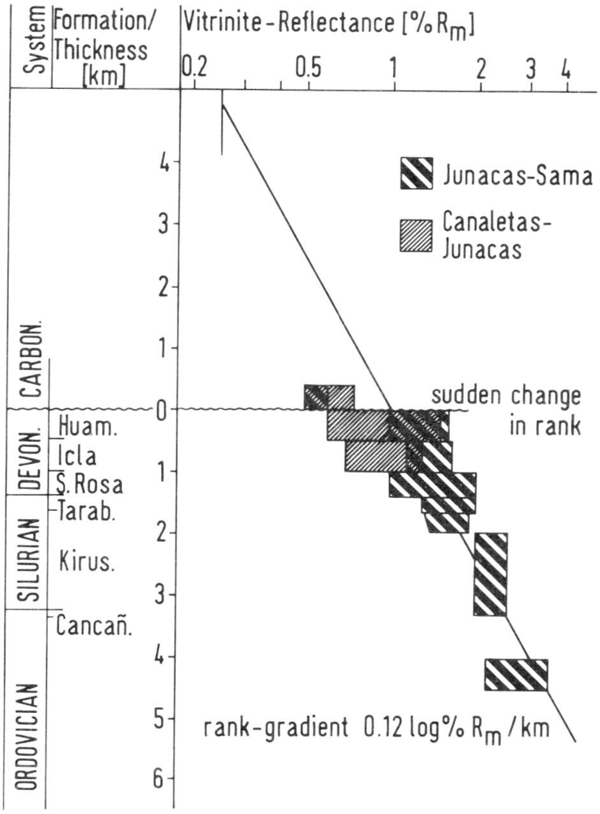

Fig. 8. Vertical vitrinite reflectance profile of the Transition Zone. Note the sudden decrease in rank from Upper Devonian to Upper? Carboniferous indicating a pre-Upper? Carboniferous coalification of the older sediments. Extrapolation of the vertical trend up to an assumed surface value of 0.25% R_m leads to the conclusion that about 4-5 km of sediment must have been deposited and eroded during the Hercynian orogeny to reach the measured sediment maturity before the Upper? Carboniferous

The east-vergent syncline is of Andean age and, as pointed out, is probably older than the Subandean structures. It exhibits components of E-W compression as well as of N-S trending dextral wrench faulting. In contrast the Ordovician rocks to the west of the syncline show upright symmetric folds with almost vertical axial planes and an unfolded vertical slaty cleavage which does not affect the Cretaceous. In many places, however, the cleavage dies out within the Palaeozoic rocks before reaching the base of the Cretaceous and is not cut by the unconformity, thus making field evidence for the age of the cleavage somewhat ambiguous. Nevertheless, the marked difference in the diagenetic/metamorphic state of Ordovician and Cretaceous rocks (see above), which can be demonstrated using illite crystallinities and infrared spectroscopy, proves a pre-Cretaceous age for the cleavage and associated heating event.

Differences at the east and west side of the Camargo structure hint at the existence of a Preandean fault which is now covered by Cretaceous rocks: (1) Ordovician rocks below the Cretaceous sediments are about 4.5 km thick in the east and only 1.5 km thick in the Impora anticline in the west. (2) The age of Ordovician sediments just below the Cretaceous is middle Arenigian in the east and early Arenigian in the west. The interval between the Lowermost Ordovician and the Lower Arenigian is markedly (ca. 1.5 km) thicker in the east, indicating synsedimentary movements (normal faulting?). (3) In the east, the Ordovician sediments show very-low-grade metamorphism, while in the Impora anticline and further west, low-grade metamorphism occurs.

These features can be explained by pre-Cretaceous block-faulting ending with an uplift of the western block by about 1-2 km, and bringing an older and higher metamorphic level to the surface without pronounced tilting. This accounts for the very weak angular unconformity. At least two different tectonic stages can be observed: (1) During sedimentation of the Ordovician rocks the eastern part was lowered at a normal fault (?) indicating a rift (extension); (2) Later, the direction of movement changed and a relative uplift of the eastern part indicates a compressional regime which already existed before the Cretaceous (possibly during the Hercynian orogeny). In addition, older strike-slip movements cannot be excluded.

6 Tectonic and Geothermal Evolution From the Cambrian to the Present

Here, an attempt will be made to summarize the tectonic and thermal history of southern Bolivia using data from this study as well as results obtained by other researchers.

After the transgression of a shallow Cambrian sea over folded Precambrian rocks, rapid subsidence starts in the early Ordovician (Tremadocian and Arenigian). This is possibly linked to a rift developing to the southwest (Sempéré 1990a). The sedimentation rate is high (at least 14 cm/1000 a over 25 Ma). Marked variations in the thickness of Ordovician strata in some places indicate synsedimentary movements.

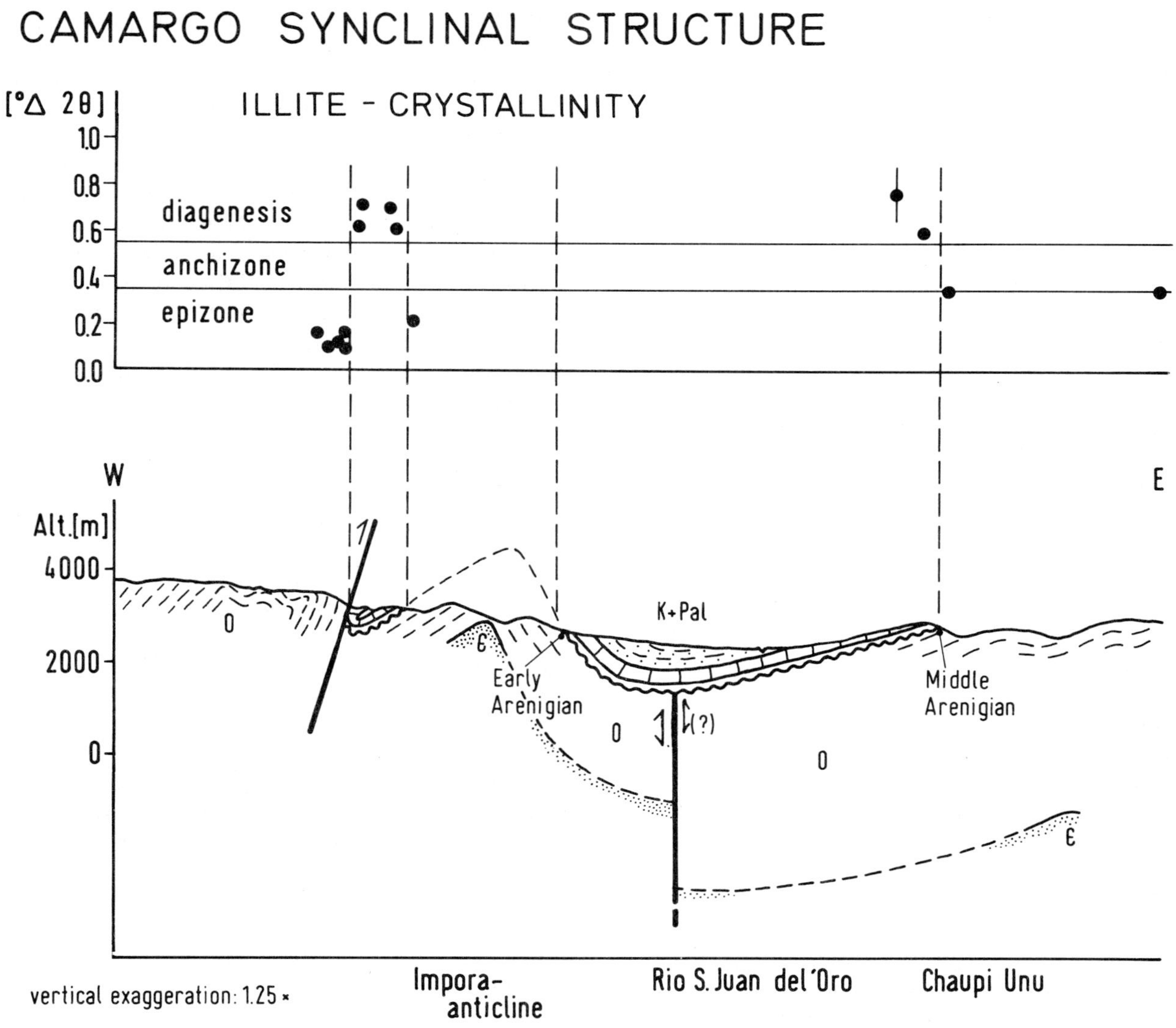

Fig. 9. Camargo synclinal structure: E-W cross section and illite crystallinity values. Abbreviations: ε Cambrian; O Ordovician; K+Pal Cretaceous and Palaeogene. Note: (1.) Diagenetic state of Cretaceous and Tertiary sediments overlying very-low-grade metamorphic Ordovician; (2.) Below the unconformity, early Arenigian rocks occur in the west (Impora anticline) whereas middle Arenigian rocks occur in the east (at Chaupi Unu)

The late Ordovician "Ocloyic" compressional phase brings about the elevation of very large areas and an overall regression. Sedimentation is resumed in the Ashgillian and continues without major interruption until the late Devonian (Frasnian).

In the period between the latest Devonian and ?early Mississippian, the "Eohercynian" orogenic phase affects the area of the later Eastern Cordillera, producing folds and slaty cleavage. The area which is now exposed in the Transition Zone acts as a foredeep while the area further east is not affected by strong subsidence or unusually high temperatures (Fig. 7). Eventually, the whole region is raised above sea level. Particularly strong uplift of the western part leads to the formation of a "Protocordillera", the precursor of today's Eastern Cordillera.

The newly formed "Protocordillera" is glaciated and eroded during the ?late Mississippian and Pennsylvanian, while predominantly continental sedimentation begins farther east (Transition Zone and Subandean Ranges). Erosion continues in the western part, possibly without interruption, up to the Cretaceous. The eastern regions are slowly subsiding during most of the Upper Palaeozoic and Mesozoic. A marine ingression in the Upper Permian, and perhaps Lower Triassic, is documented in the Transition Zone and Subandean Ranges (Vitiacua Fm.). Red continental sediments with evaporites capped by basalt, which are only found in a rather narrow N-S striking zone in the Subandean Ranges, are attributed to a mid-Triassic extensional phase by Oller and Sempéré (1990), who also assign an Upper Triassic-to-Kimmeridgian age to the overlying fluvial and aeolian sandstones. The deposition of these rocks is followed by a long period of slow erosion and weathering in the Subandean Ranges. In the Upper Oligocene, sedimentation of a thick continental foredeep fill begins and continues until the onset of Andean folding and thrusting in the late Miocene. The geothermal gradient drops from normal values of $\leq$30 °C/km in the Mesozoic to about 17 °C/km today. This is a result of thickening of the upper crust by thrusting during the Andean orogeny.

In the Eastern Cordillera, Cretaceous sediments are deposited on Ordovician rocks after a long period of erosion (possibly since the late Devonian). Sedimentation is linked in part with extensional tectonics, as indicated by basaltic intercalations (Tupiza). In the Upper Cretaceous a marine transgression reaches the area of the Eastern Cordillera. The beginning of Andean deformation (shortening) in the Palaeogene in the Eastern Cordillera contrasts with Miocene and younger shortening in the Subandean Ranges and shows a progression of folding and thrusting from west to east. As a result of this, Neogene sediments in the Eastern Cordillera are not as thick as in the Subandean Ranges.

7 Conclusions

During the Andean orogeny (Cretaceous?-Recent) the surface expression of crustal shortening is concentrated in the Subandean Ranges and Transition Zone. It is distributed among many thrust faults which show similar amounts of displacement.

In the Subandean Ranges deep subsidence and uplift took place within an extremely short period of time from the late Miocene to Pleistocene.

In the Subandean Ranges the mesozoic geothermal gradient of $\leq$30 °C/km has decreased to about 17 °C/km today as a result of a thickening of the uppermost crust.

In the eastern part of the Eastern Cordillera Preandean tectonism is related to a Hercynian (Upper Devonian/ Lower Carboniferous) orogeny producing a "Protocordillera".

In the Eastern Cordillera and Transition Zone the highest temperatures were reached during the Hercynian phase. In the Transition Zone at least 4-6 km of sediment were deposited and eroded between the Upper Devonian and Upper? Carboniferous.

Sediment maturity data show that the Cretaceous sediments of the Camargo syncline probably cover an old high-angle fault which had been active since the Ordovician.

Acknowledgements. We appreciate support from several colleagues: K.-J. Reutter initiated this project with a field trip in 1986 and laboratory investigations (Reutter et al. 1987). He encouraged us to undertake the present study and contributed to our work with many intensive discussions.We also wish to thank J. Mullis (Basel) for determination of fluid inclusions, J. Maletz (TU Berlin, graptolites) and W. Hammann (Würzburg, trilobites) for fossil dating, F. Hendriks and D. Hoffmann for illite crystallinity determinations and K. Burchard for carrying out some vitrinite reflectance measurements. Furthermore, thanks are offered to the reviewers H. Kisch and C. Martinez for useful advice.Fieldwork in Bolivia would not have been possible without generous support from YPFB (Yacimientos Petrolíferos Fiscales Bolivianos). M. Cirbian, O. Aranibar and E. Perez provided information and organizational help. A. Solíz introduced us to the geology of southern Bolivia. We are also grateful for access to unpublished data and borehole samples. H. Perez (GEOBOL) provided information on the geology of the Eastern Cordillera. This project was financially supported by the German Research Fund (DFG).

References

Baby P, Sempéré T (1989) Interpretación geológica de la parte meridional del Altiplano sur. ORSTOM en Bolivie, Mission de La Paz, Informe 15, La Paz

Baby P, Hérail G, Lopez JM, Oller J, Pareja J, Sempéré T, Tufiño D (1989) Structure de la Zone Subandine de Bolivie: influence de la géométrie des séries sédimentaires antéorogéniques sur la propagation des chevauchements. C R Acad Sci Paris 309 II: 1717-1722

Baby P, Sempéré T, Oller J, Barrios L, Hérail G, Marocco R (1990) Un bassin en compression d'âge oligo-miocène dans le sud de l'Altiplano bolivien. C R Acad Sci Paris 311 II: 341-347

Barker CE, Pawlewicz MJ (1986) The correlation of vitrinite reflectance with maximum temperature in humic organic matter. In: Buntebarth G, Stegena L (eds) Paleogeothermics, Lecture Notes in Earth Sci 5. Springer, Berlin, Heidelberg, New York, 234pp

Beltan L, Freneix S, Janvier P, Lopez-Paulsen O (1987) La faune triasique de la formation de Vitiacua dans la région de Villamontes (Département de Chuquisaca, Bolivie). Neues Jahrb Geol Paläontol Monatsh 2: 99-115

Bostick NH, Cashman SM, McCulloh TH, Waddell CT (1979) Gradients of vitrinite reflectance and present temperature in the Los Angeles and Venture Basins, California. In: Oltz DF (ed) Low temperature metamorphism of kerogen and clay minerals. The Pacific Sect SEPM, Los Angeles, pp 65-96

Castaños A, Rodrigo LA (1978) Sinopsis estratigrafica de Bolivia. Parte I de Paleozoico. Acad Nac Cienc Bol, La Paz, 144pp

Coira B, Davidson J, Mpodozis C, Ramos V (1982) Tectonic and magmatic evolution of the Andes of northern Argentina and Chile. Earth Sci Rev 18: 303-332

Dellape D, Patuel R (1987) Esquema estructural de la Sierra de Aguaragüe, Provincia de Salta, Republica Argentina. Decimo Congr Geol Argent, San Miguel de Tucuman, Actas I: 169-172

Diessel CFK, Brothers RN, Black PM (1978) Coalification and graphitization in high-pressure schists in New-Caledonia. Contrib Mineral Petrol 68: 63-78

Dunoyer de Segonzac G (1969) Les minéraux argileux dans la diagenèse. Passage au métamorphisme. Mém Serv Carte Géol Alsace Lorraine 29, Strasbourg, 320pp

Frey M (1987) Low temperature metamorphism. Blackie, London

Frey M, Teichmüller M, Teichmüller R, Mullis J, Künzi B, Breitschmid A, Gruner U, Schwizer B (1980) Very low rank metamorphism in external parts of the Central Alps: illite crystallinity, coal rank, and fluid inclusion data. Eclogae Geol Helv 73(1): 173-203

Ganz H, Kalkreuth W (1991): IR-classification of kerogen type, thermal maturation, hydrocarbon potential and lithological characterictics in potential source and reservoir rocks. J SE-Asian Earth Sci 5(1-4): 19-28

Goodarzi F, Norford BS (1989) Variation of graptolite reflectance with depth of burial. Int J Coal Geol 11: 127-141

Helwig J (1972) Stratigraphy, sedimentation, paleogeography and paleoclimates of Carboniferous ("Gondwana") and Permian of Bolivia. AAPG Bull 56(6): 1008-1033

Hendriks F (1983) Einführung in die Tongeologie. TU Berlin, Berlin, 46pp (unpubl)

Hérail G, Baby P, Sempéré T, Oller J, Barrios L, Montemurro G, Salinas R (1990) Structural cross-section in southern Bolivia. In: Forschergruppe Mobilität aktiver Kontinentalränder (ed) Final workshop, structure and evolution of the Central Andes in northern Chile, southern Bolivia and northwestern Argentina, Abstr vol. FU Berlin, Berlin, p53

Isaacson PE (1977) Devonian stratigraphy and brachiopod paleontology of Bolivia, Part A. Palaeontogr, Abt A 155: 133-192

Jacob H (1989) Classification, structure, genesis and practical importance of natural solid bitumen ("migrabitumen"). Int J Coal Geol 11: 65-79

Kisch HJ (1980) Incipient metamorphism of Cambri-Silurian clastic rocks from the Jamtland Supergroup, central Scandinavian Caledonides, western Sweden: Illite crystallinity and 'vitrinite'reflectance. J Geol Soc Lond 137: 271-288

Kübler B (1984) Les indicateurs des transformations physiques et chimiques dans la diagenèse, température et calorimétrie. In: Lagache M (ed) Thermométrie et barométrie géologiques. Soc Fr Minér Crist, Paris, pp 489-596

Martinez C (1980) Structure et évolution de la chaîne hercynienne et de la chaîne andine dans le nord de la Cordillère des Andes de Bolivie. Trav Doc ORSTOM 119, ORSTOM, Paris, 352pp

Mehl J (1982) Stratigraphische und paläontologische Untersuchungen im Silur der bolivianischen Ost-Kordillere. In: Behr H-J, Kappelmeyer O, Nicolaus H-J (eds) 8. Geowiss Lateinamerika Kolloq, Tagungsheft. Univ Göttingen, pp62-63

Oller J, Sempéré T (1990) A fluvio-eolian sequence of probable middle Triassic-Jurassic age in both Andean and Subandean Bolivia. In: ORSTOM(ed) Symp int Géodynamique Andine, Résumés des communications. ORSTOM, Collection Colloques et Séminaires, Paris, pp237-240

Pareja J, Vargas C, Suarez R, Ballon R, Carrasco R, Villaroel C (1978) Mapa geológico de Bolivia. Memoria Explicativa. Serv Geol Bolivia/Yacimientos Petrolíferos Fiscales Bolivianos, La Paz, 27pp

Perez M (1971) Geological map and cross-sections from Villamontes to Sud Lipez, 1:50.000. Yacimientos Petrolíferos Fiscales Bolivianos (unpubl)

Reinhardt M (1989) Inkohlung und Tektonik im Deckenstapel der Liguriden, Nordapennin, Italien. Berl Geowiss Abh 110: 1-108

Reutter K-J, Burchard K, Reinhardt M, Hendriks F (1987) Brief report on kerogen maturity (vitrinite reflectance and infrared spectroscopy) of surface samples from Paleozoic and Mesozoic rocks of the Andes in Southern Bolivia. With additional notes on illite crystallinity. Yacimientos Petrolíferos Fiscales Bolivianos, 9pp (unpubl)

Rivas S, Alvaro Fernandez C, Alvarez R (1969) Estratigrafía de los sistemas ordovícico-cambrico y precambrico en Tarija, sud de Bolivia. Soc Geol Bolivia 9: 27-44

Robles DE (1987) El gradiente geotermico actual en Argentina y zonas aledañas de paises vecinos. Dec Congr Geol Argent, Tucuman, Actas II: 313-316

Rodrigo LA (1973) Sedimentología del Grupo Macharetí, Sección Angosto del Río Pilcomayo, Dept. de Tarija. Soc Geol Bolivia (Anal III Conv Nal Geol), Bol 20: 199-214

Schlatter LE, Nederlof MH (1966) Bosquejo de la geología y paleogeografía de Bolivia. Serv Geol Bolivia, Bol 8: 1-49

Sempéré T (1990a) Late Cambrian to early Silurian evolution of the Central Andean Basin (10°-26°S). In: Forschergruppe Mobilität aktiver Kontinentalränder (ed) Final workshop, structure and evolution of the Central Andes in northern Chile, southern Bolivia and northwestern Argentina, Abstr vol. FU Berlin, Berlin, pp 56-57

Sempéré T(1990b) Cuadros estratigraficos de Bolivia: Propuestas nuevas. ORSTOM en Bolivie 20, La Paz, 26pp

Sempéré T, Oller J, Barrios L (1988) Evolución tectosedimentaria de Bolivia durante el Cretácico. V Congr Geol Chil III: H37-H65

Thorez J (1975) Phyllosilicates and clay minerals. A laboratory handbook for their X-ray diffraction analysis. Lelotte, Dison, 582pp

Thorez J (1976) Practical identification of clay minerals. Lelotte, Dison, 90pp

Wood DA (1988) Relationships between thermal maturity indices calculated using Arrhenius equation and Lopatin method: implications for petroleum exploration. AAPG Bull 72(2): 115-134

Appendix: List of geothermometrical data

Explanation (compare with text):

No. of location refers to fig.3
System: for letter see fig. 2. M = Mesozoic (Jurassic or Cretaceous)
Formation: Ordovician: I = Iscayachi, II = Guanacuno, III = Cieneguillas, IV = Obispo, V = Agua & Toro, VI = Pircancha (after Rivas et al. 1969)
IC (illite-crystallinity) in [Delta2Θ°] or, if indicated by #, [mm half height peak width].
IRS (infrared-spectroscopy). Vitrinite-reflectances in [%R_m] calculated with:
A = A/C-factor
W = minimal wavenumber W_{min}.
Vi. (vitrinite reflectance) in [%R_m]:
m = maximum reflectance
* = problematic value
+ = also higher values occuring
B = bitumen
G = graptolite
/ or ⊥: sample cut parallel or perpendicular to bedding plane.
n: number of measurements
ox: oxidation: 0 = none, 1 = very weak, 2 = weak, 3 = medium, 4 = strong.
Remarks: fluo. = fluorescence (not indicated continuously). Numbers indicate reflectance values for additional populations of OM or values calculated from IRS using W_{min} (e.g. .4W).

No.	System	Formation	IC	IRS	Vi.	n	ox	Rem.
1. Anticline of Aguarague								
8	P	Cangapi	.65					
10	C	Taiguati	.3					
11	C	Tupambi	.25					
12	D	Los Monos			.49	26	1	
13	D	Los Monos			.41	31	1	
14	D	Los Monos		.45A				
15	D	Los Monos			.46	7	1	
16	D	Los Monos	.2		.48	51	0	
17	C	T-3	.3					
18	C	Tupambi	.2					
19	C	Escarpm.?	.3					
20	C	Escarpm.?	.45					
21	C	S.Telmo	.3					
S3	C	Taiguati	.2					
S4	C	Itacua	.3	.35A	.53*	2	4	fluo.
S5	D	Los Monos	.35		.49	62		fluo.
S6	D	Los Monos	.3		.55	59	4	fluo.
S7	C	Itacua/T-3	.35	.38A	.63	16	2	fluo.
S8	C	San Telmo		.32A	1.32*	15	4	fluo.
S9	P	Vitiacua	.4					
2. Syncline of Isiri								
23	T	Petaca	.4					
3. Anticline of Valverde								
25	C	Escarpm.	.45					
28	M	Tacurú	1.0					
S10	P	Vitiacua	.9					
S11	P	Vitiacua		.52A				
4. Anticline of Palos Blancos								
29	C	Tarija		.4A				
30	C	Tupambi	.5					
31	C	Tarija			.7+	17	2	
S12	C	Tarija	.25	.24A	.5*	7	4	
5. Anticline of Suaruro								
33	P	Vitiacua	#.5					
35	M	Tacurú	.3					
6. Syncline of Tacuarandi								
39	P	Cangapi	.35					
42	M	Tacurú	.2					
S13	C	Tarija	.23	.4A	1.24*	43	2	fluo.
S14	P	Vitiacua	.7					
S15	P	Vitiacua	.6					
S16	T	Chaco inf	.33					
7. Syncline of Entre Rios								
44	M	Tacurú	.85					
8. Anticline of Castellon								
48	M	Tacurú	.35					
51	?	Basalt	.8					
9. Anticline of San Diego								
53	P	Vitiacua	.4	.75A				
54	C	Tarija	.4					
57	P	Cangapi	.25	.75A				
58	P	Vitiacua		.65A				fluo.

No.	System	Formation	IC	IRS	Vi.	n	ox	Rem.
S17	P	Vitiacua		.35A	.61*	3	2	
10. Syncline of Narvaez								
61	?	Basalt	#.55					
S18	P	Vitiacua	.2	.33A	1.05	8	2	fluo.
11. Anticline of Tambo Grande								
65	P	Vitiacua			.97*	1		
67	P	Vitiacua	.7					
68	P	Vitiacua		.78A				.4W
69	P	Vitiacua	.45	.55A				.35W
70	P	Vitiacua		.73A				.45
					1.2*	39	0	.9+, fluo.
77	P	Vitiacua		.75A				.35W
355	C	Tupambi			.56	32	2	fluo.
S19	P	Vitiacua			.73	21	2	fluo.
12. Canaletas to Abra Condor								
80	D	Huamamp.			1.0*	3		fluo.
81	D	Huamamp.			.8*	1		
82	D	Icla		.7A	1.3	13	0	fluo.
83	D	Icla			.95+	37	3	fluo.
87	D	Huamamp.	.45	.65A	.8+	3		
89	P	Vitiacua		.7W				
91	D	Huamamp.	.25		1.2+	20	1	
92	D	Huamamp.		1.1W	1.3*	17	1	
232	D	L.Huamamp.			.8+		2	fluo.
234	D	Icla			1.28	1	2	1.19
235	D	Icla			1.1+	14	2	.9
238	D	Huamamp.			1.19	17	2	.9
243	D	Icla	.3					
244	P	Vitiacua	.45					
245	D	Huamamp.			1.5*	50	1	
247	C	S.Telm/Es	.35					
248	C	ST/Esc?	.3					
249	C	Tupambi	.35					
254	P	Ipaguazu	.45					
256	D	L.Icla			1.2*	21	2	.9
356	D	U.Devonian			1.10	20	3	.8+ fluo.
358	D	U.Devonian			1.07	24	3	
408	D	U.Devonian			1.39	18	2	
S20	C	Tarija	.25	.73A	1.29	61	2	fluo.
S21	D	Huamamp.	.35		.90	19	4	
S22	D	Huamamp.		.7A	.76	54	2	fluo.
S23	P	Vitiacua	.3	.79A	1.27	11	2	
S25	D	Huamamp.			1.21	6	4	
S26	P	Vitiacua	.3		1.34	12	1	
S27	C	Tarija			1.18	17	4	fluo.
S28	D	Icla	.3	.8W				fluo.
13. Abra Condor to Santa Ana/Yesera Norte								
93	D	Huamamp.	.4	.8W				
94	D	Huamamp.	.25	1.2W				
95	D	S.Rosa/Tb.		1.7W	1.94B	129	2	

List of geothermometrical data (continued)

No.	System Formation	IC	IRS	Vi.	n	ox	Rem.
229	D Huamamp.			.95B	43	2	
230	D Huamamp.			1.4B	20	2	1.05+
315	Sil./Devon.?			1.33	9		
337	D Icla			1.6*B	1		
338	S Tarabuco			1.66	25	2	1.3+
340	D Icla?			1.19	10	3	1.0+
344	S Kirusillas			1.73	5	2	fluo.
351	D Huamamp.			.97	46	2	fluo.
				1.1+B			
405	S Kirusillas					2	fluo.
406	C M.Carbonif.			.50	23	2	fluo.
407	D U.Devonian?			.45	17	3	fluo.
S29	D Huamamp.	.45		1.26	60	4	fluo.
S30	D Icla	.25		1.5*	2	4	
S31	D S.Rosa			1.3*		4	fluo.
S32	S Kirusillas	.25	.6W	1.09	10	2	
14. S'& E'Tarija; Sella							
100	O Sella			7.5mB	10	3	
101	O Sella	.35		/1.98mB	41	2	
102	O Kirusillas			2.5*B	14	3	
103	S Kirusillas		2.1W	2*B	12	3	
				/4.8B	3	3	
104	S Kirusillas			1.82B	18	3	
301	S Tarabuco			1.6*	2	3	
302	S Tarabuco			1.4+*B	2		
303	S Tarabuco			1.3	12		
304	S Tarabuco			1.22	16	3	fluo.
309	S Kirusillas			1.33	6		fluo.
310	S Tarabuco			1.26	4	3	fluo.
321	D S.Rosa			.95B	12	2	fluo.
322	D S.Rosa			1.11	74	3	fluo.
323	S Tarabuco?			1.67	10	3	
403	P Vitiacua			.82	41	3	fluo.
15. Anticline of Sama							
105	O Ordovic.	.15	2.2*W				
106	O Ordov.II	.16	2.8W	1.25B	70	1	fluo.
				9mG	5	1	
107	O Ordov.II	.18		3.25B	2	3	
				6m*G	1		
108	O Ordov.II	.15					
110	O Ordov.II	.15					
112	ϵ U.Cambrian	.15					
114	O Ordov.II	.15					
S33	O Ordovic.		3.5*W				
S34	O Ordov.II	.2					
S35	O Ordov.II	.175					
16. Iscayachi to Chaupi Unu							
115	O Ordovic.		2.8W				
116	O Ordov.II			6.1mB			
				9m*G			11.2G?
117	Ord./Cambr.	.24					
118	O Ordov.II		2.3W	1?B	?	1	
119	O Ordov.II			8*mB	6	2	
120	O Ordov.II		3.4W	/12.5mG	3	1	
121	O Ord.III	.28					
122	O Ordov.		2.8W				
123	O Ordov.		2.4W				
125	O Ordov.IV,V	.25					
126	O Ordov.V,VI	.35					
129	O Ordovic.	.3	3.3W				
131	O Ordovic.		2.3W				
134	O Ordov.VI	.35					
138	O Ordovic.	.4	4.0W	/3.8B		3?	
381	O Ordov.VI			1?B		3	
387	O Ordov.VI			/5.1mG	3		
392	O Ordov.V			/12mG	26		
393	O Ordov.IV			/10.5mG			
S38	O Ordovic.	.2					
S39	O Ord.III	.25	3.5*W				
S41	O Ordov.V	.3	3.5*W				
S42	O Ordov.VI	.4					
S43	O Ordov.VI	.35					
S44	O Ordov.VI	.3					
17. Camargo Syncline (Cretaceous and Tertiary)							
140	K Chaupiunu	.6					
141	K El Molino	#.7-.9					

No.	System Formation	IC	IRS	Vi.	n	ox	Rem.
143	T S.Lucia?	.45					
144	T S.Lucia	.6					
146	K El Molino	.7					
151	T S.Lucia	.4					
154	K El Molino	.7	.4A				
S46	K El Molino	.5					
S48	K El Molino	.8					
18. Impora							
147	O Ordov.VI?	.2					
148	O Ordov.VI?			5.5mB	8	1	
150	O Ordovic.		3.9W				
152	O Ordovic.	.19					
153	O Ordovic.	.15					
382	O Ordovic.			/12.1mG			
383	O Ordovic.			/12.5mG			
386	O Ordovic.			/13mG			
19. Impora to west of Mal Paso							
159	O Ordovic.		4.4W				
160	O Ordovic.	.15					
164	O Ordovic.			6.12	35	?	
165	O Ordovic.	.13					
169	O Ordov.VI	.18					
171	O Ordovic.			/11.1mG	7	2	
172	O Ordovic.			/12mG	4	1	
173	O Ordovic.	.15					
174	O Ordovic.	.17					
176	O Ordovic.	.15					
179	O Ordovic.	.15					
181	O Ordovic.	.19					
182	O U? Ordov.	.2					
185	O Ordovic.		2.7W				
186	O Ordovic.			⊥8.5mG			
187	O Ordovic.	.15		5m*B	2		
				8.5mG	2		
188	O Ordovic.			8.5mG	7	3	
S50	O Ordov.VI	.15					
S51	O Ordov.IV	.18	3W				fluo.
S52	O Ordov.V	.15					
S53	O Ordov.V	.15					
S54	O Ordov.IV	.175					
S55	O Ordov.VI	.175					
S56	O Ordov.VI	.15					
S57	O Ordovic.	.15					
S58	O Ordovic.	.15					
S59	O Ordovic.	.25					
20. Tupiza							
193	O Ordovic.	.25					
194	O Ordovic.	.3					
212	O Ordovic.			3.6*B	1	2	
214	O Ordovic.			3.5*B	10	2	
				9.5m*G	1		
				5mB	6		
215	O Ordovic.	.2					
222	O Ordovic.	.28					
223	K Aroifillas	.3					
224	O Ordovic.	.3					
21. Tupiza to S.Vicente							
199	O Ordovic.	.13					
200	O Ordovic.	.15	3.0W	4.5*	3		
201	O Ordovic.	.13	2.8W				
206	O Ordovic.	.19					
210	O Ordovic.	.21		5*	2	2	
S60	O Ordovic.	.2					
S61	O Ordovic.	.2					
S62	O Ordovic.	.2	>4.5W				
S63	O Ordovic.	.225					
S64	O Ordovic.	.175					
22. Rio San Juan del Oro							
225	O Ordovic.	.2					
227	O Ordovic.	.15					
228	O Ordovic.	.2					
23. near Uyuni							
2	S Uncía			1.30	38	2	
3	S Uncía			1.35	19	3	
4	S Uncía			1.27	15	1	

Sedimentary and Structural Evolution of the Salar de Atacama Depression

EBERHARD WILKES and KONRAD GÖRLER

Abstract. The Salar de Atacama Depression is part of the northern Preandean Depression, a SSW-NNE striking morphological low situated between the Western Cordillera de los Andes and the Chilean Precordillera. The Cordillera de la Sal is a small SSW-NNE striking intrabasinal foldbelt within the Salar de Atacama Depression and provides the best outcrops of the sedimentary record of the Preandean Depression. It is built mainly of continental red beds (San Pedro Formation, Oligo-Miocene), locally more than 3 km thick. These sediments were deposited in a playa environment and contain large amounts of evaporites (gypsum, anhydrite, glauberite, and, especially, halite). First fossil findings (charophytes, gastropods, ostracods) indicate intervals with an oligohaline lacustrine environment in the north. Towards the west, the San Pedro Formation interfingers with an alluvial fan sequence (Tambores Formation). The sediments of the Salar de Atacama Depression overlie unconformably folded Upper Cretaceous-Eocene sediments of the Purilactis Formation. The origin of the observed complicated structural pattern is considered to be a result of the combination of compressional and sinistral strike slip movements. The most evident tectonic elements within the sediments of the Salar de Atacama Depression are: (1) NNE-SSW striking folds with doubly plunging axes; (2) a complex pattern of nearly vertical small scale normal, reverse and strike slip faults; (3) NNE-SSW striking low angle reverse faults; (3) N-S to NW-SE striking vertical tension fissures; (4) effects of gravitational lateral spread of thick halite deposits.

1 Introduction

The Salar de Atacama Depression is part of the northern Preandean Depression which describes a SSW-NNE striking morphological low situated in northern Chile between 27° and 23°S flanked by the Western Cordillera de los Andes in the east and the Cordillera de Domeyko as a part of the Chilean Precordillera in the west (Fig. 1). The Preandean Depression is marked by several salares (= playas) in endorheic basins with a low at a level of 2350 m in the Salar de Atacama contrasting with highs of 6723 m (Volcan Llullaillaco) in the Western Cordillera and 5073 m (Co. Doña Ines) in the Precordillera. Neither now nor in the past has the Preandean Depression been a uniform depositional basin. Several depocentres exist, separated from each other by structural highs acting as zones of erosion that supply sediment to the adjacent basins.

At present, the Western Cordillera is the main element of the modern Andean magmatic arc with some extensions towards the east into the Puna and the Eastern Cordillera and towards the west into the Preandean Depression. Its position at the western border of the magmatic arc is therefore unusual. But owing to the similarities in shape, structural and sedimentary pattern (e.g. Alonso et al. 1991) between the Preandean Depression - especially the Salar de Atacama Basin - and the doubtless intra-arc basins of the Argentine Puna (Jordan and Alonso 1987) it represents an intra-arc basin, too. This interpretation of the Preandean Depression is supported by geophysical data which show that the crust of the Preandean Depression has similar properties to the magmatic arc of the Western Cordillera. For example, the zone of high electrical conductivity and high attenuation of seismic waves, which is observed beneath the magmatic arc, is also present beneath the Preandean Depression (Schwarz et al. this Vol.; Wigger et al. this Vol.). The geophysical anomalies have been interpreted as a result of the presence of melts in the arc's crust. The Chilean Precordillera, east of the Preandean Depression, on the other hand does not show these anomalies and clearly belongs to the cold and rigid forearc. In the sediments and structures of the Preandean Depression most tectonic movements that affected the modern magmatic arc are recorded. The aim of this study is to decipher the

Correspondence to: K. Görler, Fachrichtung Geologie, Freie Universität Berlin, Malteserstr. 74-100, D-1000 Berlin 46

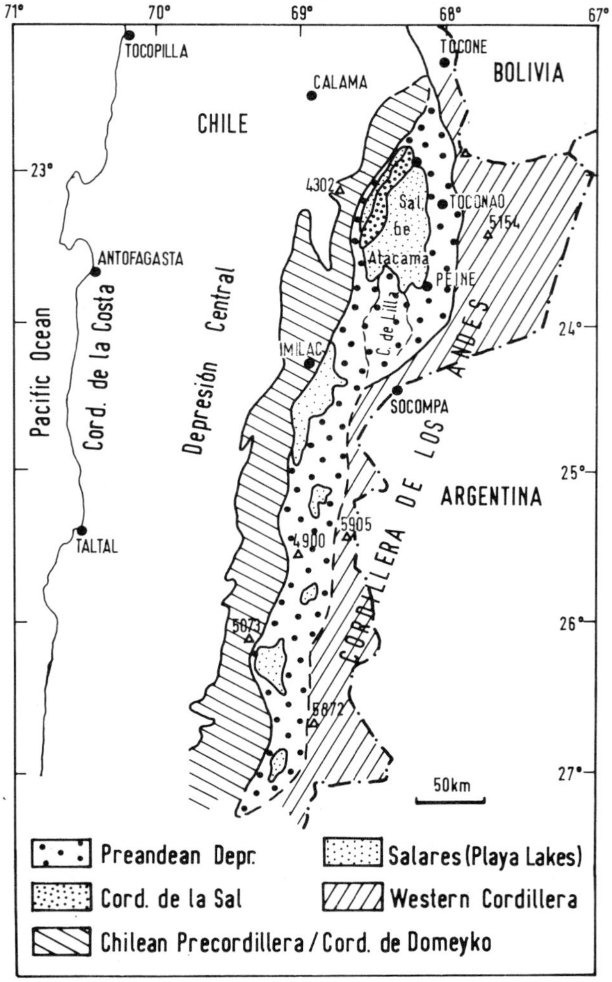

Fig. 1. Morphostructural map of northern Chile between 22° and 27°S

evolution of this basin and the adjacent magmatic arc in time and space by analyzing the sedimentary and tectonic history of the northern Preandean Depression.

2 Stratigraphic and Palaeoenvironmental Setting

Larger portions of the sedimentary fill of the Preandean Depression only crop out in the Salar de Atacama Basin. Within the basin the most favourable conditions to observe and study the rock record exist in the Cordillera de la Sal, a SSW-NNE trending foldbelt, 5-10 km wide, which traverses the Salar de Llano de la Paciencia and Salar de Atacama (Figs. 1 and 2). On average the Cordillera de la Sal rises only 200 m above the level of the surrounding playas and exposes the central parts of the Oligocene Recent sedimentary fill of the northern Preandean Depression. The fill mainly belongs to the Oligo-Miocene red beds and evaporites of the Tambores and San Pedro Formations.

North of the village of San Pedro de Atacama (localities see Fig. 2) these rocks are covered by ignimbrites (San Bartolo Group) and other volcanics of late Miocene to subrecent age and are exposed only in deeply incised valleys near San Bartolo, Rio Grande and El Tatio.

An upwarp of the recent surface of the southern Salar de Atacama measuring only a few metres, indicates the southern end of the Cordillera de la Sal. Including these outcrops, the deposition area of the Tambores and San Pedro Formations reaches a total length of about 180 km, but a further extension beneath younger rocks towards south and north is possible. Late Oligocene to Miocene continental red beds in the Lipez Basin of southern Bolivia have the same age, similar thickness, and facies development as these formations (Baby et al. 1990).

Towards the eastern and western margins of the basin continuous outcrops are absent because the Cordillera de la Sal is limited on both sides by the active playas of the Salar de Atacama in the east and the Salar de Llano de la Paciencia in the west. Seismic investigations (Townsend 1988) show that also the sedimentary basin of the Tambores and San Pedro Formations extended not far beyond the actual salares (Fig. 3).

2.1 San Pedro and Tambores Formations (Brüggen 1934)

The *Tambores Formation* crops out at the western margin of the Salar de Atacama Basin (Fig. 2). In the north a total exposed thickness in the order of up to 1300 m has been described by Wilkes (1990), who subdivided this formation into a basal lower conglomeratic unit, an evaporitic unit and an upper conglomeratic unit (Figs. 4 and 5) . In the south the Tambores Formation is built up by approximately 450 m thick conglomerates (Ramirez and Gardeweg 1982). Within the formation the following lateral trends can be observed from west towards east, i.e. from the margin towards the centre of the basin:

- decreasing grain size
- increasing thickness
- increasing proportion of evaporites.

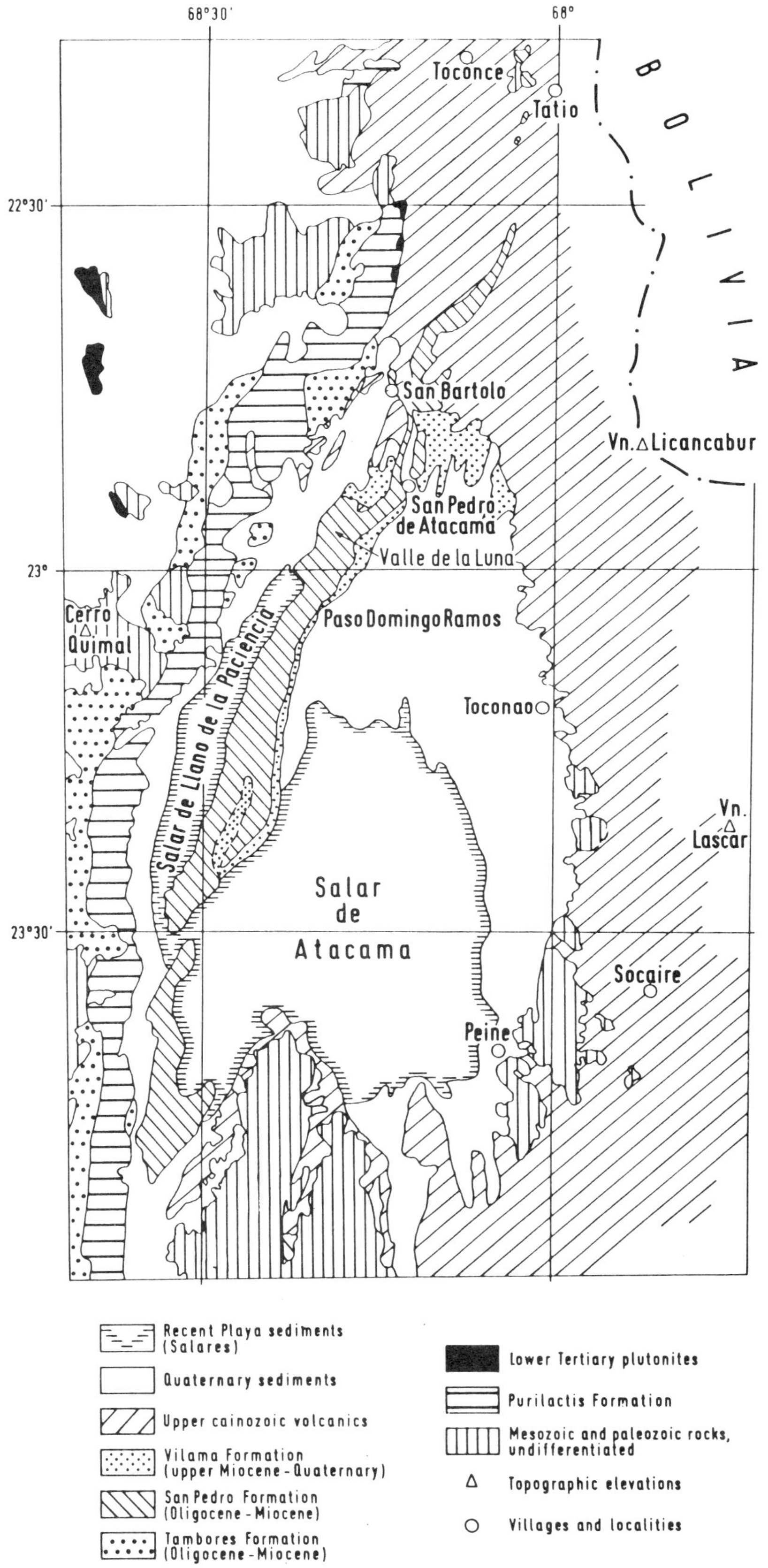

Fig. 2. Geological map of the Salar de Atacama area. (after Marinovic and Lahsen 1984; Ramirez and Gardeweg 1982)

The Tambores Formation lies with angular unconformity on top of the continental sediments of the folded and faulted about 3000 m thick Purilactis Group of Upper Cretaceous to Eocene age (Charrier and Reutter this Vol.). Locally the Tambores Formation is unconformably overlain by ignimbrites of the Upper Miocene San Bartolo Group. On the western margin of the Llano de la Paciencia the Tambores Formation interfingers with the isochronous San Pedro Formation.

The outcrops of the *San Pedro Formation* are restricted to a narrow, SSW-NNE trending strip within and beyond the Salar de Atacama Basin (Fig. 2). The maximum exposed stratigraphic thickness of the San Pedro Formation, approximately 3000 m, was measured in two sections, one situated in the area of the Valle de la Luna, southwest of San Pedro de Atacama (Figs. 4 and 5), and the other 25 km further south (Fig. 6). An even greater thickness must be assumed as the base is not exposed and the top layers are cut by erosion. In contrast, further south at 23°30'S, seismic investigations (Townsend 1988) indicate only 450 m total thickness for the San Pedro Formation (Fig. 3). Drastic changes in thickness over short distances can be observed at several locations within the Cordillera de la Sal, where layers of pyroclastics serve as marker horizons. Otherwise, a pronounced increase of thickness over short distances from the western rim towards the centre of the Cordillera de la Sal is evident (Fig. 7b). Less evidence exists for

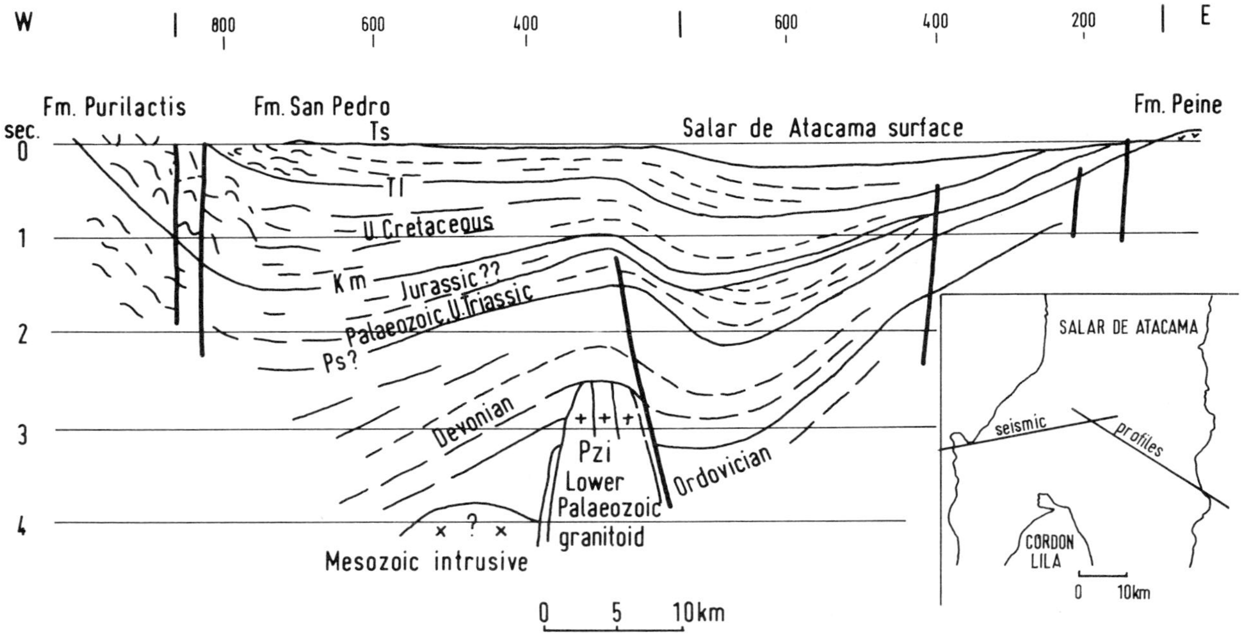

Fig. 3. Interpretation of a seismic section across the southern Salar de Atacama (Modified after Townsend 1988)

the lateral development of the San Pedro Formation east of the Cordillera de la Sal as in this region it is covered by recent sediments within the Salar de Atacama Basin and by Miocene and Pliocene ignimbrites at its eastern margin. At the few locations where older rocks are exposed beneath the ignimbrites they are, with few exceptions, of Palaeozoic or Mesozoic age. In an interpretation of a seismic profile (Townsend 1988) crossing the Southern Cordillera de la Sal, the Purilactis Group and the San Pedro Formation are wedging out towards the east beneath the Salar de Atacama (Fig. 3).

The San Pedro Formation is mainly built up by red silt- and sandstones, halites, sulphates and intercalated tuff layers.

The basal contact of the San Pedro Formation can be observed only in the extreme north and south of its outcrop area, where it unconformably overlies Cretaceous, and Palaeozoic rocks respectively.

North of the Paso Domingo Ramos the San Pedro Formation is unconformably overlain by pyroclastic and volcanic rocks of an age range between middle and late Miocene (Figs. 7 and 8). In the south the Plio-Pleistocene Vilama and Campamento Formations lie unconformably on top of the folded San Pedro Formation. In the southernmost part of the Cordillera de la Sal the evaporites of the San Pedro Formation grade without unconformity into the recent playa sediments.

The stratigraphic position of the Tambores and San Pedro Formations is not well established (Fig. 8). Two K-Ar ages of 28 ± 6 and 24.9 ± 1.0 Ma from tuff layers within the central portion of the San Pedro Formation have been published by Travisany (1978) and Marinovic and Lahsen (1984). The San Pedro Formation is therefore most probably of Oligocene to Lower Miocene age. Owing to the occurrence of *Chara* sp. and of the ostracods *Darwinula* sp., *Potamocypris* cf. *maculata* Alm (1914) and *Cyprideis* aff. *stephensoni* Sandberg (1964) in the Cementario member of the upper part of the San Pedro Formation it extends probably up to middle-late Miocene time. Considering the ages of dated rocks under- and overlying the Tambores and San Pedro Formations these must have been deposited in the north during the time interval between Oligocene and middle-late Miocene. In the south an upper time limit cannot be given. As already stated, the evaporites of the San Pedro Formation grade without interruption into the recent sedimentation of the Salar de Atacama.

The occurrence of halite beds several hundreds of metres thick indicate an arid depositional environment. The total amount of halite occurring in the sediments of the Salar de Atacama Basin is at a rough estimate 1500 km3. Trace element analysis of halite

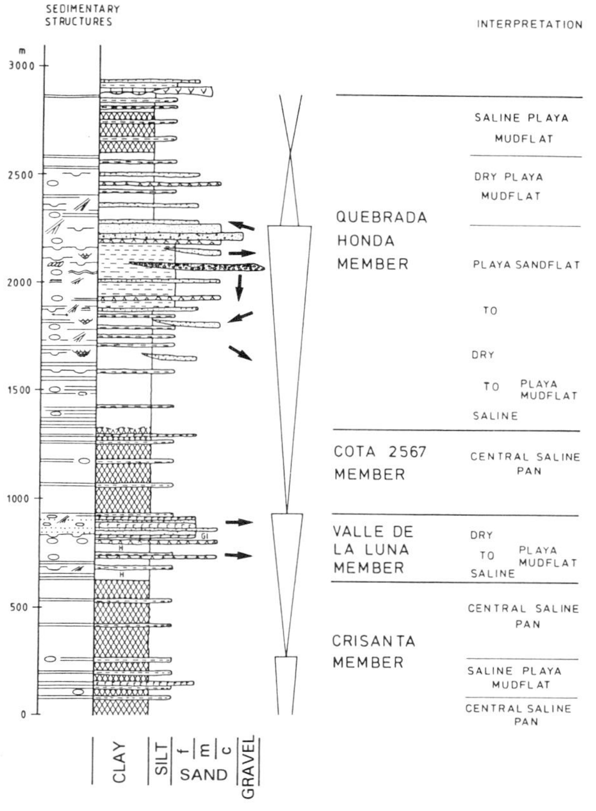

Fig. 4. Stratigraphic profile of the San Pedro Formation in the northern Cordillera de la Sal (Valle de la Luna area). (From Wilkes 1990; legend: Fig. 5)

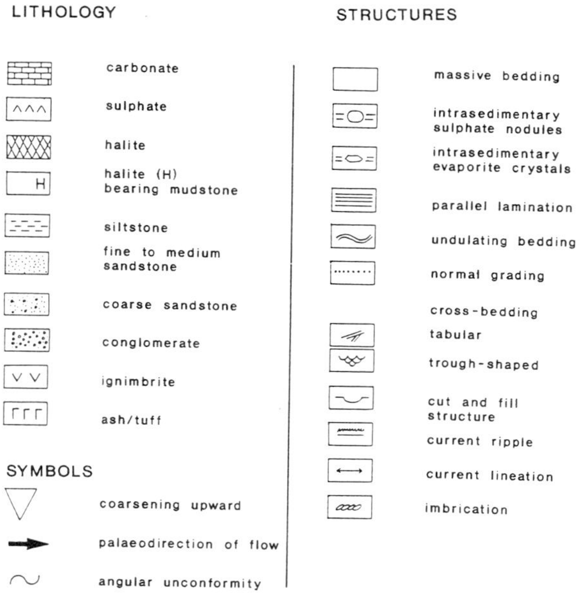

Fig. 5. Legend for the profiles of Figs. 4, 6 and 10

samples showed an extremely low content of bromine within the halite of all levels of the San Pedro Formation, which is not consistent with a marine deposition of the evaporites. It indicates a primary continental origin or at least two continental re-depositional cycles of older marine salts. Sedimentation in a basin with internal drainage under arid climatic conditions seems most probable for the complete San Pedro Formation. This view is supported by the occurrence of non-marine fossils and westward interfingering with the alluvial sequence of the Tambores Formation. Intrabasinal re-sedimentation of sulphates and chlorides is actually the most important source for evaporite sedimentation in the Salar de Atacama and the Llano de la Paciencia. Dingman (1967) proposed evaporites of the Mesozoic Tonel and Purilactis Formations west of the Cordillera de la Sal as an external sediment source. Over the time span of at least 30 Ma for which the endorheic Salar de Atacama Basin has been in existence other important additional sources may include the aerosol transport of marine salts from the coast and of chlorine of volcanic origin in combination with sodium ions from weathering of silicates. A more detailed study of the different segments of the Cordillera de la Sal will be given in the following sections.

2.1.1 El Tatio Area

In the northernmost outcrops of the San Pedro Formation, located in a deeply incised valley west of the geothermal field of El Tatio (Fig. 2), approximately 500 m of badly sorted siltstones, sandstones and conglomerates with cut and fill structures are exposed (Marinovic and Lahsen 1984). These sediments resemble those of areas of interfingering between Tambores and San Pedro Formations. The decrease of grain size and the imbrication in the conglomerates indicate a transport direction from west, and a deposition in an alluvial fan environment seems most probable (Wilkes 1990).

2.1.2 San Bartolo Area

Around the village of Rio Grande and the former copper mine of San Bartolo, the San Pedro Formation, with a total thickness in the order of

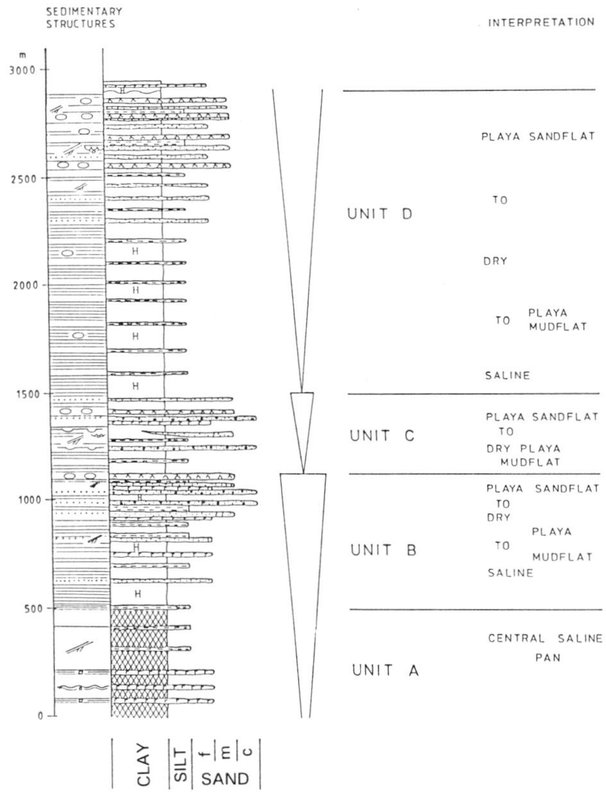

Fig. 6. Stratigraphic profile of the San Pedro Formation in the central Cordillera de la Sal (south of Paso Domingo Ramos). (From Wilkes 1990; legend: Fig. 5)

1500 m, crops out only in deeply incised valleys of the Rio Grande and the Rio Salado river systems. In the areas of great thickness, such as here, the San Pedro Formation can be divided into distinctive members, but rapid lateral changes and the absence of continuing outcrops make the correlation between different areas problematic.

Within the San Pedro Formation of the San Bartolo area (Fig. 9) Flint (1985) distinguishes and interprets from top to bottom:

-Rio Salado member: Red sandstones and evaporitic mudstone playa sandflat with fluvial system to the west (Brown to grey-green mudstones and siltstones)
-Cementario member: Brown to grey-green mudstones and siltstones (shallow ephemeral lake with fluvial feeder channels)
-Artolla member: Red evaporitic mudstones and sandstones (playa mudflat and marginal sandflat)
-Palicaye member: Brown mudstones, sandstones and bedded halite (playa saltpan-saline mudflat)

As mentioned above, WNW of San Bartolo the evaporitic San Pedro Formation and the conglomeratic Tambores Formation interfinger. The Cementario and Rio Salado members can be followed between a base and a top section of conglomerates of the Tambores Formation (Fig. 8). While the lithology does not change very much, the thickness decreases over a distance of about 2 km from approximately 800 m to less than 100 m. The lower conglomeratic unit of the Tambores Formation can therefore be correlated with the evaporite-rich beds of the Palicaye and Artolla members, probably with transition over a very short distance. From these observations we conclude that coarse alluvial fans advanced towards the east in a restricted closed basin with rapid evaporite sedimentation at the centre. The limitation of the region of great thickness to a small central part of the former sedimentation basin and strong differential synsedimentary subsidence is therefore assumed. In our interpretation, the portion of the San Pedro Formation that is isochronous with the upper conglomeratic unit of the Tambores Formation not outcropping in the San Bartolo area, seems to be either eroded or covered by younger volcanics. These portions of the San Pedro Formation outcrop south of the San Bartolo area in different sections of the Cordillera de la Sal (Fig. 9).

2.1.3 Northern Cordillera de la Sal

In the area west and southwest of San Pedro the exposed thickness of the San Pedro Formation increases to about 3000 m. Wilkes (1990) proposed a division into four members; these are from the bottom to the top (Fig. 4):

- Quebrada Honda member
- Cota 2567 member
- Valle de la Luna member
- Crisanta member.

A third of this sediment column constitutes halitic rocks which occur as massive beds at the basal, middle and top parts, thus indicating three evaporitic megacycles.

The base of the deposits of the first megacycle is not exposed. The oldest portions of the San Pedro Formation contain more than 1000 m of halite rocks (Crisanta member, Fig. 4). The second and third cycles start with red siltstones and mudstones, with little limestone but large amounts of sulphates and, especially, halite. The halitic portions of these megacycles are particularly similar and can be distinguished only by their stratigraphic position. The

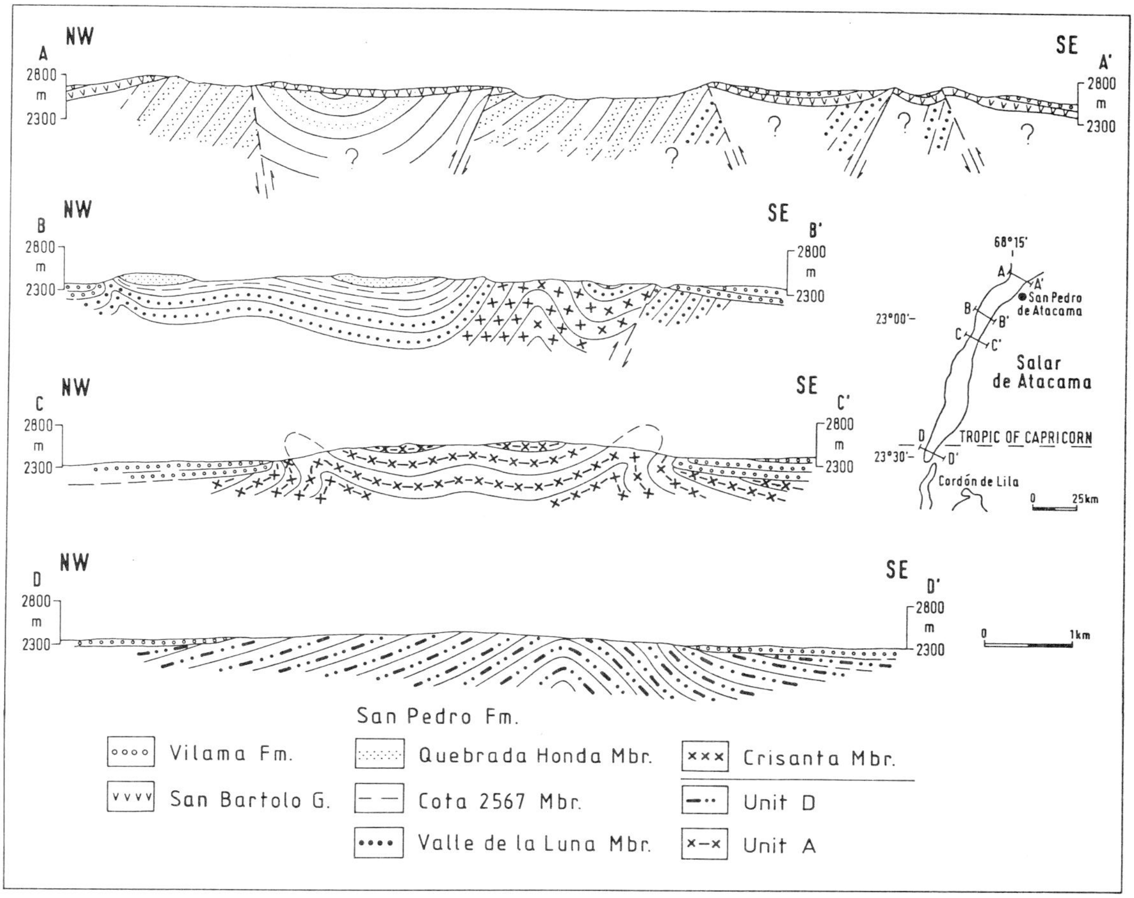

Fig. 7. Schematic tectonic profiles of the Cordillera de la Sal from north to south. Correlation between different lithostratigraphic units within the profiles is not always confirmed. (From Wilkes 1990)

description and environmental interpretation of these rocks given for the Crisanta member is therefore also valid for the halite beds of the second and third cycles.

Like other halitic members of the San Pedro Formation the *Crisanta member,* which has been named after the former Crisanta halite mine west of the Valle de la Luna, is mainly built of poorly defined beds of coarsely crystallized halite with numerous fluid inclusions. Within and between the individual halite crystals occurs red clay and silt. Thin layers of mudstone, often with displacive sulphates and pyroclastic material, are intercalated. In most cases NaCl constitutes about 90% of the 63 samples analysed. The remainder consists of clay and silt. Occasionally up to 30 % anhydrite or glauberite is included. The rare, optically pure halite is a product of recrystallization crossing the bedding planes. The measured contents of bromine within the halite rocks of the different members of the San Pedro Formation are not specific and range between 0.11 and 4.6 ppm (Wilkes and Görler 1988). Systematic lateral or vertical trends in bromine content have not been observed. Iodine ranges between 0.2 and 7.0 ppm in the halite, and this is high compared with the very low bromine values. There seems to be no correlation between the values for bromine and iodine. For cations a clear positive correlation exists between potassium (10-550 ppm) and lithium, which was detectable in only half of the samples. The highest measured value is 0.07 ppm. According to the scheme by Eugster and Hardie (1978) for the development of solutions and precipitations in saline lakes, the sequence of evaporitic megacycles in the San Pedro Formation fits type IIB: Starting in a first phase of evaporation with solutions much richer in

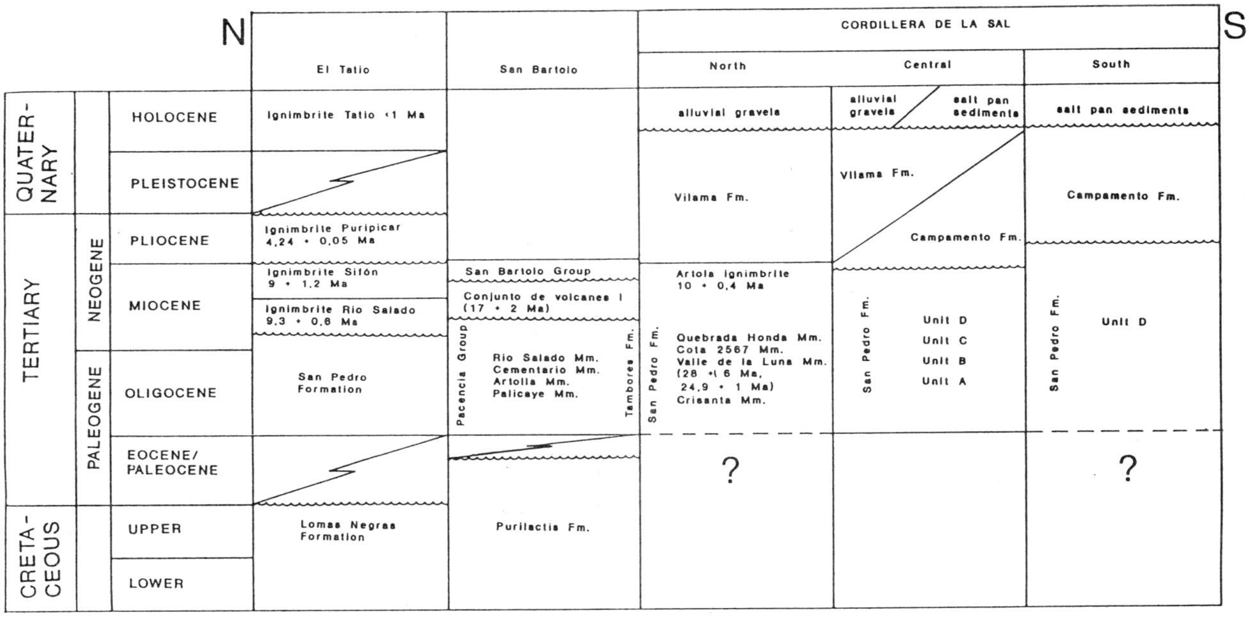

Fig. 8. Correlation of the proposed stratigraphic arrangement in the northern Preandean Depression. (After Ramirez and Gardeweg 1982; Marinovic and Lahsen 1984; Flint 1985; Wilkes 1990)

alkaline-earth metals (Ca » Mg) than in bicarbonate ions, the precipitation of calcium carbonate occurs with a changing content of magnesium until all bicarbonate ions have been consumed. In a second phase the remaining calcium ions were deposited as gypsum or anhydrite and in a third phase they were deposited as glauberite. The deposition of halite followed in the fourth and final phase. There is no indication as to the precipitation of bittern salts.

In most cases the halite rocks are contaminated by red and brown silt and clay, situated between euhedral or subhedral halite crystals and appearing as inclusions in the crystals. According to observations at Bristol Dry Lake, California (Handford 1982), this type of impure halite results from massive intrasedimentary displacive growth. As in the Bristol Dry Lake sediments, hoppered crystal cubes are common in the San Pedro Formation. They have a diameter of up to 4 cm and can be best observed within intercalated tuff layers, where halite crystals have been replaced by opal (Wilkes 1990). These features are commonly interpreted as being typical of deposition in a saline mudflat. However, according to Hardie et al. (1978) in such an environment all sedimentary structures should be destroyed by the growth of evaporite crystals, whereas in the Crisanta and Cota 2567 members of the San Pedro Formation bedding is visible and some distinctive layers with sharp contacts exist. A similar contradiction is described by Handford (1982) for Bristol Dry Lake. His proposed model for the genesis of bedded halite by repeated desiccation and reflooding of a central saltpan also seems to be applicable to the halitic members of the San Pedro Formation. An excellent model for this process is provided by present-day processes in the Salar de Atacama, where, in the central saltpan, all the stages of destruction of pure halite layers deposited from open brines can be observed (Stoertz and Ericksen 1974).

Although on the one hand there are hardly any clues pointing to the deposition of evaporites from open bodies of water in the San Pedro Formation and, on the other hand, an enormous accumulation of halite took place, it seems most likely that, as far as the Crisanta and Cota 2567 members of the San Pedro Formation are concerned, deposition was effective in large and only periodically flooded central saltpans. In such an environment the precipitation of bittern salts cannot be expected, but the most soluble components would, with time, and if not removed by other processes, be accumulated in the remaining brines; this is the case in those of the actual Salar de Atacama, where concentration took place during the last 30 Ma and large amounts of Li and K have been accumulated in the brines (Moraga et al. 1974).

Within the megacycles relapses of the evaporitic

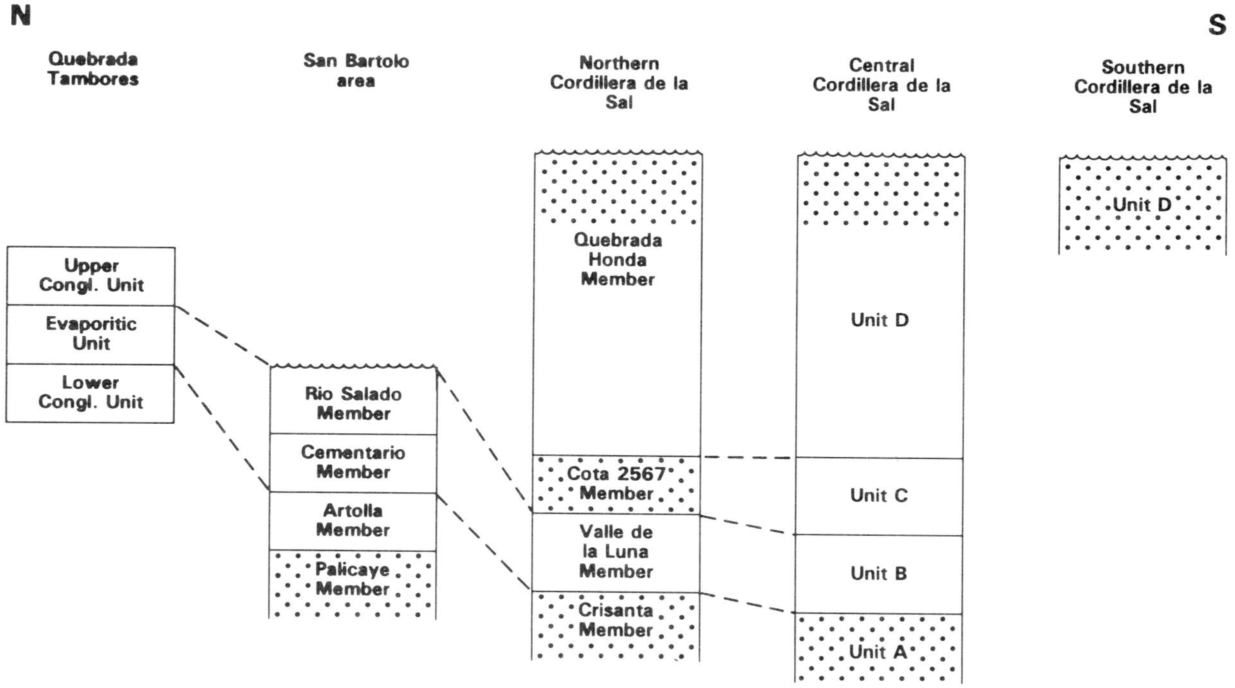

Fig. 9. Correlation between the regional stratigraphic units of the Tambores and San Pedro Formations. (*Dots:* halite beds)

succession are common and the third megacycle can be observed only locally north and south of the Valle de la Luna. The *Valle de la Luna member* and the central portion of the Quebrada Honda member are characterized by laminated brown and red mudstones and siltstones and by frequent cut and fill structures. Displacive gypsum and anhydrite within the mudstones, and nodules and layers of anhydrite and gypsum are common. Within the Valle de la Luna member, displacive glauberite crystals and locally massive impure beds of this mineral up to several metres thick are frequent. Interstitial and displacive halite is sometimes observed. These features fit a deposition of the Valle de la Luna member and the lower portion of the Quebrada Honda member in a saline to dry mudflat environment adjacent to a central saltpan (Hardie et al. 1978). The Valle de la Luna member probably corresponds to the Cementario and Rio Salado members described in the San Bartolo region by Flint (1985). His interpretation of the Cementario member as deposited in a shallow ephemeral lake has been supported by findings of lacustrine fossils (charophytes, gastropods and ostracods) at locations near Rio Grande and San Bartolo (Wilkes and Görler 1988). As these ostracods indicate an oligohaline and mixohaline environment, these lacustrine sediments were probably deposited at the northern margin of the basin in an area with a distinctive freshwater influx. This situation corresponds with the actual one, where the perennial Rio San Pedro enters the northern part of the Salar de Atacama with its oases and small lakes which, towards the south, grade into saline mudflats.

As topographical names are largely missing in the Cordillera de la Sal the *Cota 2567 member* has been named after a hill (cota) that lies north of the Valle de la Luna. It closely resembles the Crisanta member. Here, too, a deposition in a central saltpan has to be assumed.

The *Quebrada Honda member* is well exposed in the homonymous small valley north of the Valle de la Luna. In the lower and central parts the Quebrada Honda member is characterized by a thick sequence of laminated claystones and siltstones with nodular anhydrite and gypsum. Reworked nodular sulphates occur at the base of intercalated sandstone layers and channel fills. A pronounced tendency of upward coarsening can be observed, and about 500 m below the top of the profile this changes to a sequence of upward fining before beginning a new phase of upward coarsening (Fig. 4). Sedimentation in a playa mudflat and sandflat environment is assumed. Only

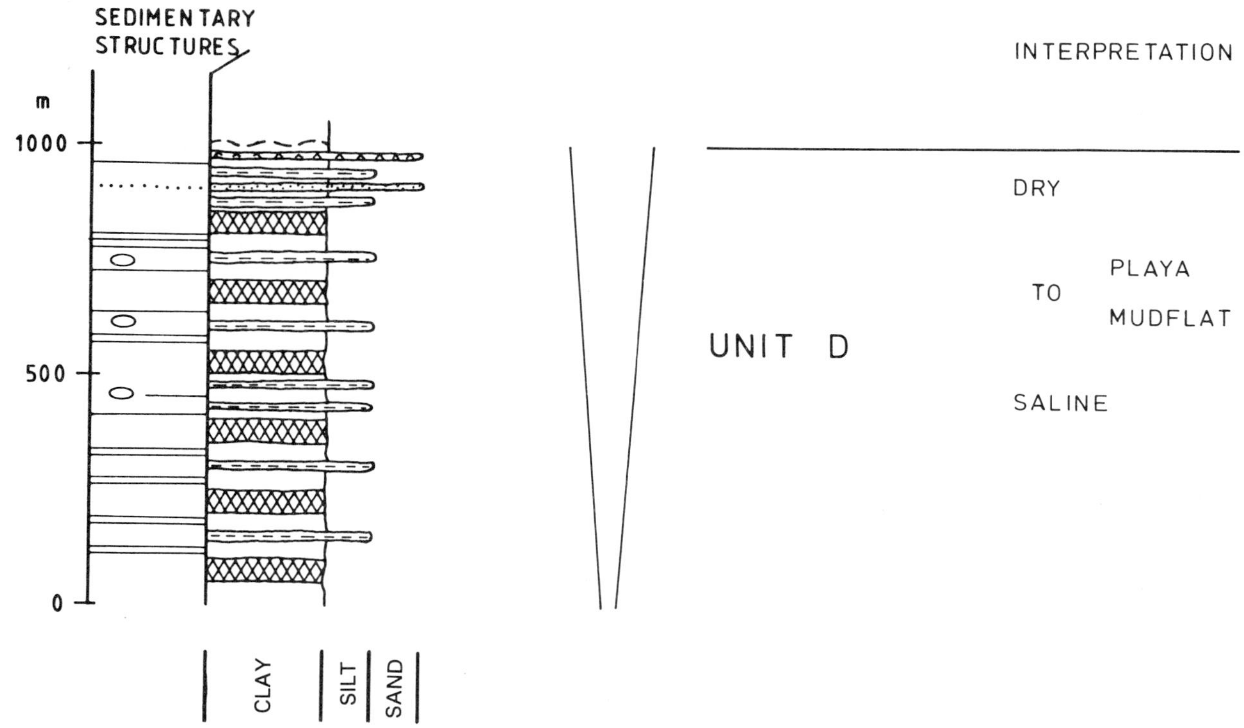

Fig. 10. Stratigraphic profile of the San Pedro Formation in the southern Cordillera de la Sal. (From Wilkes 1990; legend: Fig. 5)

for the uppermost portion does a saline mudflat or even a saline pan sedimentation seem likely.

2.1.3 Southern Central Cordillera de la Sal

South of the Paso Domingo Ramos uncertainties in the lateral correlation mean that a new informal subdivision of the San Pedro Formation has been introduced by Wilkes (1990). The correlation between the Crisanta member of the northern Cordillera de la Sal with unit A in the south seems most reliable. More problematic is the correlation between higher units. It seems probable that in the central Cordillera de la Sal in unit D the segment corresponding to the uppermost portion of the Quebrada Honda member is missing (Fig. 6).

2.1.4 Southern Cordillera de la Sal

Here the visible stratigraphic column is very restricted. Only unit D is outcropping (Fig. 10) as a dissected upwarp of the actual Salar de Atacama, and the sedimentary record continues into the most recent deposits.

2.2 Miocene to Quaternary volcanics and sediments

The sediments of the San Pedro Formation are, mostly unconformably, overlain by younger rocks of different types and ages.

2.2.1 Ignimbrites

North and west of San Pedro a sequence of locally more than 100 m thick, ignimbrite sheets of an andesitic to rhyodacitic composition and an Upper Miocene to Pleistocene age cover the intensively folded San Pedro Formation (Fig. 2). Only north of Rio Grande do still older volcanics occur directly on top of the San Pedro Formation, i.e. the "Dacita Machuca" (17 ± 2 Ma; Ramirez 1979b) at the base of the "Conjunto de Volcanes I" of Marinovic and Lahsen (1984). On the 1:250 000 official geological map Marinovic and Lahsen (1984) distinguish for the area between Rio Grande and the Valle de la Luna a sequence of six ignimbrite sheets with intercalated fluvial gravels:

- Ignimbrita Chaxas (Pliocene-Pleistocene)
- Ignimbrita Puripicar (4.2 Ma, Pliocene)
- Ignimbrita Pelon (7.0 Ma, Upper Miocene)
- Ignimbrita Yerba Buena (Upper Miocene)
- Ignimbrita Sifon (Upper Miocene)
- Ignimbrita Artola (10.0 Ma, Upper Miocene).

2.2.2 Vilama Formation (Moraga et al. 1974) and Campamento Formation (Brüggen 1942)

West of San Pedro de Atacama, on top of ignimbrites of the San Bartolo Group, sediments of the Vilama Formation are deposited. South of San Pedro de Atacama, the San Pedro Formation is only locally covered by ignimbrites but is mostly directly overlain by the Vilama Formation. However small local erosional remainders of ignimbrites have been found between the San Pedro Formation and the Vilama Formation. The up to 80-m-thick Vilama Formation is characterized by conglomerates and red sandstones and mudstones. Towards the north more and more tuffitic layers and volcanics are intercalated and eventually volcanic rocks dominate. Layers of white volcanic ash and diatomites occur. Towards the south (west of San Pedro) the volcanic influence decreases, the grain size diminishes and lacustrine limestones are intercalated in siltstones and sandstones. In some places the siltstones contain ostracods (*Cyprideis* aff. *salebrosa* van den Bold 1963 and *Ilyocypris* sp.) and at the summit of the Cerro Mármol gastropods have been found in coarse sandstones near the top of the Vilama Formation. The ostracods are of Pliocene or younger age.

South of the Paso Domingo Ramos the Vilama Formation is characterized by increasing layers of evaporites. Further south evaporites prevail and the term "Campamento Formation" is used on the official geological 1:250 000 map for sediments of the same stratigraphic position and of probably the same age (Ramirez and Gardeweg 1982). The Campamento Formation and the San Pedro Formation can only be distinguished where they are separated from each other by an angular unconformity. This is not the case in the southernmost parts of the Cordillera de la Sal. Both the Vilama Formation and Campamento Formation grade into the recent sediments of the Salar de Atacama. Their lower age limit (late Miocene) is given in the north by the superimposition of the Vilama Formation on top of late Miocene ignimbrites.

According to Moraga et al. (1974) the brine beneath the central saltpan of the Salar de Atacama contains up to: 34.42 g/l Mg, 34.50 g/l Na, 37.00 g/l K, 6.40 g/l Li, 0.119 g/l Rb, 0.0083 g/l Cs. For several years the brines which are probably most concentrated and with a high lithium content have been exploited in the southwestern part of the Salar de Atacama. The high content of Mg, K and Li is probably a consequence of saltpan sedimentation in an endorheic basin over about 30 Ma with precipitation of large amounts of calcium sulphate, glauberite and halite occurring. For halite alone, a volume of approximately 1500 km3 has been calculated for the sediments of the Salar de Atacama and the Cordillera de la Sal. Concentration of the most soluble cations in the remaining brine is the result of this ongoing process.

3 Sedimentary and Palaeogeographic Development

Profiling and lithostratigraphic correlation between the profiles, especially with help from intercalated tuff layers used as marker horizons in the field and on aerial photographs, have demonstrated that with time the size and position of the centre of this evaporitic basin have changed markedly in a north-south direction (Fig. 9).

Determinations of palaeocurrent directions (Flint 1985; Wilkes and Görler 1988; Wilkes 1990) demonstrated a prevailing sediment transport from the north and west. However, a shift to easterly directions has also been observed for the highest portions of the San Pedro Formation.

The central saltpan was at its largest during the time of the deposition of the Crisanta member, and this probably corresponds with the Palicaye and Artolla members in the San Bartolo area, the lower conglomeratic unit of the Tambores Formation and unit A in the southern Cordillera de la Sal (Fig. 8). The centre of this initial basin was situated in the central portion of the present-day Cordillera de la Sal in the area between Valle de la Luna and the Paso Domingo Ramos, where the Crisanta member is especially poor in siliciclastic contaminations and where more than 1000 m of halite rocks were deposited. The saltpan of the Crisanta member must have measured at least 100 km from north to south.

During the deposition of the Valle de la Luna member in the northern part of the Cordillera de la Sal, halite precipitation continued in a central portion as indicated by the informal unit B (Fig. 9). Oligohaline to mixohaline lacustrine sediments of the Cementario member and the playa sandflat and fluvial sediments of the Rio Salado member (Flint 1985), deposited further to the north in the Rio Grande-San Bartolo area, are probably isochronous too. This would indicate a marked reduction and a southward shift of the central saltpan.

The maximum halite deposition in the Cota 2567 member was again situated in the Valle de la Luna-Paso Domingo Ramos area, but did not reach the thickness and regional extension of the Crisanta member. The Cota 2567 member can be interpreted as isochronous to the alluvial fan sediments of the upper conglomeratic unit of the Tambores Formation northwest of San Bartolo, but a time equivalent unit in the San Pedro Formation of the San Bartolo area seems to be missing (Fig. 9).

The Quebrada Honda member seems to be without a time equivalent in the north and shows saltpan sedimentation only in its upper portion; it can probably be correlated with the informal unit D in the south (Fig. 8). In this case a further shift of the depocentre towards the south must be assumed. The same trend exists for the Pliocene-Pleistocene time interval, as shown in the facies distributrions in the Vilama and Campamento Formations. Again the present-day position of the saltpan of the Salar de Atacama is a result of a further shifting towards the southwest and has an excentric position due to a freshwater influx coming mainly from the north and east.

Changes in the Quaternary-recent sediment dispersal pattern due to eastwards migrating compressional tectonics are evident and have been well documented for the northwestern margin of the Llano de la Paciencia by Jolley et al. (1990). This effect is even more spectacular at the southeastern margin of the Salar de Atacama south of Tilomonte. The morphology of this area is characterized by the steep up to 100 m high eastern limbs of N-S trending asymmetric folds of the Pliocene Tucúcaro ignimbrite. Locally (e.g. 6 km southwest of Tilomonte) remnants of a drainage pattern predating the folding of the ignimbrite and its cover of unconsolidated gravels and almost perpendicular to the actual drainage pattern are well preserved.

4 Structural Setting and Development

As far as can be seen, the sediments of the Preandean Depression have been deposited on top of older sediments, folded during the late Eocene Incaic phase (38-39 Ma, Döbel et al. 1992). Especially along the western border of the Salar de Atacama Depression, thick sequences of red continental clastics with volcanic intercalations of the Purilactis Formation (Upper Cretaceous-Eocene) are unconformably covered by the Tambores Formation. But, according to the interpretation (Townsend 1988) of seismic profiles across the Salar de Atacama (Fig. 3), there is no angular unconformity between the Purilactis Formation and the San Pedro Formation towards the east beneath the Salar de Atacama. Otherwise, the sediments of the Preandean Depression themselves are often affected by younger tectonics which is generally compressive or transpressive with a sinistral sense of displacement. The concentration of deformation on the Preandean Depression shows that the Preandean depression is a zone of crustal weakness probably due to the presence of melts at mid-crustal levels (Schwarz et al. this Vol.; Wigger et al. this Vol.). Furthermore the eastern margin of the Preandean Depression is the border between the rigid forearc and the more ductile magmatic arc. From this rheological contrast a concentration of deformation on the weak Preandean Depression can be expected. For example, on the southern border of the Salar de Atacama basin the N-S-striking Tucúcaro Fault crops out, which is one of the major structures of this region. According to Niemeyer (1984) the fault originated in Palaeozoic times and was reactivated as a reverse fault with a vertical displacement of >2000-3000 m during the late Miocene-early Pliocene. The last vertical (reverse) movements with a displacement of 3 m occurred during the Pleistocene. The Cordillera de la Sal has been also strongly deformed since the early Miocene. Probably due to the presence of thick halite layers the uppermost part of the crust was weaker than other parts of the Salar de Atacama Basin and, thus, deformation concentrated on the Cordillera de la Sal. Here at least five different types of deformation have been observed. The most evident tectonic elements are:

NNE-SSW trending folds with doubly plunging axes. The fold axes display sinoidal bendings, evolving locally towards an "en échelon" grouping (Ramirez and Gardeweg 1982; Hooper and Flint 1987, Forsythe and Toskos 1988; Wilkes 1990). SE-vergence prevails, but there is also a tendency towards bivergent structures (Wilkes 1990). This type of deformation can be observed mainly in the Cordillera de la Sal (Figs. 11 and 12). Seismic investigations (Townsend 1988) have revealed that the fold structures decline rapidly eastwards beneath

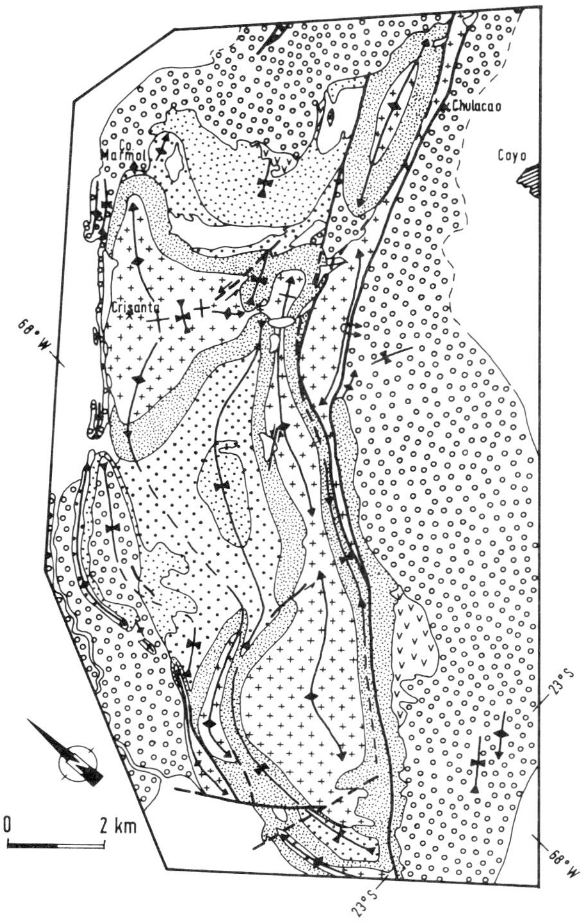

Fig. 11. Geological map of the northern Cordillera de la Sal between San Pedro de Atacama and Paso Domingo Ramos. (From Wilkes 1990)

the Salar de Atacama (Fig. 3) while continuing westwards beneath the Llano de la Paciencia to become visible at the eastern margin of the Cordillera de Domeyko where the San Pedro Formation or the Tambores Formation crop out (Ramirez 1977, 1979a; Hooper and Flint 1987). The main fold belt coincides with the zone of maximum thickness and prevailing evaporitic sedimentation of the San Pedro Formation. As mentioned, another area of large scale monoclinal folds exists at the southeastern margin of the Salar de Atacama south of Tilomonte. In this area and especially around the northwestern margin of the Llano de la Paciencia the asymmetric folds evolve over short distances to thrust faults. At least two phases of folding exist, one before the emplacement of the middle Miocene and Pliocene volcanics, which in the Rio Grande and San Pedro area cover the intensively folded San Pedro Formation and another one after the deposition of these volcanics, as the ignimbrite sheets themselves show a large scale fold pattern.

NNE-SSW striking thrust faults. These are mostly with an ESE thrust direction, but also show the opposite vergency. Such low angle reverse faults, sometimes with visible overthrusts of more than 100 m on top of unconsolidated gravels, are common outside the Cordillera de la Sal near the northwest (Hooper and Flint 1987, Jolley et al. 1990) and southeast margins of the Salar de Atacama basin where ignimbrites are outcropping. They evolve over short distances from monoclinal folds. Jolley et al. (1990) have interpreted the structural pattern of the northwestern part of the Salar de Atacama Basin as products of an east vergent dynamic thrust system with deformation progressively propagating towards the east.

A very complex pattern of nearly vertical small scale faults (normal, reverse and strike slip, mostly sinistral). This type of deformation is typical for the Cordillera de la Sal too and is easily visible in vertical walls of the evaporites and siltstones of the Valle de la Luna member (Wilkes 1990). This type of deformation has not been detected in the ignimbrites and the Vilama Formation on top of the San Pedro Formation. Sinistral strike-slip faults are not restricted to the fold belt and occur elsewhere in the region. For example Macellari et al. (1991) inferred sinsitral movements from the interpretation of seismic sections of the Salar de Atacama. In Pliocene ignimbrite beds from east of the Salar de Atacama a N-S trending system of right-stepping Riedel shears occurs which also indicates sinistral displacement. Quaternary sinistral strike-slip movements have been reported by Armijo & Thiele (1990) for the N-S trending Atacama Fault in the Coastal Cordillera near Antofagasta. This fault, which originated as a sinistral strike-slip fault in Jurassic times (Scheuber & Andriessen 1990), has been reactivated since the Neogene. Armijo & Thiele (1990) attributed the sinistral movements, which are in contradiction to the present minor dextral component of convergence between the Nazca and the South American plates, to a clockwise rotation of the Central Andes south of the Bolivian orocline (Arica bend).

Vertical tension fissures. These are well preserved especially in the Oligocene-Miocene conglomeratic Tambores Formation (N-S), the late Miocene ignimbrites of the San Bartolo Group (NW-SE), the Pleistocene ignimbrites east of San Pedro (WNW-

ESE) and in the Crisanta member (NW-SE) within the Valle de la Luna area. Especially within the ignimbrites systems of open cracks are widespread and well visible in aerial photographs. Their arrangement, the geometry of the observed small scale faults and the sinoidal bending of the fold axes within the Cordillera de la Sal are consistent with the assumption of a SW-NE striking strike-slip movements.

Effects of gravitational lateral spreading. These are manifested as recumbent folds and salt glaciers on both sides of the Cordillera de la Sal in zones where halite rocks were uplifted.

The fold structures observed within the Cordillera de la Sal become increasingly complicated from south to north (Wilkes and Görler, 1988; Wilkes, 1990). A simple, small east-vergent anticline in the south (Fig. 7d; shortening 7%) develops towards the north into a complicated divergent fold sequence at the Paso Domingo Ramos. Further to the north, SSW-NNE striking reverse faults with vergencies towards the WNW and ESE occur within and at the margins of the Cordillera de la Sal (Fig. 7a; shortening 30%). Quaternary alluvial gravels of Cordillera de Domeyko material within the cores of synclines of the folded San Pedro Formation and even along the eastern side of the Cordillera de la Sal imply a young uplift and related folding of the central Cordillera de la Sal. In the northern Cordillera de la Sal the uplift is younger than the antecedent Rio San Pedro which, north of San Pedro de Atacama, follows the axis of the Cordillera de la Sal, avoiding the depression of the Llano de la Paciencia in the west and the Llano Vilama in the east (Abele 1988).

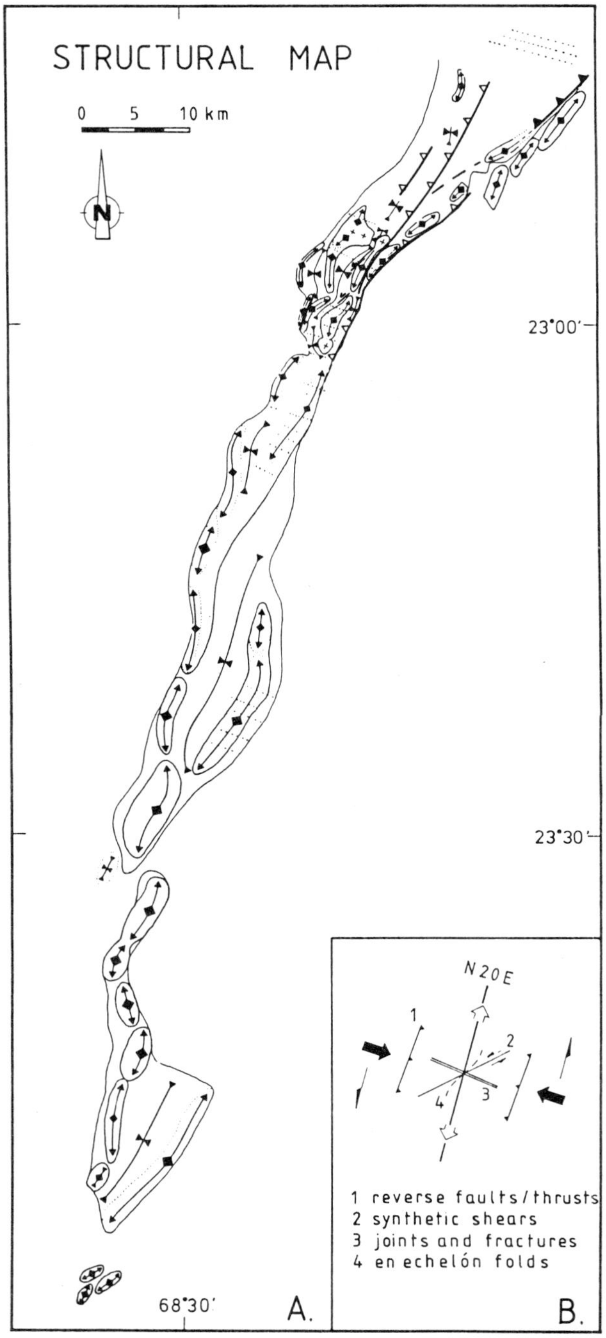

Fig. 12. *A* Structural map of the Cordillera de la Sal. *B* Structural scheme (after Harding 1974) with directions of shortening (*filled thick arrows*) and extension (*open arrows*). (From Wilkes 1990)

There is good evidence for repeated tectonic syndepositional movements from the Oligocene-Miocene to the present, but the timetable of deformational events in the Salar de Atacama Depression can only be estimated. These estimates are, however, consistent with the ages of unconformities published for the Western Cordillera (Lahsen 1982) and the SW Bolivian Altiplano (Kussmaul et al. 1975). According to the few relevant radiometric dates (Travisany 1978; Ramirez 1979b; Marinovic and Lahsen 1984), a first folding in the north (Rio Grande area) took place in early Miocene time between 24.9 and 17 Ma, (Quechua 1 phase; Mégard et al. 1984). In the Altiplano and the Western Cordillera shortening took place at ~23 Ma (Kussmaul et al. 1975; Lahsen 1982). Otherwise, the above-mentioned findings of fossils of probably late Miocene age in the folded San Pedro Formation of the San Bartolo area are an indication of younger folding. According to Marinovic and Lahsen (1984) the unconformably overlying Sifón Ignimbrite has an age of 9.0 ± 1.2 Ma, a late Miocene (Quechua 2)

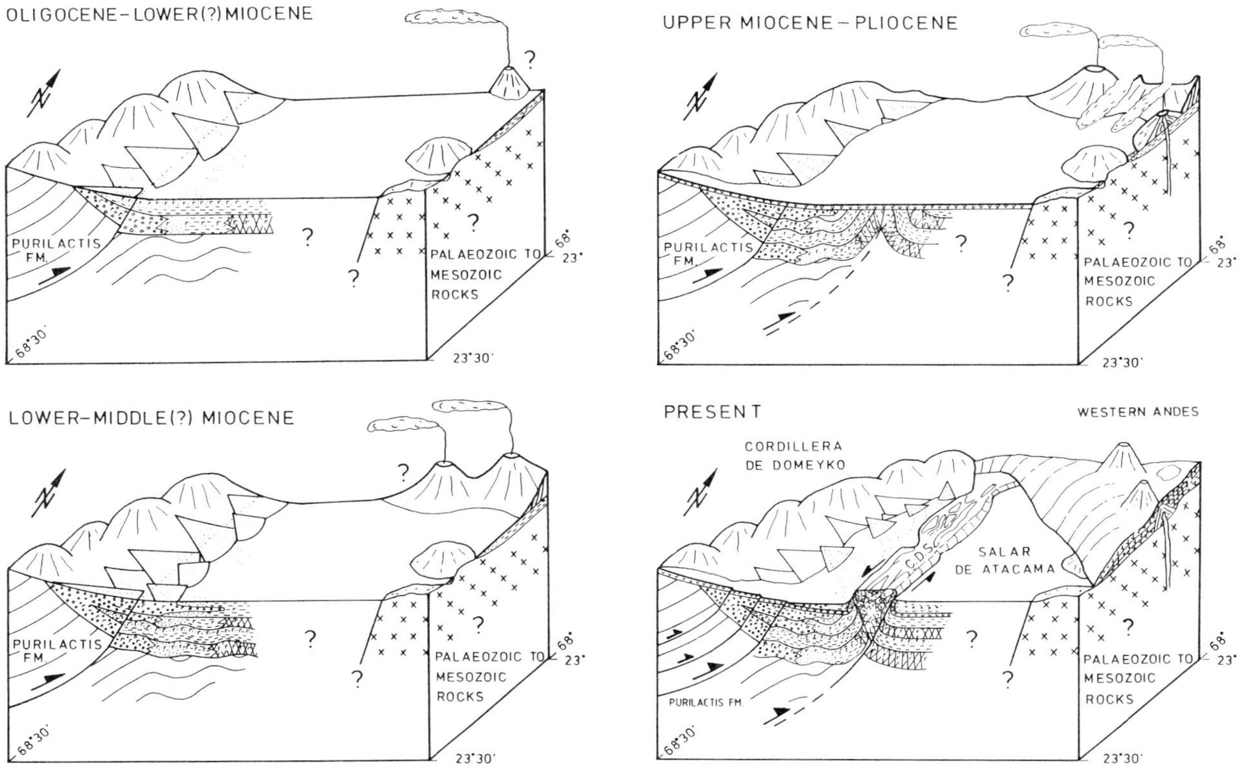

Fig. 13. Stages in the development of the Preanden Depression since the late Oligocene. Stage I (*late Oligocene to early(?) Miocene*) After the Incaic folding, faulting and uplift of the area of the late Cretaceous-early Palaeogene magmatic arc in Oligocene times, severe erosion started. Most of the eroded material has been transported towards the east via alluvial fans - represented by the Tambores Formation - into the Preandean Depression. In the centre of a N-S directed basin, at least 100 km long but not very wide, uniform deposition of halite took place. Stage II (*early to middle(?) Miocene*) sedimentation within a large alluvial fan/playa system continues. The extent and position of the central saltpan is frequently shifting owing to climatic changes and tectonic activity in the source area. Local angular unconformities within the San Pedro Formation and significant changes of thickness over short distances are provoked by initial transpressional movements within the basin. Stage III (*late Miocene to Pliocene*). In the northern Salar de Atacama Basin intensive folding was followed by erosion. The folded and eroded sediments of the San Pedro Formation were covered by different ignimbrite sheets of the San Bartolo Group. In the southernmost Salar de Atacama Basin evaporite sedimentation continues without interruption. Stage IV (*Pleistocene and Holocene*). In the northern and central Salar de Atacama Basin further folding and the uplift of the Cordillera de la Sal, probably as a "pop up" structure occurred. In the southernmost Salar de Atacama this process has hardly begun. Evaporite sedimentation continues west and east of the Cordillera de la Sal in the actual Llano de la Paciencia and the Salar de Atacama. In the areas of greatest thickness the San Pedro Formation can be divided into distinctive members, but rapid lateral changes and the absence of continuing outcrops make the correlation between different areas problematic

phase of folding is therefore likely. Deformations of ~10 Ma have been reported by Jordan & Alonso for the Argentine Puna (Jordan & Alonso 1987) and also for the Altiplano and Western Cordillera (Kussmaul et al. 1975; Lahsen 1982). Deformations also occurred after the late Miocene as the ignimbrites of the Upper Miocene San Bartolo Group are also affected by ample folding and by local involvement of unconsolidated gravels of an uncertain age on top of the ignimbrites. Further to the south the angular unconformity between the folded San Pedro Formation and the younger Campamento Formation is reduced to an increasingly concordant position. In the southernmost portion of the Cordillera de la Sal, only an upwarp of the present-day halitic surface of the Salar de Atacama not yet affected by karstification is a last indication of a very young folding that is ongoing. Otherwise, indications of older phases of folding are missing in the southern Cordillera de la Sal. These observations suggest that the folding shifted with time and started later in the south than in the north. The post-Miocene deformations in the Preandean Depression may be attributed to the Pliocene-Pleistocene Diaguita Phase (Jordan & Alonso 1987); however, widespread angular unconformities and slumping horizons within the San Pedro Formation of the northern Cordillera de la Sal make frequently repeated synsedimentary deformations, uplift and erosion more probable than a single tectonic phase. A more accurate timetable of deformation cannot be established until more radiometric dating has been carried out within the thick continental red beds.

The causes of the deformation and uplift of the Cordillera de la Sal have been discussed by various authors (Gerth 1955; Dingman 1962; Zeil 1964; Hollingworth and Rutland 1968; Guest 1969; Thomas 1970; Hooper and Flint 1987; Forsythe and Toskos 1988; Jolley et al 1990; Wilkes 1990; Williams et al. 1990; Macellari 1991). However, the older models of diapirism by buoyant salt movement or gravity-gliding folding do not coincide with field and seismic observations. According to calculations by Lerche and O'Brien (1987), diapirism by a contrast in density between salt and the surrounding sediments seems less probable in this case. Salt movements and lateral gravitational spreading may be triggered by compressional stress but are presumably not of great importance. The origin and the observed sedimentary and structural patterns of the Salar de Atacama Depression fit best with the assumption of compression with a dominant tectonic transport direction towards the southeast (Jolley et al. 1990; Williams et al. 1990) combined with episodic sinistral strike-slip displacements (Macellari et al. 1991).

5 Conclusions - a Brief Outline of the Development of the Preandean Depression

The development of the Preandean Depression since the Oligocene is illustrated by the sketches of Figs 13. During the late Cretaceous to early Oligocene the later Preandean Depression was part of the backarc basin of the magmatic arc situated in the Chilean Precordillera. The basin formed during an early Oligocene period of low convergence rate (Pardo-Casas & Molnar 1987) and decreased magmatic activity (Scheuber et al. this Vol). The crust which had thickened during the late Eocene Incaic Phase probably reacted to the decreasing convergence rate by a readjustment of crustal thickness to the lowered stresses; this readjustment may have led to the formation of basins such as the Preandean Depression. In Oligocene times compressional thrust faulting and uplift began within the magmatic arc at the western margin of the Preandean Depression Sedimentation in the N-S striking small basin was characterized by alluvial fans in the west, and massive evaporite deposition in the centre. The asymetrical position of the central saltpan near the western margin was probably caused by active thrusting and tectonic subsidence. Rare intercalated pyroclastic layers within the sedimentary record are the first signs of new volcanic activity to the east of the Preandean Depression.

During the early to middle Miocene, increasing compressional tectonics at the western margin of the Salar de Atacama Basin caused a shift in the centres of erosion and deposition, together with intraformational unconformities and significant lateral changes in facies and thicknesses over short distances. The intraformational angular unconformities are indicative of a cannibalistic resedimentation, especially of the evaporites within the Salar de Atacama Basin. The increased occurrence of pyroclastics and a change in the direction of alluvial fan transport hint at the buildup of the modern magmatic arc towards the north and east of the Salar de Atacama Basin during this period.

In late Miocene to Pliocene times the zone of main tectonic activity shifted from the western rim of the Salar de Atacama Basin towards the central saltpan. A "Palaeocordillera de la Sal" evolved and this was essentially eroded before the deposition of the Upper Miocene ignimbrite sheets. The ignimbrites are indicative of a very active magmatic arc east of the Salar de Atacama Basin, and coincide with the Quechua 2-phase of Mégard et al. (1984).

The uplift of the actual Cordillera de la Sal, which started during the Pleistocene, coincides with the Diaguita-compressional phase of Jordan and Alonso (1987). The position of the Cordillera de la Sal probably marks the outcrop of the main thrust of an eastward-migrating thrust belt (Jolley et al. 1990). Regarding its steep reverse fault at its eastern margin and back thrust faults in the west, the Cordillera de la Sal can be interpreted as a "pop-up"-structure. As a consequence of the uplift the Salar de Atacama Basin was divided into two depocentres by the Cordillera de la Sal which acts as the main source for the actual intensive cannibalistic resedimentation of evaporites.

Acknowledgements. We thank Dr. F.-F. Helmdach for determining the ostracod faunas and Mrs. E. Oeff for carrying out the chemical analyses.

References:

Abele G (1988) Geomorphological west-east section through the northern Chilean Andes near Antofagasta. In: Bahlburg H, Breitkreuz C, Giese P (eds). The southern Central Andes. Lecture Notes in Earth Sciences, 17. Springer, Berlin Heidelberg New York, pp 153-168

Alonso RN, Jordan ET, Tabutt KT, Vandervoort DS (1991) Giant evaporite belts of the Neogene central Andes. Geology 19: 401-404

Armijo R, Thiele R (1990) Active faulting in northern Chile: ramp stacking and lateral decoupling along a subduction plate boundary. Earth Planet Sci Lett 98: 40-61

Baby P, Sempéré T, Oller J, Barrios L, Hérrail G (1990) The southern Altiplano of Bolivia: An Oligo-Miocene intermontane foreland basin. Forschergruppe Mobilität aktiver Kontinentalränder, Final Workshop, Abstr Vol Berlin, p 69

Brüggen J (1934) Grundzüge der Geologie und Lagerstättenkunde Chiles. Heidelberger Akademie der Wissenschaften

Brüggen J (1942) Geología de la Puna de San Pedro de Atacama y sus formaciones de areniscas y arcillas rojas. Primer Congr Panam Ing Minas Geol An II, 1: 342-367

Dingman RJ (1962) Tertiary salt domes near San Pedro de Atacama. US Geol Surv Prof Pap 450D: 92-94

Dingman RJ (1967) Geology and ground-water resources of the northern part of the Salar de Atacama, Antofagasta province, Chile. US Geol Surv Bull 1219, 49 pp

Döbel R, Friedrichsen H, Hammerschmidt K (1992) Implication of $^{40}Ar/^{39}Ar$ dating of early Tertiary volcanic rocks from the north Chilean Precordillera. Tectonophysics 202:55-81

Eugster HP, Hardie LA (1978) Saline lakes. In: Lerman A.(ed) Lakes - chemistry, geology, physics. Springer, Berlin Heidelberg New York, pp 237- 293

Flint S (1985) Alluvial fan and playa sedimentation in an Andean arid closed basin: The Pacencia Group, Antofagasta province, Chile. J Geol Soc Lond 142: 535- 546

Forsythe RD, Toskos T (1988) Late Cenozoic tectonics along the southwestern border of the Altiplano/Puna, Salar de Atacama Depression, Antofagasta region, Chile. VII. Congr. Latinamericano de Geologia Belem Abstr, p 315

Gerth H (1955) Der geologische Bau der südamerikanischen Kordillere. Bornträger, Berlin, 264 pp

Guest JE (1969) Upper Tertiary ignimbrites in the Andean Cordillera of part of the Antofagasta province, northern Chile. Geol Soc Am Bull 80: 337-362.

Handford CR (1982) Sedimentology and evaporite genesis in a Holocene continental-sabkha playa basin - Bristol Dry Lake, California. Sedimentology 29: 239-253.

Hardie LA, Smoot JP, Eugster HP (1978) Saline lakes and their deposits: A sedimentological approach. In: Matter, A., Tucker, M. E. (eds.). Modern and ancient lake sediments. Int Assoc Sedimentol Spec Publ 2: 7-41

Harding TP (1974) Petroleum traps associated with wrench faults. AAPG Bull 58: 1290-1304

Hollingworth SE, Rutland RWR (1968) Studies of Andean uplift Part I - Post-Cretaceous evolution of the San Bartolo area, north Chile. Geol J 6(1): 49-61

Hooper B, Flint S. (1987) Miocene-Recent tectonic evolution of the San Bartolo Area, northern Chile. Zentralbl Geol Paläontol Teil I (7/8): 967-981

Jolley EJ, Turner P, Williams GD, Hartley A, Flint S (1990) Sedimentological response of an alluvial system to Neogene thrust tectonics, Atacama desert, northern Chile.- J Geol Soc Lond 147: 769-784

Jordan TE, Alonso RN (1987) Cenozoic stratigraphy and basin tectonics of the Andes mountains, 20°-28° south latitude. AAPG Bull 71(1): 49-64

Kussmaul S, Jordan L, Ploskonka E (1975) Isotopic ages of Tertiary volcanic rocks of southwest Bolivia. Geol Jahrb B14: 111-120

Lahsen A (1982) Upper Cenozoic volcanism and tectonism in the Andes of northern Chile. Earth Sci Rev 18: 285-302

Lerche I, O'Brien JJ (1987) Modelling of buoyant salt diapirism. In: Lerche I, O'Brien JJ (eds) Dynamical geology of salt and related structures. Academic Press, New York, pp 129-162

Macellari CE, Su MJ, Townsend F (1991) Structure and seismic stratigraphy of the Atacama basin, northern Chile. VI Congr. Geol Chil Resumenes ampliados: 133-137

Marinovic SN, Lahsen A (1984) Hoja Calama. Serv. Nac. Geol Min, Santiago, Carta Geol de Chile 58, 140 pp

Mégard F, Noble DC, McKee EH, Bellon H (1984) Multiple pulses of Neogene compressive deformation in the Ayacucho intermontane basin, Andes of central Peru. Geol Soc Am Bull 95: 1108-1117

Moraga A, Chong G, Fortt MA, Henriquez H (1974) Estudio geologico del Salar de Atacama, provincia de Antofagasta. Bol Inst Invest Geol Santiago 29: 56

Niemeyer H (1984) La Megafalla Tucúcaro en el extremo sur del Salar de Atacama: una antigua zona de cizalle reactivada en el Cenozoico. Comunicaciones 34: 37-45

Ramirez CF (1977) Geologia de la parte norte de los cuadrangulos San Pedro de Atacama y Cordillera de la Sal, provincia El Loa, II. Región. Taller de Titulo I y II, Univ de Chile, Santiago 67 pp (unpubl)

Ramirez CF (1979a) Geologia de cuadrangulo Rio Grande y sector sur oriental de cuadrangulo Barros Arana, provincia El Loa, II. Región. Tesis de Grado, Univ de Chile, Santiago 139 pp (unpubl)

Ramirez CF (1979b) Edades potassio-argón de rocas volcánicas cenozoicas en la zona de San Pedro de Atacama-El Tatio, región de Antofagasta. II Congr. Geol Chil Arica 1: F31-F41

Ramirez CF, Gardeweg PM (1982) Hoja Toconao. Serv Nac Geol Min, Santiago, Carta Geol de Chile 54, 121 pp

Scheuber E, Andriessen PAM (1990) The kinematic and geodynamic significance of the Atacama Fault Zone, northern Chile. J Struct Geol 12: 243-257

Schwab K (1985) Basin formation in a thickening crust - the intermontane basins in the Puna and the Eastern Cordillera of NW-Argentina (Central Andes). IV. Congr Geol Chil, Antofagasta 1: 138-158

Stoertz GE, Ericksen GE (1974) Geology of salars in northern Chile. US Geol Surv Prof Pap 811: 65 pp

Thomas A (1970) Beitrag zur Tektonik Nordchiles. Geol Rundsch 59: 1013-1027

Townsend F (1988) Exploracion petrolera en la cuenca del Salar de Atacama, región de Antofagasta, Chile. Vertiente, Antofagasta 4: 45-55

Travisany VA (1978) Mineralización cuprifera en areniscas de la Formación San Pedro en el distrito San Bartolo. Tesis de Grado, Univ de Chile, Santiago 71 pp (unpubl)

Wilkes E (1990) Die Geologie der Cordillera de la Sal, Nordchile. Berl Geowiss Abh (A) 128: 145 pp

Wilkes E, Görler K (1988) Sedimentary and structural evolution of the Cordillera de la Sal, II. Regíon, Chile. V. Congr Geol Chil, Santiago 1: A173-A188

Williams GD, Turner P, Flint S, Stimpson I, Jolley EJ, Hartley AJ (1990) Andean basin dynamics in northern Chile. Symp int Geodynamique andine, Résumés des communications. ORSTOM, Grenoble p 231

Zeil W (1964) Geologie von Chile. Bornträger, Berlin, 233 pp

The Purilactis Group of Northern Chile: Boundary Between Arc and Backarc from Late Cretaceous to Eocene

REYNALDO CHARRIER and KLAUS-J. REUTTER

Abstract. The approximately 2000- to 4000-m-thick detritic Purilactis Group exposed for ≈150 km along the western escarpment of the tectonic depression of the Salar de Atacama is subdivided from bottom to top into: (1) Tonel Fm. (≈1000 m of fine sandstones and gypsiferous mudstones), (2) Purilactis Fm. (750-3000 m of conglomerates with volcanic intercalations), and (3) Yesífera Superior Fm. (≈400 m of gypsiferous sandstones). The age of the Purilactis Group comprises the time interval from the latest Cretaceous to the Eocene-Oligocene boundary. Its depocentre bordered, to the west, probably along normal faults, on the relatively elevated area of a contemporaneously active magmatic arc in the Precordillera of Northern Chile (Cordillera Domeyko), and, hence, it is considered to be the westernmost part of an epicontinental backarc basin. To the east this basin was connected with the rift-induced depocentre of the Salta Group in northwestern Argentina. The vertical transition from the fine-grained sediments of the Tonel Fm. to the conglomeratic facies of the Purilactis Fm. at about 44 Ma coincides with a shift of magmatic activity towards the east. The Incaic orogenic phase at about 38 Ma led to general crustal shortening which affected the arc and especially the backarc area. Thrusts and folds developed at the former basin boundary which also was subjected to dextral strike-slip displacements as revealed by local vertical folds. Also, younger tectonics related to the development of the Salar de Atacama Depression affected the western escarpment, characterizing it as an important tectonic lineament.

1 Introduction

The western escarpment of the tectonic depression of the Salar de Atacama shows the red continental sediments of the Purilactis Group which form a narrow belt about 5-10 km wide between the younger sediments of the Salar and the quite different, and mostly older, rocks of the Cordillera de Domeyko to the west. The intense fold and fault structures of the Purilactis sediments suggest that this western escarpment may represent an important tectonic zone. This is underlined by the small and evidently younger fold belt of the Cordillera de la Sal, which developed within the Salar de Atacama in front of the western escarpment (Fig. 1), and by gravity anomalies which show, in the residual field, a steep slope running parallel to the escarpment and separating a gravity minimum to the west from a maximum to the east (Götze et al. this Vol.; Map 3 this Vol.).

Stratigraphy and tectonics of the well-exposed Purilactis Group have attracted the attention of geologists over many years (e.g. Brüggen 1934, 1942; Dingman 1963; Moraga et al. 1974; Ramírez and Gardeweg 1982; Salfity et al. 1985; Hartley et al. 1988; Flint et al. 1989; Charrier and Reutter 1988, 1990). It was recognized that the Purilactis Group rests unconformably on the late Palaeozoic Agua Dulce and El Bordo Fms. and is unconformably covered by the gravels of the Oligocene-Miocene Tambores Fm. (Brüggen 1934), an equivalent of the San Pedro Fm. which is involved in the structures of the Cordillera de la Sal. The age of the Purilactis sediments, constrained only by the ages of the overlying and underlying rocks, was especially a matter for discussion.

Brüggen (1934, 1942) divided the Purilactis sequence in the area to the east of P. (Portezuelo) Barros Arana (Fig. 1: Barros Arana syncline) into three subunits (from the bottom): Formación Salina de Purilactis (fine grained unit with gypsiferous beds), Formación Porfirítica de Purilactis (1000 m of dark brown sandstone with conglomeratic intercalations) and Conglomerados de Purilactis (1000 m of conglomerates). He assigned an early Cretaceous age to the second subunit and also considered the rim of the western escarpment of the Salar de Atacama to be a fault.

Correspondence to: R. Charrier Departamento de Geología y Geofísica, Universidad de Chile, Casilla 13518, Correo 21, Santiago, Chile

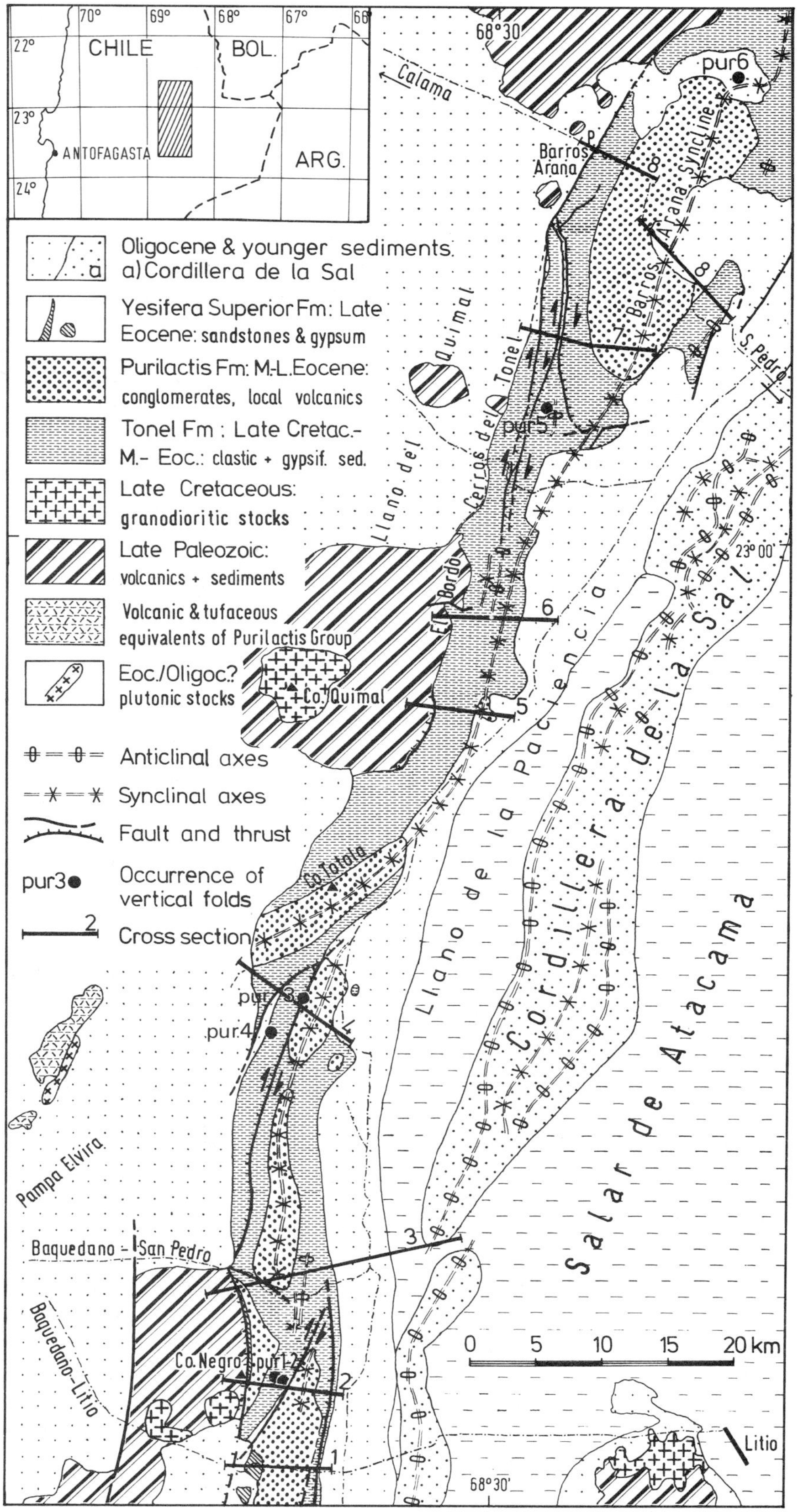

Fig. 1. Geological map of the western escarpment of the Salar de Atacama based on Landsat-MSS imagery, aerial photographs, and field mapping. Heavy lines indicate locations of *cross sections 1-8*, and points *pur1-pur6* indicate sites of vertical folds due to longitudinal dextral strike-slip movements

Dingman (1963) proposed new formal names for these units: Tonel Fm. for Brüggen's Formación Salina de Purilactis, and, separated by a disconformity, Purilactis Fm. composed of the other two formations and subdivided into three members (from the bottom): (1) fine to coarse sandstones, (2) sandstones and gypsiferous siltstones with conglomeratic intercalations, and (3) coarse conglomerates with a volcanic level near the base. He assigned a Jurassic age to the Tonel Fm. (a view still shared by Macellari et al. 1991) and a Cretaceous age to the Purilactis Formation. This division was followed by Moraga et al. (1974).

Ramírez and Gardeweg (1982) integrated Dingman's two formations in a unique Purilactis Fm. of late Jurassic-Cretaceous age. In the area of the map Hoja Toconao (23°-24°S), however, they used the name "Cinchado Fm." for a Palaeocene-Eocene conglomeratic and volcanic sequence which was supposed to overlie unconformably the Purilactis Formation.

According to Hartley et al. (1988) the Tonel Fm. is separated by an unconformity from the overlying Purilactis Fm. On the basis of facies analyses, they subdivided the Purilactis Fm. of their definition into eight members. From a 40 m thick andesitic lava, overlying the fifth member, Flint et al. (1989) obtained a ^{40}Ar-^{39}Ar age of 63.7 ± 10 Ma. Hartley et al. (1991), revising earlier stratigraphic concepts, include both formations in the Purilactis Group and attribute to it a late Cretaceous-early Tertiary age. Such a stratigraphic concept had also been proposed by Kriz and Cherroni (1966) and Salfity et al. (1985).

Field observations by the present authors over almost the whole N-S extension of the Purilactis rocks, make it necessary to revise the stratigraphic concept yet again. In this respect, the fact must be stressed that the Purilactis Fm., as defined by Charrier and Reutter (1990, present chapter), corresponds only to the upper part of the Purilactis Fm. defined by Hartley et al. (1988) and Jolley et al. (1990). The present study is based on stratigraphic and tectonic research work along cross sections as well as on reconnaissance mapping in the western escarpment of the Salar de Atacama between 22°45' and 23°45' S (five campaigns of 10 to 15 days each, 1984-1989; Fig. 1). These data, and previous work by these and other authors, together with radiometric datings by Döbel (1989; Döbel et al. 1992) on samples taken by the present authors now permit a tentative synthesis of the stratigraphic, lithological, and structural features of the Purilactis rocks and of their palaeotectonic setting.

2 Stratigraphic Subdivision

The definition of the Purilactis Group and its subdivision into three formations was introduced by Charrier and Reutter (1990). The group includes sequences described by preceding authors under the names Tonel, Purilactis, and Cinchado Fms. We conserve the well known name of Purilactis and use it sensu lato for the whole sequence, attributing to it the rank of a group, and sensu stricto for its middle part, which is considered to represent a lithostratigraphic formation corresponding lithologically to Brüggen's (1942) type locality. The three lithostratigraphic units distinguished in the Purilactis (sensu lato) Group are, from bottom to top (Table 1, Fig. 3): the fine grained, partly evaporitic Tonel Fm., the mainly conglomeratic Purilactis (sensu stricto) Fm., and the arenaceous and evaporitic Yesífera Superior Formation. Along the segment studied the Purilactis Group overlies disconformably the Agua Dulce and El Bordo Fms., both of late Palaeozoic age (Breitkreuz et al. 1992), and is unconformably covered by the white, scarcely consolidated coarse clastics of the Tambores Fm. (Oligocene-middle Miocene) and younger detritic units.

The following description is based on a cross section (section 2 in Figs. 1 and 2) located in the southern part of the study area near Cerro Negro, where the stratigraphic sequence of the Purilactis Group is almost complete, including the basal contact with the Palaeozoic Agua Dulce Formation. Only the upper part of the uppermost formation is absent because of a thrust which brought the Palaeozoic lavas to the surface again.

Tonel Formation

The lower, mostly fine-grained and partly evaporitic formation of the Purilactis Group is, in this section, approximately 1300-1400 m thick. It rests with a disconformity or very low angle unconformity on dark grey vesicular lavas of the late Palaeozoic Agua Dulce Formation. The transition to the overlying conglomeratic Purilactis Fm. occurs gradually over some tens of metres. The lithological sequence is given in Table 1.

The main features of the Tonel Fm. are maintained throughout the study area, although variations in thickness and lithology occur. So, along the trace of section 3 (Fig. 1), frequent conglomeratic intercalations dominate the upper part of the formation beneath the uppermost gypsiferous sandstones. In the

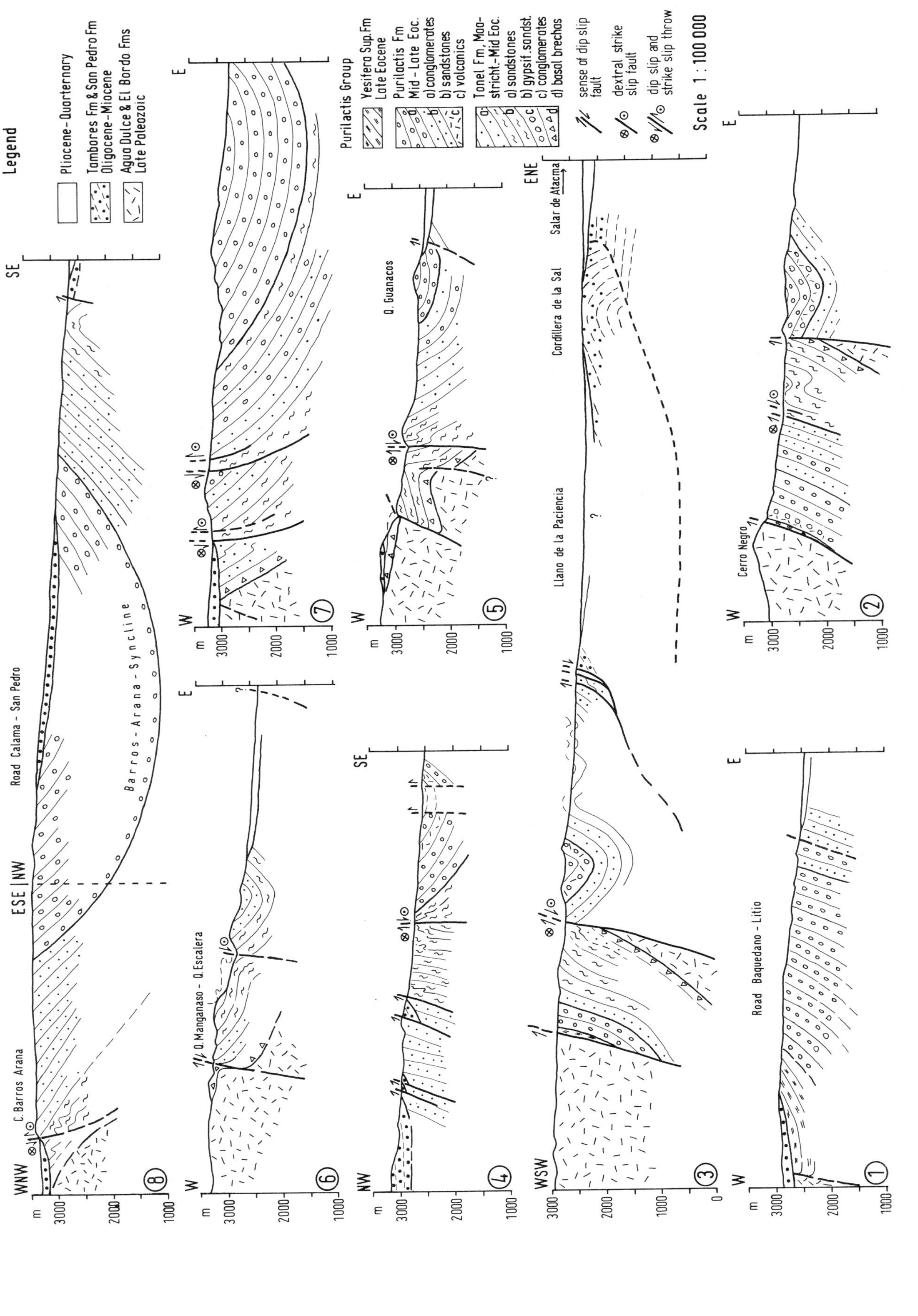
Legend
Pliocene-Quarternary
Tambores Fm & San Pedro Fm Oligocene-Miocene
Agua Dulce & El Bordo Fms Late Paleozoic
Purilactis Group
Yesifera Sup. Fm Late Eocene
Purilactis Fm Mid-Late Eoc.
a) conglomerates
b) sandstones
c) volcanics
Tonel Fm, Maastricht.-Mid Eoc.
a) sandstones
b) gypsif. sandst.
c) conglomerates
d) basal brechas
sense of dip slip fault
dextral strike slip fault
dip slip and strike slip throw
Scale 1 : 100 000
WNW
C. Barros Arana
ESE | NW
Road Calama - San Pedro
Barros - Arana - Syncline
SE
W
Q. Manganaso - Q. Escalera
E
W
E
NW
SE
W
Q. Guanacos
E
WSW
Llano de la Paciencia
Cordillera de la Sal
Salar de Atacma
ENE
W
Road Baquedano - Litio
E
W
Cerro Negro
E
m
3000
2000
1000
0
8
6
7
4
5
3
1
2

area of the Barros Arana syncline (Fig. 1) these conglomerates probably correspond to the "isolated conglomerate packets" which Hartley et al. (1988) attributed to the "Members 3 and 4" of the Purilactis Fm. of their definition. According to our subdivision of the Purilactis Group (Charrier and Reutter 1990), however, these "Members 3 and 4" should be included in the Tonel Fm., together with the fine-grained "Members 1, 2, and 5" of Hartley et al. (1988).

In the lowermost part of the Cerros del Tonel area (Fig. 1) the basal limestones found in the areas of section 2 (Table 1) and Fig. 4 are lacking, but medium- and fine-grained carbonate-cemented sandstones, about 200 m thick, occur. The upwards-following members probably contain greater quantities of gypsum than other localities to such a degree that diapirism occurred as a result of shortening. Subvolcanic stocks and dykes, which possibly pertain in age to the tuffs and lavas in the lowermost part of the Purilactis Fm., are frequent in the area of locations pur3-pur4 (Fig. 1) and influence the local structural pattern.

Purilactis Formation

In the area of section 2 (Fig. 1) fine-grained clastics with sporadic intercalations of conglomerates near the top of the Tonel Fm. are gradually replaced further up by thick massive conglomeratic layers, containing only very few fine-grained intercalations, of the lowermost Purilactis Formation. In some places the first conglomerates (matrix supported) alternate with orange mudstones of the uppermost portion of the Tonel Formation. About 2.5 km to the northeast, the basal alternating conglomerates and mudstones of the Purilactis Fm. are followed by nearly 20 m of non-fossiliferous carbonate-cemented conglomeratic breccias and limestones, and further up by brown conglomerates with a green tuffaceous intercalation (lower volcanic level). The main part of the Purilactis Fm. is a rather monotonous sequence of thick beds of coarse conglomerates. In section 2 and to the south of section 1 (Fig. 1) the top of the formation is characterized by lava flows and pyroclastic rocks which alternate with the conglomerates (upper volcanic level).

The conglomerates that form the syncline immediately north of the San Pedro-Baquedano road (section 3 in Figs. 1 and 2) and the alternation of conglomerates and lava flows of Cerro Totola (Fig. 1) were considered by Ramírez and Gardeweg (1982) to be an independent unit ("Cinchado Fm.", a term introduced by Montaño [1976] for a sequence of lavas and conglomerates at Cerro Cinchado situated within the magmatic arc domain, more than 50 km to the west of the study area) separated from the underlying Purilactis sediments by an angular unconformity. Indeed, from aerial photographs an apparent unconformity may be interpreted for the marked differences between short wavelength folds in the mudstones and fine sandstones of the upper Tonel Fm. and the wide syncline of the overlying conglomerates. Field observations, however, reveal a perfect stratigraphic continuity at this contact, which lithologically corresponds to the transition from the Tonel Fm. to the Purilactis Fm. in other places.

Supposing that, in the area of the Barros Arana syncline, the Purilactis Fm. also starts with compact conglomerates (Fig. 1; section 8 in Fig. 2), it can only embrace the upper three ("Members 6-8") of the eight members of the Purilactis Fm. of Hartley et al. (1988). Consequently, the dimensions and the thickness of the Purilactis Fm. of Charrier and Reutter (1990, this chapter) are smaller in this area than those indicated by Hartley et al. (1988).

The lithostratigraphic sequence of the Purilactis Fm. represented in Table 1 amounts to a thickness of $\approx$750 m. This value is extraordinarily low when compared with that of the Purilactis Fm. in the Barros Arana syncline, where it is more than 2000 m thick (section 8 of Fig. 2). Great variations in sedimentary thickness even occur over much shorter distances. Conglomerates about 3000 m thick are exposed along the Litio-Baquedano road (section 1 in Figs. 1 and 2), 8 km to the SE from the location of the sequence represented in Table 1, but separated from it by a major thrust fault. These variations in thickness indicate an irregular, possibly fault-controlled subsidence of the Purilactis depocentre, although an influence of post-depositional tectonics on the exposed thickness cannot be ruled out in all cases.

◀ **Fig. 2.** *left* Eight cross sections through the Purilactis Group at the western escarpment of the Salar de Atacama. For location of sections see Fig. 1. The lithostratigraphic column shown in Table 1 is based on field work along cross section 2

Table 1. Lithostratigraphic section through the Purilactis Group in the area of section 2 (Figs. 1 and 2)

Tambores Formation, Oligocene, overlying Purilactis Group down to basement with angular unconformity	
	Yesifera Superior Formation
50 m	Coarse sandstones and impure gypsum beds; partly gypsum-filled fractures
	Purilactis Formation
30 m	Volcanic member with dark grey, slightly porphyric lavas intercalated with conglomerates
10 m	Thick yellowish-red conglomerates with well rounded igneous pebbles
200 m	Thick red conglomeratic breccias containing clasts of tuff, limestone, sandstone, and different sorts of porphyric rocks
500 m	Thick massive brown, fine and coarse, laminated, cross-bedded, channelized conglomerates in 0.30-to 1.50-m-thick layers, frequently fining upwards, with red laminated sandstone fragments; maximum diameter of boulders 0.5 m; stratification is better developed near the base than farther up
	Tonel Formation
100 m	Alternating laminated medium to coarse red sandstones and gypsiferous sandstones with cross-bedding
3 m	Red, channelized, upward-fining conglomeratic sandstone bench
100 m	Orange mudstones and gypsiferous sandstones with isolated, well rounded pebbles and intraformational sandstone clasts
200 m	Alternating gypsiferous mudstones and sandstones with parallel lamination and cross bedding
40 m	Dark brown sandstone layer
250 m	Red and brown sandstones and occasional mudstone intercalations with parallel laminations and cross-bedding; mud cracks can be observed
60 m	Aeolian gypsum-arenitic sandstone sequence with large cross-bedding, indicating north-northeastward transport
50-60 m	Yellow mudstones and thin, fine-grained conglomerates
60 m	Yellowish-white limestones with isolated granules and pebbles, and sandstones with calcareous cement and organic tracks
100 m	Red sandstone and limestone sequence with thin grained conglomerate intercalations
70-80 m	Sandstone sequence with gypsiferous cement, showing parallel laminations and tabular cross-bedding
100 m	Conglomerate sequence with cobbles of 10 cm maximum diameter
150 m	- Conglomeratic level containing limestone fragments - Alternation of finer breccias, grey micritic limestones and fine red sandstones and siltstones - Massive sandy breccia (5 m) with dark volcanic and calcareous rock fragments
Agua Dulce Formation, late Palaeozoic, underlying Purilactis Group with parallel unconformity	

Yesífera Superior Formation

In the stratigraphic column described for section 2 (Figs. 1 and 2) the lavas and conglomerates belonging to the top of the Purilactis Fm. are overlain by a sequence of coarse white to yellowish sandstones with intercalations of impure gypsum beds. These steeply dipping sediments form a narrow band, about 50 m wide, between the top of the Purilactis Fm. on the eastern side and the upthrust Palaeozoic Agua Dulce Fm. on the western side.

Coarse gypsiferous sandstones and fine breccias, commonly with gypsum-filled fractures, appear again in section 1 (Fig. 2) on top of the 3000-m-thick sequence of conglomerates of the Purilactis Formation. Ramirez and Gardeweg (1982) attributed these gypsiferous sediments to the San Pedro Fm. (Oligocene-Miocene) and, hence as an evaporitic equivalent of the Tambores Fm. (Dingman 1963). However, as no angular unconformity exists between the Purilactis Fm. and these sediments, they are considered here as the uppermost formation of the Purilactis Group, i.e. the Yesífera Superior Formation. Its outcropping thickness is more than 300 m; a stratigraphic top of this sequence is not exposed.

3 Regional Stratigraphic Correlations

Although variations in lithology and thickness are common in the Purilactis Group, the main subdivision is easily recognizable throughout the area studied. As no fossils of stratigraphic significance were found in the Purilactis Group, the attribution of ages depends on lithostratigraphic correlations with similar formations outside the western border of the Salar de Atacama and on some radiometric dating of volcanic intercalations.

For section 2 (Figs. 1 and 2) it was shown (Table 1) that limestone lumps and beds occur in the basal 150 m of the Tonel Fm. above its disconformable contact with the underlying late Palaeozoic lavas. A

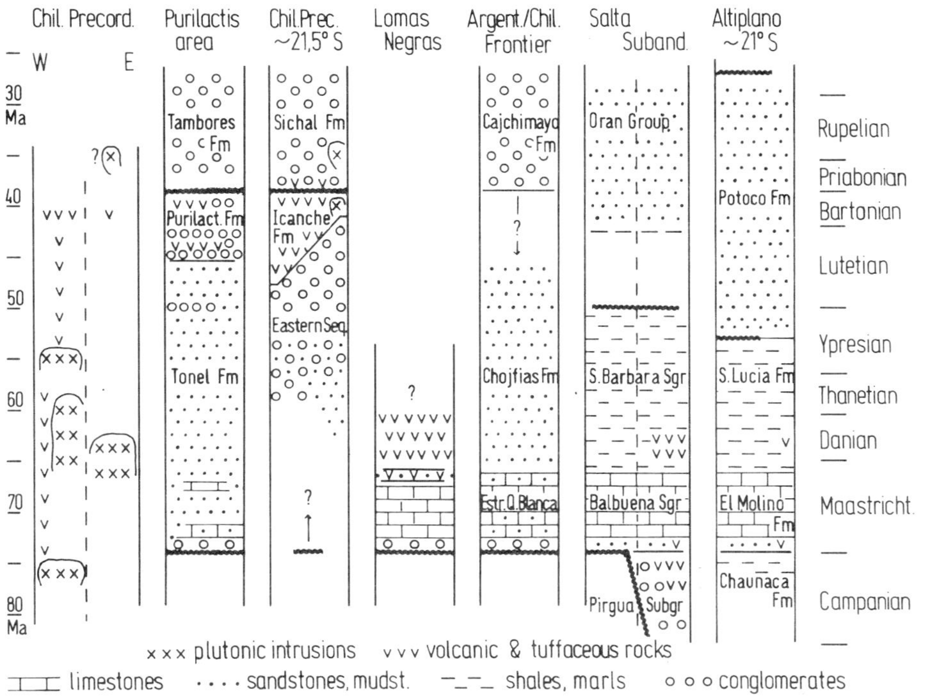

Fig. 3. Chronostratigraphic correlation corresponding to a W-E section through columns 1, 2, 4, 5 and 6 (from left to right), i.e. from the Chilean Precordillera to northwestern Argentina. Columns 3 and 7 are located farther to the N. Columns have been drawn after Bogdanic (1990: col. 3), Boric et al (1990: col. 1), Döbel (1989: col. 2 to 4), Ramirez and Gardeweg (1985: col. 5), Marinovic and Lahsen (1984. col. 5), Marquillas and Salfity (1988: col. 6), Sempéré (1986: col. 7), and YPFB (1990: col. 7). Question marks suggesting other stratigraphic possibilities were added by the present authors

similar sequence was found farther south, outside the study area, near Estación Pan de Azucar (Fig. 4). Here, the basement, consisting of a late Palaeozoic granitoid rock, grades into a reddish basal breccia. It is overlain by red stratified breccias which fine upwards to red sandstones and mudstones. A 3 m thick impure limestone bed which shows stromatolitic structures in its upper part is found 18 m above the base. After a further 85 m of mudstones, characteristic gypsiferous beds crop out. Except for the evaporites, the basal sequence of the Tonel Fm. of section 2 and of the exposure shown in Fig. 4 resembles strongly that of the Yacoraite or El Molino Fms. in northwestern Argentina and Bolivia, although the thickness and the stratigraphic extent of the limestone beds of these formations are much larger. This correlation, however, remains hypothetical because there is no proof of the marine origin of the limestones at the base of the Tonel Fm. and the nearest exposures of calcareous sediments that can undoubtedly be attributed to the Yacoraite Fm. are situated more than 100 km to the east, at the frontier between Chile and Argentina near the Huaitiquina Pass (Donato and Vergani 1987), and near the triple junction of the frontiers between Chile, Argentina and Bolivia (Quebrada Blanca de Poquis; Gardeweg and Ramírez 1985).

Nevertheless, this hypothetical stratigraphic relationship is supported by data from the Lomas Negras area (68.05°W, 22.31°S, NNE of the study area), situated only 15 km to the east of the northeasternmost outcrops of the Purilactis Group. Marinovic and Lahsen (1984) described foraminifera of presumed late Cretaceous age from limestones in the lower part of their Lomas Negras Formation. The age was confirmed by Döbel (1989; Döbel et al. 1992) who, by means of Ar-^{39}Ar dating, obtained a plateau age of 66.6 ± 1.2 Ma from biotites of a tuffaceous level overlying the marine sequence. The upper part of the Lomas Negras Fm., above the dated tuffite, consists of a some hundred metres thick sequence of andesitic lavas. These volcanics have no equivalent in the Tonel Fm. but in northwestern Argentina, where magmatic rocks which intruded into the Salta Group yielded an age of 63 ± 2 Ma (Galliski and Viramonte 1988; Fig. 3).

The age of the base of the Purilactis Fm. (sensu stricto) has been determined by ^{40}Ar-^{39}Ar datings carried out by Döbel (1989; Döbel et al. 1992) on samples of a pyroclastic layer from the lower volcanic level in the central part of the study area. At two different locations (pur3 and NE of pur2, Fig. 1) concordant ^{40}Ar-^{39}Ar plateau ages of 44.0 ± 0.5 Ma were obtained from two biotite samples and one hornblende sample. This agrees closely with the K-Ar age determination of 39.9±3.0 Ma and 41.0±3.6 Ma reported by Ramirez and Gardeweg (1982) from a sample collected in the tuff level intercalated near the base of the conglomerates of the Purilactis Fm. that forms the ample syncline north of the San Pedro-Salar Elvira road (Fig. 1).

In the area of the Barros Arana syncline (Fig. 1) Hartley et al. (1988) found a 40 m thick andesitic lava between fine-grained sediments (their "Member 5") and an overlying thick sequence of compact conglomerates (their "Member 6"), i.e. at the base of the Purilactis Fm. in the sense of this paper. Flint et al. (1989), who analyzed a clinopyroxene crystal from that strongly altered lava using the ^{40}Ar-^{39}Ar method, proposed an effusion age of 63.7±10 Ma corresponding to the integrated age from the obtained ^{39}Ar spectrum. Except for the known difficulties of ^{40}Ar-^{39}Ar dating on pyroxene (McDougall and Harrison 1988: p. 28), the following facts raise some doubts about this interpretation of the effusion age: (1) the first extraction steps show excess ^{40}Ar (or loss of ^{39}Ar), and (2), only with the extraction of the last 36 % of ^{39}Ar (2 steps in Run 890), a plateau at a level of about 48 Ma is reached. It is most remarkable that this relatively small plateau is very close to the Döbel values and that, moreover, the age of the last extraction step (Run 889: 47.4±4.3 Ma; Run 890: 45.2±2.9 Ma) is perfectly concordant with his plateau ages. Thus, it may be concluded that, with some probability, the lava of the Barros Arana syncline extruded more or less contemporaneously with the eruptions that produced the pyroclastic rocks at 44 Ma (Döbel 1989; Döbel et al. 1992) in the same lithostratigraphic level (lower volcanic level of the Purilactis Fm.) of more southern parts of the Purilactis Group.

No radiometric datings exist for the upper volcanic level of the Purilactis Formation. Its age can only be inferred by correlating it with a sedimentary and volcanic sequence occurring in the north Chilean Precordillera to the north of Calama (Sierra de Moreno, ≈ 21°-22°S). In that area, Jurassic sediments are disconformably overlain by red continental conglomerates and sandstones whose thickness varies between 0 and 2300 m (Bogdanic 1990: "Eastern Sequence"), and above this by a volcanic (andesitic) formation up to 2000 m thick (Maksaev 1978: Icanche Fm.; Bogdanic 1990). For the base of the Icanche Fm., ^{40}Ar-^{39}Ar ages vary between 47.8±0.8 Ma and 41.2±0.6 Ma, and for the top a minimum age of 38.5±0.9 Ma was obtained (Döbel 1989; Döbel et al. 1992).

Döbel (1989; Döbel et al. 1992) also succeeded in dating an angular unconformity at the top of the Icanche Fm. as 38.5±0.5 Ma (Scheuber et al. this Vol.). This tectonic event corresponds to the Incaic phase whose effects are observed in many parts of the Central Andes. Therefore, it is probable that at 38.5 Ma also the sedimentation of the Purilactis Group came to an end and compressional tectonics commenced.

4 Depositional environment

The sediments of the Purilactis Group accumulated in a basin subsiding at the back, i.e. to the east, of an active magmatic arc which was located in the Precordillera (Cordillera Domeyko) and in the adjacent parts of the Chilean Longitudinal Valley. This arc, which came into being 72 Ma ago (Scheuber et al. this Vol.) and whose activity was terminated during the early Oligocene, was elevated above sea level and had a positive topography with respect to the basins of the Purilactis, Salta, and Puca Groups (Salfity et al. 1985). The volcanic activity was intermittent and irregularly distributed in the arc area, and the volcanic rocks covered only incompletely the older substratum, especially towards the backarc area. All the clastic material contained in the sediments of the Purilactis group can be derived from the arc area and consists either of the arc volcanics or rocks underlying them.

Erosion in the source area and deposition of the Purilactis Group took place under arid conditions. The fine grained carbonate cemented sandstones and impure limestones of the Tonel Fm., which overlie the basal breccia, may have been deposited close to the shore of a shallow epicontinental sea, although these are not clearly marine and there is no direct proof that the Maastrichtian transgression leading to the deposition of the Yacoraite (in Argentina) and El Molino (in Bolivia) Fms. reached the Purilactis depocentre. An environmental change to an inland sabkha domain is indicated by the subsequent fine-grained sediments and a partly gypsum-rich facies of the Tonel Formation. The restricted extension and reduced thickness of lacustrine limestone deposits indicate that the water bodies necessary for their deposition were small and of short duration. Layers of gypsum-arenites, 5 to 20 m thick, with large cross-bedding evidence aeolian deposition. Red and yellow mudstones with gypsum nodules and fine grained sandstones with parallel lamination and cross-bedding suggest interfingering with distal wadi deposits.

Distal wadi fan deposits are represented by fine-grained and thin-layered conglomerates intercalated in the mudstone sequence. Their close association with limestone deposits suggests the existence of periods with stronger precipitation, which favoured not only the sedimentation of relatively coarse deposits but also the development of water bodies with some kind of organic activity. Cross-bedding polarities indicate a sediment supply by rivers flowing from west to east, while aeolian deposits show a northward direction of transport.

The predominantly conglomeratic facies of the Purilactis Fm. represents deposition by an eastward prograding wadi fan system. Clastic material was supplied by rivers flowing to the east as evidenced by cross-bedding and imbrication polarities. These intermittent braided rivers had their origin in the elevated area of the magmatic arc, located to the west of the Purilactis basin, which was the source of clasts consisting of late Palaeozoic volcanics, granitoids and sediments, fossiliferous Jurassic limestones, late Cretaceous granodiorites (Cerro Quimal) and early Tertiary arc volcanics. Red sandstone clasts may have been derived from late Jurassic-early Cretaceous deposits or from reworked western parts of the Tonel Formation.

The onset of conglomeratic sedimentation of the Purilactis Fm. coincides with the onset of magmatic activity in the Purilactis depocentre. Pebbles and boulders of intermediate volcanic rocks in the conglomerates and tuffaceous layers (Döbel 1989: 44.0 ± 0.5 Ma) in the basal part of the Purilactis Fm. are evidence of nearby active volcanoes. The local interstratification of lava flows up to 10 m thick with the conglomerates of Cerro Totola (Fig. 1) points to fissure eruptions of basaltic andesites. Also, the intrusion of subvolcanic bodies and sills and dykes into the Tonel Fm., especially in the area comprised between sections 4 and 5 (Fig. 1) may be contemporaneous with the sedimentation of the Purilactis Formation. Its coarse polymictic sediments reveal a strong accentuation of the relief by accelerating vertical tectonics not only in the western source area but also in the basin. These tectonics were accompanied by an activation of arc magmatism and a shift of its axis towards the east (Scheuber et al. this Vol.). Finally, in the Yesífera Superior Fm. less energetic transport conditions were re-established.

5 Structures

In contrast to the highly uplifted Palaeozoic rocks to the west and the depressed Neogene and Quarternary sediments to the east, the relatively narrow strip of sediments of the Purilactis Group, cropping out along the western escarpment of the Salar de Atacama, shows the effects of intense deformation. Structures include longitudinal folds and high-angle dipping thrusts, indicating E-W or ESE-WNW shortening, as well as asymmetrical vertical folds due to N-S directed strike-slip movements (Fig. 1: pur1-pur6).

Two different types of structural setting can be recognized along the escarpment. In the northern part (e.g. along the El Bordo ridge, Fig. 1), the Palaeozoic rocks were uplifted along a high-angle reversed fault in such a way that sort of compressional flexure developed, in which the basal members of the Tonel Fm. are found near the fault (or flexure) and in some places even upon the uplifted Palaeozoic, while the upper parts of the Tonel Fm. descend stepwise towards the east with several east-vergent folds and minor thrusts. In this case, the Purilactis Fm. crops out, if at all, only in the lower part of the scarp (Fig. 2: sections 5 and 6). In the southern part of the study area, the Palaeozoic rocks are thrust upon the Yesífera Superior Fm., i.e. upon the top of the Purilactis Group (Fig. 2: sections 1-3). Also in this second case, several eastward verging folds and secondary thrusts affect the Purilactis Group so that the Tonel Fm., and even slices of the Palaeozoic basement, may crop out farther to the east at some distance from the main thrust (Fig. 2: sections 2 and 3). The situation in the extreme north of the study area (Fig. 2: sections 7, 8), is different in that the Barros Arana syncline was back-thrust to the west over the Palaeozoic rocks and their Neogene sedimentary cover in the course of post-Incaic shortening (Jolley et al. 1990).

The structures of the Purilactis Group sediments formed at a shallow tectonic level. No cleavage developed and no cohesive cataclastic, or even mylonitic, rocks were developed along the thrusts and strike-slip faults. A shallow tectonic level is also indicated by some fold structures which show the influence of down-slope gravitational deformation (Fig. 2: section 6). Diapirism, caused by tectonic compression, developed in gypsum-rich parts of the Tonel Fm., especially in the Cerros del Tonel area. This is evidenced by irregular bodies of breccias, several or even hundreds of metres wide, which consist of sandstone fragments cemented by an impure gypsum matrix. Normally stratified Tonel

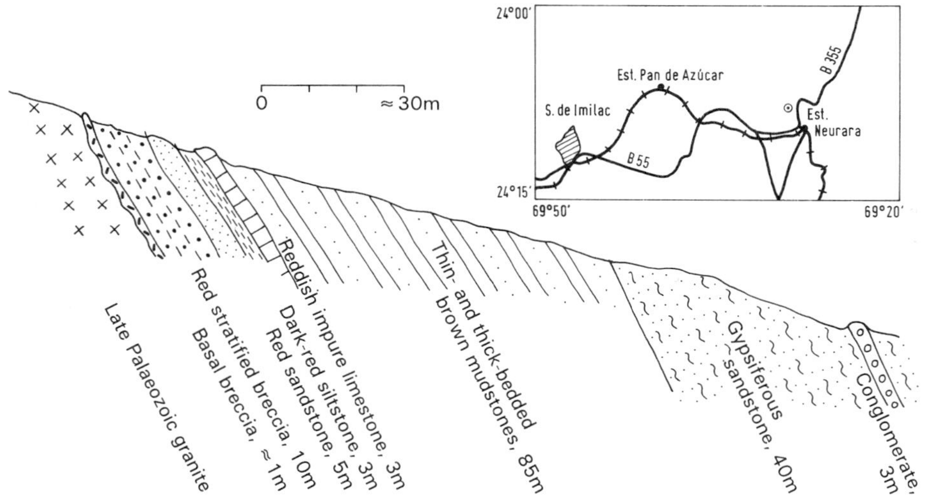

Fig. 4. Basal part of Tonel Fm. in an exposure outside the study area to the south of Salar de Atacama, near the Antofagasta-Salta railway (69°30'45"W, 24°09'30" S). The bed of extraclastic limestone, 18 m above the basement, is possibly a marine or non-marine equivalence of the Yacoraite limestones of the Argentine Puna

Fm. sediments lie between the breccia bodies and show gradational contacts to the broken rocks.

An important feature is the vertical folds which were generated by strike-slip faults running parallel or almost parallel to vertical strata which had been rotated to vertical by preceding compressional tectonics. In plan view the asymmetrical geometry of these folds shows Z-shapes, in which the middle limb may involve sediments 5 to 200 m thick. The biggest vertical fold structures occur in the Cerros del Tonel area, but smaller structures were found all over the study area (locations pur1-pur6 in Fig. 1). The uniform Z-fold asymmetry indicates dextral strike-slip movements. Reutter et al. (1991) showed that such and other structures of dextral orogen-parallel strike-slip tectonics are present elsewhere within the north Chilean Precordillera.

The age of the structures mentioned is constrained by the age of the uppermost sediments of the Purilactis Group and by the imprecisely known age of the base of the unconformably overlying Oligocene-Miocene Tambores Formation. As pointed out above, first strong shortening probably occurred at 38.5±0.5 Ma, which is the age of the unconformity in the Chilean Precordillera to the north of Calama (Döbel 1989; Döbel et al. 1992), but these exposures also show that shortening, and possibly also the orogen-parallel strike-slip movements, still continued during the Oligocene. While the strike-slip tectonics probably affected only the arc and the arc-backarc boundary, Incaic phase shortening advanced into the backarc area (or tectonic foreland) affecting also the Salta Group and its Bolivian equivalents (Fig. 3; Scheuber et al. this Vol.; Andriessen and Reutter this Vol.).

As evidenced by the age and the great thickness of the San Pedro Fm. in the Cordillera de la Sal (Wilkes and Görler this Vol.: 2850 m), the Salar de Atacama Depression started its strong subsidence during the Oligocene, when Incaic phase shortening had come to an end. During that time the western escarpment may have been affected by tensional flexuring. It was superseded by Neogene (Quechua phase) to Quaternary compressional tectonics which, as a main effect, caused folding of Oligocene and Neogene sediments in the Cordillera de la Sal. This young structure is considered to have formed in the hanging wall of a thrust which extended, from the western escarpment, eastwards beneath the folded Purilactis Group and the unconformably overlying younger sediments of the Llano de la Paciencia (section 3 in Figs. 1 and 2; Jolley et al. 1990; Wilkes and Görler this Vol.).

Neogene shortening also affected the western escarpment itself and was superimposed upon the Palaeogene structures. Thus, in the area of section 4 (Figs. 1 and 2) high-angle dipping reversed faults caused a repetition of the exposures of the unconformity between the Tonel Fm. and the Tambores Fm. by tectonic imbrication. In the lower part of the scarp, Purilactis Group sediments border on the San Pedro Fm. or younger Neogene sediments with east-vergent reversed faults (section 3 in Fig. 2), which correspond to the "Frontal Domeyko Thrust" of Jolley et al. (1990). Aerial photographs also reveal that the alluvial footplains of the escarpment, covered

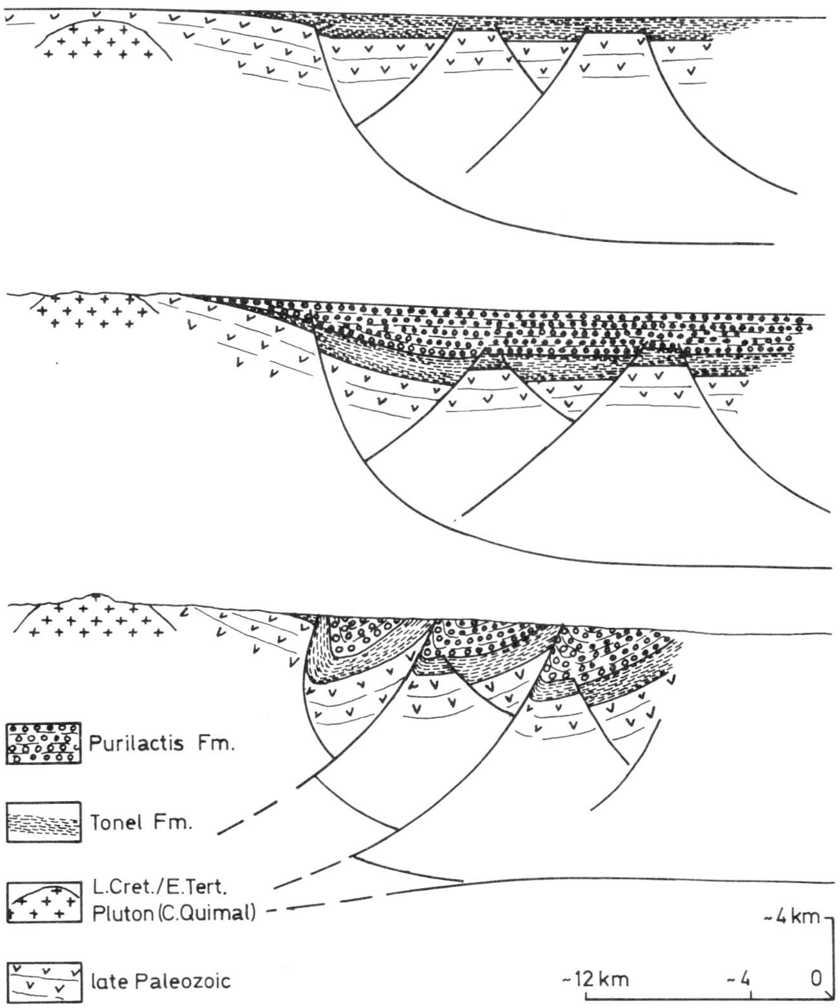

Fig. 5. Sketch of the tectonic and palaeogeographic evolution of the depocentre of the Purilactis Group as the westernmost part of an epicontinental backarc basin at the boundary with the adjacent magmatic arc from late Cretaceous to early Oligocene times. During deposition, this boundary was presumably a divide between backarc extensional tectonics and slightly compressional tectonics of the magmatic arc. In the course of the overall compression at the end of the Eocene (Incaic phase), the normal faults were rotated and inverted, and the boundary became a site of thrusting and longitudinal dextral strike-slip movements

by sediments of Pleistocene or even Holocene age, have been affected by faults and this proves that both uplift of the western rim of the Salar de Atacama and shortening are still active.

6 Geotectonic Setting

If the assumptions about the age of the base of the Tonel Fm. are correct, the tectonic and sedimentary activity in the Purilactis depocentre started contemporaneously with the magmatic arc activity in the north Chilean Precordillera and adjacent parts of the Longitudinal Valley at about 72 Ma. This late Cretaceous-Palaeogene arc system was emplaced to the east of a preceding mid-Cretaceous arc system, centred in the Longitudinal Valley, whose activity had ended during the late Cretaceous in the course of the compressional tectonics of the Peruvian phase (Mégard 1987: Santonian; Muñoz et al. 1989; Scheuber et al. this Vol.: Fig. 9) which affected not only the Longitudinal Valley, but also the western part of the Precordillera. Thus, the late Cretaceous-Palaeogene lavas and volcaniclastics (Chile-Alemania Fm. and parts of Augusta-Victoria Fm.) were deposited with angular unconformity on top of older rocks on elevated ground. On the other hand, the Purilactis Group overlies the late Palaeozoic rocks with a very low angle unconformity, or a disconformity, which shows that this area had not undergone Peruvian phase folding but only uplift and erosion.

The thickness of the arc volcanics is greatest (max. ≈ 2 km) in the westernmost parts of the arc area, close to the Chilean Longitudinal Valley, while in the ranges west of the Purilactis area, it is strongly reduced (Marinovic and Lahsen 1984: 100 m at Cerro Los Mellizos, ≈ 50 km west of El Bordo; about 200 m in the exposure north of Salar Elvira, see Fig. 1) or not preserved at all. Thus, there is no reason to assume subsidence for the eastern part of this area. On the contrary, fission track data by Andriessen and Reutter (this Vol.: apatite age of 50 Ma in Palaeozoic granite located ≈35 km west of Cerro Negro, and within their limits of error almost concordant K-Ar biotite and FT (fission track) zircon, FT sphene, and FT apatite ages of ≈63 Ma from the granodiorite of Cerro Quimal, Fig. 1) prove, at least locally, rapid uplift of this area. This contrasts with the high amount of subsidence in the adjacent Purilactis area, where an average thickness of ≈3000 m can be assumed. Furthermore, as typical Purilactis sediments have not been found beyond the western escarpment (exposures indicated in Hoja Toconao by Ramirez and Gardeweg [1982] turned out to consist of volcanic material [north of Pampa Elvira] and late Palaeozoic sediments [surroundings of Cerro Quimal]; Fig. 1), it must be concluded that the Purilactis depocentre, as the westernmost part of the backarc area, bordered with normal faults on the arc area (Fig. 5).

There is no direct proof of synsedimentary normal faulting at the western border of the Purilactis depocentre and within it, but there are further indications which favour this assumption. For example, great differences in thickness of the Purilactis Fm. in neighbouring areas separated by faults (compare Fig. 2: sections 1 and 2) hint at block tectonics during deposition. E-W tension is indicated also by some dykes up to 50 m thick in the Cerros del Tonel area, but this argument is not unequivocal as, farther to the south in the reaches of section 5 (Fig. 1), a set of E-W directed dykes crosscuts Tonel sediments. More evidence comes from the Cerros de Tuina, outside the study area, about 15 km to the north of P. Barros Arana (Fig. 1), where Marinovic and Lahsen (1984) mapped a series of N-S trending high-angle reversed faults which cut, with W- as well as E-vergencies, the late Palaeozoic volcanic and sedimentary basement rocks and sediments attributed to the Purilactis Group. The opposite dip directions of these faults suggest that they were generated as normal faults and subsequently inverted during a period of crustal shortening.

Distributed tensional tectonics and consequent crustal thinning, also governed subsidence of the different subbasins of the Salta Group in northwestern Argentina and of the Puca Group in southwestern Bolivia. These tectonics, however, had already started during the early Cretaceous and were accompanied by a characteristic magmatic activity between 97±5 and 123±5 Ma (Galliski and Viramonte 1988), i.e. prior to the emplacement of the late Cretaceous-Palaeogene arc system in the north Chilean Precordillera. Nor can the early and mid-Cretaceous crustal extension in Bolivia and Argentina be attributed to subduction during the preceding mid-Cretaceous arc system in the Longitudinal Valley, because there is no evidence of a contemporaneous backarc depocentre in Chile. The question posed by Grier et al. (1991) whether the subbasins of the Salta Group are "subduction related back-arc basins, a failed rift system associated with the opening of the South Atlantic, or a combination of the two", therefore has to be answered in a differentiated way: During the early and mid-Cretaceous the development seems to have been independent of subduction, but since the Campanian at least the western parts of these Cretaceous basins were involved in backarc spreading processes to which the magmatic events in Argentina at about 78 Ma and 63 Ma may also be attributed (Galliski and Viramonte 1988; Marquillas and Salfity 1988; Reyes et al. 1976; Semperé 1986).

During the sedimentation of the Purilactis Fm., volcanic arc activity affected not only the elevated western area but also the depocentre of the Purilactis Formation. According to the radiometric data compiled by Maksaev (see Scheuber et al. this Vol.: Fig. 2), arc volcanism became more intense at about 50 Ma and shifted to a position farther eastward from that occupied from 72 to 55 Ma. So, the source area of young volcanic rocks came closer to the depocentre than during deposition of the Tonel Fm., and also the relief probably became more accentuated, thus causing the sudden onset of the psephitic sedimentation at 44 Ma.

With the beginning of the Incaic phase the tensional regime of the whole epicontinental backarc basin was superseded by a strong compressional regime. The Purilactis Group was deformed into elongate fold structures striking N-S in the southern part of the escarpment and NNE-SSW in the northern part (Fig. 1). The postulated normal faults between arc and backarc area must have been either inverted or rotated to east-vergent reversed faults and thrusts. Thus, the complex compressional flexure of the late Palaeozoic basement at its boundary with the basal Tonel Fm., in the northern part of the study area, cannot be conceived as a simple compressional structure; it was formed probably by the rotation of an originally east-dipping normal fault to a west-dipping reversed fault. Inversion of originally normal faults may have occurred along some of the thrusts in the southern part of the study area (Fig. 5).

Magmatic activity, mostly in the form of granodioritic intrusions, reached a maximum during and shortly after the Incaic phase (Scheuber et al. this Vol.: Fig. 2). The rather small granodioritic stocks which intruded the Purilactis Group, to the north and the south of the study area, are possibly the easternmost intrusions of this age, but no respective radiometric data are available. As a consequence of heating, the strength of the crust was reduced and its tectonic mobility increased. This may be the reason for the development of the longitudinal dextral strike-slip faults, which, according to Reutter et al. (1991), are a characteristic feature of the tectonics of the late Cretaceous-Palaeogene magmatic arc and are due to oblique subduction. In the Chuquicamata area, 70 km to the northwest of the Barros Arana syncline, longitudinal strike-slip movements along the West Fissure were active at least until 34 Ma (Scheuber et al. this Vol.). Thus, it is reasonable to postulate a similar age for the strike-slip movements along the western escarpment of the Salar de Atacama.

The compressional tectonics of the Incaic phase also affected the Salta and Puca Groups in the Argentine Puna and the Bolivian Altiplano, but did not reach beyond the Eastern Cordillera. It can be supposed that shortening started in the Purilactis Group, and then, similar to a fold-and-thrust belt, advanced eastwards converting the former epicontinental backarc area to a foreland. The normal faults related to subsidence of the backarc area or earlier rifting must have been inverted to thrusts, in a similar way as indicated by Grier et al. (1991) for Neogene compression in the Eastern Cordillera and the Subandean ranges.

7 Conclusions

Stratigraphic studies and age determinations (Döbel 1989; Döbel et al. 1992) suggest that, in accordance with a proposal made by Salfity et al. (1985), the Purilactis Group is a western, partly heteropic, equivalent of the Balbuena and Santa Bárbara Subgroups of the Salta Group in Argentina (Marquillas and Salfity 1988) and of the El Molino and Santa Lucía Fms. of the Puca Group in Bolivia. Deposition took place in a basin situated at the continentward side of a coeval magmatic arc, which was located in the north Chilean Precordillera. On the basis of this geotectonic configuration, which persisted from the late Cretaceous to the late Eocene (72-38 Ma), the Purilactis depocentre and at least the western parts of its eastern equivalents in Argentina and Bolivia are here considered to have been an epicontinental backarc basin. There is evidence that the boundary between the uprising magmatic arc domain to the west and the subsiding Purilactis depocentre to the east was determined by syndepositional normal faults. From that time this boundary has remained a zone of crustal weakness and a tectonically important lineament up to the present.

During the compressive tectonics, which started with the Incaic phase at about 38 Ma, the arc area to the west was pushed upon the Purilactis Group by rotation and inversion of earlier normal faults. From the arc-backarc boundary, Incaic phase shortening advanced into the foreland, i.e. the former backarc area, where tectonic inversion of faults related to the development of an earlier rift basin or later backarc basin is supposed to have generated a complex, basement-involving fold-and-thrust belt. Crustal weakness along this lineament is also revealed by structures caused by longitudinal dextral strike-slip movements which took place still during the Incaic phase and simultaneously to other orogen-parallel strike-slip faults within the magmatic arc area (Reutter et al. 1991).

The depression of the Salar de Atacama developed during the Oligocene, after the drop of the strong compressional stresses of the Incaic phase (Scheuber et al. this Vol.: Fig. 14), perhaps as a consequence of relaxation of the newly thickened crust. During this time the western escarpment of the Salar came into being along the former arc-backarc boundary, possibly first as a zone of extensional flexuring. When, finally, during the Miocene, compressional stresses increased again (Quechua phase), renewed tectonic shortening originated from this lineament and caused the fold structures of the Cordillera de Sal (Jolley et al. 1990; Wilkes and Görler this Vol.).

Acknowledgements. This work is part of the project "Mobility of Active Continental Margins" supported by the German Research Council (DFG). It also is a contribution to project No. 242: "Cretaceous of Latin America" of the International Geological Correlation Program. Fieldwork by R.C. was financed by research grant E-2456 of the Departamento Técnico de Investigación (DTI) of the Universidad de Chile, and that by K-J.R. by the DFG. R.C. received grants from the Alexander von Humboldt Stiftung for short stays in Berlin.

References

Bogdanic T (1990) Kontinentale Sedimentation der Kreide und des Alttertiärs im Umfeld des subduktionsbedingten Magmatismus in der chilenischen Präkordillere. Berl Geowiss Abh A123, 117 pp

Boric R, Diáz F, Maksaev, V (1990) Geología y yacimientos metaliferos de la región de Antofagasta. Servicio Nacional de Geología y Minería, Santiago, Bol 40, 246 pp

Breitkreuz C, Helmdach F-F, Kohring R, Mosbrugger V (1992) Late Carboniferous intra-arc sediments in the north Chilean Andes: Stratigraphy, paleogeography and paleoclimate. Facies, Erlangen, 26: 67-80

Brüggen J (1934) Grundzüge der Geologie und Lagerstättenkunde Chiles. Math-Naturwiss Klasse Heidelberger Akademie der Wissenschaften, Tübingen, 362 pp

Brüggen J (1942) Geología de la Puna de San Pedro de Atacama y sus formaciones de areniscas y arcillas rojas. Primer Congreso Panamericano Ing Minas Geol, Santiago, An 2: 342-367

Charrier R, Reutter K-J (1988) La Formación Purilactis en el borde occidental del Salar de Atacama, 23°-23°45' de latitud sur, Chile. Comunicaciones 39: 211 (Abstr)

Charrier R, Reutter K-J (1990) The Purilactis Group of Northern Chile: Link between arc and backarc during late Cretaceous and Paleogene. Symp Int Géodyn Andine, 15-17 May 1990 Grenoble, ORSTOM, Paris, pp 249-252

Dingman RJ (1963) Cuadrángulo Tulor. Inst Invest Geol, Santiago, Carta geológica de Chile 1:50000 11, 35 pp

Döbel R (1989) Geochemie und Geochronologie alttertiärer Vulkanite aus der Präkordillere Nordchiles zwischen 21° und 23°30'S. Ph D, Freie Univ Berlin, 152 pp (unpubl)

Döbel R, Friedrichsen H, Hammerschmidt K (1992) Implication of ^{40}Ar-^{39}Ar dating of early Tertiary volcanic rocks from the north-Chilean Precordillera. Tectonophysics 202: 55-81

Donato E, Vergani G (1987) Estratigrafía de la Formación Yacoraite (Cretácico) en Paso Huaytiquina, Salta, Argentina. 10 Congr Geol Argent Tucumán Actas 2: 263-266

Flint SS, Hartley AJ, Rex DC, Guise P, Turner P (1989) Geochronology of the Purilactis Formation, northern Chile: an insight into late Cretaceous-early Tertiary basin dynamics of the Central Andes. Rev Geol Chile 16: 241-246

Galliski M, Viramonte JG (1988) The Cretaceous paleorift in northwestern Argentina: a petrologic approach. J S Am Earth Sci 1: 329-342

Gardeweg M, Ramirez CF (1985) Hoja Rio Zapaleri, Región de Antofagasta. Serv Nac Geol Minería, Santiago, Carta geológica de Chile 1:250000 66, 89 pp

Grier ME, Salfity JA, Allmendinger RW (1991) Andean reactivation of the Cretaceous Salta rift, northwestern Argentina. J S Am Earth Sci 4: 351-372

Hartley AJ, Flint S, Turner P (1988) A proposed lithostratigraphy for the Cretaceous Purilactis Formation, Antofagasta province, northern Chile. 5 Congr Geol Chil Santiago Actas 3: H83-H99

Hartley AJ, Jolley E, Turner P (1990) Paleomagnetic and structural constraints of Mesozoic-Recent thrust sheet rotation in the Precordillera of northern Chile. Int Symp Géodyn Andine 15-17 May 1990 Grenoble, ORSTOM, Paris, pp 61-62

Hartley AJ, Jolley E, Turner P, Flint S (1991) Preliminary paleomagnetic results from the late Cretaceous Tonel Formation (Purilactis Group), Precordillera of northern Chile, constraints on thrust sheet rotation. 6 Congr Geol Chil Viña del Mar Actas 1: 1-5

Jolley E, Turner P, Williams GD, Hartley AJ, Flint S (1990) Sedimentological reponse of an alluvial system to Neogene thrust tectonics, Atacama Desert, northern Chile. J Geol Soc London 147: 769-784

Kriz SJ, Cherroni C (1966) Diagramas correlativos de formaciones cretácicas del sudoeste de Bolivia. Serv Geol Bolivia La Paz, Hoja informativa 2

Macellari CE, Su MJ, Townsend F (1991) Structure and seismic stratigraphy of the Atacama basin, northern Chile. 6 Congr geol. chil Viña del Mar Actas 1: 133-137

Maksaev V (1978) Geología de los Cuadrángulos Chitigua y Cerro Palpana. Inst Invest Geol, Santiago, Carta geológica de Chile 1:50000 31, 55 pp

Marinovic N, Lahsen A (1984) Hoja Calama. Serv Nac Geol Minería, Santiago, Carta geológica de Chile 1:250000 58, 140 pp

Marquillas R, Salfity JA (1988) Tectonic framework and correlations of the Cretaceous-Eocene Salta Group, Argentina. In: Bahlburg H, Breitkreuz C, Giese P (eds) The Southern Central Andes, Lecture Notes in Earth Sciences 17 Springer, Berlin, Heidelberg, New York, pp 119-136

McDougall I, Harrison TM (1988) Geochronology and thermochronology by the ^{40}Ar-^{39}Ar Method. Oxford Monogr Geol Geophys 9: 212 pp

Mégard F (1987) Cordilleran Andes and marginal Andes: a review of Andean geology north of the Arica elbow (18°S). In: Monger JW and Francheteau J (eds) Circum-Pacific orogenic belts and evolution of the Pacific Ocean Basin, Am Geophys Union Geodyn Ser 18: 71-95

Montaño JM (1976) Estudio geológico de la zona de Caracoles y áreas vecinas, con énfasis en el sistema Jurásico, provincia de Antofagasta, II. Región, Chile. Mem Univ. Chile Santiago, 168 pp

Moraga A, Chong G, Fortt MA, Henriquez H (1974) Estudio geológico del Salar de Atacama, provincia de Antofagasta. Inst Invest Geol., Santiago, Bol 29, 59 pp

Muñoz N, Charrier R, Pichowiak S (1989) Cretácico Superior volcano-sedimentario (Formación Quebrada Mala) en la Región de Antofagasta, Chile, y su significado tectónico. Contr. Simp Cretácico América Latina, Buenos Aires, Parte A: Eventos y Registro Sedimentario, pp 133-148

Ramirez CF,Gardeweg M (1982) Hoja Toconao. Servicio Nacional de Geología y Minería, Santiago, Carta geológica de Chile 1:250000 54, 122 pp

Reutter K-J, Scheuber E, Helmcke D (1991) Structural evidence of orogen-parallel strike slip displacements in the Precordillera of northern Chile. Geol Rundsch 80: 135-153

Reutter K-J, Scheuber E (1988) Relation between tectonics and magmatism in the Andes of northern Chile and adjacent areas between 21° and 25°S. 5 Congr Geol Chil Santiago Actas 1: A345-A363

Reyes FC, Salfity JA, Viramonte JG, Gutierrez W (1976) Consideraciones sobre el volcanismo del Subgrupo Pirgua (Cretácico) en el noroeste argentino. 6 Congr Geol Argent Bahía Blanca Actas 1: 205-223

Salfity JA, Marquillas RA, Gardeweg M, Ramirez C, Davidson J (1985) Correlaciones en el Cretacico Superior de Argentina y Chile. 4 Congr Geol Chil Antofagasta Actas 1: 654-667

Semperé T (1986) Contribución a la estratigrafía del Mesozoico boliviano en el dominio andino. Publ Misión ORSTOM La Paz 1, 34 pp

YPFB (Yacimientos Petrolíferos Fiscales Bolivianos), Gerencia de Exploración (1990) Cuadro cronoestratigráfico de Bolivia. La Paz, (Chart)

Early Jurassic to Early Cretaceous Magmatism in the Coastal Cordillera and the Central Depression of North Chile

SIEGFRIED PICHOWIAK

Abstract. Central Andean mountain-building processes in an active continental margin setting can be traced back to late Proterozoic times. In the evolution of the Central Andes two major periods of different trend can be distinguished, the Preandean development which is characterized by continental growth, and the Andean development where destructive processes such as subduction erosion and crustal thickening prevailed. Both periods were associated with important magmatic cycles. The Preandean development is documented by a westward migration of magmatic arcs, starting from the Brazilian Shield and reaching the Coastal Cordilleran area with final magmatic pulses during the late Palaeozoic. The late Preandean magmatism was triggered by subduction processes, and initiated remelting of overlying relatively young crust (e.g. the Pampa Elvira complex in the Precordillera: 285 Ma, Sr_i ca. 0.7050). The change from accretionary to destructive processes is marked by the culmination of magmatic activity in the Coastal Cordillera during the Jurassic and early Cretaceous. This period coincides with maximum continental growth. Since the mid-Cretaceous magmatic belts have migrated back to the east. The first cycle of the Andean development is characterized by magmatic conditions that differ fundamentally from those of the late Palaeozoic. Mobilization and differentiation of primitive upper-mantle partial melts dominate magma genesis. The very first pulses of initial basics (HIB ca. 180 Ma, Sr_i 0.7033) with continental basalt affinities do not show subduction components and originate from upper-mantle decompressional melting during an extensional regime, presumably due to extremely oblique subduction and/or a greatly reduced subduction rate. The volcanic succession La Negra (JV ca. 183 Ma, Sr_i 0.7030), contemporaneous gabbroic intrusives like Coloso Coastal Gabbro Complex (CCG, ca. 183 Ma, Sr_i 0.7030) and post-dating granitoid intrusives (G I ca. 155 Ma, Sr_i 0.7031) are products of various differentiation processes from the same upper mantle source magma, but also show subduction components by the rise in LIL-element concentrations. This cycle terminated as magmatic activities moved into the Central Depression; here granitoid intrusives (G II ca. 80-74 Ma, Sr_i 0.7040) are structurally and compositionally similar to the G I granitoids, but some geochemical data suggest that their melts were slightly contaminated with crustal components or were even derived from arc-crust material which had been consolidated before (200 to 150 Ma).

1 Introduction

The geoscientific literature on the history of the Andes commonly distinguishes between a Preandean and an Andean development. The former is a sum of several orogenic cycles occurring before the early Jurassic (>200 Ma); the latter means mountain-building of a young orogen since the early Jurassic (<200 Ma). With the progress of modern geodynamic investigations (e.g. Pitcher et al. 1985) the importance of cycles in the Preandean and Andean developments has been recognized.

The Preandean development comprises various long and complex orogenic cycles affecting the continental margin of Gondwana with the oldest dating back to the mid-Proterozoic (Baeza and Pichowiak 1988a). These orogenies seem to have one stable trend: They were associated with a westward migration of magmatic arcs from the Brazilian Shield in the late Proterozoic to the Coastal Cordillera in the latest Palaeozoic. This observation has been interpreted in the recent literature as successive accretionary growth of the continental crust due to active continental margin processes (Mahlburg Kay et al. 1989; Ramos et al. 1986). A contradictory view of dominantly intracontinental processes without subduction components is favoured by other authors (Damm et al. this Vol.). It is established that, in contrast to the Preandean development, the subduction processes of the Andean development triggered continental margin destruction and crustal thickening, leading to backward-migration of the magmatic arcs since the early Cretaceous. The first

Correspondence to: S. Pichowiak, HPC Consult, Oraniendamm 64-72, D-1000 Berlin 46

Andean magmatic arc was emplaced in the Coastal Cordillera which has only a few remnants of Preandean rocks. Prevailing rocks of this new arc are Jurassic volcanic series and numerous plutonic complexes (Fig. 1). Magmatic activity of the mid Cretaceous occurred farther to the east in the Central Depression. Age relations, compositional features and some structural aspects of Jurassic to Cretaceous, i.e. early Andean, magmatic rocks of the Coastal Cordillera and adjacent regions of the Central Depression will be described in the following sections.

2 Rocks of the Jurassic-Cretaceous Magmatic Arcs

2.1 General Remarks

2.1.1 Previous Work on Radiometric Dating

Coastal Cordilleran magmatic rocks in north Chile have been dated by Halpern (1978), Damm and Pichowiak (1981), Berg and Breitkreuz (1983), Diaz et al. (1985, Hervé et al. (1985), Rogers (1985), Damm et al. (1986) and Hervé and Marinovic (1989). Most of the data presented resulted from K-Ar whole rock and mineral determinations on plutonic rocks. Other approaches have been published, for example by Rogers (1985) who reports a Rb-Sr age of 186.5 ± 13.6 Ma for La Negra volcanic rocks near the town of Tocopilla. Damm et al. (1986) report U-Pb datings that show continuous magmatic activity from 196 ± 4 to 152 ± 10 Ma for gabbroic to granitoid intrusives of the early Andean Cycle and a K-Ar age of 126 ± 3.8 Ma for a final pulse of the granitoid rocks in the Coastal Cordillera. Apatite fission track data from Scheuber and Andriessen (1990) reveal the uplift history of 118 ± 12 Ma ago (see also Andriessen and Reutter this Vol.).

2.1.2 Rb-Sr Isotope Investigations of Intrusive Complexes and Volcanic Rocks

Isotope ratios of the samples are listed in Table 1. Quoted errors are at the 95% confidence level. The initial $^{87}Sr/^{86}Sr$ ratio was calculated with the age given in the table. Isochrons were calculated using the isotope ratios, and the error was obtained from the external reproducibility. In some samples the statistical error of the measurement exceeds that of the external reproducibility. In such cases the error of the measurement statistics was used as an overall error. The isotope composition of the samples was measured on a MAT 261 (Finigan) thermionic mass spectrometer at the Institut für Geologie, Freie Universität Berlin. Detailed information on sample treatment, measurement techniques and accuracy was reported by Fiechtner (1991).

2.2 Regional Aspects and Age Relations

2.2.1 Magmatic Rocks of the Coastal Cordillera

The Initial Basics (Gabbros) of the Hornitos Type (HIB)

Rocks of the earliest pulse, i.e. the Hornitos type (HIB), occur in several small outcrops as relics of elongated stocks or dykes, with tectonic contacts to strongly fractured younger plutonic rocks or magmatic contacts to still younger plutonic rocks. The most important sites are: (1) the Hornitos composite stock (location I in Table 2), (2) the Caleta Coloso fragmented dykes and stocks (location II), (3) the Peninsula de Mejillones stock, and (4) the Caleta El Cobre fragmented stock. Samples for Rb-Sr analysis were only taken from locations I and II. Most of the rocks suffer from highly variable secondary alteration by hydrous phases (documented for sample 87-0-1 with a high $^{87}Sr/^{86}Sr$ intercept of 0.704567) and the spread of compositions is not sufficient for age calculations. However, an initial Sr ratio of 0.703164-0.703380 is obtained by analysing samples from different sites and, tentatively, using a U-Pb zircon age of ≈190 Ma from Damm et al. (1986) for basic plutonic rocks of the Mejillones Peninsula.

The La Negra (JV) Formation Volcanics and the Coloso Coastal Gabbro Complex (CCG)

The La Negra volcanic rocks are commonly basalts to andesites with glomerophyric to ophitic textures and a vitrophyric to microlithic (doleritic) groundmass, where plagioclase and pyroxene (pyroxene-hornblende intergrowth) occur typically as phenocrysts (Garcia 1967; Buchelt and Tellez 1988). La Negra rocks were sampled along the section Cerro Cristales - Caleta El Cobre (Coastal Cordillera, location III in Table 2). Only fresh samples from the centres of lava flows were taken for analysis in order to exclude the widespread alteration effects along flow stratifation and other structures. In order to avoid a great stratigraphic spread a set of six samples from a 10-m-thick sequence of lava flows was taken at location III (Table 2), although a small

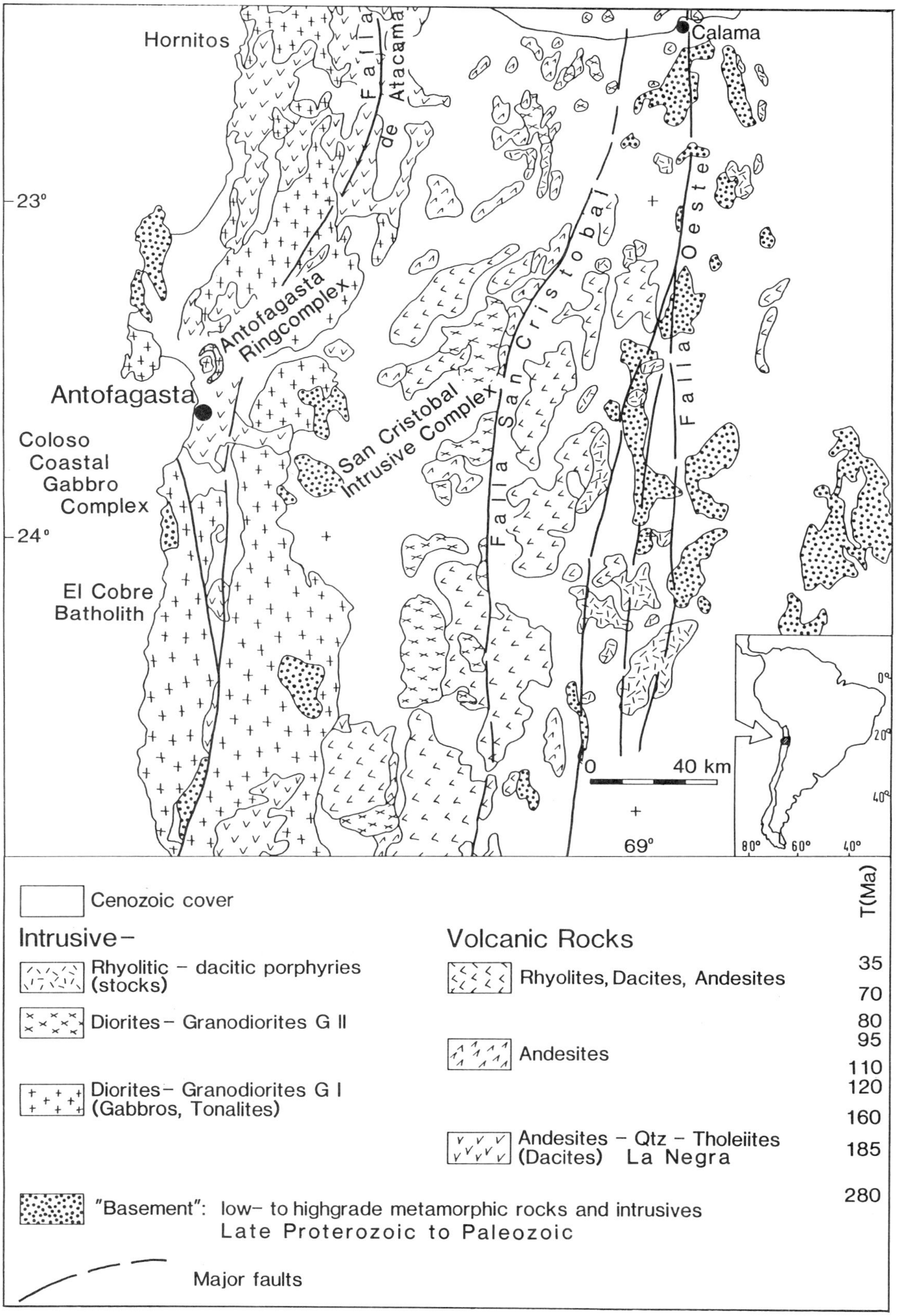

Fig. 1. Distribution of Preandean rocks and Andean magmatic rocks in north Chile

Table 1. Early Andean magmatic rocks from the Coastal Cordillera and Central Depression. Rb-Sr concentrations and isotope ratios

Sample	Sr(ppm)	Rb(ppm)	$^{87}Rb/^{86}Sr$	$^{87}Sr/^{86}Sr$	± 2 σ	Sr_i
Hornitos composite stock (190 Ma)						
87-1-1	436.6	5.02	0.0333	0.703354	0.000035	0.703264
87-1-3	500.0	2.84	0.0164	0.703422	0.000169	0.703378
87-1-4	542.0	1.77	0.0095	0.703304	0.000011	0.703278
87-1-7	298.2	1.11	0.0108	0.703192	0.000049	0.703163
87-0-1	330.1	35.16	0.3080	0.705355	0.000015	0.704553
87-0-2	255.4	70.80	0.8019	0.705292	0.000129	0.703206
Caleta Coloso fragmented dyke (190 Ma)						
86-2-8	320.3	16.30	0.1471	0.703651	0.000147	0.703253
La Negra volcanics (183 Ma)						
ln-1	270.0	62.63	0.6709	0.704822	0.000038	0.703076
ln-2	272.5	63.90	0.6781	0.704815	0.000016	0.703051
ln-3	276.1	63.62	0.6665	0.704794	0.000043	0.703060
ln-4	283.7	64.55	0.6579	0.704731	0.000012	0.703019
ln-5	275.6	62.18	0.6525	0.704708	0.000041	0.703010
ln-6	279.7	63.54	0.6569	0.704786	0.000045	0.703077
Coloso coastal gabbro complex (183 Ma)						
86-2-1	316.6	16.32	0.1491	0.703380	0.000034	0.702992
86-2-4	320.2	13.79	0.1245	0.703446	0.000053	0.703122
86-2-5	396.3	7.09	0.0517	0.703168	0.000015	0.703033
86-2-6	302.9	22.25	0.2124	0.703597	0.000043	0.703044
86-2-7	254.2	67.49	0.7678	0.704815	0.000020	0.702817
Antofagasta ring complex (155 Ma)						
86-4-1	364.8	27.40	0.2172	0.703695	0.000055	0.703216
86-4-3	348.9	32.74	0.2713	0.703990	0.000016	0.703392
86-4-4	315.5	48.30	0.4428	0.704162	0.000037	0.703186
86-4-6	282.8	49.73	0.5085	0.704403	0.000025	0.703282
86-4-7	324.8	44.81	0.3990	0.704205	0.000013	0.703326
86-4-9	323.1	57.25	0.5124	0.704497	0.000079	0.703368
86-4-10	345.1	50.87	0.4263	0.704124	0.000043	0.703185
86-4-11	349.2	39.90	0.3304	0.703776	0.000036	0.703048
86-4-12	391.8	56.08	0.4139	0.703783	0.000035	0.702871
86-3-2	170.9	78.72	1.3321	0.705981	0.000035	0.703046
86-3-4	132.0	223.27	4.8967	0.714002	0.000051	0.703212
86-3-7	252.9	68.94	0.7883	0.704816	0.000036	0.703079
86-3-8	289.1	10.59	0.3744	0.704072	0.000025	0.703247
86-3-9	203.8	48.58	0.6892	0.704658	0.000022	0.703139
86-3-13	167.0	112.75	1.9523	0.707234	0.000083	0.702932
86-3-14	168.1	107.46	1.8487	0.707008	0.000014	0.702935
San Cristobal intrusive complex (80 Ma)						
86-1-3	247.1	204.75	2.3970	0.706734	0.000035	0.704009
86-1-5	243.6	227.16	2.6970	0.706841	0.000045	0.703775
86-1-6	308.6	160.01	1.4998	0.705738	0.000074	0.704033
86-1-8	348.1	155.70	1.2937	0.705463	0.000053	0.703993
86-1-10	275.3	182.78	1.9204	0.706000	0.000080	0.703817
86-1-12	391.4	91.61	0.6769	0.704690	0.000025	0.703921
86-1-19	405.9	86.84	0.6188	0.704688	0.000019	0.703985

Table 2. Early Andean magmatic rocks from the Coastal Cordillera and Central Depression. Locations and petrographical features of samples

I: Hornitos composite stock (HIB) (22°25'30''S, 70°16'10''W)	
87-0-1	Coarse-grained (pegmatitic) marginal facies of stock, autometasomatized and hydrothermally altered (not included in Nicolaysen diagram), schlieren-type with contact to:
87 0-2	Medium-fine marginal facies of stock, minor alteration close to (?)tectonic zone (former contact to basement host rocks?)
87-1-3/-4/-7	Medium-grained cumulate gabbro, neighbouring 87-0-2 facies to the centre of stock, lacking sharp contact but with some minor tectonic disruptions
87-1-3	Average type
87-1-4	Plagioclase cumulate type
II: Caleta Coloso fragmented dyke(?) or stocks (HIB) (3 outcropping isolated localities: 1,2 megaxenolite/inclusion in CCG layered gabbro (?) 3 stock in Bolfin complex) (23°50'00''S, 70°30'05''W)	
86-2-8	Fine-grained gabbroic stock in Bolfin rocks slightly plagioclase-cumulated suite
III: Profile La Negra: Cerro Cristales - Caleta Cobre (LN) (24°10'05''S, 70°30'07''W)	
LN-1	(Top 11 m) lava flow, microporphyric andesitic basalt with plagioclase phenocrysts
LN-2	Lava flow, porphyric andesitic basalt with plagioclase and pyroxene phenocrysts
LN-3	Lava flow, strongly porphyric andesitic basalt with plagioclase and pyroxene phenocrysts
LN-4	Sill(?) doleritic basalt weathered and sheared contacts to upper and lower layers
LN-5	Lava flow, microporphyric andesitic basalt poor in plagioclase phenocrysts
LN-6	(Base 0 m) lava flow, porphyric andesitic basalt with plagioclase and pyroxene phenocrysts
IV: Coloso coastal gabbro complex (CCG) (23°50'30''S, 70°30'07''W)	
86-2-1	Medium-grained olivine-bearing gabbro, layer central part
86-2-4	Medium-grained gabbro (plag-cumulate trend), layer central part
86-2-5	Medium-grained hornblende-gabbro, layer central part
86.2.6	Medium-grained hornblende-gabbro, layer central part
86-2-7	Coarse-grained (pegmatitic) hornblende-gabbro, vein with nebulitic contact (from layer transition, autometasomatic pegmatitic facies not included in age calculation)
V: Antofagasta ring complex (G I) (23°29'30''S, 70°22'20''W)	
86-4-1	Outer ring, fine-grained monzodiorite (aplitic facies)
86-4-9	Outer ring, medium-grained porphyritic granodiorite (aplitic facies, strongly autometasomatically influenced)
86-4-10	Outer ring, medium-grained porphyritic granodiorite (subvolcanic facies, some LN-xenoliths)
86-4-11	Outer ring, fine-grained porphyritic granodiorite (subvolcanic facies, xenolith-free)
86-4-12	Outer ring, fine-grained porphyritic granodiorite (subvolcanic facies, some LN-xenoliths)
86-4-3	Centre, medium-grained quartz-rich granodiorite (plutonic facies, autometasomatically influenced)
86-4-4	Centre, medium-grained granodiorite (plutonic facies, autometasomatically influenced)
86-4-6	Centre, medium-grained leucogranodiorite (dyke facies, autometasomatically influenced)
86-4-7	Centre, medium-grained granodiorite (plutonic facies)
VI: Caleta el Cobre batholith (G I) (24°14'15''S, 70°30'25''W)	
86-3-2	Medium-grained quartz-rich granodiorite (dyke facies)
86-3-4	Fine-grained leucogranite (aplitic facies)
86-3-7	Medium-grained quartz-monzodiorite (plutonic facies)
86-3-8	Medium-grained monzodiorite (plutonic facies)
86-3-9	Medium-grained monzodiorite (plutonic facies)
86-3-13	Medium-grained leucogranodiorite (plutonic facies)
86-3-14	Medium-grained leucogranodiorite (plutonic facies)
VII: San Cristobal intrusive complex (G II) (23°25'15''S, 69°31'47''W)	
86-1-3	Central part, fine-grained leucogranodiorite (homogeneous plutonic facies)
86-1-6	Central part, fine-grained granodiorite (homogeneous plutonic facies)
86-1-8	Central part, fine-grained granodiorite (homogeneous plutonic facies)
86-1-5	Peripheral part, fine-grained quartz-rich granodiorite (homogeneous plutonic facies)
86-1-10	Peripheral part, medium-grained granodiorite (slightly foliated plutonic facies)
86-1-12	Peripheral part, fine-grained diorite (mafic schlieren facies, hbl-bi segregates)
86-1-19	Peripheral part, fine-grained diorite (mafic schlieren facies, hbl-bi segregates)

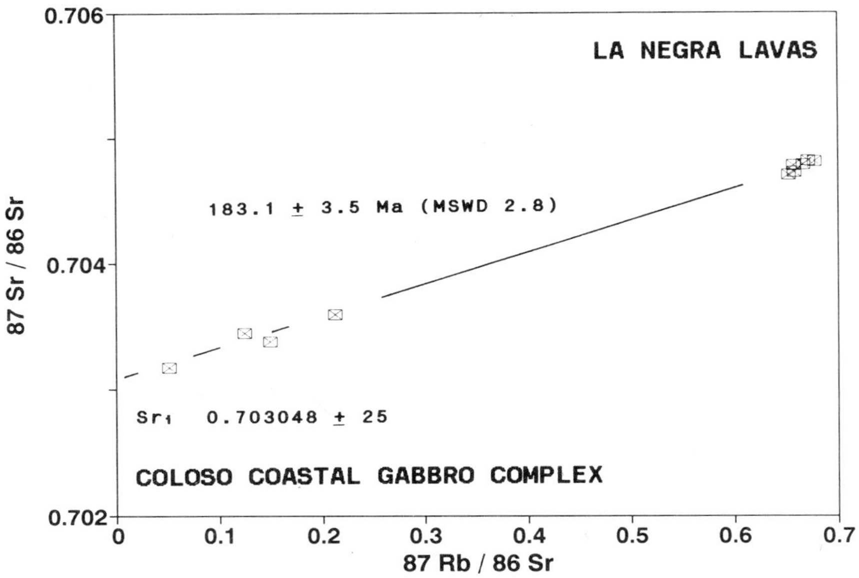

Fig. 2. Isochron diagram of Coloso Coastal Gabbro Complex and La Negra rocks

compositional spread had to be considered.

The rock types of the Coloso Coastal Gabbro Complex (CCG) occur at two sites: (1) The Coloso Coastal Gabbro Complex shows classical structures such as well-developed layered series. Its contacts with the host rocks, the Bolfin complex (Rössling 1989; Lucassen 1991), are rare, whereas tectonic and intrusive contacts with younger granitoids are common. The analysed samples (location IV of Table 2) are from the central parts of the thick (1-3m) cumulate layers of the CCG. (2) The Caleta El Cobre gabbro stock is a small gabbroic body with indistinct layering and contacts to the Caleta El Cobre batholith at the mouth of the Quebrada El Cobre.

The La Negra rocks were compared with the layered series of the Coloso Coastal Gabbro (see Tables 1 and 2). Lavas and intrusive rocks were found to have similar chemical compositions and systematic trends in trace element diagrams (see Sect. 3), as well as similar mineral phases of zoned calcic plagioclase and complex pyroxene-hornblende-biotite aggregates. Thus, it is supposed that these rocks are consanguineous, and, on the basis of this assumption, they have been tentatively combined in order to calculate a common isochron (Fig. 2) which corresponds to an age of 183.1 ± 3.5 Ma (MSWD = 2.8) with an initial Sr ratio of 0.703048 ± 25 (2δ). This age agrees with that of 186 ± 13 Ma (Sr_i: 0.70315 ± 0.00023) reported by Rogers (1986) for the La Negra in the region of Tocopilla (Coastal Cordillera at 22°S). The only sample, that does not fit the isochron of Fig. 2 is no. 86-2-7 (Table 1 and 3), a pegmatitic hornblende gabbro, with hornblende megacrysts of up to 5 cm length but which extend to up to 1 m in other places, so that the homogeneity of this sample is not ensured.

The Caleta El Cobre Batholith and the Antofagasta Ring Complex - G I Granitoids

The rocks of these complexes (G I) are mostly of granodioritic to monzodioritic compositions but with granitic, tonalitic or dioritic varieties. The trend to monzonitic compositions may occur in connection with autometasomatic transformation of granodioritic to dioritic rocks. The G I rocks represent the culmination of intrusive activity and also the final pulse of the magmatism of the early Andean development (Fig. 1). They intruded into La Negra lavas or pre-Mesozoic rocks and occur typically as shallow-level plutons such as the Antofagasta Ring Complex (location V, Table 2). In some cases they were eroded to deeper parts of the plutonic bodies which have tectonic contacts to older plutonic or metasedimentary series along steeply dipping fractures; an example is the Caleta El Cobre batholith (location VI, Table 2). Samples were taken from both complexes but were treated separately.

The Caleta El Cobre batholith consists of a series of differentiated rock types starting with medium-grained monzodiorites and quartz-bearing monzodiorites, followed by leucogranodiorites and dykes of quartz-rich granodiorites, and finally aplitic leucogranites. Taking into account all the data points the calculated isochron age is 155.4 ± 2.5 Ma (MSWD = 3.8) and there is an $^{87}Sr/^{86}Sr$ intercept of 0.703082 ± 77 (2δ) (Fig. 3).

Fig. 3. Isochron diagram of G I rocks (Caleta El Cobre batholith)

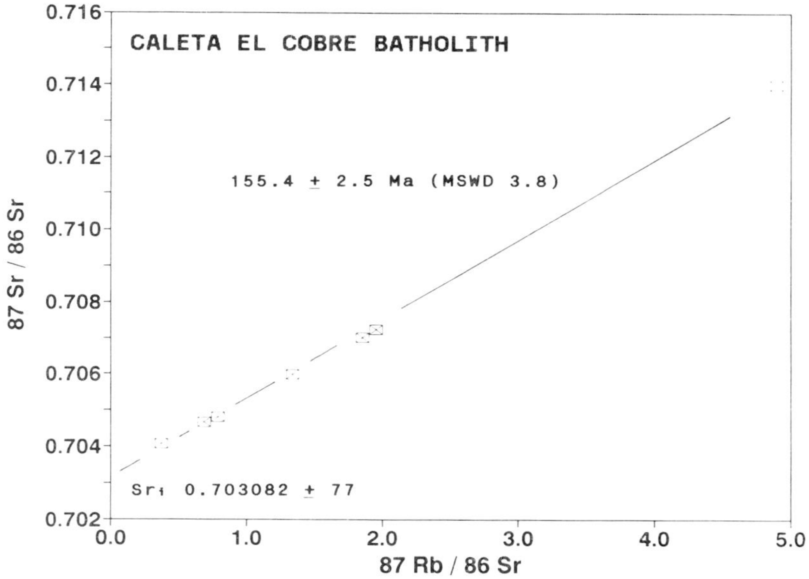

The Antofagasta Ring Complex consists of an outer subvolcanic ring and a central intrusive body. Most of the samples from the central part are altered by autometasomatism due to fluid enrichment in the roof of the intrusive, and the samples from the ring are additionally contaminated with La Negra xenoliths. Taking into account an age of 155 Ma, the initial ratios vary between 0.7029 and 0.7034.

2.2.2 Magmatic Rocks of the Central Depression

The San Cristobal Intrusive Complex - G II Granitoids

Like the G I rocks, the granitoid intrusive rocks (G II) of the Central Depression range in composition from granodiorites to monzodiorites, but are generally more quartz-rich. Metasomatism is common and even more intens than in the case of the G I rocks. Samples of the San Cristobal intrusive complex were collected from the central part of the plutonic body (location VII, Table 2) with homogeneous structural and textural patterns of fine-grained rocks, and additionally from a peripheral zone where rocks show more inhomogeneities like flow structures, mafic schlieren of generally fine-to medium-grained textures and partial foliation.

The three samples from the central core yield a straight line (MSWD = 0.16) in the isochron diagram of Fig. 4 where the slope corresponds to an age of 80.3 ± 2.3 Ma. The intercept was calculated to be 0.704003 ± 60 (2δ). In contrast, the four samples from the peripheral zone of the complex give an isochron age of 73.7 ± 1.3 Ma (MSWD = 0.46) with a corresponding intercept of 0.704009 ± 32 (2δ). To avoid over-interpretation of the sparse data a tentative age-calculation on the base of all measured data from this complex results in an errorchron of 78.5 ± 5.5 Ma and an Sr initial ratio of 0.7040 respectively; this age corresponds to the $^{40}Ar/^{39}Ar$ age of 78.6 ± 0.4 Ma published by Maksaev et al. (1988). The age range of 80-73 Ma gives a frame for interpretation of the G II magmatic pulse (cooling ages of this pluton: Andriessen and Reutter this Vol.).

2.3 Conclusions from Age Determinations for the Magmatic Succession and Time-Space Relationships

A comprehensive survey of time and space relations of dated magmatic rocks from the Antofagasta segment is given in Fig. 5, where Sr_i ratios vs. time are plotted. Those data points marked with location abbreviations result from Rb-Sr whole rock determinations by the present author. The difference between the Preandean and Andean magmatic rocks is striking, as Palaeozoic granitic rocks show Sr_i ratios of 0.7050 or higher, but at the beginning of the Mesozoic, around 200 Ma, the Sr_i ratios decrease abruptly to a lowest value of about 0.7030. Minimum ratios occur frequently and are stable over the period 200 to 120 Ma but then increase to the highest ratios of modern Andean magmatic rocks; this implies the occurrence of processes like significant crustal contamination, or source enrichment (melts derived

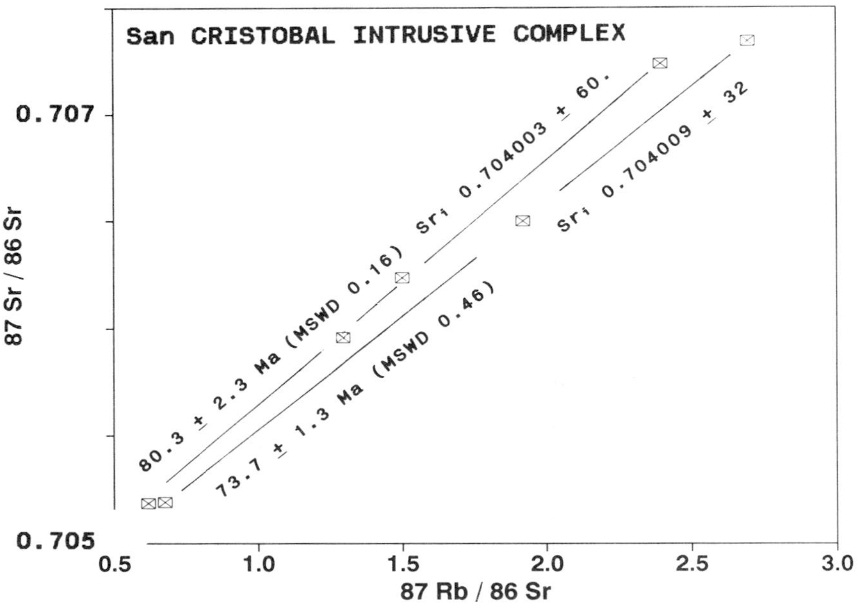

Fig. 4. Isochron diagram of G II rocks (San Cristobal intrusive complex)

from metasomatically enriched mantle sources), or derivation by anatexis from young arc crust or oldest basement (see also Sect. 3).

The post-Palaeozoic magmatism starts with HIB, small gabbroic stocks and dykes which are precursors of CCG-type gabbroic pulses. The HIB are followed by the La Negra volcanic series (the age of 183.1 ± 3.5 Ma is restricted to the oldest parts of the La Negra sequence in the area) and related gabbros of the Coloso Coastal Gabbro Complex (CCG). Granitoids similar to those of the Caleta El Cobre batholith and Antofagasta Ring Complex followed and intruded into earlier rocks of the same magmatic arc. This G I unit started intruding around 150 Ma and intrusions of this type continued until 120 Ma, when the uplift history of the Coastal Cordillera began. The next important magmatic unit - G II granitoids like the San Cristobal intrusive complex at 73.7 ± 1.3 to 80.3 ± 2.3 Ma - is displaced towards the Central Depression. This is the first step of the migration of magmatic activity to the east and also shows a change in conditions for magmagenesis, indicated by the rise in Sr_i ratios.

3 Geochemistry

The geochemical methods employed here comprise whole rock major and trace element analyses by XRF techniques in the institutes of Geology (Geochemistry section) and Mineralogy at the Freie Universität Berlin, using international standard reference rocks (USGS: BHVO-1, RGM-1, STM1, QLO-1; CRPG: GA, GH, BR; SABS: NIM-D, NIM-P, NIM-S; CANMET: MRG1). REE concentrations were determined at the CRPG in Nancy (France) by optical emission spectrochemical analysis utilizing microwave plasma excitation (Govindaraju et al. 1976). The results are given in Table 3.

3.1 Classification and Differentiation Trends

Compositional features of the modal and chemical components are demonstrated in the Streckeisen triangles (Fig. 6) and the R_1/R_2 De La Roche diagram (Fig. 7). A more detailed description of petrographical and structural aspects of these rocks is given by Pichowiak et al. (1990). The Streckeisen triangle for intrusive rocks (mineral modes) plots HIB and other gabbroic rocks nearest to the plagioclase corner. These rocks range from subordinate peridotite/pyroxenite to dominant olivine-bearing gabbro and gabbronorite. The plagioclase content for the gabbroic rocks generally varies from 30 to 50 vol% and is never less than 10 vol%. The granitoid G I and G II rocks are generally granodiorites to diorites with some tonalite or even monzonite tendencies. They have fine- to medium-grained, homogeneous textures, but also some compositional variations within individual intrusions which result from internal differentiation or multiple intrusion events. Volcanic rocks occur more variably, mostly with basaltic to andesitic compositions and different degrees of silica or alkali enrichments (Streckeisen triangle: Rittmann norm components; the La Negra volcanics have

Table 3. Early Andean magmatic rocks from the Coastal Cordillera and Central Depression.; major- and trace-element geochemistry

Sample	86-2-8	87-0-1	87-0-2	87-1-1	87-1-3	87-1-4	87-1-7	LN 1	LN 2	LN 3	LN 4	LN 5	LN 6	86-2-1	86-2-4	86-2-5	86-2-6	86-2-7	86-4-1	86-4-9
Location	II	I	I	I	I	I	I	III	III	III	III	III	III	IV	IV	IV	IV	IV	V	V
	HIB								*LN and CCG*										*G I rocks*	
SiO_2	43.78	50.12	51.87	47.47	44.95	44.45	44.90	54.03	52.00	52.01	51.86	52.71	53.85	50.33	45.84	54.11	52.26	57.97	51.28	58.40
Al_2O_3	22.19	18.48	16.92	14.88	18.17	28.66	18.41	14.69	13.80	13.45	15.56	16.75	17.53	17.39	17.24	18.43	16.84	16.55	18.14	15.72
FeO	7.16	5.33	5.50	7.31	9.50	2.41	5.06	7.55	8.73	8.82	7.38	5.58	5.86	4.73	9.85	6.25	6.65	5.43	6.54	4.92
Fe_2O_3	2.13	3.10	2.91	2.47	2.76	1.75	1.78	3.59	3.19	3.42	3.25	2.91	2.60	1.91	2.86	2.56	2.77	2.47	2.92	2.99
MgO	6.49	4.80	5.57	10.91	6.86	3.25	11.16	3.18	4.39	4.13	4.02	3.68	3.42	8.87	5.81	3.71	5.56	2.41	4.23	2.85
MnO	0.14	0.15	0.16	0.16	0.20	0.05	0.12	0.42	0.38	0.34	0.35	0.40	0.17	0.13	0.20	0.19	0.17	0.11	0.16	0.11
CaO	14.16	9.86	7.34	11.08	12.38	15.91	15.55	5.40	6.10	6.62	6.70	6.53	8.42	11.53	12.22	8.50	9.85	5.43	7.09	5.52
Na_2O	1.24	2.82	3.55	2.27	1.87	1.48	0.93	4.02	3.71	3.17	4.20	3.76	3.59	1.71	1.95	3.65	3.18	4.09	3.81	4.73
K_2O	0.18	1.01	1.83	0.44	0.22	0.02	0.00	2.49	2.38	3.10	2.65	2.81	1.71	0.37	0.56	0.79	0.75	2.12	1.21	2.16
TiO_2	0.63	1.6	1.41	0.97	1.26	0.25	0.28	1.70	1.73	1.89	1.76	1.50	1.15	0.41	1.36	1.06	0.77	0.97	0.92	0.61
P_2O_5	0.00	0.41	0.3	0.22	0.34	0.11	0.10	0.47	0.44	0.57	0.45	0.25	0.42	0.11	0.07	0.09	0.13	0.26	0.13	0.16
H_2O	0.98	1.4	1.42	0.68	0.73	1.20	1.41	1.55	1.77	1.50	1.30	1.85	0.95	1.49	0.60	0.64	0.33	0.60	2.02	1.01
TOTAL	99.08	99.08	98.78	98.86	99.24	99.54	99.70	99.19	98.62	99.02	99.48	98.73	99.67	98.98	98.56	99.98	99.26	98.41	98.45	99.18
La	1.69	12.81	11.26	7.99	7.41	2.07	1.08	25.20	26.20	16.80	19.05	18.30	26.20	4.40	3.68	6.10	4.72	nd	8.15	nd
Ce	5.25	30.50	30.53	21.97	20.79	4.01	5.33	59.44	62.50	40.21	44.90	46.05	59.30	11.65	12.30	17.30	15.46	nd	18.98	nd
Nd	2.10	17.14	18.58	15.11	14.72	2.45	2.76	35.49	38.30	25.70	26.83	26.86	37.85	5.90	6.10	8.70	6.93	nd	11.50	nd
Sm	0.78	4.15	4.80	3.96	4.06	0.63	1.00	8.98	9.42	5.78	6.75	6.74	8.73	1.80	1.98	2.80	1.85	nd	3.40	nd
Eu	0.44	1.35	1.64	1.13	1.17	0.38	0.39	1.92	2.31	1.58	1.84	1.85	2.09	0.50	1.00	1.10	0.59	nd	1.10	nd
Gd	0.70	4.28	5.16	3.96	4.47	0.69	1.26	9.41	11.22	5.94	6.47	7.44	9.31	1.70	2.00	2.40	1.76	nd	2.90	nd
Dy	0.78	3.78	5.15	3.45	3.88	0.64	1.29	9.72	10.41	5.10	6.53	7.43	10.19	1.90	2.10	2.70	1.93	nd	3.10	nd
Er	0.47	1.97	2.84	1.89	2.17	0.45	0.78	4.80	5.20	3.04	4.00	4.64	5.12	1.10	1.20	1.60	1.20	nd	1.70	nd
Yb	0.46	1.68	2.62	1.60	1.71	0.28	0.58	4.46	4.54	2.86	3.42	4.05	4.40	1.20	1.20	1.80	1.40	nd	2.00	nd
Lu	0.06	0.32	0.45	0.28	0.33	0.05	0.08	0.59	0.62	0.41	0.49	0.54	0.58	0.20	0.20	0.30	0.20	nd	0.30	nd
Rb	16.00	35.00	71.00	5.00	3.00	2.00	5.00	62.00	63.00	64.00	65.50	62.18	64.00	16.00	14.00	7.00	22.00	67.00	27.00	57.00
Sr	320.00	330.00	255.00	436.00	500.00	542.00	298.00	255.00	272.00	276.00	283.00	275.00	280.00	317.00	320.00	396.00	303.00	254.00	65.00	323.00
Ba	66.00	204.00	168.00	109.00	77.00	36.00	17.00	515.00	480.00	522.00	522.00	466.00	391.00	77.00	145.00	228.00	183.00	330.00	188.00	290.00
Zr	33.00	59.00	143.00	53.00	30.00	7.00	12.00	251.00	210.00	240.00	227.00	210.00	171.00	72.00	65.00	88.00	160.00	237.00	120.00	158.00
Nb	nd	5.00	5.00	5.00	5.00	5.00	5.00	8.00	8.00	8.00	8.00	9.00	6.00	3.00	2.00	3.00	4.00	24.00	15.00	27.00
Th	nd	5.00	5.00	5.00	5.00	5.00	6.00	9.00	7.50	5.00	5.50	7.50	7.50	nd	nd	nd	nd	nd	nd	nd
Y	5.49	22.77	32.58	20.45	23.27	3.57	7.83	50.00	49.00	48.00	49.00	38.00	35.00	14.00	14.80	20.00	17.00	nd	22.10	26.00
Sc	nd	36.09	24.50	40.90	43.79	16.60	58.59	32.00	40.00	28.00	30.00	37.00	30.00	nd	nd	nd	nd	nd	nd	nd
V	374.00	618.00	162.00	290.00	431.00	119.00	134.00	310.00	270.00	390.00	355.00	261.00	275.00	118.00	502.00	289.00	207.00	130.00	26.00	114.00
Cr	79.00	46.00	148.00	621.00	73.00	21.00	202.00	26.00	47.00	40.00	45.00	57.00	40.00	203.00	80.00	60.00	151.00	120.00	176.00	192.00
Co	nd	64.00	73.00	82.00	66.00	54.00	62.00	22.00	30.00	20.00	22.00	25.00	21.00	nd	nd	nd	nd	nd	nd	nd
Ni	nd	23.00	72.00	215.00	36.00	25.00	71.00	10.00	11.00	12.00	13.00	29.00	12.00	98.00	30.00	15.00	20.00	34.00	34.00	29.00
Cu	nd	29.00	49.00	133.00	137.00	188.00	25.00	nd	nd	nd	nd	nd	nd	nd	nd	nd	nd	nd	nd	nd
Zn	52.00	51.00	67.00	67.00	83.00	23.00	35.00	nd	nd	nd	nd	nd	nd	36.00	69.00	76.00	63.00	35.00	143.00	60.00
Ga	nd	5.00	5.00	6.00	21.00	24.00	14.00	nd	nd	nd	nd	nd	nd	nd	nd	nd	nd	nd	nd	nd

Sample	86-4-10	86-4-11	86-4-12	86-4-3	86-4-4	86-4-6	86-4-7	86-3-2	86-3-4	86-3-7	86-3-8	86-3-9	86-3-13	86-3-14	86-1-3	86-1-6	86-1-8	86-1-5	86-1-10	86-1-12	86-1-19
Location	V	V	V	V	V	V	V	VI	VI	VI	VI	VI	VI	VI	VII	VII	VII	VII	VII	VII	VII
		G I rocks (cont.)														*G II rocks*					
SiO_2	57.96	59.11	57.83	62.95	60.98	65.33	64.49	64.36	72.12	58.25	55.31	68.25	60.74	65.91	65.04	63.28	61.78	66.24	63.64	60.58	57.76
Al_2O_3	15.94	17.11	17.53	16.52	16.23	15.35	16.37	13.41	13.10	15.27	17.10	15.03	16.65	13.35	15.25	15.92	16.24	15.02	16.02	15.78	16.24
FeO	4.65	4.09	4.68	3.29	3.71	1.13	3.74	4.63	0.87	5.88	4.96	2.45	4.39	3.81	2.64	3.11	2.97	2.55	2.99	3.33	3.60
Fe_2O_3	3.75	2.38	2.74	2.09	2.50	1.91	1.63	2.82	1.84	2.56	2.58	1.98	2.56	2.66	2.34	2.41	2.39	2.37	2.61	2.70	2.55
MgO	2.89	2.67	2.48	1.75	2.39	1.02	2.36	1.87	0.61	3.39	4.89	1.33	2.58	1.61	1.79	2.24	2.14	1.54	2.09	2.43	2.75
MnO	0.24	0.11	0.16	nd	0.12	0.04	0.11	0.10	0.01	0.13	0.13	0.06	0.12	0.05	0.08	0.08	0.07	0.07	0.08	0.08	0.10
CaO	5.75	5.73	5.84	4.47	4.80	4.13	4.33	4.32	1.87	6.53	8.29	3.66	5.19	3.83	3.64	4.16	4.86	3.32	4.25	6.35	7.02
Na_2O	4.30	4.52	4.33	4.27	3.90	4.52	4.08	3.48	2.43	3.23	2.90	4.10	3.79	3.27	3.51	3.37	4.14	3.11	3.55	3.42	3.53
K_2O	1.40	1.66	1.51	1.99	2.07	2.44	2.31	2.69	5.27	1.92	1.17	2.18	2.78	3.21	3.74	3.43	2.88	4.27	3.38	2.35	2.19
TiO_2	0.75	0.88	0.84	0.59	0.60	0.41	0.13	1.13	0.34	0.87	0.51	0.32	0.77	0.97	0.62	0.68	0.69	0.62	0.71	0.75	0.78
P_2O_5	0.13	0.18	0.20	0.14	0.17	0.09	0.13	0.25	0.11	0.14	0.01	0.12	0.18	0.22	0.08	0.14	0.14	0.13	0.14	0.16	0.15
H_2O	1.46	0.70	1.50	0.50	0.87	1.92	0.10	0.17	0.77	0.76	1.50	1.10	0.01	0.90	0.66	0.86	1.05	0.51	0.45	1.65	2.22
TOTAL	99.22	99.14	99.64	98.56	98.34	98.29	99.78	99.23	99.34	98.93	99.35	100.58	99.76	99.79	99.39	99.68	99.35	99.75	99.91	99.58	98.89
La	10.90	15.30	nd	14.60	12.50	10.90	13.62	21.50	15.65	10.70	5.10	6.79	22.30	20.28	25.35	22.59	13.6 0	22.00	21.53	17.85	15.91
Ce	27.80	37.50	nd	32.70	28.70	25.10	33.20	53.60	34.50	27.10	13.30	16.33	48.90	49.76	54.80	50.83	34.80	52.80	48.12	41.05	38.05
Nd	13.90	17.50	nd	16.02	14.70	10.60	16.70	28.50	16.81	14.10	6.00	7.52	23.70	26.15	23.40	22.50	19.40	22.60	22.30	20.92	19.50
Sm	3.80	4.30	nd	4.20	3.80	2.90	4.30	8.40	3.99	4.20	1.90	2.07	5.70	7.56	5.70	5.48	5.30	5.50	5.56	5.09	4.81
Eu	0.90	1.50	nd	1.20	0.70	0.90	0.60	1.60	0.67	0.97	0.60	0.64	1.50	1.45	1.00	1.16	1.30	1.10	1.01	1.18	1.22
Gd	3.10	3.70	nd	3.50	2.90	2.20	3.50	7.26	4.18	3.70	1.70	2.23	4.20	6.64	4.50	4.06	4.20	4.20	4.30	4.12	3.92
Dy	3.30	3.70	nd	3.50	3.10	2.20	3.60	7.60	4.25	4.10	2.00	2.41	3.60	6.94	4.40	4.01	4.20	4.10	4.21	3.89	3.70
Er	1.80	2.10	nd	2.00	1.80	1.30	2.00	4.20	2.68	2.40	1.20	1.54	1.96	3.90	2.40	2.23	2.30	2.30	2.30	2.22	2.11
Y	2.20	2.50	nd	2.40	2.20	1.70	2.30	4.70	2.76	2.80	1.48	1.71	2.40	4.28	2.80	2.54	2.60	2.80	2.60	2.33	2.19
Lu	0.30	0.40	nd	0.40	0.30	0.20	0.30	0.70	0.49	0.40	0.20	0.31	0.30	0.64	0.40	0.34	0.30	0.40	0.40	0.31	0.31
Rb	51.00	40.00	56.00	33.00	48.00	50.00	45.00	79.00	223.00	69.00	11.00	49.00	113.00	107.00	247.00	160.00	155.00	227.00	183.00	92.00	87.00
Sr	345.00	349.00	392.00	348.00	315.00	283.00	325.00	171.00	132.00	253.00	289.00	204.00	167.00	168.00	204.00	308.00	348.00	243.00	275.00	391.00	406.00
Ba	nd	290.00	253.00	335.00	317.00	352.00	351.00	83.00	305.00	303.00	65.00	300.00	554.00	357.00	610.00	637.00	585.00	717.00	617.00	577.00	572.00
Zr	144.00	166.00	176.00	144.00	132.00	229.00	104.00	188.00	211.00	169.00	65.00	98.00	209.00	150.00	250.00	267.00	279.00	318.00	320.00	198.00	201.00
Nb	27.00	27.00	24.00	12.00	27.00	27.00	8.00	10.00	8.00	7.00	2.50	6.00	27.00	nd	8.00	5.00	13.00	10.00	10.00	nd	nd
Th	nd	nd	nd	nd	nd	nd	nd	nd	37.00	nd	nd	5.00	nd	nd	nd	nd	nd	nd	nd	nd	nd
Y	24.00	27.40	20.00	27.13	23.50	17.10	26.69	55.46	31.34	30.00	14.70	17.74	27.00	50.00	33.80	29.73	30.60	31.00	30.17	25.00	19.00
Sc	nd	nd	nd	nd	nd	nd	nd	nd	5.50	nd	nd	11.19	nd	18.30	nd	nd	nd	nd	nd	nd	nd
V	260.00	160.00	155.00	89.00	127.00	67.00	115.00	156.00	47.00	217.00	90.00	71.00	171.00	135.00	84.00	113.00	97.00	78.00	100.00	137.00	141.00
Cr	140.00	180.00	142.00	181.00	222.00	293.00	191.00	170.00	13.00	190.00	220.00	171.00	230.00	120.00	160.00	189.00	225.00	199.00	189.00	45.00	50.00
Co	nd.	nd	nd	nd	nd	nd	nd	nd	262.00	nd	nd	90.00	nd	155.00	nd	nd	nd	nd	nd	nd	nd
Ni	29.00	29.00	25.00	25.00	29.00	29.00	30.00	25.00	17.00	10.00	105.00	32.00	29.00	nd	11.00	10.00	15.00	10.00	20.00	29.00	23.00
Cu	nd	nd	nd	nd	nd	nd	nd	nd	45.00	nd	nd	123.00	nd	52.00	nd	nd	nd	nd	nd	nd	nd
Zn	252.00	73.00	104.00	69.00	66.00	23.00	54.00	70.00	14.00	61.00	45.00	26.00	110.00	50.00	98.00	45.00	38.00	46.00	55.00	70.00	nd
Ga	nd	nd	nd	nd	nd	nd	nd	nd	16.00	nd	nd	19.00	nd	11.00	nd	nd	nd	nd	nd	nd	nd

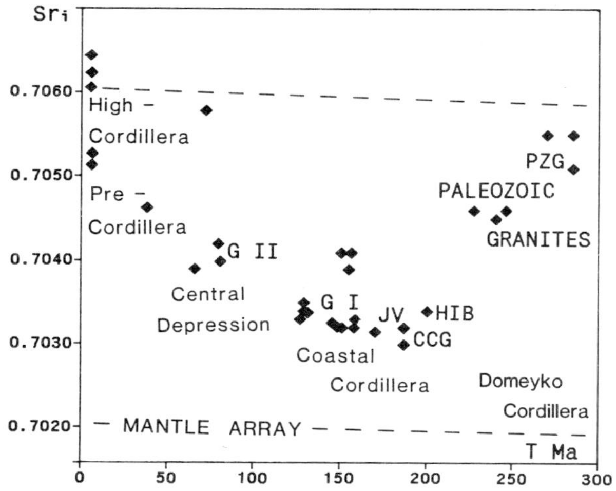

Fig. 5. Sri-values vs. time. Note the obvious fall from Palaeozoic granites (*PZG* - Pampa Elvira, 285 Ma, Baeza and Pichowiak 1988b) to the lowest values of Coastal Cordilleran rocks. This is followed by a steady increase in values and a movement of the magmatic loci until the High-Cordillera is reached (*HIB* Hornitos Initial Basics, 190 Ma; *JV* La Negra and *CCG* Coloso Coastal Gabbro complex, 183 Ma; *G I* Caleta El Cobre batholith and Antofagasta Ring Complex granitoids, 155 Ma; *G II* San Cristobal intrusive complex: 74-81 Ma; Rb-Sr data from this programme; additional data from Halpern 1978, Hawkesworth et al. 1982, Diaz et al. 1985, Rogers 1985, Hervé et al. 1985)

plagioclase, pyroxene and minor olivine or amphibole as the principal phenocrysts). The chemical characteristics and trends of plutonic and volcanic rocks are shown in the De La Roche diagram of (Fig. 7). Mafic plutonic rocks range from ultramafic lithologies (including cumulate rocks) to olivine gabbro and gabbronorite. Dioritic to granitic plutonic rocks of the G I and G II cycles suggest calc-alkaline differentiation processes. The volcanic rocks are compositionally intermediate between the rocks of the initial and late plutonic cycles.

3.2 Magma Sources and Fractionation Processes

Based on the low Sri ratios and the pattern of compositional variations of the rocks as mentioned above, a homogeneous mantle is assumed to be the main source for magmagenesis. The differentiation behaviour of similar magmatic suites, where tholeiitic fractionation was obvious for the primary gabbroic rocks (HIB type) despite the differentiation behaviour with calc-alkaline affinities of the La Negra and granitic rocks, was discussed in Pichowiak et al. (1990).

An attempt to demonstrate the relationship between different magmatic units was made by investigating element partition behaviour. Based on a selection of compatible and incompatible elements as well as the most important REEs, Fig. 8 displays distribution patterns for characteristic rocks of each unit and illustrates the systematic changes from one unit to the next.

Part 1 of Fig. 8 represents the HIB rock normalized against Anderson's (1983) mantle model. In this diagram only nickel shows depletion, a great part of it remaining captured in the source; all other elements are progressively enriched. Using this pattern to estimate melt-source proportions on the basis of Rb concentration (5 ppm Rb of the sample) and with the assumption that all LIL elements of the melt-batch are concentrated in the liquid, the result gives about 15% derived melt from the assumed source. As La Negra volcanics, related CCG gabbros, and the granitoids may not be derived directly from mantle sources, but may represent different stages of differentiation from assumed primary mantle melt-batches of a HIB-type composition. The latter composition was used as hypothetical parental magma for the calculation in part 2 of Fig. 8 (which means that the normalization baseline here corresponds to the HIB composition in part 1).

The patterns of all units in part 2 exhibit a similar shape, i.e. enrichment at one end and depletion towards the other end, with increasing slopes from the CCG to the G II pattern. Systematic step-by-step changes in element behaviour are evident from the columns in part 2. Here, Rb and K, as LIL elements, show progressive enrichment from CCG through JV and G I, to G II (a behaviour which is sometimes interpreted as reflecting subduction components during magma generation). Ce for the LREE exhibits a one-step enrichment similar to that for the JV to G II, due to a concentration in melts after removal of early crystallizing ol- and px-phases. The Eu trend initially exhibits an enrichment for the JV during ol/px fractionation, but this turns to progressive depletion for G I and G II by complementary plg-fractionation. This coincides with the Ca development and its progressive depletion for the CCG and JV by cpx-fractionation (at the same time Eu shows a trend of enrichment!). Finally, the additional plg-fractionation (probably also hbl-fractionation) causes the Eu depletion for the G I and G II. Yb for the HREE shows enrichment for the JV by concentration in the liquid phase, and this progresses to depletion for G I by enhanced removal of cpx/hbl components and accessories. A re-enrichment occurs for the G II and

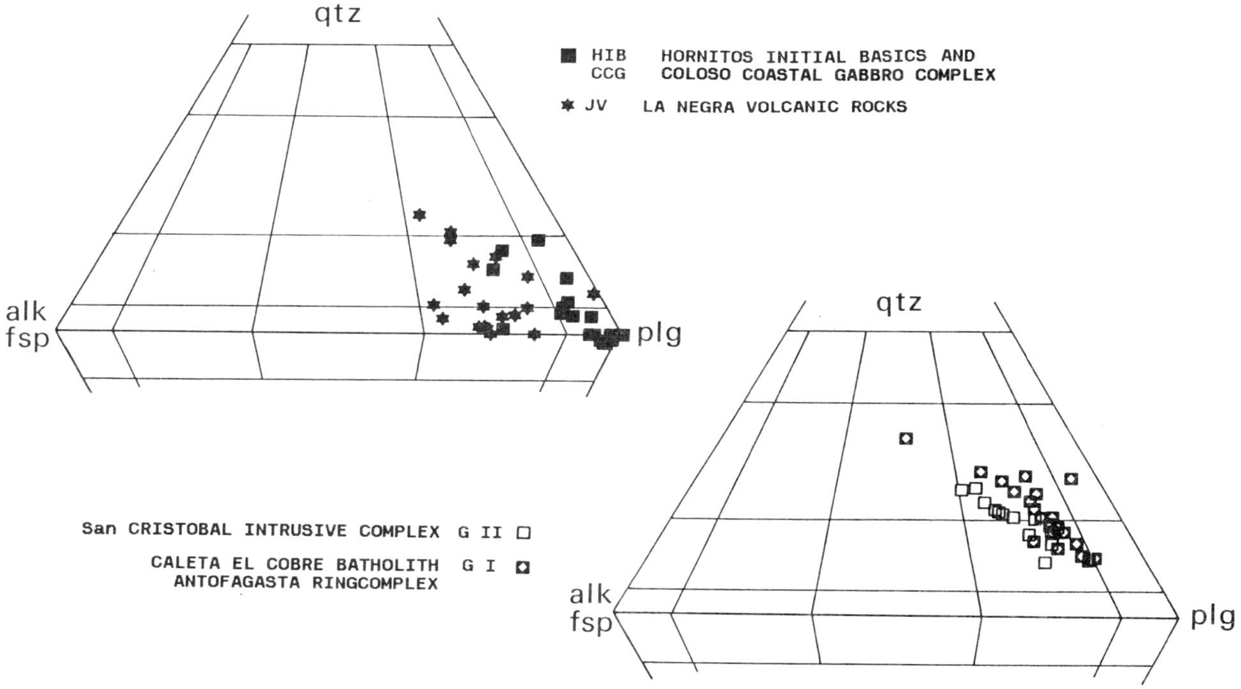

Fig. 6. Streckeisen triangles. *Above* volcanic rocks and initial basic intrusives; *below* granitoid rocks

indicates a change in conditions for magmagenesis of this unit. Probably the incorporation of HREE enriched components to the magma is due to the same mechanism which may have raised the Sr_i ratios for these rocks. Finally, Ni demonstrates a moderate to complete depletion from the CCG to the G I and G II, caused by ol- and px-removal.

4 A Hypothetical Model for Magmatism and Tectonic Setting

4.1 Initial Andean Development From 220 to 190 Ma: Magmatic Activity Controlled by Extensional Tectonics

This period marks the start of the Andean Cycle. In an extensional regime gabbroic bodies and dyke-swarms with continental basalt affinities and without subduction components penetrated the crust (HIB) due to decompressional upper-mantle melting with basaltic ponding in the lower crust (Fig. 9a). Presumably the extensional conditions were supported by a plate convergence which was too oblique to induce a subduction-related magmatism (cf. Scheuber et al. this Vol.).

4.2 Andean Development From 190 to 130 Ma: Maximum Activity of the Jurassic Magmatic Arc Controlled by Subduction

The start of the La Negra volcanism and the first culmination in magmatic activity corresponds to a change in conditions for magma generation. This has been interpreted by various authors as subduction input (e.g. Rogers 1985; Buchelt and Tellez 1988; Pichowiak et al. 1990). Thus, at the south American continental margin the kinematics had changed to stronger plate convergence leading to subduction of the oceanic plate. Arc-parallel motion still continued and this was recognized by Scheuber (1987) in a prominent sinistral shear system along the Atacamafault system which was active during the Jurassic and early Cretaceous. This system of oblique subduction in combination with arc-parallel motion (Fig. 9b) caused extension normal to the plate boundary and allowed the formation of pull-apart structures. Magma generation may then have been controlled by mantle wedge melting, triggered by subduction fluids, and also the continuation of upper mantle decompression. Under the influence of subduction the production of magma increased enormously and built up La Negra basalts and

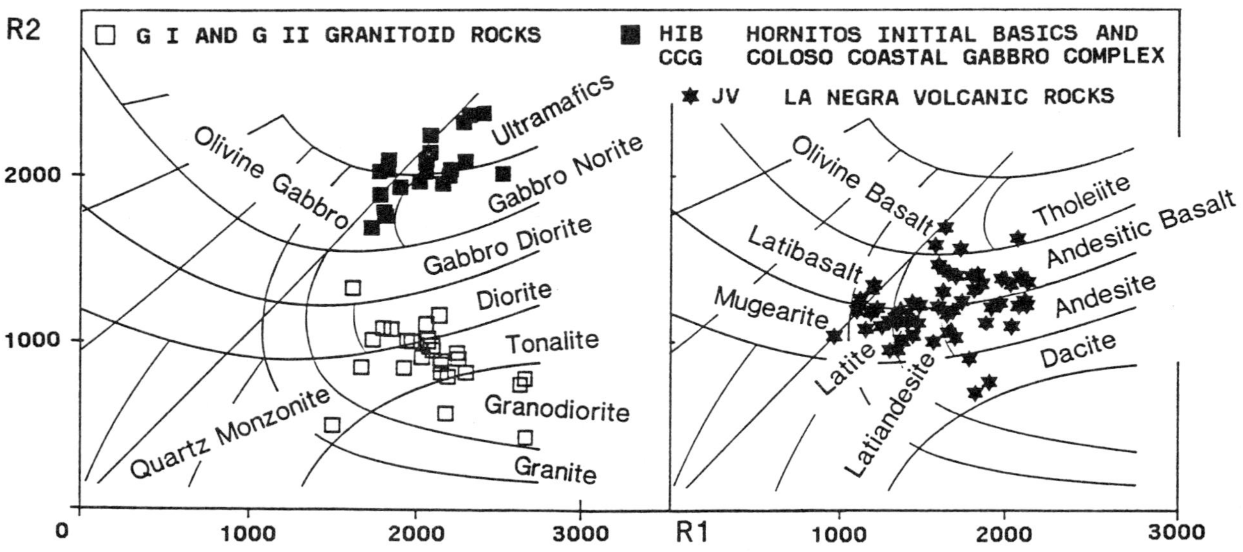

Fig. 7. De La Roche classification. Left gabbroic and granitoid plutonic rocks (left) show a bimodal distribution; right volcanic rocks plot between these two groups

andesites to a thickness of 10 km; it also produced minor contemporaneous CCG gabbros. Conditions may have been similar for the mobilization of the huge G I complexes which intruded into the arc between 150 and 120 Ma. But it may also be assumed that the problem of space for the emplacement of G I complexes was exclusively solved by oblique subduction kinematics, as it was described by Glazner (1991) for the formation of Mesozoic plutons in California.

An important aspect of this hypothesis is that those units which dominate in the Coastal Cordillera, i.e. the La Negra volcanics and the G I, may have left behind a similar amount of residual material from fractionation processes in intracrustal magma-chambers as volcanic material extruded and plutonic material was emplaced. This means, in fact, that most of the arc must have been built up by magmatic material of this cycle only.

The question remains as to what happened to the host rock basement. As there is no evidence for any incorporation of older basement crust material into

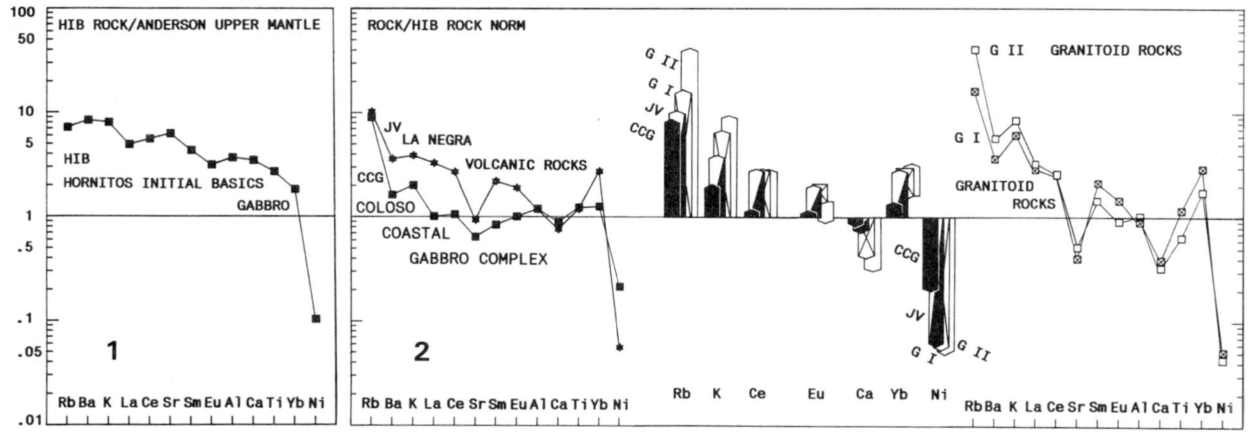

Fig. 8. Spider diagrams. 1 HIB-type composition, normalized against Anderson's(1983) mantle model. 2 CCG, JV, G I and G II-type compositions normalized against HIB-source. Arrow columns show different fractionation behaviour for elements indicated. For explanation, see text

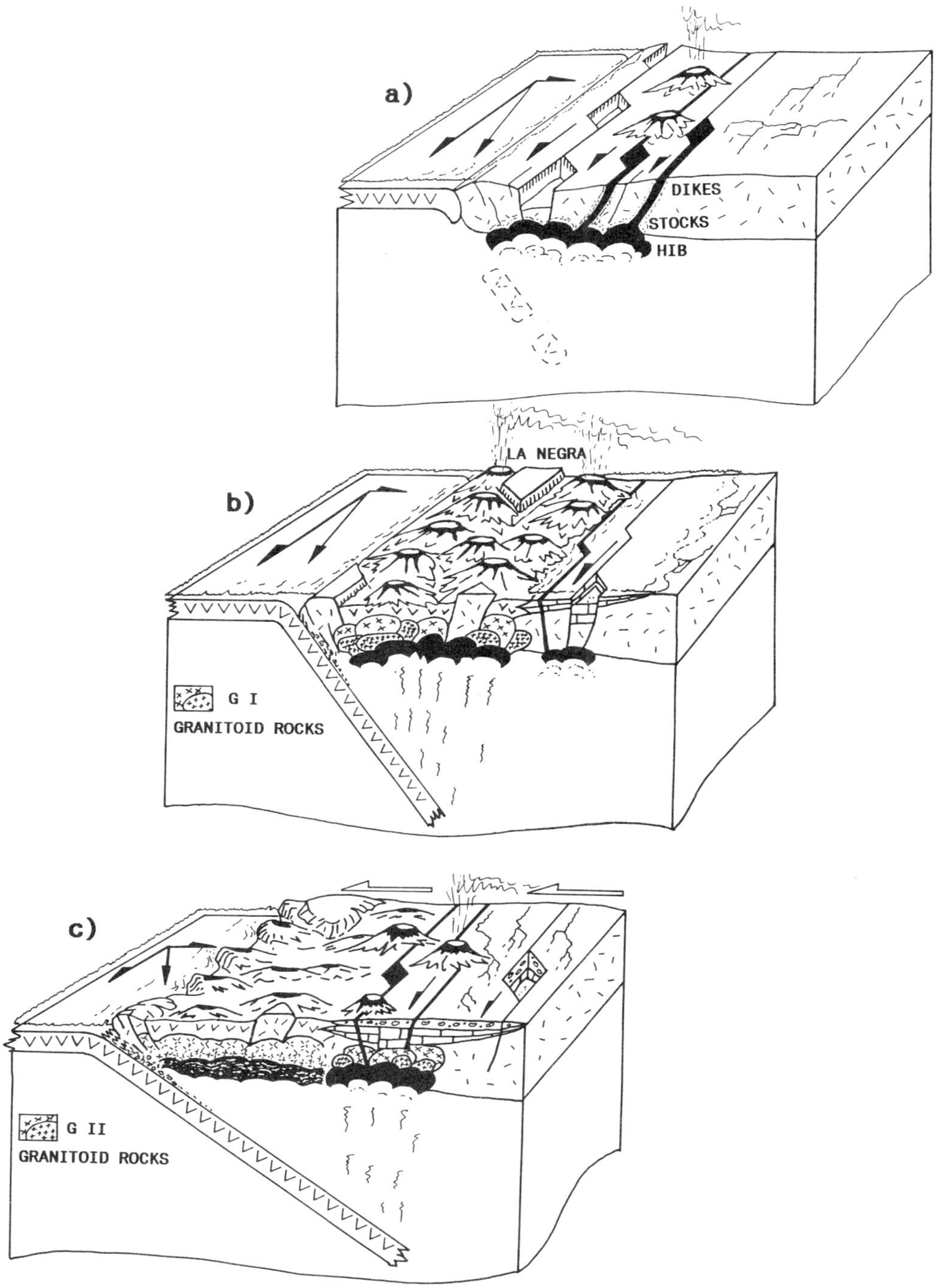

Fig. 9. Model for magmatic and tectonic setting during the time-span 200-80 Ma
a 220-190 Ma. Initial Andean magmatic activity controlled by tensional tectonics in the continental margin area: pull-apart type crustal stretching by lateral plate motions without subduction; magmatic accretion by decompressional upper-mantle melting, basaltic ponding. Gabbro complexes and related dyke-swarms with continental basalt affinities (HIB initial basics)
b 190-130 Ma. Maximum of the Jurassic arc formation: high subduction rate, oblique subduction triggers parallel-margin shear with pull-apart structures - the magmatic arc is built up by La Negra volcanics, contemporaneous gabbros (CCG) and huge granitoid plutonic complexes (G I)
c 130-80 Ma. Decrease in magmatic activity in the Coastal Cordillera, beginning of west to east magmatic migration: changes in subduction rate and angle of dip, tensional tectonics by oblique plate movements with shear systems; magmatic activity with Central Depression G II intrusives shows first signs of change in magmagenesis indicated by rise in Sr_i-values (see text)

magma generation, because of low and stable Sr_i ratios (0.7031 ± 0.0002 for all units from 200 to 120 Ma), it is assumed that the continental margin crust was dismembered and the fragments were dispersed and isolated by the masses of mobilized magmatic rocks (Scheuber et al. this Vol.). In general they did not contribute to magmagenesis as contaminants. Further support for this explanation comes from coincidence with geophysical data, as the investigations of Götze et al. (this Vol.) showed a system of maximum gravity anomalies in the residual field for the La Negra volcanics, granitoids, plus underlying heavy residual magmatites but contrasting minimum values for the isolated basement regions which cover only some dispersed patches in the Coastal Cordillera.

4.3 Andean Development From 130 to 80 Ma - Decrease of Magmatic Activity in the Coastal Cordilleran Area and Beginning of West to East Migration of the Magmatic Arc

Extensional conditions, due to oblique plate movements continued from 130 to 90 Ma although there were possible changes in the configuration of the continental margin, such as changes in subduction rate and angle of dip (probably due to the now rapid opening of the South Atlantic). The contemporaneous intrusive activity is represented by the G II granitoid intrusives which occur mainly in the Central Depression (Fig. 9c). This intrusive cycle marks the beginning of the eastward migration of the magmatic arc and the termination of the early Andean magmatic phase; it also provides the first signs of changed conditions for magma generation, as indicated by the rise in Sr_i ratios (ca. 0.7040). Thus magma generation may comprise various conditions that, on the one hand, may still include components of HIB composition but, on the other hand, may involve the increasing influence of contamination, as well as processes such as melting from an enriched mantle source or remelting of the youngest arc crust, blended and homogenized in a mantle-crust transition zone and probably enriched by upper crustal alkalic-silicic melts before reaching the final emplacement position.

Such processes were quoted in the MASH model of Hildreth and Moorbath (1988) and also in a model of successive time-related processes like that proposed by Rogers and Hawkesworth (1989) who distinguish a first trend of mobilization of old Proterozoic mantle lithosphere and increasing crustal contamination combined with subduction components under the eastwards migration of Mesozoic magmatism, from a second trend of younger magmatism with intracrustal melting and subduction components interchanging with the thickening crust.

Acknowledgements. Analytical investigations of major- and trace elements and isotope analyses were processed by the laboratories of the Institutes of Geology and Mineralogy of the Freie Universität Berlin under the direction of Prof. H. Friedrichsen. For the measurement and evaluation advice and constructive discussions I wish to thank Dr. K. Hammerschmidt. I also wish to thank Ms. M. Feth for taking care of all laboratory procedures. Pre-investigations were carried out at the Zentrallaboratorium für Geochronologie Münster (FRG) under the direction of Prof. B. Grauert. Thanks are due to A. Baumann and laboratory personnel for careful work and fruitful discussions. REE analyses were performed by the CRPG Nancy, France. To all the Chilean colleagues, especially G. Chong and T. Bogdanic from the Universidad Católica del Norte in Antofagasta, Chile, I wish to address special thanks for their contribution to my fieldwork.

References

Anderson DL (1983) Chemical composition of the mantle.- J Geophys Res 88 Supplement: B41-B52

Baeza L, Pichowiak S (1988a) Ancient crystalline basement provinces in the north Chilean Central Andes - Relics of continental crust development since the mid Proterozoic. In: Bahlburg H, Breitkreuz C, Giese P (eds): The southern Central Andes. Lecture Notes in Earth Sciences 17: Springer, Berlin Heidelberg New York, pp 3 - 24

Baeza L, Pichowiak S (1988b) Complejos plutonicos controlados por estructuras en la Precordillera del Norte de Chile: Geoquimica y geochronologia de Limon Verde y Catorce de Febrero. V Congreso Geologico Chileno Santiago Actas III/I: 91-107

Berg, K, Breikreuz C (1983) Mesozoische Plutone in der nordchilenischen Küstenkordillere: Petrogenese, Geochronologie, Geochemie und Geodynamik mantelbetonter Magmatite. Geotekt Forsch 66: 1-107

Breitkreuz C, Bahlburg H, Delakowitz B, Pichowiak S (1989) Paleozoic volcanic events in the Central Andes. J S Am Earth Sci 2(2): 171-189

Buchelt M, Tellez C (1988) The Jurassic La Negra Formation in the area of Antofagasta, north Chile (lithology, petrography, geochemistry) In: Bahlburg H, Breitkreuz C, Giese P (eds): The southern Central Andes. Lecture Notes in Earth Sciences 17: Springer, Berlin Heidelberg New York, pp 171-182

Damm KW, Pichowiak S (1981) Geodynamik und Magmengenese in der Küstenkordillere Nordchiles zwischen Taltal und Chañaral. Geotekt Forsch 61: 1-166

Damm KW, Pichowiak S, Todt W (1986) Geochemie, Petrologie und Geochronologie der Plutonite und des metamorphen Grundgebirges in Nordchile. Berl Geowiss Abh (A) 66: 73-146

Damm KW, Pichowiak S, Harmon RS, Todt W, Kelley S, Omarini R, Niemeyer H (1990) Pre-Mesozoic evolution of the Central Andes; The basement revisited. In: Mahlburg Kay S, Rapela CW (eds) Plutonism from Antarctica to Alaska. Geol Soc Am Spec Pap 241: 101-126

De La Roche H, Leterrier J, Grandclaude P, Marchal M (1980) A classification of volcanic and plutonic rocks using R_1R_2-diagram and major element analyses - Its relationships with current nomenclature. Chem Geol 29: 185-210

Diaz M, Cordani UG, Kawashita K, Baeza L, Venegas R, Hervé F, Munizaga F (1985) Preliminary radiometric ages from the Mejillones Peninsula, north Chile. Comunicaciones 35: 59-67

Farrar E, Clark AH, Haynes SJ, Quirt GS, Conn H, Zentilli M (1970) K-Ar evidence for the post-Paleozoic migration of granitic intrusion foci in the Andes of northern Chile. Earth Planet Sci Lett 10: 60-66

Fiechtner, L (1990) Geochemie und Geochronologie frühmesozoischer Tholeiite aus Zentral Marokko. Berl Geowiss Abh (A) 118: 76 pp

García F (1967) Geologia del Norte Grande de Chile. Simposio Geosinclinal Andino, Sociedad Geológica de Chile Publicaciones 3: 138 p

Glazner AF (1991) Plutonism, oblique subduction and continental growth: an example from the Mesozoic of California. Geology 19: 784-786

Govindaraju K, Mevelle G, Chouard C (1976) Automated optical emission spectrochemical bulk analysis of silicate rocks with microwave plasma excitation. Anal Chem 48: 1325-1331

Halpern M (1978) Geological significance of Rb-Sr isotopic data of northern Chile crystalline rocks of the Andean orogen between 23° and 27° S. Geol Soc Am Bull 89: 522-532

Harmon RS, Barreiro B, Moorbath J, Hoefs J Francis PW, Thorpe RS, Déruelle B, McHugh J, Viglino JA (1984) Regional O-, Sr- and Pb-isotope relationships in the late Cenozoic calc-alkaline lavas of the Andean Cordillera. J Geol Soc Lond 141: 803-822

Hawkesworth CJ, Hammill J, Gledhill AR, van Clasteren P, Rogers G (1982) Isotope and trace element evidence for late-stage intra-crustal melting in the High Andes. Earth Planet Sci Lett 58: 240-254

Hervé M, Marinovic N (1989) Geocronologia y evolucion del batolito Vicuña Mackena, Cordillera de la Costa, sur de Antofagasta. Rev Geol Chile 16 (1): 31-49

Hervé M, Marinovic N, Mpodozis C, Perez DE, Arce C (1985) Geocronologia K-Ar de la Cordillera de la Costa entre los 24° y 25° latitud sur. Antecedentes prelimnares. IV Congr Geol Chil Antofagasta, Resumenes 4.19: 158

Hildreth W, Moorbath S (1988) Crustal contributions to arc magmatism in the Andes of central Chile. Contrib Mineral Petrol 98: 455-489

Lucassen F (1991) Geologie, Metamorphosegeschichte und Geochemie neugebildeter basischer Kruste im jurassischen magmatischen Bogen der Küstenkordillere Nordchiles Region Antofagasta 23°25'-24°20'S. PhD Thesis, Berlin, 128 pp (unpubl)

Mahlburg Kay S, Ramos AV, Mpodozis C, Sruoga P (1989) Late Paleozoic to Jurassic silicic magmatism at the Gondwana margin: analogy to the middle Proterozoic in North America?- Geology 17: 324-328

Maksaev V, Boric R, Zentilli M, Reynolds PH (1988a) Metallogenetic implications of K-Ar, ^{40}Ar-^{39}Ar, and fission track dates of mineralized areas in the Andes of northern Chile V Congr Geol Chil Actas 1: B65-B86

McNutt R, Crocket JH, Clark AH, Caelles JC, Farrar E, Haynes S, Zentilli M (1975) Initial $^{87}Sr/^{86}Sr$ ratios of plutonic and volcanic rocks of the Central Andes between latidudes 26° and 29°S. Earth Planet Sci Lett 27: 305-333

Pichowiak S, Buchelt M, Damm KW (1990) Magmatic activity and tectonic setting of the early stages of the Andean cycle in North Chile. In: Mahlburg Kay S, Rapela CW (eds) Plutonism from Antarctica to Alaska. Geol Soc Am Spec Pap 241: 127-144

Pitcher WS, Atherton MP, Cobbing EJ, Beckinsale RD (1985) Magmatism at a plate edge - The Peruvian Andes. Wiley, New York, 328 pp

Ramos AV, Jordan TE, Allmendinger RW, Mpodozis C, Kay SM, Cortes JM, Palma M (1986) Paleozoic terranes of the central Argentine-Chilean Andes. Tectonics 5: 855-880

Reutter. KJ, Giese. P, Götze HJ, Scheuber E, Schwab K, Schwarz G, Wigger P (1988) Structures and crustal development of the Central Andes between 21° and 25°S. In: Bahlburg H;,Breitkreuz C, Giese P (eds): The southern Central Andes. Lecture Notes in Earth Sciences 17: Springer, Berlin Heidelberg New York, pp 231-258

Rogers, G (1985) Geochemical traverse across the north Chilean Andes. Thesis, Department of Earth Sciences, Open University, Milton-Keynes, 333 pp

Rogers G, Hawkesworth CJ (1989) A geochemical traverse across the north Chilean Andes: Evidence for crust generation from the mantle wedge. Earth Planet Sci Lett 91: 271-285

Rössling R (1989) Petrologie in einem tiefen Krustenstockwerk des jurassischen magmatischen Bogens in der nordchilenischen Küstenkordillere südlich von Antofagasta. Berl Geowiss Abh A 112: 73 pp

Scheuber E (1987) Geologie der nordchilenischen Küstenkordillere zwischen 24°30' und 25°S - unter Berücksichtigung duktiler Scherzonen im Bereich des Atacama-Störungssystems. Thesis, Institut für Geologie, Freie Universität Berlin, 170 pp

Scheuber E, Andriessen PAM (1990) The kinematic and geodynamic significance of the Atacama Fault Zone, northern Chile. J Struct Geol 12: 243-257

Streckeisen A (1975) To each plutonic rock its proper name. Earth Sci Rev 12: 1-33

Thompson RN, Morrison MA, Dickin AP, Hendry GL (1983) Continental flood basalts ... arachnids rule OK? In: Hawkesworth CJ, Norry MJ (eds) Continental Basalts and Mantle Xenoliths. Shiva, Nantwich, pp 158-185

Sediment Accumulation and Subsidence History in the Mesozoic Marginal Basin of Northern Chile

PETER PRINZ, HANS-G. WILKE and AXEL von HILLEBRANDT

Abstract. In connection with the break up of Pangaea, a marine cycle started with the ingression of the late Triassic sea in Colombia, Peru and Chile. To describe the basin evolution, we have produced computer isopach maps of the marine sediments for the stages from Norian to Kimmeridgian. The basin was essentially formed during the Sinemurian, i.e. at the same time as the volcanic activity (La Negra Formation) of the magmatic arc in the Coastal Cordillera began. The marine sediment accumulation and subsidence rates in this active backarc basin did not exceed those typical of passive continental margins, and the basin subsidence was primarily produced by sedimentary loading of the underlying crust.

1 Introduction

In order to compare the thicknesses of the marine sediments of the Mesozoic marginal sea of northern Chile, as many sections as possible were examined. In this chapter the area studied extends between 20° and 29°S in northern Chile. Most data were collected by personal field observations, but an additional contribution was provided by Quinzio (1987); also, other literature was reviewed and improved where necessary. The next step was to change the thickness of each Jurassic stage, independent of lithology and facies, directly into sedimentation rates. Compaction, erosion and interruption of sedimentation, as well as resedimentation, have not been included in this simplified model. When biostratigraphic data were insufficient to subdivide the complete pile into several sequences, the total thickness was distributed proportionally over the time range of each stage.

Computer isopach maps were produced with the kind help of Dr. S. Schmidt and Prof. Dr. H.-J. Götze (FU Berlin) (Figs. 3-8) using a large quantity of reliable data (Table 1) collected from the N-S trending Precordillera and from its eastern margin (Fig. 1). From the adjacent area to the west, i.e. the Longitudinal Valley, data are scarce. The isolines are influenced by additional zero points measured on the eastern border, and partly estimated for the western border. Therefore, calculated isopachs are never accurate, and the shape and extension of depocentres must be regarded as a first attempt to quantify subsidence processes. A N-S orientated longitudinal section (Fig. 2) encompasses the best density of data and is therefore the most accurate, and the total amount of sediments deposited up to the Kimmeridgian in the Precordillera provide sufficient data for a discussion of the geodynamic process.

The thickness of a sediment sequence can be influenced by several factors, some of which include: Sedimentary input, subsidence or uplift, sea level changes and compaction; these have all been taken into account. The total amount of strata deposited during marine sedimentation is a good approximation of the subsidence throughout this period. The determination of subsidence at a particular stage would provide us with a highly sensitive indicator for crustal movements.

The distribution of localities of high subsidence is not uniform, despite the fact that rates of sedimentation coincide with those of a quiet epicontinental sea. Furthermore, the thickness is determined by the duration of the stage. The time scale and eustatic curve are from Posamentier and Vail (1988). With the exception of volcanic rocks, including tuffic material, the sedimentary input depends on tectonic, climatic and morphological processes in the hinterland and is thus determined by corresponding sediment transport e.g. by rivers into the basin as well as by chemical and biological processes within the basin itself. Sedimentary input and even a eustatic sea-level rise result in basin subsidence caused by loading (Ziegler 1982).

Correspondence to: A. v. Hillebrandt Technische Universität Berlin, Straße des 17. Juni 135, D-1000 Berlin 12

Table 1. Measured thicknesses (m)

	Trias	Hett.	Sine.	Plien.	Toarc.	Aal.	Baj.	Bath.	Call.	Oxf.	Kimm.
1 Q. Copaquiri E								—	150	25	
2 Q. Copaquiri W										300	
3 Cº Challacollo										300	
4 Q. Pinchal		100	350	350	100	—	—	—	—	—	
5 Adamito			50								
6 Cº. de la Mica			10								
7 Q. Arcas					100	20	—	—	—	—	
8 Q. Quinchamale										50	
9 Camino			30								
10 SW' Quillagua									30	70	
11 W' Cº Jaspe							85	35	145	50	—
12 SW' Cº Jaspe							25	70	100	50	94
13 N' Q. Chug-Chug		20	—	—	—	20	—	—	140	150	
14 Sª. San Lorenzo										—	100
15 Cºˢ. San Lorenzo			80	220	90	20	50	25	40	40	40
16 Moctezuma			40	40	40						
17 Cts. Bayos			75	120	100	25	30	30	15	150	125
18 Caracoles							100	50	65	75	60
19 Rencoret / Cº Difícil		10	30				—	—	—	—	
20 Q.S. Pedro W											47
21 Marmol Selco								49	115	70	
22 Q.S. Pedro								64	155	96	125
23 Q.S. Pedro S										180	
24 Q. Primorosa									140	35	
25 S' Cºˢ de Cuevitas	20	20	70								
26 Cº Amarillo S	50	10	50	100	60	—	—	70	100	270	30
27 Pzo. Azabache						20	120	65	212	200	80
28 A. Victoria										100	30
29 NE' Cº. Pascua	5			70	30			30	—		
30 Mina Magellanes	—	—	—	70				70	120	85	
31 Cº. Pascua									80	75	
32 NE' Cº. del Arbol	7	10									
33 SE' Cº Sombrero Chino											60
34 Imilac									20		
35 W' Cº. Bayo	230	—	—								
36 N' Cº. Chinchilla		10									
37 W' Cº Campamento		100	300	180	180	50	200	200	285	225	100
38 Ag. El Oro	10						130	105	65	175	
39 Sª. Argomedo	150	80	150					40	60		
40 Ag. de Varas		10	35								
41 Q. Granate	60					30					
42 Cº Islote										20	—
43 Cº. Yumbes		50	130								
44 E' Ptzo. de la Sal	140	—							—	—	90
45 Q. Profeta E							130	130	275	—	

	Trias	Hett.	Sine.	Plien.	Toarc.	Aal.	Baj.	Bath.	Call.	Oxf.	Kimm.
46 Q. Profeta W							120	175	225	300	—
47 Ag. El Minero	40	45	80	160	180	45					
48 N' Q. Bonita				140							
49 Q. Profeta										—	50
50 Ag. Vizcachas		20	100	100	175						
51 Q. Las Mulas		15	200	150	100	—	225				
52 Q. Punta del Viento	35										
53 Sª. Candeleros								170	350	280	40
54 Cºˢ. Plomo del Corral de Alambre		20	300	350							
55 Q. Chaco Sur	5	8	60	100	100						
56 Sª. Cenizas			400	300							
57 Ag. del Carretón			400							400	
58 Q. del Medio								90	110	300	
59 Q. Incaguasi		18	200	20	75	—	100	125	125		
60 Cº. Buena Esperanza							80				
61 P. Los Hidalgos		35	210								
62 E' Q. de la Encantada		40	40	180	390	—	190				
63 Q. de Los Burros		140	140	170	360	—	230	240	480	580	
64 Doña Ines Chica	20	45	210	65							
65 Q. Pan de Azúcar		80	100								
66 Q. El Asiento				100	60	—	5	5	150		
67 Q. Caballo Muerto				20			10		110		
68 Sª. Fraga							250				
69 Q. El Peñon				30	70	20			70		
70 Q. Paipote				100	140	50			80		
71 Q. El Bolito			10	40	15	15	25		85		
72 Q. El Paton				45	40	5	10				
73 Yerbas Buenas				80	100						
74 Q. Vaca Muerta/Potrerillos			70	110	70	15					
75 Q. Larga			80	80	120	25					
76 Q. Noria			470	120							
77 Q. Llareta					60						
78 Rio Figueroa				45	30	10	10		70		
79 La Guardia				70	60	7					
80 M. del Carrizo				160	60	50					
81 Vegas de Chañar				200					25		
82 Q. Calquis				80	120						
83 Manflas			10	110	260	20			15		
84 Río Manflas							140				
85 Q. La Totora			160	110	90	10	5				
86 Q. Acevedo			260	130	100	10	25				
87 Q. Chanchoquin				35	35						
88 Q. La Plaza			80	80	50	—					
89 Q. La Plata			12	13	15	20	20				
90 Q. Pinte			220	110	20	10	70				

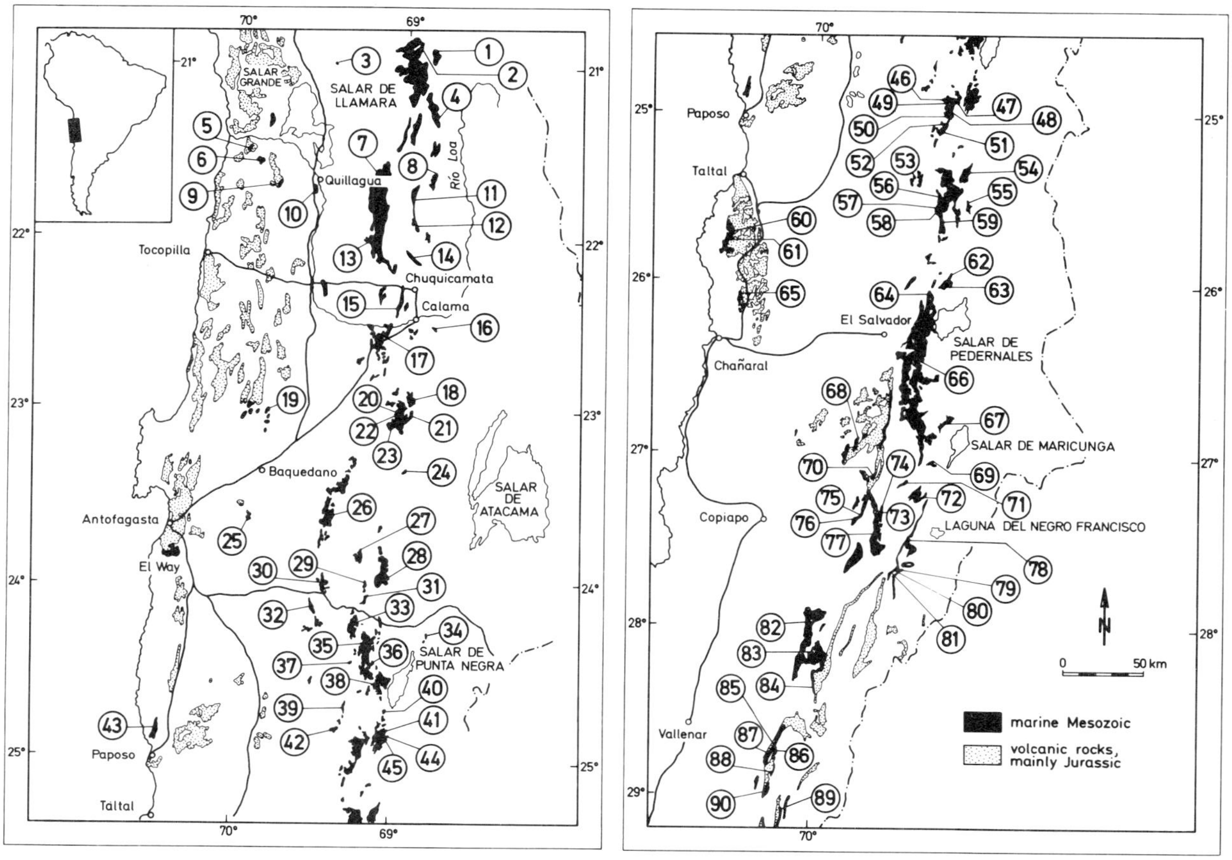

Fig. 1. Map of marine Mesozoic outcrops (*black*) in northern Chile. *1* to *90* Sediment thickness data derived from measured sections (see also Table 1)

2 Palaeogeography and Sedimentation in the Mesozoic Marginal Sea

Stratigraphy and palaeogeography of the marine Mesozoic sedimentation area in northern Chile between 21° and 26°S were described and summarized by von Hillebrandt et al. (1986) and Gröschke et al. (1988).

In this area, marine late Triassic sedimentation starts in the Norian, at the same time as in Colombia, Peru (Geyer 1980) and central Chile (Corvalan 1982). As this transgression contradicts Posamentier's and Vail's eustatic curve, plate tectonics are considereed to be responsible for the event. The late Triassic sediments in northern Chile overlie unconformably volcanic rocks and terrestrial clastic rocks.

They crop out in the Precordillera and easternmost Coastal Cordillera (i.e. easternmost outcrops of Cerros de Cuevitas) [25; see Fig. 1] in the north and in the southern Cordillera Domeyko in the south. The easternmost outcrops occur at the western border of Salar de Punta Negra (Chong and Hillebrandt 1985). The extension of the late Triassic marginal basin and its connection with the ocean to the west are not known. The late Triassic sea has not reached locations where early Jurassic trangressively overlies terrestrial Triassic or older basement, i.e. the western part of Sierra de Moreno [4, 13], the western outcrops of Cerros de Cuevitas [25] and the Coastal Cordillera between Paposo and Chañaral [43, 61, 65]. The late Triassic sediments indicate mainly shallow marine facies as is shown by frequent reef-building corals (Chong and Hillebrandt 1985, Prinz 1991), however, a pelagic influence with few benthic fauna was observed mainly in the western part of the Triassic belt and towards the top of the sequence and this implies a connection with the ocean south of 26°S.

With the exception of the easternmost part of Cerros de Cuevitas [25], the marine Triassic is uncertain in the Coastal Cordillera (Scheuber et al. 1986). Marine Mesozoic sediments in the Coastal Cordillera are only Hettangian, Sinemurian (part), exceptionally Bajocian (intercalated in the volcanic La Negra Formation) and early Cretaceous, so that together with the mainly magmatic rock composition we assume that a peninsula or island arc chain existed. We interpret the late Triassic transgression was facilitated by a graben-like structure in the region of the Precordillera. Volcanic activity occurred as early as the beginning of the Jurassic (Hettangian, Sinemurian) in the west of the present Coastal Cordillera (Quinzio 1987).

During the Jurassic the marginal sea expanded in a north- and eastward direction. North of 26°S, the centre of the basin lay in the western part of the Precordillera, and follows the N-S trend of the present-day Andes. The current Coastal Cordillera formed a barrier, probably almost completely closed during the late Triassic but during the Hettangian was transgressed by the sea between Chañaral and Paposo (Quinzio 1987), most of Cerros de Cuevitas [25], Rencoret [19] and west of Quillagua [6]. Abundant ammonites and radiolarians confirm a pelagic influence and good connection with the ocean.

In the Coastal Cordillera intense volcanic activity (La Negra Formation) started in the late Sinemurian. The volcanic rocks, partly submarine and mostly subaerial, were deposited on a preexisting high to the west of the continent. Numerous gabbroic intrusions of the Coastal Cordillera and of the peninsula of Mejillones are of early Jurassic age (Rogers 1985; Damm et al. 1986; Hervé and Marinovic 1989). In the eastern part of the basin, especially north of 24°S, sediments are transgressive to the east, up to the Callovian (von Hillebrandt et al. 1986). In the area between Salar de Punta Negra and 26°S the sea transgressed more to the east and the facies also suggests a continuous transgression in this area.

Starting with the transgressive Mesozoic and continuing partly until the Kimmeridgian the backarc-basin was exclusively marine but connection modalities are not well known. In the area investigated the Jurassic marginal sea and the Panthalassa must have been separated for most of the time by a barrier or by islands of a magmatic arc (e.g. in the Bajocian). Further to the north between Iquique and Arica, the magmatic arc is interrupted by a basin filled with mostly volcanoclastic sediments and in this area there must have existed a better connection with the ocean.

This hypothesis conjures up the image of a structure similar to that of the Californian Gulf with

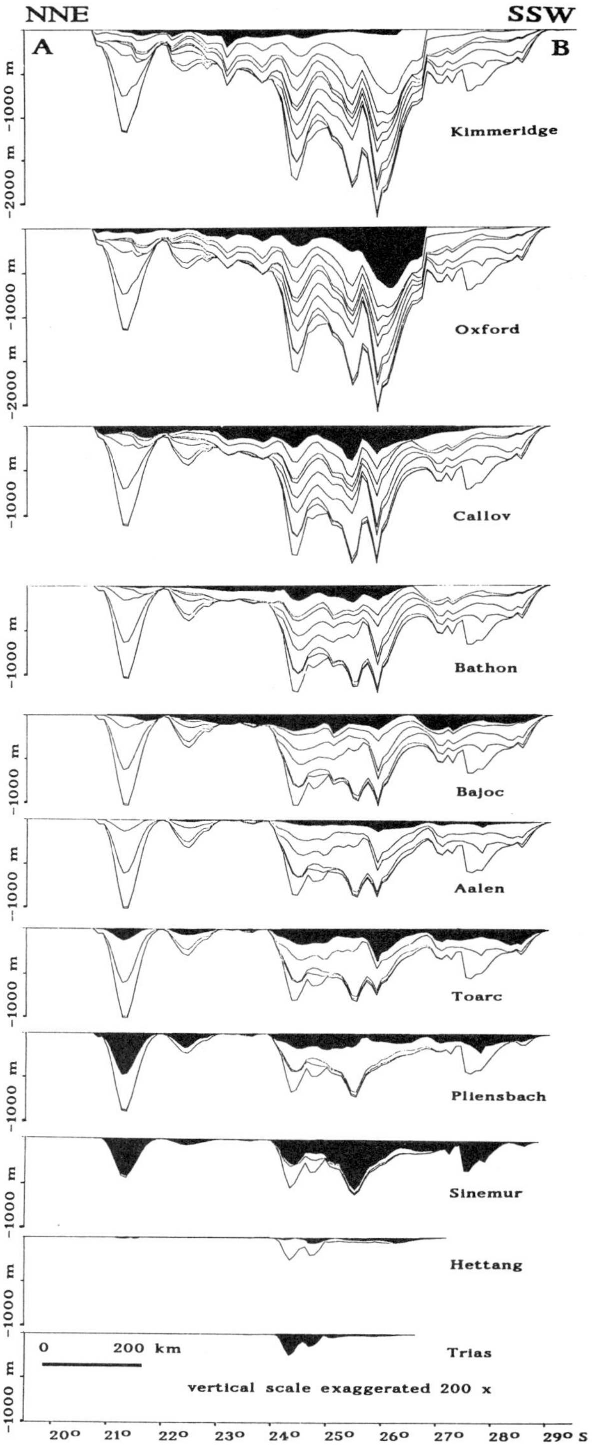

Fig. 2. Evolution of sediment thicknesses in a NNE to SSW longitudinal section (A and B in fig. 7)

an opening towards the north, but with a peninsula consisting of volcanoes (Riccardi 1983). Further support for this assumption is provided by sedimentary developments during the late middle Jurassic and late Jurassic, which is incomplete in the Precordillera south of 26°30'S, while a facies with calcareous concretions predominates in the centre of the northern part and indicates deeper water. The separation of the marginal sea from the ocean, caused by increasing magmatic/tectonic activity or by eustatic sea-level fall, culminated for a short time during the late Oxfordian, when widespread evaporation and gypsum sedimentation occurred. Compared to older Jurassic deposits, Kimmeridgian ammonite-bearing sediments are rare and incomplete. They are found only in the Precordillera between 22°30' and 25°30'S. Towards the end of the Jurassic and during the early Cretaceous, a gradual regression of the marginal sea away from northern Chile can be observed. Some hundreds of metres of partially marine sediments (south of 23°S) and some thousands of metres of mainly deltaic to fluviatile deposits (north of 23°S) in the western Precordillera, of late Jurassic to early Cretaceous age, indicate continuing subsidence of the basin (Bogdanic 1990). Outcrops of marine Cretaceous sediments are rare in the Precordillera. In the Coastal Cordillera south of Antofagasta, near El Way, shallow marine carbonates were deposited in the Hauterivian and Barremian. This area probably bordered on a highland in the west (Rössling 1989). An early Cretaceous basin with high sediment thicknesses developed between Copiapó and Vallenar (Jurgan 1977). The Cretaceous system was not considered in the evaluation of the isopach maps.

3 Depocentres and Sedimentation Rates

Late Triassic marine sediments (Fig. 3) transgressively overlie the terrestrial volcanics of Permo-Triassic age. As the Triassic transgression is not well dated, it is not possible to calculate subsidence rates from the isopachs. This isopach map shows various areas with greater thicknesses in the centre of the Triassic marginal sea, whose position corresponds to the area of the posterior higher subsidence. In the northern and southern parts the thickness is not great.

In the Hettangian, the sea transgresses mainly towards the north and over the Coastal Cordillera between 21°45' and 26°15'S; however, the configuration of the isopachous lines (Fig. 2) remains basically the same as they are for the Triassic. North and south Quebrada Pinchal an area of greater thickness sets in, and this is of increasing significance at a later stage.

The Sinemurian is a stage of high sediment accumulation in some areas, partly with turbidites. The sediments are deposited in well defined, en-echelon arranged centres (Fig. 4). In between 22°, 23°15' and 26°30'S, the thickness is reduced, a fact that applies to all the following Jurassic stages. This arrangement does not necessarily indicate basin or uplift areas. Nevertheless, an area southeast of Taltal is remarkable for its low thickness throughout the Jurassic.

In the early Pliensbachian (Fig. 4), peripheral marine areas of a lower thickness than those in the area west of Calama can be observed in the southeast (von Hillebrandt 1973) and in the northeast in Moctezuma [16]. Along the N-S axis thicknesses are fairly regular, with the exception of the area east of Taltal and areas that remained unaffected by sedimentation.

In the Toarcian the distribution of thicknesses (Fig. 5) is quite similar to that of the Sinemurian. The depocentre of Taltal moves to the south in the Quebrada de Los Burros area [63], where it subsequently gains further significance.

The Aalenian is characterized by very small thicknesses throughout the area (Fig. 5) even when the short duration of this stage is taken into account. Very great isochronous differences in facies (e.g. transgressive clastic sediments in Portezuelo Azabache [27] and bituminous calm-water sediments in Cerros San Lorenzo [15]) exist, and no important diachronous changes can be observed, apart from the areas of transgression. The low thickness must therefore be due to low subsidence and mainly to low sedimentary input. The location of the depocentres remains the same as during the Toarcian.

In the Bajocian (Fig. 6) thickness amounts are similar to those before the Aalenian. The sea transgressed at the area of Cerro Jaspe [11, 12] and Caracoles-Quebrada San Pedro [18, 20]. In the Coastal Cordillera north and south of Iquique (20°20'S), a basin with marine, mainly volcaniclastic sediments arose and between Taltal and Chañarel at different localities marine sediments are intercalated into the volcanic La Negra-Formation.

The thicknesses of the Callovian sediments vary widely over short distances (Fig. 7). In the regions of Manflas [83,84], Imilac [34] with shallow water limestones and Cerritos Bayos [17] with calcareous concretions, the sediments are a mere 15-20 m thick, whereas in the Quebrada de Los Burros [63] they are up to 450 m thick. The distribution suggests a N - S

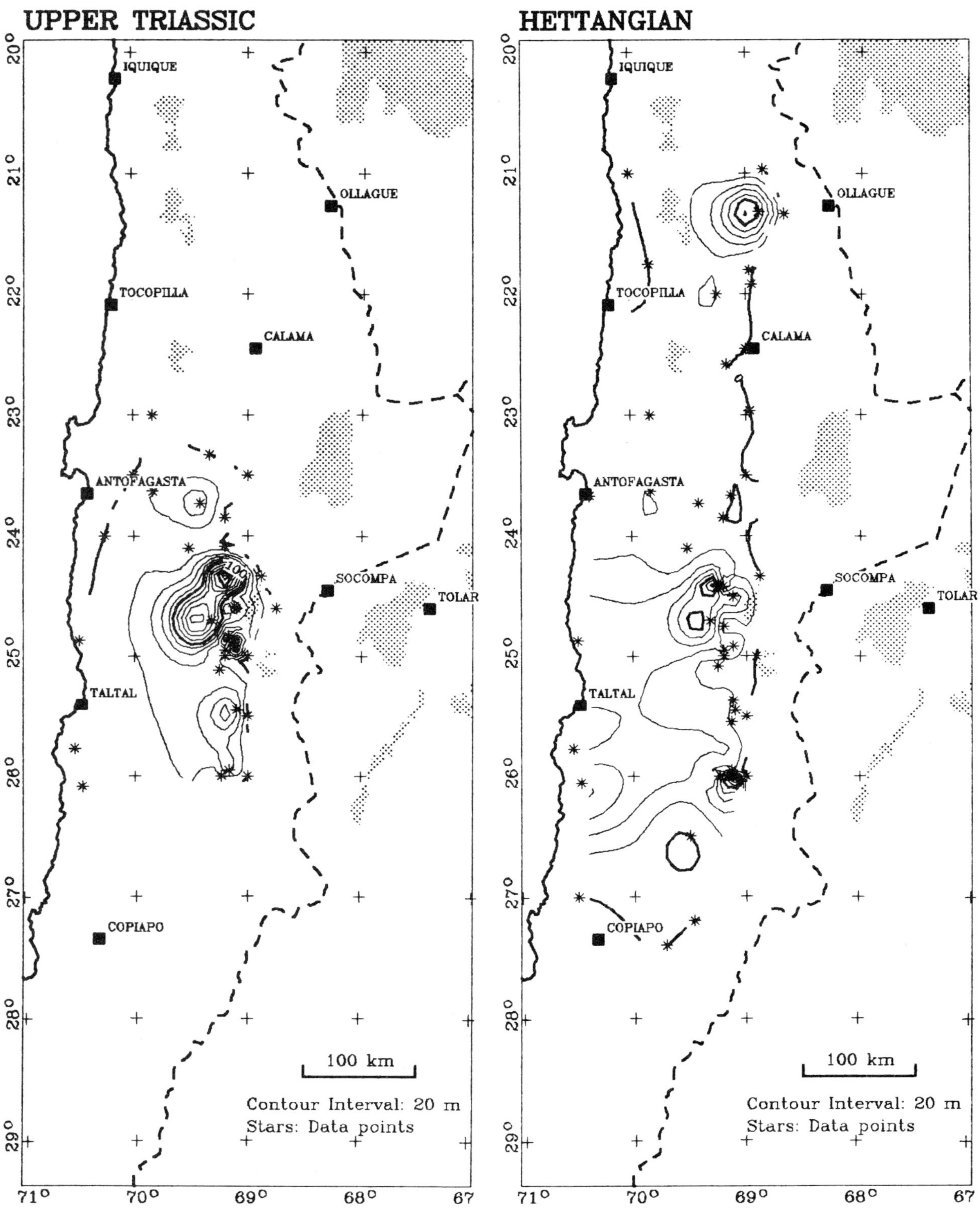

Fig. 3. Isopach maps of the Norian/Rhaetian and the Hettangian stages

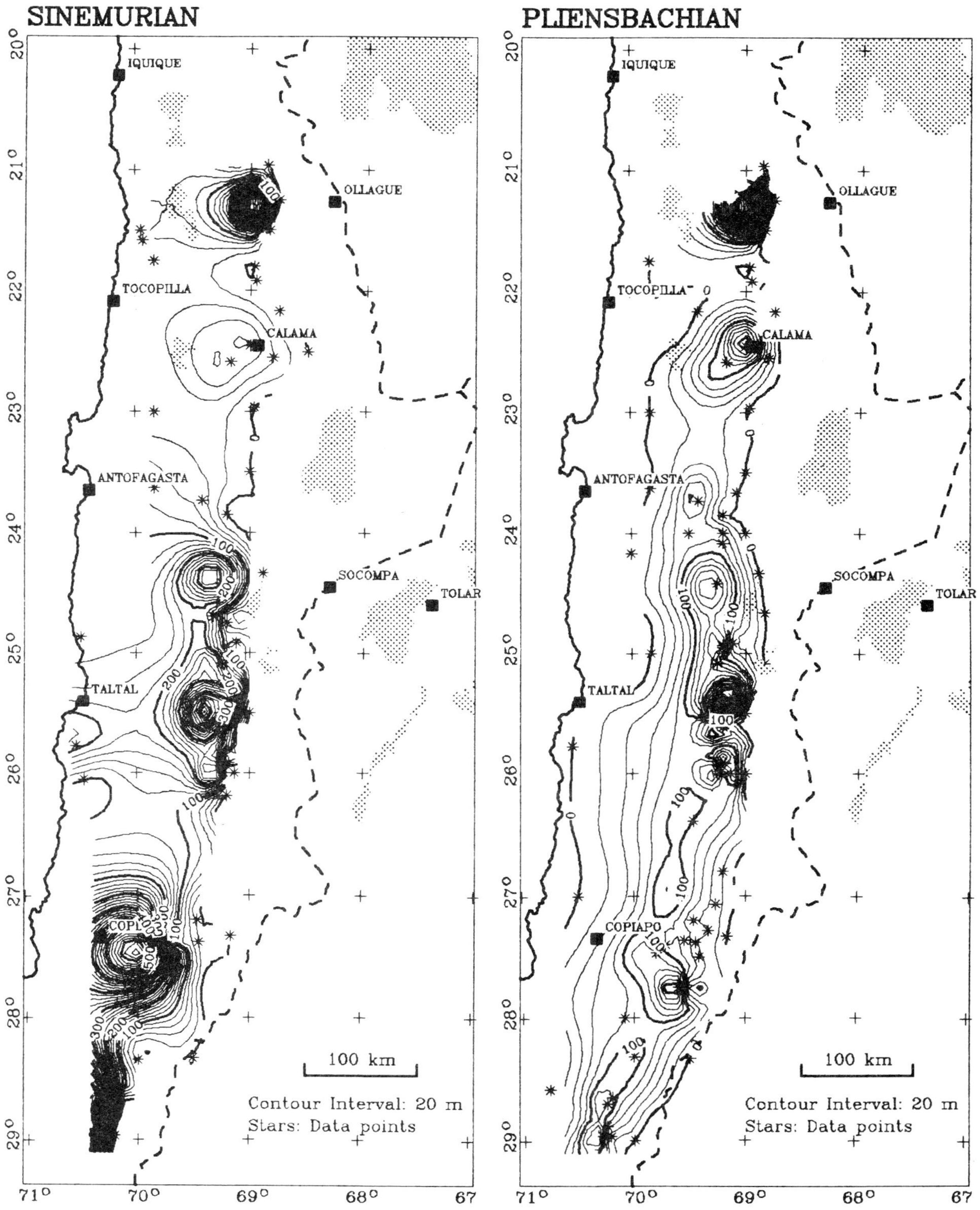

Fig. 4. Isopach maps of the Sinemurian and the Pliensbachian stages

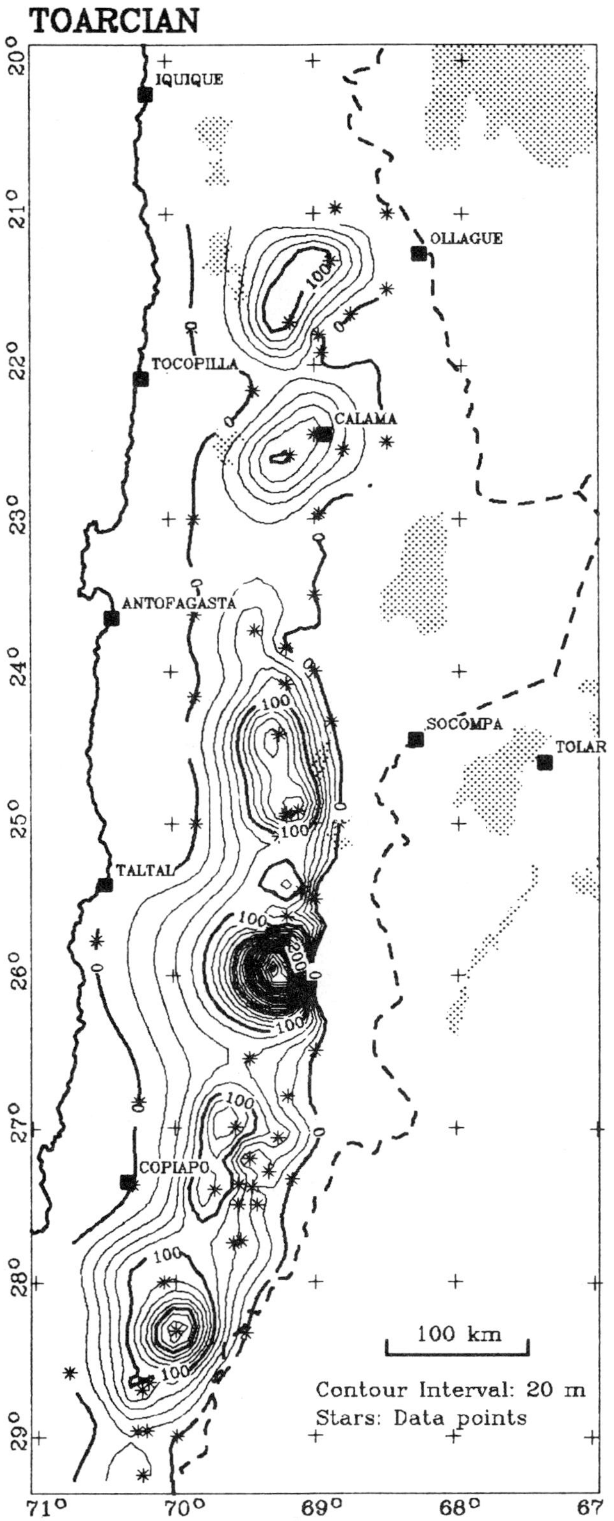

Fig. 5. Isopach maps of the Toarcian and the Aalenian stages

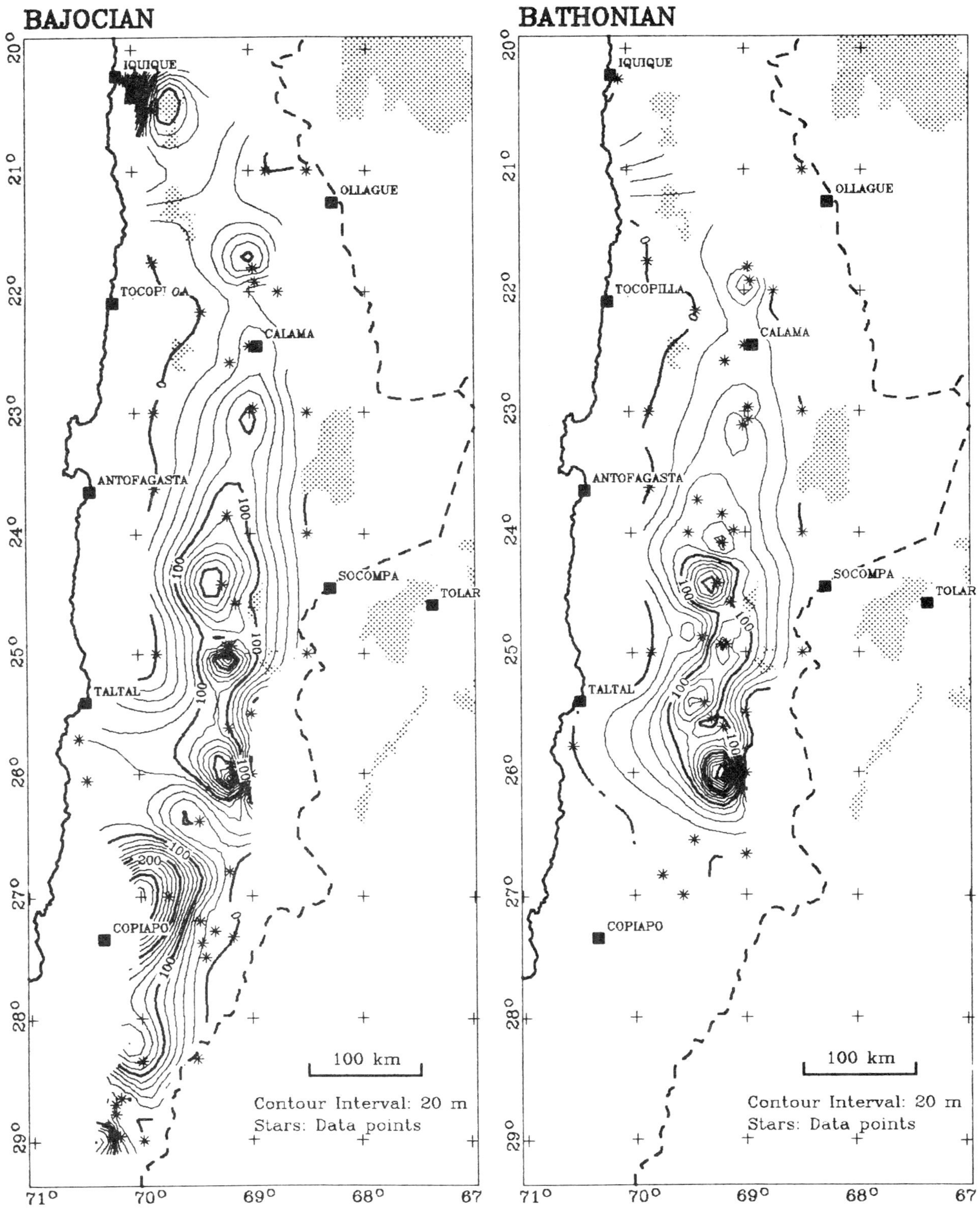

Fig. 6. Isopach maps of the Bajocian and the Bathonian stages

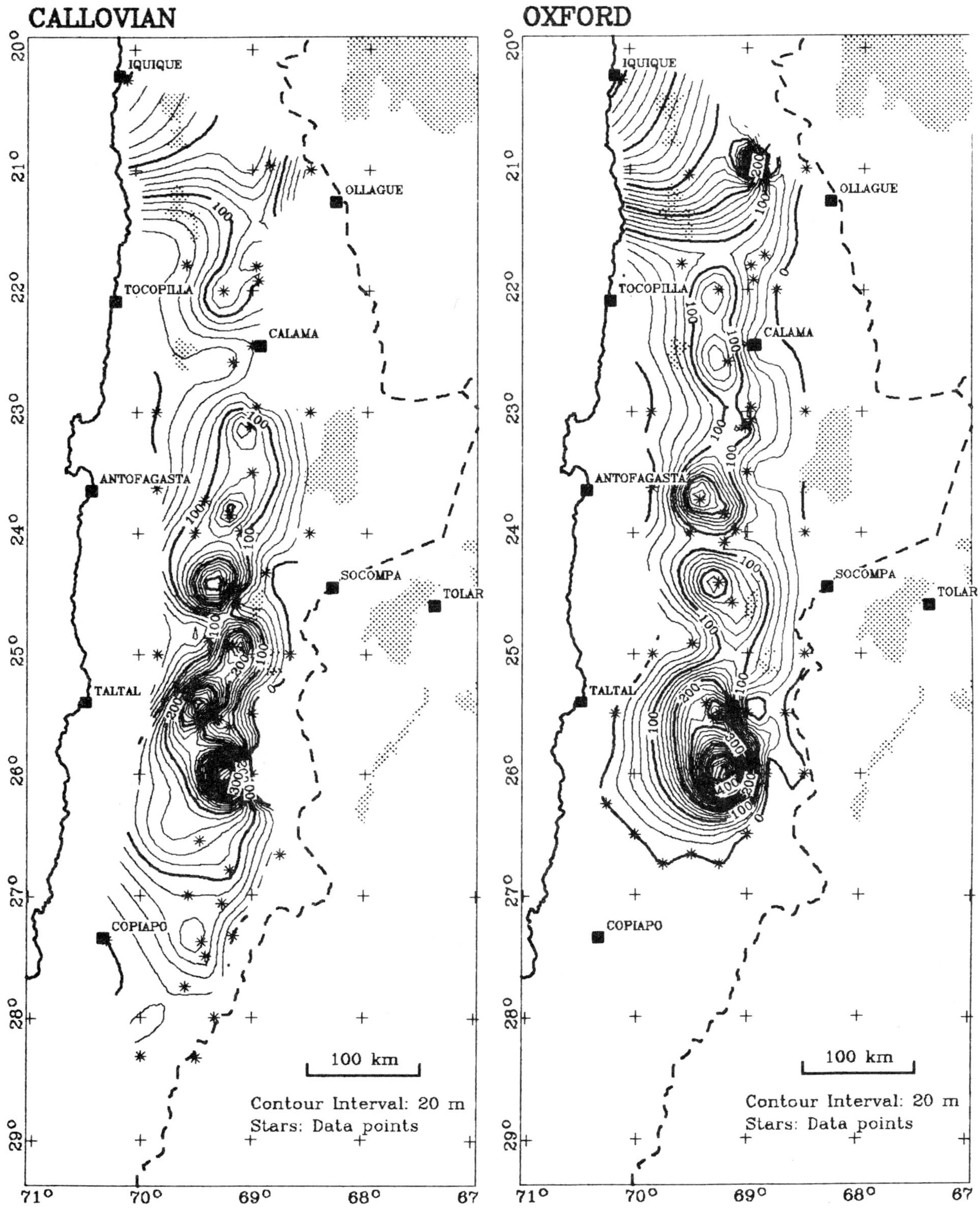

Fig. 7. Isopach maps of the Callovian and the Oxfordian stages

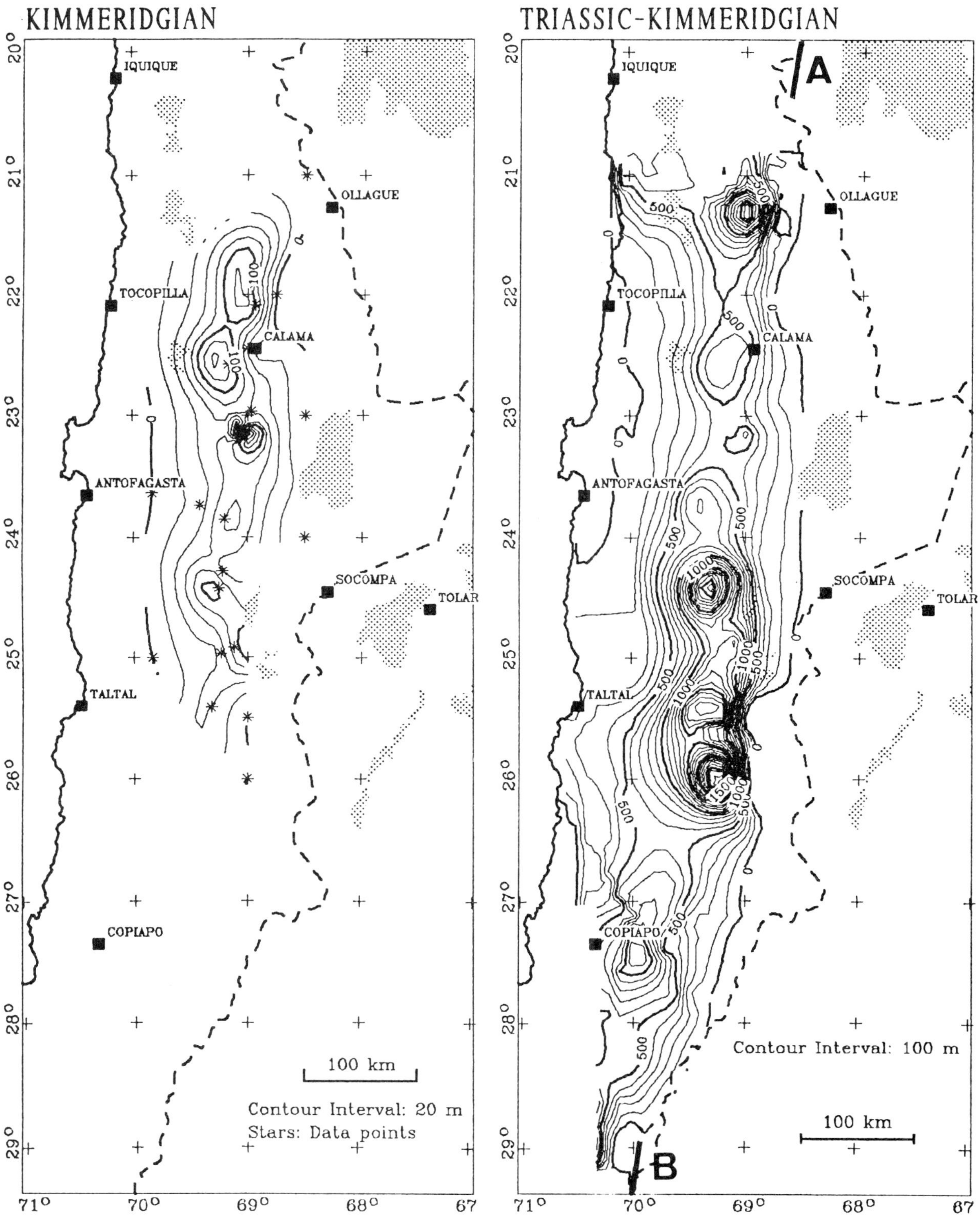

Fig. 8. Isopach maps of the Kimmeridgian stage and the total sediment thickness from the Upper Triassic to the Kimmeridgian

oriented axis with larger sediment accumulation, whereas in the profiles to the west and east lower thicknesses occur. The profiles in the Cerros San Lorenzo [15]-Cerritos Bayos [17] area with their low thicknesses are notable; they were also prominent in the Bajocian and Bathonian. In the region east of Copiapó, there is a recurrence of sediments, but most of them are remarkably thin. Along the axis Antofagasta-Socompa, an area of low thickness was formed and can be observed up to the Kimmeridgian. In the Oxfordian, the differences in thicknesses of sediments are enormous and generally there is a high sedimentation rate (Fig. 7). The area west of Calama is characterized by low thicknesses. The greatest thickness in the Oxfordian is 580 m in the Quebrada de Los Burros [63] area, and this is the same as in the Callovian. South of 26°30'S, sediments of the Oxfordian are not known; this is a repeat of what happened in the Bathonian. Again, the areas north of 21°S and at Iquique gain importance.

Marine sediments of the Kimmeridgian are of relatively small thickness and they have been confirmed only in the Precordillera between 22° and 25°30'S (Fig. 8). With the exception of the Quebrada San Pedro [22] area, where a Callovian depocentre is reactivated, the structure of the isopachs is the same as in the Oxfordian.

4 Interpretation

Figure 8 is an accumulated isopach map of the late Triassic and the Jurassic marine sediments. In this map, extended areas with important sediment accumulation are interrupted by areas with lower thicknesses. In the Quebrada Pinchal area [4] at 21°15'S an important depocentre is situated at the eastern margin of the basin. To the south, up to 24°S there follows an area of low thickness that is structured by minor variations. The southern Cordillera Domeyko between 24°15' and 26°30'S is a striking area which has the highest accumulation rates. Within this area, three different depocentres can be observed. Between 26°30' and 27°S an area with lower thickness again occurs and further to the south there is another depocentre. These are separated by a NW-SE striking lineament of at least Cenozoic age (Salfity 1985). Altogether, this map (Fig. 8) supplies a definitive picture of the Mesozoic marine basin at the end of the Kimmeridgian.

The formation of the areas of different thickness was not a continuous process, as is shown in a N-S oriented longitudinal section through the area of highest data density (Fig. 2). The following course of events can be observed: As early as during the Triassic a sedimentation area developed in the Precordillera at 24°30'S and this continuously received sediments of relatively high thickness throughout all stages. The Hettangian has no influence on the sedimentation area, whereas in the Sinemurian several structures at 21°15', 22°20', 25°20'and 27°30'S were initiated. The area of highest thickness at Quebrada de los Burros (25°50'S) was formed during the Toarcian. The middle Jurassic is a phase of continuous and uniformly distributed sedimentation that is controlled more by eustasy than by structural changes in the basin. The lack of Bathonian and Oxfordian sediments south of 26°30'S should probably be viewed in connection with a new NW-SE striking structure which is part of a greater tectonic lineament. Finally during the Kimmeridgian, a smaller maximum occurred in the Quebrada San Pedro section [22] at 23°08'S, which had already been insinuated during the Callovian. Greater tectonic lineaments, as depicted by Salfity (1985) for the Central Andes, are connected with arrangements of the isopachs within the total isopach map. Our observations lead to the conclusion that these lineaments had been active at least since the Jurassic or partly since the Triassic.

The Mesozoic marginal sea was initiated as early as during the late Triassic. With the beginning of the subduction-related volcanism in the lower part of the late Sinemurian, the course of developments in the Coastal Cordillera changed completely, whereas in the Precordillera north of 26°S, this important event is not clearly reflected in the sediments. Further to the south, east of Copiapó, the influence of volcanism is documented during sedimentation (von Hillebrandt 1973). The formation of depocentres during the Sinemurian, i.e. at the same time as the beginning of Jurassic volcanism, suggests relations between the arc-volcanism and backarc basin. The zone of highest volcanic activity remains in the Coastal Cordillera and ranges from the Sinemurian to the Oxfordian. The strip of highest observable sediment accumulation throughout the Jurassic in the Precordillera is located in the northern part at 69°W and in the southern part at 70°W. In the Bajocian, marine shallow water sediments with intensive volcanic influence are intercalated in the volcanic series of the Coastal Cordillera between Taltal and Chañaral. This means that the magmatic arc was not always a closed barrier. Volcanic influence within the backarc basin is small and is limited to the middle Jurassic specially in the western part of the basin where also clasts of volcanic rocks and corresponding heavy minerals support the volcanic activity of the Coastal Cordillera. After Pichowiak et al. (1990) the area studied since the Triassic was part of an area with ductile shear zones and left lateral strike-slip faults which originated as the result of a collision between the continental crust and the ocean crust drifting from northwest to southeast. A strike-slip system of this type could have led to the formation of pull-apart basins (Rogers 1980; Aydin and Nur 1982). The formation of the depocentres cannot have involved the Atacama Fault System because it is situated too far to the west. If these en-echelon arranged depocentres of Jurassic age really represent pull-apart basins, then another N-S striking fault system further to the east must be postulated.

It has been shown that the Jurassic marginal sea contained areas of higher and lower subsidence. The relatively uniform facies of the sediments and the continuously neritic evolution corresponds to global sea-level curves and does not suggest important tectonic events within the backarc basin during this time span. The relation between the evolution of the volcanic arc and the backarc basin is less evident than we previously believed. There is a great similarity between the isopach map of the Sinemurian and that of the whole Mesozoic. The initial basin forming process was related to the onset of eastward subduction, as evidenced by the initiation of arc magmatism. The en-echelon arrangement of the depocentres may indicate pull-apart basins. However, marine sediment accumulation and subsidence rates in this active margin basin did not exceed those typical of passive continental margins, and subsidence was primarily produced by sedimentary loading of the underlying crust.

References

Auboin J, Borello AV, Cecioni G, Charrier R, Chotin P, Frutos J, Thiele R, Vicente J (1973) Esquisse paléogéographique et structurale des Andes Méridionales. Rev Géogr phys Géol dyn 15: 11-72

Aydin A, Nur A (1982) Evolution of pull-apart basins and their scale independence. Tectonics 1: 91-195

Bogdanic T (1990) Kontinentale Sedimentation der Kreide und des Alttertiärs im Umfeld des subduktionsbedingten Magmatismus in der chilenischen Präkordillere (21°-23°S). - Berl geowiss Abh (A) 123: 1-117

Chong G, von Hillebrandt A.(1985): El Triásico Preandino de Chile entre los 23°30' y 26° de Lat. Sur. 4. Congr. Geol. Chileno. Actas 1: 1-162 - 1-210, 4 figs., 1 tab., 4 pls.; Antofagasta.

Corvalan J (1982) El límite Triásico-Jurásico en la Cordillera de la Costa de las provincias de Curicó y Talca. - 3. Congr Geol Chil Actas 3: F/63-F/85

Dalziel IWD (1986) Collision and Cordilleran orogenesis: An Andean perspective. In: Coward, MP, Ries AC (eds): Collision tectonics. Geol Soc Spec Publ., Oxford 19: 389-404

Damm K-W, Pichowiak S, Todt W (1986) Geochemie, Petrologie und Geochronologie der Plutonite und des metamorphen Grundgebirges in Nordchile. Berl Geowiss Abh (A) 66: 73-146

Ellison R, Klinck B, Hawkins M (1989) Deformation events in the Andean orogenic cycle in the Altiplano and Western Cordillera, southern Peru. J. S Am Earth Sci 2: 263-276

Geyer OF (1980) Die mesozoische Magnafazies-Abfolge in den nördlichen Anden (Peru, Ekuador, Kolumbien). Geol Rundsch 69: 875-891

Gröschke M, von Hillebrandt A, Prinz P, Quinzio LA, Wilke H-G (1988) Marine Mesozoic paleogeography in northern Chile between 21°-26°S. - In Bahlburg H, Breitkreuz C, Giese P (eds): The Southern Central Andes. Lecture Notes in Earth Sciences 17. Springer, Berlin Heidelberg New York, pp 105-117

Hervé M, Marinovic N (1989) Geocronología y evolución del Batolito Vicuña Mackenna, Cordillera de La Costa, Sur de Antofagasta (24-25°S). Rev Geol Chile 16: 31-49

Jurgan H (1977) Strukturelle und lithofazielle Entwicklung des andinen Unterkreide-Beckens im Norden Chiles (Provinz Atacama). Geotekt Forsch 52: 1-138

Kay SM, Ramos VA, Mpodozis C, Sruoga P (1989) Late Paleozoic to Jurassic silicic magmatism at the Gondwana margin: Analogy to the middle Proterozoic in North America? Geology 17: 324-328

Mégard F (1984) The Andean orogenic period and its major structures in central and northern Peru. J geol Soc Lond 141: 893-900

Mégard F (1989) The evolution of the Pacific Ocean margin in South America north of Arica Elbow (18°S). In Ben-Avraham Z (ed): The evolution of the Pacific Ocean margins. Oxford University Press, Ney York, pp 208-230

Pichowiak S, Buchelt M, Damm K-W (1990) Magmatic activity and tectonic setting of the early stages of the Andean cycle in northern Chile. - In Kay SM, Rapela CW (eds): Plutonism from Antarctica to Alaska. Geol. Soc. Am. Spec. Pap. 241: 127-144

Posamentier HW, Vail PR (1988) Eustatic controls on clastic deposition II - sequence and systems tract models. - In Wilgus CK, Hastings BS, Kendall CG, Posamentier HW, Ross CA, Van Wagoner JC (eds): Sea level changes: an integrated approach. Spec Publs Soc Econ Paleontol Miner 42: 125-154

Prinz P (1991) Mesozoische Korallen aus Nordchile. Palaeontographica (A) 216: 147-209

Quinzio LA (1987) Stratigraphische Untersuchungen im Unterjura des Südteils der Provinz Antofagasta in Nord-Chile. Berl geowiss Abh (A) 87: 1-100

Riccardi AC (1983) The Jurassic of Argentina and Chile. - In Moullade M, Nairn AEM (eds): The Phanerozoic geology of the world II. The Mesozoic. pp 201 - 236

Riccardi AC (1986) Cretaceous paleogeography of southern South America. - Palaeogeogr Palaeoclimatol Palaeocol 59: 169-195

Riccardi AC (1988) The Cretaceous System of southern South America. Geol Soc Am Mem 168: 161 pp

Rössling R (1989) Petrologie in einem tiefen Krustenstockwerk des jurassischen magmatischen Bogens in der nordchilenischen Küstenkordillere südlich von Antofagasta. Berl geowiss Abh (A) 113: 1-73

Rogers DA (1980) Analysis of pull-apart basin development produced by en-echelon strike-slip faults. Spec Publ Int Assoc Sedimentol (1980) 4: 27-41

Rogers G (1985) A geochemical traverse across the north Chilean Andes. PhD Thesis, Dep Earth Sci Open Univ Milton-Keynes, 333 pp (unpubl)

Salfity JA (1985) Lineamentos transversales al rumbo andino en el noroeste Argentino. 4 Congr Geol Chil Actas 2: 2/119-2/137

Scheuber E, Rössling R, Reutter K-J (1986) Strukturen in der chilenischen Küstenkordillere zwischen Paposo und Antofagasta. Berl geowiss Abh (A) 66: 209-224

von Hillebrandt A (1973) Neue Ergebnisse über den Jura in Chile und Argentinien. Münster Forsch Geol Paläontol 31/32: 167-199

von Hillebrandt A, Gröschke M, Prinz P, Wilke H-G (1986) Marines Mesozoikum in Nordchile zwischen 21° und 26°S. - Berl geowiss Abh (A) 66: 169-190

Zeil W (1979) The Andes - a geological review. Borntraeger, Berlin, 260 pp

Ziegler PA (1982) Geological atlas of western and central Europe. Shell Int Petrol Maatschappij BV, Den Hague

Jurassic Cretaceous Palaeogeographic Evolution of the Chilean Andes at 23°-24°S Latitude and 34°-35°S Latitude: A Comparative Analysis

REYNALDO CHARRIER and NELSON MUÑOZ

Abstract. A comparative palaeogeographic analysis between the northern (23°-24°S) and central (34°-35°S) Chilean Andes during the Jurassic and the Cretaceous indicates that both regions had the following general palaeogeographic pattern: Successive volcanic arcs shifted to the east of the former arc, and backarc domains with a marine backarc basin in the latest Triassic to early Cretaceous, a continental backarc basin in the late Cretaceous, and a possible marine backarc basin, connected to marine forearc basins in the latest Cretaceous-earliest Tertiary. The particular evolution of these palaeogeographic domains shows some interesting differences. The evolution of the backarc basin in northern Chile corresponds to a single transgression-regression cycle, while the evolution of the backarc in central Chile corresponds to two such cycles. In the volcanic arc domain in central Chile an extensive intra-arc basin developed in the early Cretaceous. This basin was separated from the backarc basin by a volcanic ridge and extended from the Coastal Cordillera to the present international border, with a width of several tens of kilometres. The Neocomian marine sediments, present in the Coastal Cordillera at Antofagasta (El Way Fm.) and Iquique (Blanco Fm.), correspond to forearc deposits related to the commencement of the early to late ("middle") Cretaceous Quebrada Mala volcanic arc. The similar palaeogeographic pattern which developed in both regions indicates an essentially similar tectonic regime along the continental margin during the late Triassic-Jurassic and Cretaceous. The differences detected in the evolution of some of the palaeogeographic domains were caused by local, though significant, variations in the dominant tectonic conditions and clearly demonstrate the existence of a palaeosegmentation of the mountain range at that time.

1 Introduction

Morphostructural and seismological variations along the Andean Cordillera as well as interruptions along the volcanic arc provide the most important evidence supporting the present segmented nature of this mountain range. Palaeogeographic variations along the Argentine-Chilean Andes also demonstrate a segmented nature of the orogen during earlier periods.

A comparative analysis of the palaeogeographic evolution of two well known regions of the Chilean Andes (Fig. 1), one in northern Chile at the latitude of Antofagasta (23° and 24°S) and the other in central Chile south of Santiago (34° and 35°S), show for both regions essentially the same palaeogeographic pattern during the Jurassic and Cretaceous, although important differences in the particular evolution of some of the palaeogeographic domain are also evident. We describe here the palaeogeographic evolution of both regions, underline the main differences, and finally discuss briefly the significance of the similarities and differences detected.

Correspondence to: R. Charrier Departamento de Geología y Geofísica, Universidad de Chile, Casilla 13518, Correo 21, Santiago-Chile

2 Stratigraphy

A brief outline of the stratigraphy of both regions is given in order to facilitate the understanding of the palaeogeographic comparison between them. In northern and central Chile the Mesozoic sedimentary sequence begins in the Triassic and lies unconformably on a Palaeozoic basement. In northern Chile the Triassic deposits consist of continental acidic volcanic deposits covered by fossiliferous marine sandstones and conglomerates. In central Chile a thick Ladinian acidic volcanic unit is intercalated between detritic marine deposits of Anisian age, below, and Rhetian age, above (see von Hillebrandt et al. 1986; Muñoz 1989, for the Antofagasta region; Muñoz-Cristi 1942; Thomas 1958; Cecioni and Westermann 1968;

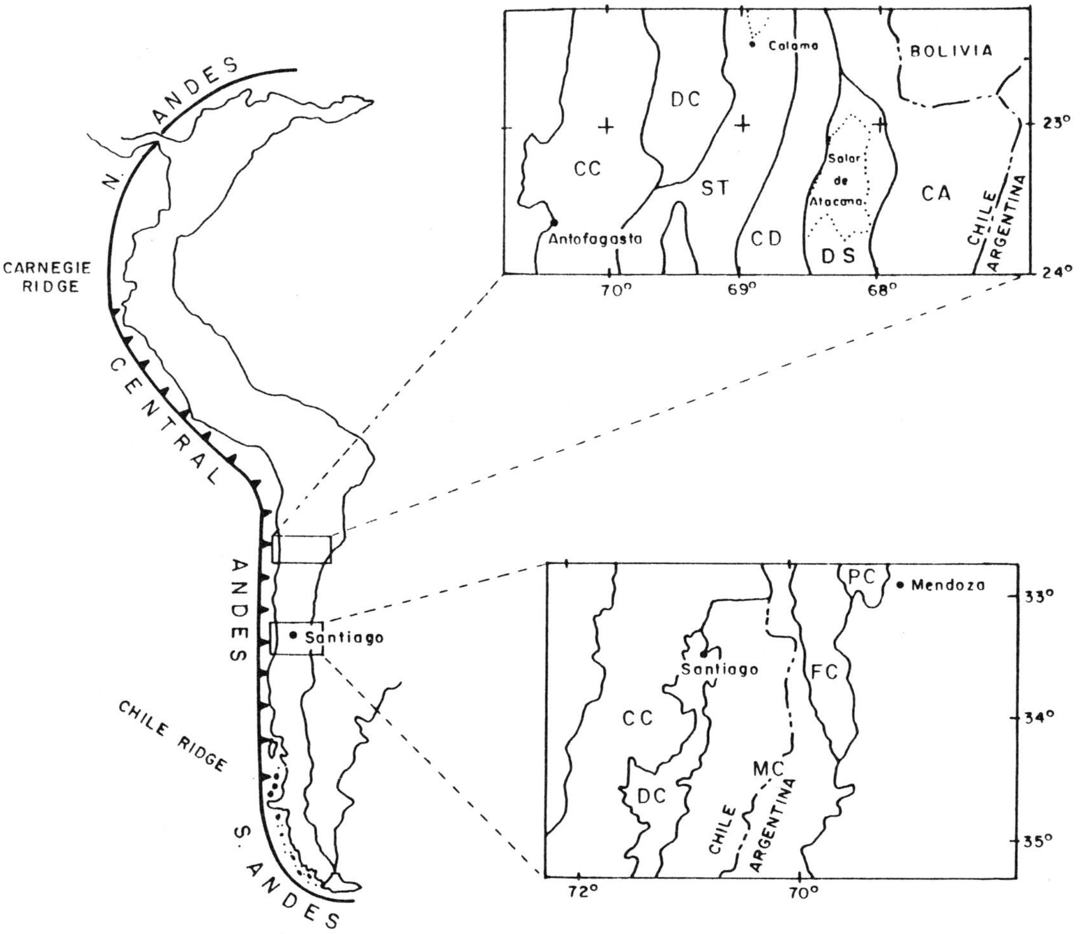

Fig. 1. Main subdivisions of the Andes and location of the analysed regions. Morphostructural units for each region are indicated: *CC* Coastal Cordillera; *DC* Central Valley; *ST* Transversal Cordillera; *CD* Domeyko Cordillera; *DS* Salar Basins; *CA* Altiplano; *MC* Main Cordillera; *FC* Frontal Cordillera; *PC* Precordillera

Corvalán 1976; Charrier 1979; Thiele and Morel 1981, for central Chile).

The Anisian-Ladinian deposits of central Chile and western Argentina accumulated in NW trending grabens (Rolleri and Criado-Roque 1966; Stipanicic 1972; Stipanicic and Bonaparte 1972; Charrier 1979). The sediments transported towards the northwest along these grabens formed extensive submarine fans, presently located along the Chilean coastal area. The trend of these grabens, which is clearly oblique to the dominant N-S Andean trend, and the prevailing extensional regime in the Triassic indicates that this period had a totally different tectonic setting. This period is considered to be transitional between the Palaeozoic and the post-Palaeozoic orogenic cycles (see Aubouin et al. 1973; Ramos 1988). In northern and central Chile there is a gradual transition between the Triassic and the Jurassic marine deposits (von Hillebrandt et al. 1986; Muñoz 1989; Fuenzalida 1938; Cecioni and Westermann 1968; Thiele and Morel 1981).

2.1 Stratigraphy of the Region Between 23° and 24°S Latitude

Along an East-West profile between the Coastal Cordillera and the Domeyko Cordillera (Fig. 2) it is possible to differentiate between areas with different stratigraphic evolutions (Fig. 3).

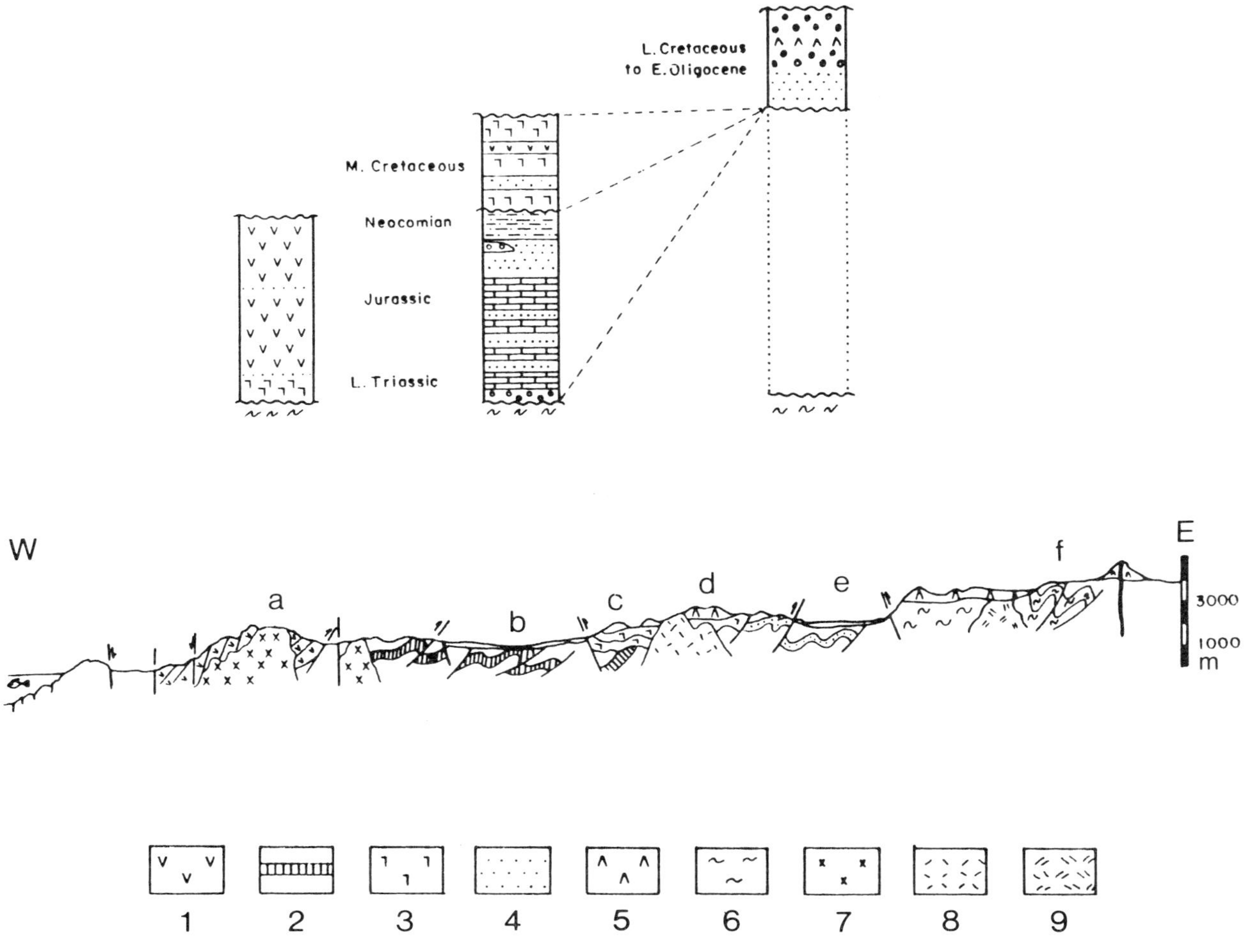

Fig. 2. Geological profile across northern Chile at 23°-24°S lat. and stratigraphic columns for the Coastal Cordillera, Central Valley-Transversal Cordillera, and Salar basins. After Muñoz (1989) and Charrier and Reutter (1990). *1* Jurassic andesites; *2* Jurassic marine deposits; *3* Ignimbrites; *4* Sandstones; *5* Cenozoic volcanics intercalated in Eocene-early Oligocene conglomerates; *6* Palaeozoic units; *7* Jurassic plutonic rocks; *8* Cretaceous plutonic rocks; *9* Tertiary plutonic rocks; *a* Coastal Cordillera; *b* Central Valley; *c* Transversal Cordillera; *d* Domeyko Cordillera; *e* Salar Basins; *f* Altiplano

To the west latest Triassic-Jurassic calc-alkaline, mainly andesitic volcanic sequences, which are several kilometres thick, form the bulk of the Coastal Cordillera (e.g. La Negra Fm. of Jurassic age; Harrington 1961; García 1967).

East of the Coastal Cordillera these volcanics, in some parts with an acidic character at the base, interfinger with Jurassic detritic and calcareous marine equivalents of the Sierra del Cobre (Muñoz 1989) or Cerritos Bayos (Biese 1961; Baeza 1979; Lira 1989). The marine sequence is developed on top of late Triassic detritic continental deposits (von Hillebrandt et al. 1986; Muñoz 1989; Prinz et al. this Vol.; Scheuber et al. this Vol.) and corresponds to a transgression-regression cycle developed from the latest Triassic to the early Cretaceous. Oxfordian evaporites are succeeded by thick red siltstones and sandstones and by local conglomeratic deposits (Lira 1989; Muñoz 1989; Bogdanic 1990). These are conformably covered by a coarse detritic and volcanic, continental, early Cretaceous unit (Muñoz 1989). The continental volcanic and sedimentary early to late ("middle") Cretaceous Quebrada Mala Fm. (Muñoz 1989; Muñoz et al. 1989) covers unconformably older Mesozoic units.

Farther east in the Domeyko Cordillera the Mesozoic sequence is unconformably deposited on Palaeozoic volcanics (Fig. 3). A thick, early to

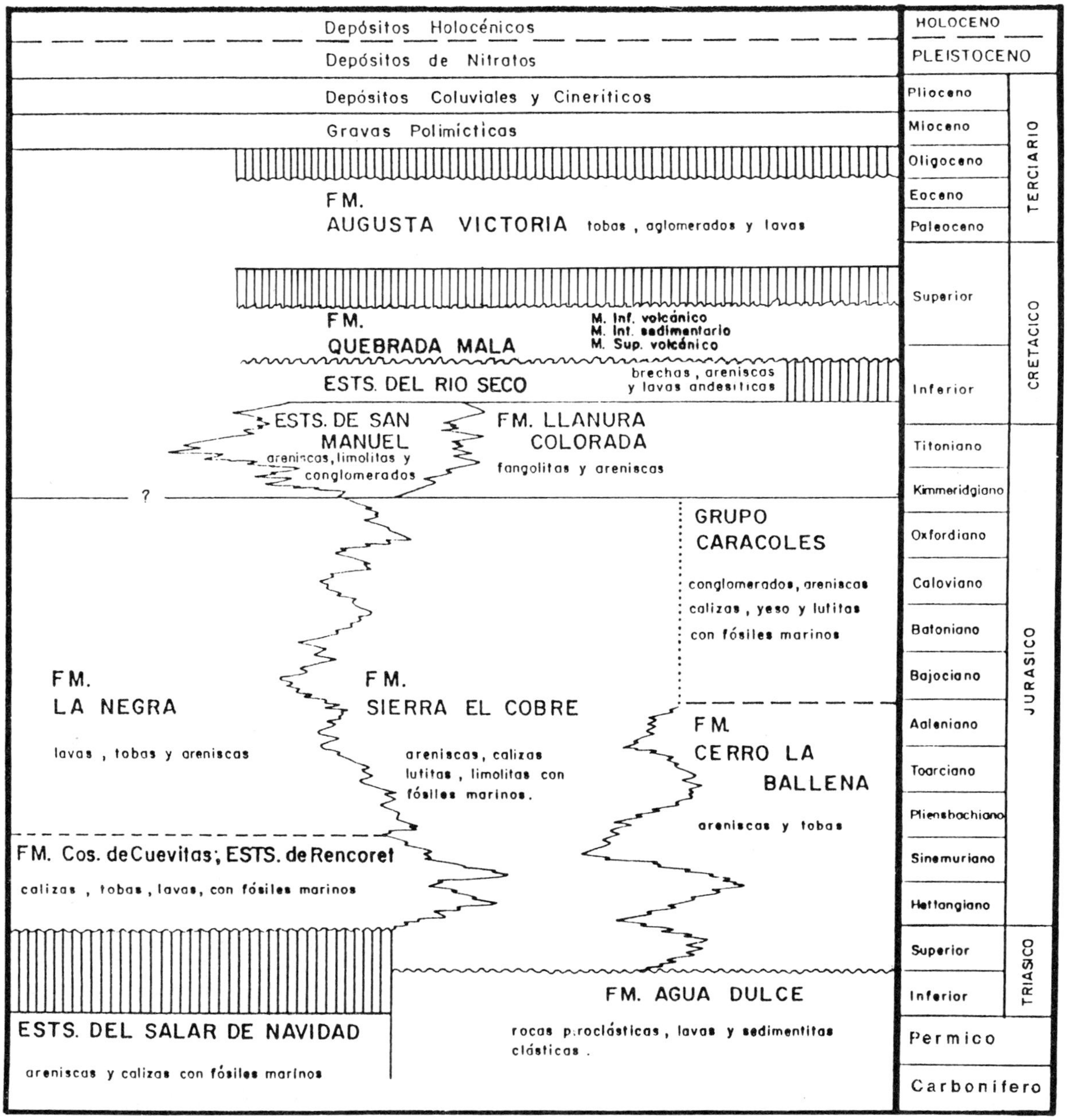

Fig. 3. Stratigraphic and palaeogeographic relationships between the Mesozoic and Cenozoic deposits of the Coastal Cordillera, Central Valley and Domeyko Cordillera in the Antofagasta region. After Muñoz (1989)

middle Jurassic detritic and tuffaceous unit with fossiliferous marine intercalations is followed by the marine detritic and calcareous Caracoles Group.

Possible marine, latest Cretaceous deposits are known in the Main Cordillera (Lomas Negras Fm.; Marinovic and Lahsen 1984) and on the west border of the Salar de Atacama (Tonel Fm. of the Purilactis Group; Charrier and Reutter 1990 this Vol.). In this latter locality the gypsiferous Tonel deposits are conformably covered by the thick conglomeratic Eocene-early Oligocene Purilactis Fm. sensu stricto (Fig. 2).

2.2 Stratigraphy of the Region Between 34°-35°S Latitude

In this region the Mesozoic and Cenozoic deposits form a wide synclinorium, with Palaeozoic rocks outcropping on both sides. Here, it is possible to identify two areas with different stratigraphic evolutions: one in the Coastal Cordillera and the other in the Main Cordillera (Fig. 4).

The Mesozoic sequence of the Coastal Cordillera is more than 10 000 m thick. It is formed mainly by volcanic and volcaniclastic deposits of Triassic (La Ligua = Pichidangui Fm.), late Jurassic (Horqueta Fm.) and early Cretaceous age (Veta Negra Fm.) and by thick, coarse detritic deposits of late early Cretaceous age (Las Chilcas Fm.). Marine sedimentary and fossiliferous deposits are known from the middle Triassic (El Quereo Fm.), late Triassic (basal Los Molles Fm.), early to middle Jurassic (La Calera Fm.) and earliest Cretaceous (Lo Prado Fm.). The Upper Cretaceous is represented by the essentially rhyolitic Lo Valle Fm. (Drake et al. 1976).

Although the Mesozoic sequence of the Main Cordillera is thinner than that of the Coastal Cordillera, it is still nearly 8000 m thick. It is composed mainly of red detritic deposits of the Río Damas (Kimmeridgian) and Colimapu (late early Cretaceous) Formations. As in the Coastal Cordillera, there are in this sequence two intercalated marine formations: the Nacientes del Teno (Callovian-Oxfordian) and the Baños del Flaco (Tithonian-Neocomian) Formations. These deposits correspond to two transgression-regression cycles: (1) late Liassic-Bajocian to Kimmeridgian and (2) Tithonian to Neocomian-Aptian-Albian. The latest Cretaceous is possibly represented by the eastern outcrops of the Coya-Machalí (= Abanico) Formation.

3 Palaeogeography

Here we discuss the main features and differences in the palaeogeographic evolution of the two considered regions during the Jurassic and the Cretaceous. The palaeogeographic settings of both regions during the latest Triassic-middle Jurassic, Kimmeridgian-Tithonian, Tithonian-Neocomian-(Albian), late Neocomian-Albian-Cenomanian, and latest Cretaceous, are given in Fig. 5.

In both regions the latest Triassic to late early Cretaceous magmatic arc was located in the present Coastal Cordillera. In northern Chile the western margin of the marine backarc basin was located immediately east of the present Coastal Cordillera. In central Chile, however, the early Cretaceous backarc basin had its western margin in the Main Cordillera. The evolution of the backarc domain in both regions was parallel until the Kimmeridgian. In northern Chile the withdrawal of the sea from the backarc basin occurred at this point and only local and thin marine deposits are intercalated in the thick red continental late Jurassic to early Cretaceous sequence exposed along the Precordillera (Muñoz et al. 1989, p. A140). At the same time, in central Chile, the marine sedimentation in the backarc basin was interrupted by a widely extended regressive episode. After this episode the marine conditions lasted until the late Neocomian and sedimentation continued in the backarc basin until the Aptian-Albian with thick red detritic deposits and some thin gypsum intercalations at the base of the detritic sequence.

The Kimmeridgian regression is marked in both backarc basins by the presence of gypsum deposits followed by thick sequences of red continental sediments (Llanura Colorada and Río Damas Fms., respectively). In northern Chile the monotonous 2000 m thick red siltstones of the Llanura Colorada Fm., which were deposited in a tidal flat environment during the Kimmeridgian to earliest Cretaceous, indicate rapid subsidence of the backarc basin floor (Muñoz 1989). In central Chile, during the final regressive episode of the backarc basin in the late Neocomian a more than 2000 m thick sequence of calcareous external platform sediments (Baños del Flaco Fm.) and red tidal flat deposits (Colimapu Fm.) accumulated in a rapidly subsident area close to the west margin of the basin (Charrier 1981b, 1984).

The formation of local and rapidly subsident areas seems to be characteristic of the backarc basins and apparently occurs during the generalized emergence of the continental margin and the beginning of the basin inversion. These local subsident areas could, therefore, correspond to "compressional grabens" caused by a change in the dominant tectonic regime while passing from extensional to compressional stress conditions (Scheuber et al. this Vol.).

In central Chile during the early Cretaceous a great amount of volcanic deposits and conglomerates accumulated in an intra-arc basin (Charrier 1984). Some volcanic deposits contain marine fossiliferous limestone intercalations of early Cretaceous age (Lo Prado Fm.). These are covered by a thick basaltic sequence (Veta Negra Fm.) with low Sr initial ratios (Aberg et al. 1984) this deviates from the typical calc-alkaline character of the Andean volcanic rocks. On this basis, it was concluded that there was strong

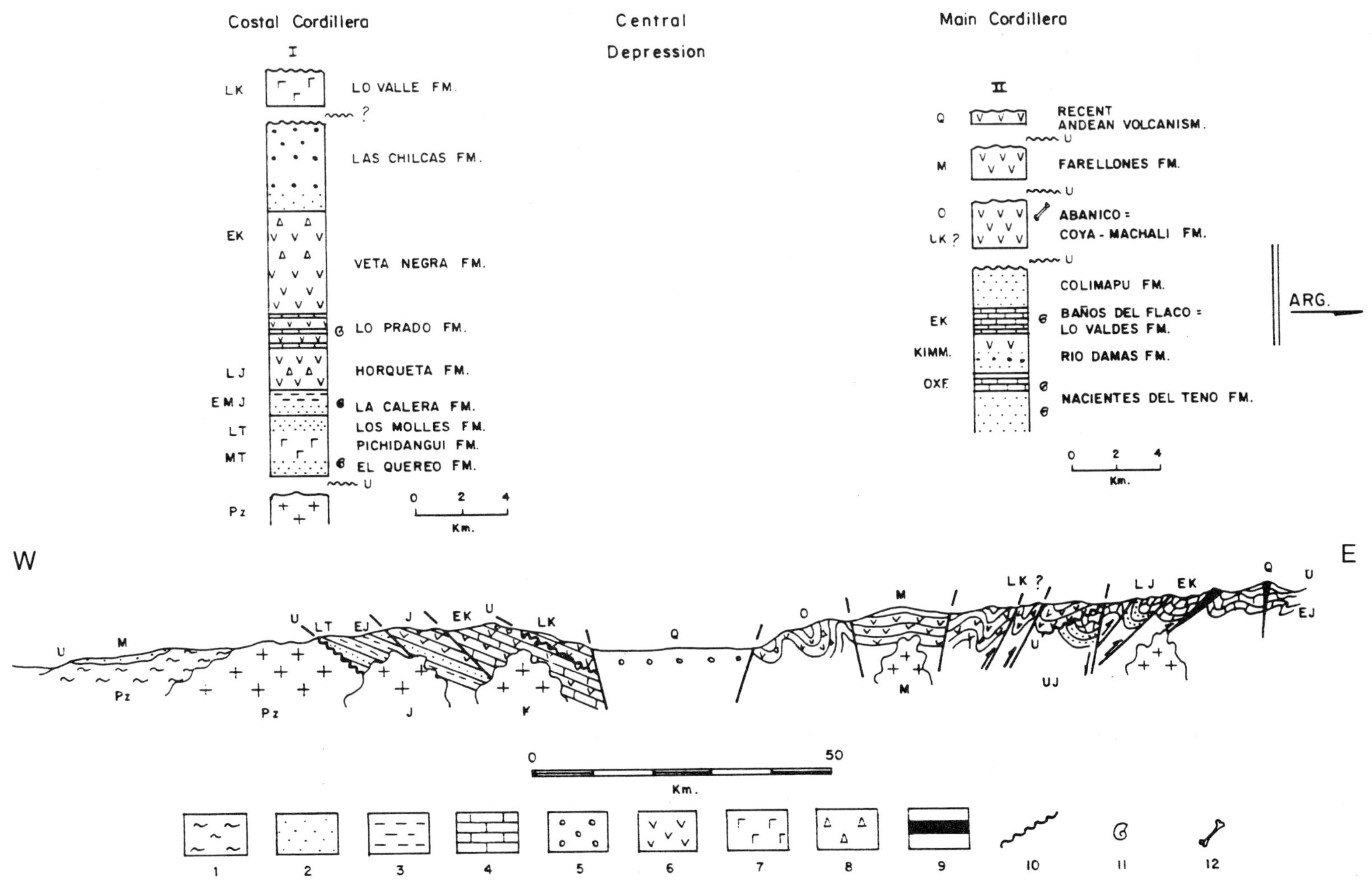

Fig. 4. Schematic geological profile across central Chile at 34°-35°S lat. and stratigraphic columns for the Coastal Cordillera and the Main Cordillera. Profile after Charrier (1973, 1981b) and Thiele and Morel (1981). Stratigraphic column I after Thomas (1958), Drake et al. (1976), Vergara and Drake (1978), Thiele and Morel (1981), Thiele and Nasi (1982), Rivano et al. (1986). Stratigraphic column II after Klohn (1960), Charrier (1973, 1981a,b), Davidson and Vicente (1973), Piracés (1976), Vergara et al. (1988), Valencia et al. (in press), Wyss et al. (1991). *1* Foliated metamorphic rocks; *2* sandstone; *3* lutite; *4* limestone; *5* conglomerate; *6* andesite; *7* rhyolite; *8* breccia; *9* gypsum; *10* unconformity; *11* ammonite; *12* vertebrate fossil; + + intrusive complexes; *U* unconformity; *Pz* Palaeozoic; *MT* middle Triassic; *LT* late Triassic; *J* Jurassic; *EJ* early Jurassic; *EMJ* early to middle Jurassic; *LJ* late Jurassic; *EK* early Cretaceous; *LK* late Cretaceous; *M* Miocene; *O* Oligocene; *Q* Quaternary

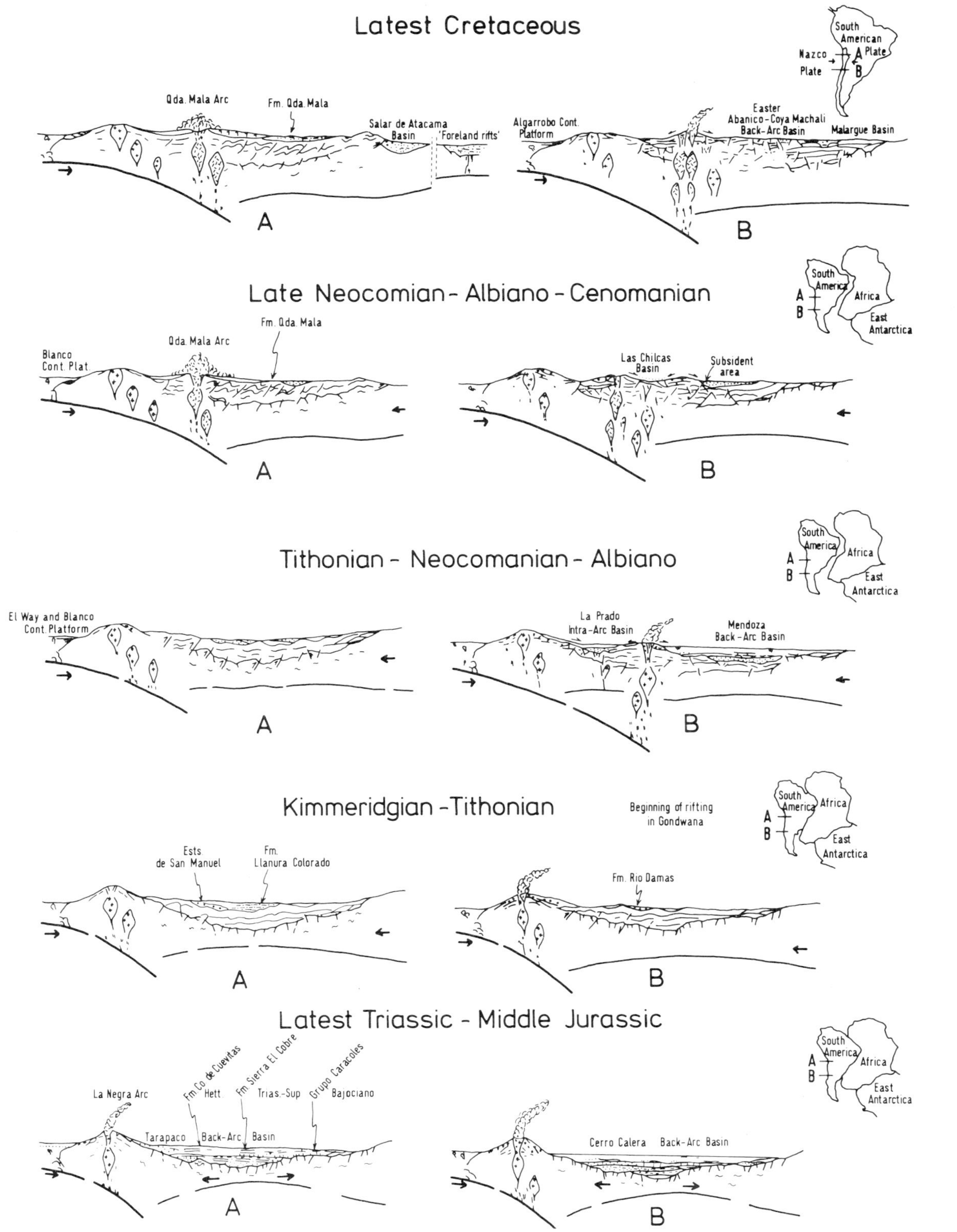

Fig. 5. Comparative palaeogeographic evolutions for the northern (23°-24°S lat.) and central (34°-35°S lat.) Chilean Andes for the latest Triassic middle Jurassic, Kimmeridgian-Tithonian, Tithonian-Neocomian-Albian, late Neocomian-Albian-Cenomanian, and late Cretaceous

crustal extension at this time, and the development of an ensialic marginal basin in the backarc domain was postulated by Levi and Aguirre (1981). However, we suggest that the extension occurred not in the backarc domain, but in the arc domain.

In the Main Cordillera in central Chile, thick volcanic and coarse clastic deposits of early Cretaceous age, supplied to the backarc basin from the west by an emergent active calc-alkaline volcanic arc, are intercalated in the marine fossiliferous backarc deposits of the Baños del Flaco (= Lo Valdés) Fm. (Charrier 1981b). These deposits are presently exposed along the international border, or very close to it, on the Chilean versant of the Main Cordillera (Charrier 1981b; Ramos 1988). On this basis we suggest, following Charrier (1984), that the mainly volcanic deposits of early Cretaceous age, present in central Chile, were accumulated in an intra-arc basin. This basin was located west of the active volcanic arc, which supplied thick volcanic and coarse detritic deposits to the backarc basin, and it occupied the area, at least 100 km wide, between the Coastal Cordillera and the western versant of the present Main Cordillera.

In northern Chile an angular unconformity separates the, mainly clastic, early Cretaceous backarc sediments from the early to late ("middle") Cretaceous volcanic deposits (Quebrada Mala Fm.; Muñoz 1989; Muñoz et al. 1989). In the central Chilean region under consideration there is no evidence for the existence of a late Cretaceous volcanic arc. North of this region between 29° and 32°S the remains of this arc are well exposed, overlying unconformably older Mesozoic units (Rivano and Sepulveda, in press). This arc was located to the east of the former Jurassic arc, but still in an area which was formerly part of the intra-arc domain. This episode represents the most important and abrupt change in the palaeogeographic evolution of the Argentine-Chilean Andes (Charrier and Vicente 1972; Vicente et al. 1973; Mpodozis 1984) and coincides with the increased spreading rate of the Atlantic and Pacific oceans between 110 and 85 Ma associated with the break-up of Gondwana (Larson and Pitman 1972).

In the Coastal Cordillera in northern Chile the Neocomian marine calcareous El Way (Antofagasta) and Blanco (Iquique) Fms. conformably cover the continental detritic Coloso and Atajaña Fms. Because of the age of these marine units and their western location, covering at Antofagasta the volcanic deposits of the arc and near Iquique the Jurassic plutonic rocks of the arc's root, we consider them to be forearc deposits related to the "middle" (early to late) Cretaceous Quebrada Mala volcanic arc (Scheuber et al. this Vol.: Fig. 5).

In the Chilean regions under consideration there are again common features during the latest Cretaceous. In northern Chile a new broad volcanic arc formed which partly overlapped the Quebrada Mala volcanic arc and its continental backarc domain (Scheuber et al. this Vol.). To the east of it a basin was formed which was possibly occupied by a marine transgression during the Maastrichtian-Danian. Marine sediments of this age are well documented in northwestern Argentina (Marquillas and Salfity 1988). Charrier and Reutter (this Vol.) postulate a possible connection between this backarc basin and the Andean foreland rifts described for NW-Argentina by Reyes et al. (1976) and Gallisky and Viramonte (1988). In central Chile a marine ingression is also postulated for the backarc domain during the deposition of the sedimentary and slightly calcareous intercalations of the eastern outcrops of the lower Coya-Machalí (= Abanico) Fm., which is considered to be of latest Cretaceous age (Drake et al. 1982; Thiele and Nasi 1982). This area was possibly connected to the marine ingression, coming from the Atlantic ocean, whose sediments are present in the Andean foreland at this latitude (Volkheimer 1978).

4 Conclusions

It is concluded that in the regions analysed the general features of the palaeogeographic pattern were similar during the Jurassic and Cretaceous: (1) a volcanic arc and a marine backarc basin, in the latest Triassic to early Cretaceous; (2) a volcanic arc, shifted to the east of the former arc, and a continental backarc domain in the early to late ("middle") Cretaceous; and (3) a backarc domain, probably connected to marine basins of the Andean foreland, in the latest Cretaceous-earliest Tertiary.

However, a more detailed analysis based on the development of the backarc basin suggests that both regions underwent a parallel palaeotectonic evolution from the latest Triassic to the late Jurassic (Kimmeridgian) but had different features from the latest Jurassic to the late Cretaceous. The palaeogeographic conditions became similar again in the latest Cretaceous-earliest Tertiary.

In northern and central Chile the contemporaneous regression in the backarc basin at the end of the Jurassic coincides with the onset of the rifting stage of Gondwana in the Kimmeridgian. This process seems to have caused a strong coupling along the continental margin and an increase in the compressive stress, acting on the arc and backarc basin palaeogeographic pair, which forced the region to emerge. Apparently, this effect was maintained until the Neocomian in northern Chile, but was relaxed in central Chile. The cause for this is not clear to us, however, we suspect that the northward opening of the Atlantic could have relaxed this coupling earlier in southern Chile than in northern Chile. The marine El Way and Blanco Fms. in the Coastal Cordillera in northern Chile cannot be related directly to this Tithonian-Neocomian transgression-regression cycle in central Chile as they were deposited in a forearc domain.

During the early Cretaceous an extensive ensialic marginal intra-arc basin was formed in central Chile, comprising the Coastal Cordillera and the present western or Chilean versant of the Main Cordillera. This intra-arc basin was separated from the Mendoza-Neuquén backarc basin by an active calc-alkaline volcanic arc. The intra-arc domain, almost exclusively composed of volcanics, some of which have an alkalic character, and later intruded by a late Cretaceous batholith, can be considered an isolated palaeogeographic, petrologic and, possibly also, metallogenic province in the central Chilean Andes.

Acknowledgements. This chapter is a contribution to project No. 242 "Cretaceous of Latin America" of the International Geological Correlation Program (IGCP; IUGS-UNESCO).

References

Aberg G, Aguirre L, Levi B, Nystrom JO (1984) Spreading-subsidence and generation of ensialic marginal basins: an example from the early Cretaceous of central Chile. In: Kokelaar BP, Howells MF (eds) Volcanic and associated sedimentary and tectonic processes in modern and ancient marginal basins. Geol Soc Lond Spec Publ 16: 185-193

Aubouin J, Borrello AV, Cecioni G, Charrier R, Chotin P, Frutos J, Thiele R, Vicente J-C (1973) Esquisse paléogéographique et structurale des Andes Meridionales. Rev Geogr Phys Geol Dyn 15 (1-2): 11-72

Baeza A (1979) Distribución de facies sedimentarias marinas en el Jurásico de Cerritos Bayos y zonas adyacentes, norte de Chile. 3 Congr Geol Chil Santiago Actas 2 : H45-H61

Biese W (1961) El Jurásico de Cerritos Bayos. Inst Geol Univ de Chile, Santiago, Publ 19: 66 pp

Bogdanic T (1990) Kontinentale Sedimentation der Kreide und des Alttertiärs im Umfeld des subduktionsbedingten Magmatismus in der chilenischen Präkordillere (21°-23°S). Berl Geowiss Abh A 123: 117 pp

Cecioni G, Westermann G (1968) The Triassic/Jurassic marine transition of coastal central Chile. Pac Geol 1 (1): 41-75

Charrier R (1973) Geología regional de las provincias O'Higgins y Colchagua. Inst Inves Rec Nat, Santiago, (IREN) 7: 69 pp

Charrier R (1979) El Triásico en Chile y regiones adyacentes de Argentina: una reconstrucción paleogeográfica y paleoclimática. Dep Geol Geofís Univ de Chile, Santiago, Comunicaciones 26: 1-37

Charrier R (1981a) Mesozoic and Cenozoic stratigraphy of the central Argentinian-Chilean Andes (32°-35°S) and chronology of their tectonic evolution. Zentralbl Geol Paläontol 1 1981 (3-4): 344-355

Charrier R, (1981b) Geologie der chilenischen Hauptkordillere zwischen 34° und 34°30' südlicher Breite und ihre tektonische, magmatische und paläogeographische Entwicklung. Berl Geowiss Abh A 36: 270 pp

Charrier R, (1984) Areas subsidentes en el borde occidental de la cuenca de tras-arco jurásico-cretácica, Cordillera Principal chilena entre 34° y 34°30'S. 9 Congr. Geológico Argentino Bariloche Actas 2: 107-124

Charrier R, Reutter K-J (1990) The Purilactis Group of Northern Chile: Link between arc and backarc during late Cretaceous and Paleogene. Symp Int Géodyn Andine, 15-17 May 1990 Grenoble, ORSTOM, Paris, pp 249-252

Charrier R, Vicente JC (1972) Liminary and geosynclinal Andes: Major orogenic phases and synchronical evolutions of the Central and Magellan sectors of the Argentine-Chilean Andes. Solid Earth Probl Conf Buenos Aires Upper Mantle Proj 2: 451-470,

Corvalán J (1976) Triásico-Jurásico de Vichuquén-Tilicura y de Hualañé, Provincia de Curicó, implicaciones paleogeográficas. 1 Congr Geol Chil Santiago Actas 1: A137-A154

Davidson J, Vicente JC (1973) Características paleogeográficas y estructurales del área fronteriza de las Nacientes del Teno (Chile) y Santa Elena (Argentina) (Cordillera Principal, 35°-35°15' de latitud sur). 5 Congr Geol Argent Buenos Aires Actas 5: 11-55

Drake RE, Curtis G, Vergara M (1976) Potassium-argon dating of igneous activity in the Central Chilean Andes - latitude 33°S. J Volc Geotherm Res 1: 285-295

Drake RE, Charrier R, Thiele R, Munizaga F, Padilla H, Vergara M (1982) Distribución y edades K/Ar de volcanitas post-neocomianas en la Cordillera Principal entre los 32° y 36° L. S. Implicaciones estratigráficas y tectónicas para el Meso-Cenozoico en Chile Central. 3 Congr Geol Chil Concepción Actas 1: D42-D78

Fuenzalida H (1938) Las capas de Los Molles. Museo Nac. Historia Natural, Santiago, Bol 16: 67-92

Gallisky M, Viramonte JG (1988) The Cretaceous paleorift in northwestern Argentina: A petrologic approach. J South Am Earth Sci 1: 329-342

García F (1967) Geología del Norte Grande de Chile. Soc Geol Chile, Santiago, Symposio sobre el Geosinclinal Andino 3, 138pp

Harrington HH (1961) Geology of parts of Antofagasta and Atacama provinces, northern Chile. AAPG Bull 45: 169-197

Klohn C (1960) Geología de la Cordillera de los Andes de Chile Central, Prov. de Santiago, Colchagua y Curicó. Bol Inst Investig. Geológicas, Santiago, 8: 95 pp

Larson RL, Pitman III WC (1972) World-wide correlation of Mesozoic magnetic anomalies, and its implications. Geol Soc Am Bull 83: 3645-3662

Levi B, Aguirre L (1981) Ensialic spreading-subsidence in the Mesozoic and Paleogene Andes of central Chile. J Geol Soc Lond 138: 75-81

Lira G (1989) Geología del área pre-andina de Calama, con énfasis en la estratigrafía y paleogeografía del Mesozoico, 22° a 22°40' Latitud sur; Región de Antofagasta, Chile. Thesis, Dep Geol Geofís Univ de Chile, Santiago, 211 pp

Marinovic N, Lahsen A (1984) Hoja Calama, Región de Antofagasta. Serv Nac Geología y Minería, Santiago, Carta Geol de Chile 58, 144 pp

Marquillas RA, Salfity JA (1988) Tectonic framework and correlations of the Cretaceous-Eocene Salta Basin, Argentina. In: Bahlburg H, Breitkreuz C, Giese P (eds) The southern Central Andes, Lecture Notes in Earth Sciences 17. Springer, Berlin Heidelberg New York, pp 119-136

Mpodozis AC (1984) Geodinámica de los márgenes continentales activos. Seminario de actualización de la geología de Chile, Serv Nac Geología y Minería, Santiago, pp 4-8

Muñoz N (1989) Geología y estratigrafía de las hojas Baquedano y Pampa Unión, II Región, Antofagasta, Chile. Thesis, Dep Geol Geofís Univ de Chile, Santiago, 152 pp

Muñoz N, Charrier R, Pichowiak S (1989) Cretácico Superior volcánico-sedimentario (Formación Quebrada Mala) en la Región de Antofagasta, Chile, y su significado geotectónico. In: Contrib. Simp Cretácico de América Latina, Parte A Eventos y Registro Sedimentario. Spalletti L (ed) Buenos Aires, pp 133-148

Muñoz-Cristi J (1942) Rasgos generales de la constitución de la Cordillera de la Costa especialmente en la provincia de Coquimbo. 1 Congr Panam Ing Minas y Geología Santiago An 2: 285-318

Piracés R (1976) Estratigrafía de la Cordillera de la Costa entre la Cuesta Melón y Limache, Provincia de Valparaiso, Chile. 1 Congr. Geol Chil Santiago Actas 1: A65-A82

Ramos VA (1988) The tectonics of the Central Andes; 30° to 33°S latitude. Geol Soc Amer Spec Paper 218: 31-54

Reyes FC, Salfity JA, Viramonte J G, Gutiérrez W (1976) Consideraciones sobre el volcanismo del Subgrupo Pirgua (Cretácico) en el Norte Argentino. 6 Congr Geol Argent Bahía Blanca Actas 1: 205-223

Rivano S, Sepulveda P (1990) Hoja Illapel, Región de Coquimbo. Serv Nac Geol Min Carta Geol de Chile, (in press)

Rivano S, Sepulveda P, Boric R, Hervé M, Puig A (1986) Antecedentes radiométricos para una edad cretácica inferior de la Formación Las Chilcas. Rev. Geol Chile 27: 27-32

Rolleri EO, Criado-Roque P (1966) La cuenca triásica del norte de Mendoza. 3 Jorn Geol Argent Actas 1: 1-76

Stipanicic PN (1972) Cuenca triásica de Barreal. In: (Leanza AF (ed) Geología Regional Argentina, Academia Nac Ciencias, Córdoba, pp 537-566

Stipanicic PN, Bonaparte JF (1972) Cuenca triásica de Ischigualasto-Villa Unión, In: Leanza AF (ed) Geología Regional Argentina, Academia Nac. Ciencias, Córdoba, pp 507-536

Thiele R, Morel R (1981) Tectónica triásico-jurásica en la Cordillera de la Costa, al norte y sur del Río Mataquito (34°45'-35°15' Lat. S.), Chile. Rev. Geol Chile 13-14: 49-61

Thiele R, Nasi C (1982) Evolución tectónica de los Andes a la latitud 33° a 34° Sur (Chile Central) durante el Mesozoico-Cenozoico. 5 Congr Latinoam Geol Buenos Aires Actas 3: 403-426

Thomas H (1958) Geología de la Cordillera de la Costa entre el valle de La Ligua y la cuesta Barriga. Bol Inst Investig Geol 2: 86 pp

Valencia J, Covacevich V, Marshall LG, Rivano S, Charrier R, Salinas P (1992) Oldest record of Teiid lizard from the Colimapu Formation (early Cretaceous) at Termas del Flaco, central Chile. Rev Geol Chile, (in press)

Vergara M, Charrier R, Munizaga F, Rivano S, Sepulveda P, Thiele R, Drake RE (1988) Miocene volcanism in the central Chilean Andes (31°30'S-34°35'S). J South Am Earth Sci 1: 199-209

Vicente JC, Charrier R, Davidson J, Mpodozis AC, Rivano S (1973) La orogenesis Subhercínica: fase mayor de la evolución paleogeográfica y estructural de los Andes argentino-chilenos centrales. 5 Congr Geol Argent Buenos Aires Actas 5: 81-98

Volkheimer W (1978) Descripción geológica de la Hoja 27b, Cerro Sosneado, Prov. de Mendoza, (1:200.000). Bol Minist Econ Secret Estado Min 151: 83 pp

von Hillebrandt A, Gröschke M, Prinz P, Wilke HG (1986) Marines Mesozoikum in Nordchile zwischen 21° and 26°S. Berl Geowiss Abh A 66: 169-190

Wyss AR, Norell MA, Flynn JJ, Novacek MJ, Charrier R, McKenna MC, Swisher III CC, Frassinetti D, Salinas P, Meng Jin (1991) A new early Tertiary mammal fauna from central Chile: implications for stratigraphy and tectonics. J Vertebr Paleontol 10 (4): 518-522

The Southern Andes Between 39°and 44°S Latitude: The Geological Signature of a Transpressive Tectonic Regime Related to a Magmatic Arc

FRANCISCO HERVÉ

Abstract. Only the active stratovolcanoes of the main range are more than 2000 m high, in this segment of the Andes where the submerged Central Valley separates the coastal ranges from the main range, south of 41°S lat. A late Palaeozoic accretionary subduction complex, mainly composed of turbiditic sequences with subordinate greenstones and cherts, with low grade metamorphism and deformation increasing to the west, underlies most of the coastal ranges and archipelagos. Isolated occurrences of similar rocks in the main range, which have yielded a Rb-Sr age of ca. 300 Ma, suggest that the same complex forms the basement there. The main range is mostly composed of the North Patagonian Batholith (NPB). Except for minor late Palaeozoic and early Jurassic bodies near its northern end, it has yielded middle Jurassic to Neogene intrusion ages. It is calcalkaline, with a lithologic variation from gabbro to monzogranite, with tonalite and granodiorite as the predominant rock types. The younger plutons appear to be located near the central axis of the batholith, which has a well defined early Cretaceous monzogranitic eastern margin. A backarc basin developed to the east of the batholith during the middle Jurassic to the early Cretaceous, and was filled with marine sediments and volcanic sequences. From the late Cretaceous onwards, only subaerial volcanism is recorded in the backarc position. These rocks show very little deformation. The Liquiñe-Ofqui Fault Zone (LOFZ), a dextral trench-linked strike slip structure runs near the centre of the NPB. It has been active at least since the Eocene, as is indicated in the sedimentary record by pull-apart basins which developed to the west of the fault zone. The LOFZ has controlled the emplacement of syntectonic neogene plutons and the presently active volcanic belt. The described scenario, i.e. the intrusion of the Meso-Cenozoic batholith into a late Palaeozoic accretionary complex, the development of an important transcurrent fault parallel to the continental margin, a lack of migration of the magmatic foci with time and very mild backarc deformation, have resulted in a low mountain range with a relatively thin continental crust. These relations could be useful for interpreting similar situations in the past development of other segments of the Andes.

1 Introduction

The Andes are a typical example of a modern orogenic belt developed in an active continental margin with subduction of the adjacent ocean floor as the main cause of magmatic and tectonic processes. However, the mountain range has a longitudinal segmentation, which gives rise to differences in topography, magmatism and geological evolution.The present topographic height of the Andes in Chile gradually decreases from north to south, as does the thickness of the continental crust and the absolute values of the Bouguer gravity anomalies. In the segment of the belt considered here, only the summits of the active stratovolcanoes in the main range are more than 2000 m high. This is in strong contrast with the southern Central Andes (northern Chile), where the stratovolcanoes rise to more than 6500 m and are built on a pre-volcanic substratum more than 4000 m high.

Through a brief account of the main rock units in this part of the southern Andes and by highlighting the geological evolution, this chapter seeks to explain the strong difference between this segment of the Andes and the southern Central Andes, which form the main subject of this book.

Correspondence to: F. Hervé, Departamento de Geología y Geofísica, Universidad de Chile, Casilla 13518-Correo 21, Santiago-Chile

2 The Palaeozoic Basement

The oldest rocks of the main range in this region (Fig. 1) are low grade metamorphic slates and greenstones with occasional well preserved pillow structures. A higher metamorphic grade is observed where these rocks are intruded by the Meso-Cenozoic NPB. They are considered to be part of the wide late Palaeozoic accretionary complex which is well exposed in the coastal ranges of Chile at the same latitudes (Godoy et al. 1984).

The age of the regional metamorphism is probably indicated by a ca. 300 Ma isochron in low grade metasediments at Huinay (Hervé et al. 1990). Similar radiometric ages have been obtained in the accretionary complex of the coastal ranges (Davidson et al. 1987). At Buill, Devonian trilobites occur in slate boulders, similar in lithology to the slates which have been dated at Huinay (Levi et al. 1966).

There is no evidence known of older continental crust or Palaeozoic granitoids from this part of the Andes. This means that the Palaeozoic magmatic arc and the old continental crust of Gondwana lay east of the present mountain belt. This is in contrast with the situation of the Central Andes, where Precambrian and early Palaeozoic units (Baeza and Pichowiak 1988) are found in the main range.

3 The Mesozoic-Cenozoic Development of a Magmatic Arc

During this period subduction of the Pacific crust beneath the South American continent produced a magmatic arc which grew over the late Palaeozoic accretionary complex.

The magmatic arc was located along the present outcrops of the NPB. A backarc basin with marine episodes is represented by the middle to late Jurassic Futaleufu Group and by the early Cretaceous Palena Group (Thiele et al. 1979), which presently crop out east of the NPB.

In the early Cretaceous, 110 to 120-Ma monzonite plutons, which form the easternmost 20 km of the NPB (Hervé et al. 1990) intruded the mentioned backarc sequences. Dioritic, granodioritic and tonalitic plutons constitute the bulk of the more central portions of the NPB. They have yielded late Cretaceous ages (Drake et al. 1990) immediately to the east of the LOFZ (see below). Miocene and Pliocene intrusion ages are present only around this main structure which, in this region, lies near to the western margin of the NPB.

These data imply that the locus of subduction related arc magmatism at these latitudes remained nearly stationary from the Jurassic to the Miocene. On the other hand, a 150 to 200-km eastward migration of magmatic arcs during the Meso-Cenozoic is a common feature of the geological development of the southern Central Andes, as has been indicated, for example, by Reutter et al. (1988).

The present day magmatic arc, characterized by a chain of andesitic to basaltic stratovolcanoes, is located along the LOFZ (Fig. 1), indicating that no significant E-W lateral displacement of the magmatic foci has taken place since the Miocene.

4 The Liquiñe-Ofqui Fault Zone (LOFZ)

This major trench-linked dextral strike slip structure extends from the vicinity of Istmo de Ofqui along the Andes for more than 1100 km into the lake region. Its origin has been related to oblique subduction beneath the continental margin (M. Hervé 1976; Beck 1988; García et al. 1988) or to the indentor effect of subduction of the Chile Rise beneath the continent (Forsythe and Nelson 1985). The history of the fault is not known in detail. M. Hervé (1977) mentions a 3-km-wide belt of fault rocks near its northern end and these are cut by a 28 Ma dacite porphyry which is not affected by mylonitization.

Forsythe and Nelson (1985) indicate that at Golfo de Penas, south of the Ofqui isthmus, which is considered to be a pull-apart basin related to the dextral movement of the LOFZ, the sedimentary record starts in Eocene. The same age is recorded by Levi et al. (1966) for the Ayacara Formation, a folded volcanoclastic marine sequence which outcrops sporadically west of the LOFZ, which may also have been deposited in short lived pull-apart basins. In the Reloncaví area, the fault zone separates late Cretaceous plutons to the east, from Miocene plutons to the west. According to Cembrano (1990), at mainland Chiloe a syntectonic tonalitic pluton was emplaced in a ductile shear zone near the main trace of the LOFZ. The age of the pluton (Drake et al. 1990; Hervé et al. 1990) ranges between 4.7 ± 0.5 Ma (Rb-Sr whole rock isochron) and 3 Ma (biotite and hornblende Ar-Ar dates). South of Lago Yelcho, a horizontal lacustrine sedimentary sequence deposited along the topographic depression carved over the main branch of the LOFZ, is topped by a 1.2 Ma basaltic flow (Hervé et al. 1990) which seals the activity of the fault. Muir Wood (1989) proved that near the southern end of the LOFZ, the fault has been active during the Quaternary. Furthermore he believes that there was 1200 m of relative vertical uplift of the eastern block. Thus, the LOFZ appears to be a structure which has been active at least since the late Oligocene in its northern end and which is still active near its southern end. Since the Miocene,

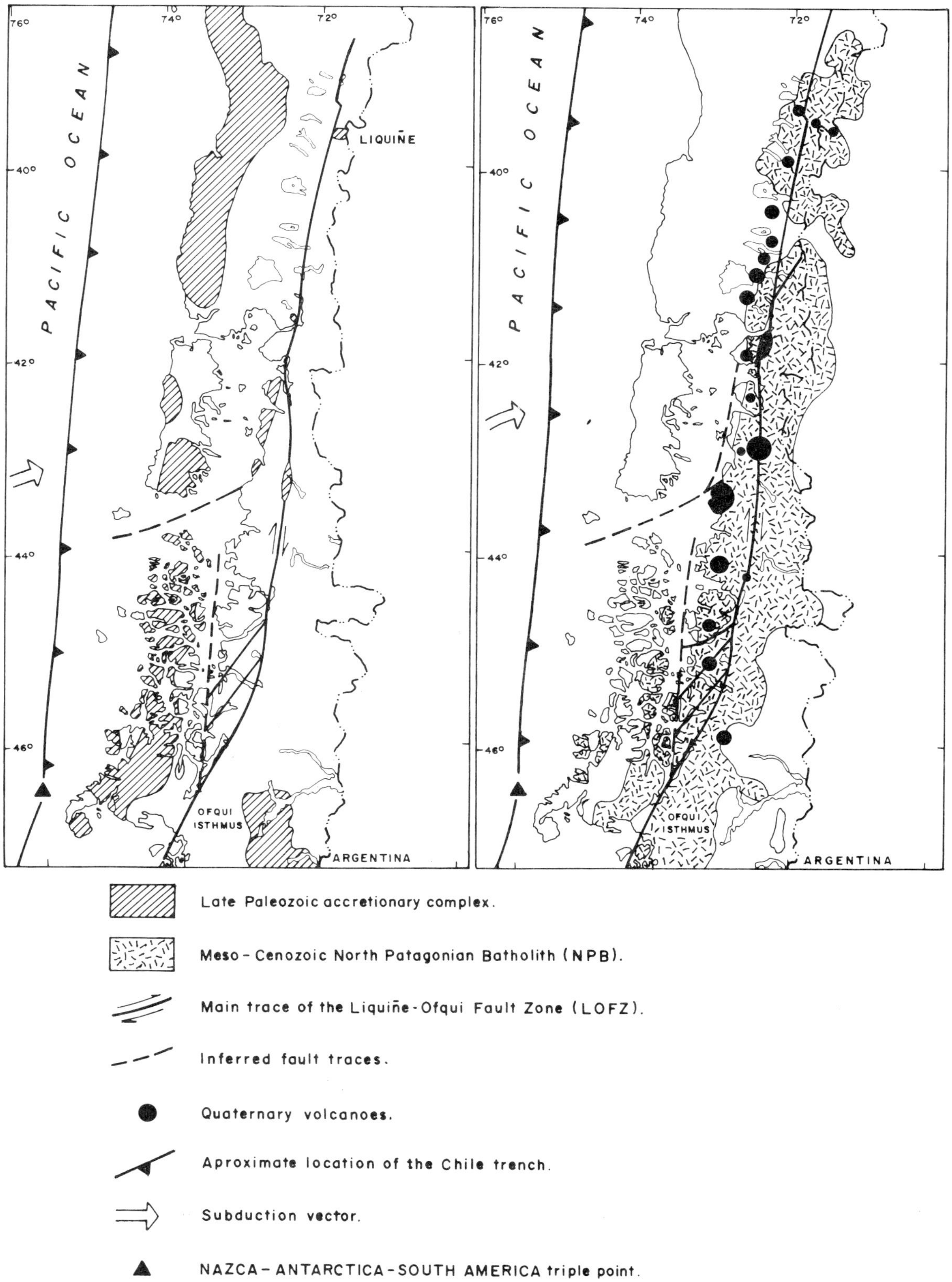

Fig. 1. Relation of the Liquiñe-Ofqui Fault Zone to the main geological units of the southern Andes between 39° and 44°S

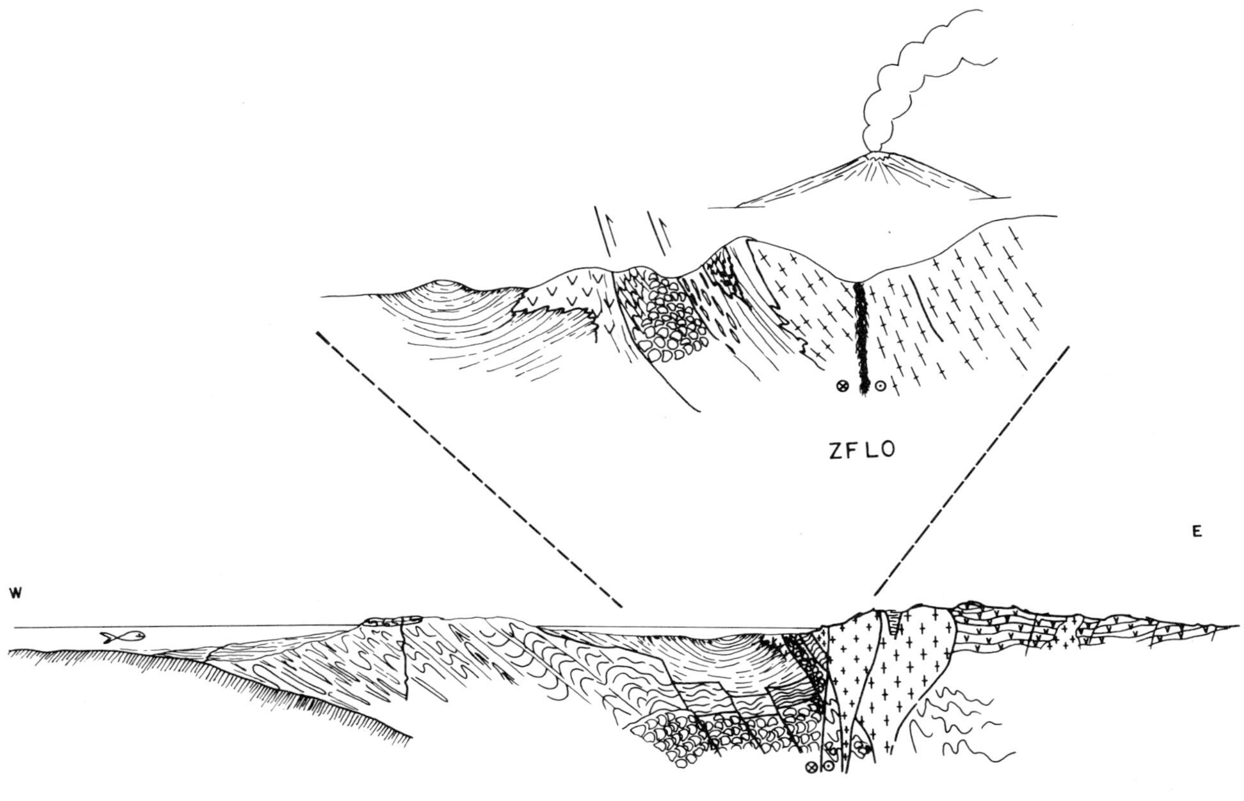

Fig. 2. Generalized schematic cross section of the southern Andes at 42°S

it has controlled the location of plutonic intrusions and is still controlling the access of magmas to the surface in the present chain.

Palaeomagnetic evidence (García et al. 1988) allows 440 to 550 km of northward displacement of the coastal block, with some kilometres of differential uplift of the eastern block.

5 Tectonics of the Southern Andes

After the build-up of the late Palaeozoic accretionary complex, a Meso-Cenozoic magmatic arc and a short-lived Mesozoic marine backarc basin were developed over the deformed and metamorphosed rocks of the complex.

Since the onset of the Mesozoic subduction regime which has persisted to the present, the continental margin has experienced deformational events which, in contrast to the Central Andes, did not lead to a significant crustal thickening. It appears that a considerable portion of the strain induced by the subduction on the continental margin has been dissipated through the transcurrent movements along the LOFZ, generated along the thermally weakened crust of the magmatic belt.

The deformation in the forearc can be summarized as follows. The western portions of the subduction complex register a late Jurassic-early Cretaceous reactivation (Davidson et al. 1987) which resulted in the generation of a penetrative S2 crenulation cleavage and a metamorphic phase that reset Rb-Sr and K-Ar systems. The eastern parts of the complex were apparently not affected by this event. The Cretaceous magmatic arc was later uplifted and deeply eroded before the unconformable deposition of marine Eocene sediments (La Cascada Formation, Thiele et al. 1979) remnants of which are preserved in tectonic depressions in the NPB.

The onset of activity along the LOFZ probably occurred as early as the Eocene, as indicated by the 28 Ma dyke crosscutting mylonites at Liquiñe, and by the sedimentary record in the pull apart basins west of it which started in the Eocene (Levi et al. 1966; Forsythe and Nelson 1985). Dalla Salda and Franzese (1987) describe the formation of pull apart basins to the east of the main range, starting in the Eocene and lasting until the late Miocene; these were accompanied by intense acid to basic magmatism. Acid volcanism took place near the LOFZ during the Miocene and Pliocene. The deposits of these basins were folded during this period.

No fold-and-thrust belt developed to the east of the arc behind the span of LOFZ. The Jurassic-early Cretaceous sequences are affected only by open broad folds and by normal faulting of unknown age (Ramos and Mpodozis 1989) (Fig. 2). Other segments of the Andes, including the Central Andes, develop very conspicuous backarc fold-and-thrust belts, a mechanism indicative of crustal shortening and thickening. The LOFZ and the forearc region seem thus to have absorbed most of the Meso-Cenozoic deformation in this segment of the Andes. Palaeomagnetic data (García et al. 1988) for the Lake District indicate clockwise rotation of rocks as young as Pliocene, west of the LOFZ.

6 Conclusions

The Andes are a young orogenic belt which has been built by magmatic and tectonic processes related to the subduction of the oceanic crust beneath the leading edge of the South American plate. The segment of the Andes between 39° and 44°S lat. has low elevations and a thinner continental crust than that of the Central Andes. These characteristics can be explained by a different geological evolution of the segments.

In the southern Andes considered here, the Meso-Cenozoic magmatic arc was built over a late Palaeozoic accretionary prism with no Precambrian continental basement. The magmatic arc had a rather fixed position inland of the continental margin, when compared with the eastward migration observed in the Central Andes. The magmatic arc was built in the vicinity of a major dextral strike-slip fault, which has been active at least since the Eocene. Since that time this fault zone has controlled the access of magmas to upper levels in the crust. The forearc region, and the LOFZ which limits it to the east appear to have absorbed most of the strain at the leading edge of the continental margin, inhibiting the development of a backarc fold-and-thrust belt. A similar situation might have existed in the late Jurassic-early Cretaceous evolution of the southern part of Central Andes, when a major arc related fault zone, the Atacama Fault Zone, was active.

The different features of the segments of the Andes are thus dependent on the way the strain, related to the compressive forces generated by subduction, is partitioned into components either more parallel or more perpendicular to the continental margin.

Acknowledgements. Field work was supported by FONDECYT grants 0568/88 and 1181/90. Discussions with M. Beck, E. Godoy, F. Munizaga, R.J. Pankhurst and R. Drake among others, have helped in the development of the ideas presented here. An invitation from the FU and TU, Berlin, to the "Final Workshop on Structure and Evolution of the Central Andes" greatly assisted with the writing of this chapter. Revisions by M. Atherton and an anonymous referee significantly improved the text.

References

Baeza L, Pichowiak S (1988) Ancient crystalline basement provinces in the north Chilean Central Andes - Relics of continental crust development since the mid Proterozoic. In: Balburg H, Breitkreutz Ch, Giese P (eds): The Southern Central Andes. Lecture Notes in Earth Sciences, 17. Springer, Berlin Heidelberg New York, pp 3-24

Beck M (1988) Analysis of late Jurassic-Recent paleomagnetic data from active plate margins of South America. J S Am Earth Sci 1: 39-52

Cembrano J (1990) Geología del batolito norpatagónico y rocas metamórficas de su margen occidental (41°50'-42°10'S), Chile. Thesis, Dep de Geología, Univ de Chile (unpubl)

Dalla Salda L, Franzese J (1987) Las megaestructuras del Macizo y la Cordillera Norpatagónica argentina y la génesis de las cuencas volcano-sedimentarias terciarias. Rev Geol Chile 31: 3-14

Davidson J, Mpodozis C, Godoy E, Hervé F, Pankhurst RJ, Brook M (1987) Late Paleozoic accretionary complexes on the Gondwana margin of southern Chile: Evidence from the Chonos archipelago. In: McKenzie E (ed) 6th Gondwana Symp, AGU Monogr, 40: 221-227

Drake R, Hervé F, Munizaga F, Pankhurst RJ (1990) Radioisotopic ages of the North Patagonian batholith and their relation to the Liquiñe-Ofqui Fault zone, Chile (40°-42°S lat.). VIIth Int Conf Geochronology, Cosmochronology and Isotope Geology. Abstr vol 29. Canberra, Australia

Forsythe R, Nelson E (1985) Geological manifestations of ridge collision: Evidence from the Golfo de Penas-Taitao Basin, Southern Chile. Tectonics 4: 447-495

García A, Beck M, Burmester R, Munizaga F, Hervé F (1988) Paleomagnetic reconnaissance of the Region de los Lagos, southern Chile and its tectonic implications. Rev Geol Chile 15: 13-30

Godoy E, Davidson J, Hervé F, Mpodozis C, Kawashita K (1984) Deformación sobreimpuesta y metamorfismo progresivo en un prisma de acreción paleozoico: Archipiélago de Chonos, Aysén, Chile. Actas IX Cong Geol Argent, San Carlos de Bariloche T4, 211-232.

Hervé F, Pankhurst RJ, Cembrano J, Munizaga F (1990) Magmatism and tectonics in the Andes of Chiloe (42° -45°S), Chile. In: Symp Int "Geodynamique Andine". ORSTOM, Colloques et Seminaires, Grenoble, pp 305-308

Hervé M (1976) Estudio geológico de la falla Liquiñe-Reloncaví en el área de Liquiñe: Antecedentes de un movimiento transcurrente (Province de Valdivia). Actas 1er Congr Geol Chil, Santiago, 1: B39-B56

Hervé M (1977) Geología del área al este de Liquiñe, Provincia de Valdivia, Xa Región, Chile. Thesis, Dep Geología, Univ. de Chile, Santiago, 111 pp (unpubl)

Levi B, Aguilar A, Fuenzalida R (1966) Reconocimiento geológico en las Provincias de Llanquihue y Chiloé. Bol 19, IIG, Santiago, 45pp

Muir Wood R (1989) Recent normal faulting at the Laguna de San Rafael, Aisen province, southern Chile. Comunicaciones 40: 57-68

Ramos V, Mpodozis C (1989) The Andes of Chile and Argentina. In Geology of the Andes and its relation to hydrocarbon and mineral resources. In: Ericksen G, Cañas, M, Reinemund JA (eds) Circum Pacific Council for Energy and Mineral Resources. Earth Sci Series 11: 59-90

Reutter K, Giese P, Götze HJ, Scheuber E, Schwab K, Schwarz G, Wigger P (1988) Structures and crustal development of the Central Andes between 21°S and 25°S. In Bahlburg H, Breitkreuz C, Giese P (eds) The Southern Central Andes. Lecture Notes in Earth Sciences 17, Springer, Berlin Heidelberg New York, pp 231-261

Thiele R, Castillo JC, Romero G, Ulloa M (1979) Geología del sector fronterizo de Chiloé Continental entre los 43° y 43°45'lat. S, Chile. Actas VII Congr Geol Argent Neuquén 1: 577-591

Terranes of Southern Gondwanaland and Their Control in the Andean Structure (30°-33°S Latitude)

VICTOR A. RAMOS

Abstract. A revision of the present knowledge of the late Proterozoic-Palaeozoic accretionary history in this segment of southwestern Gondwanaland enables us to discuss its role in the control of the geometry of Andean tectonics. The previous terrane boundaries were reactivated by extension in early Mesozoic times during the break-up of Gondwanaland. The asymmetric rift systems that formed were located in the hanging wall of the Proterozoic-Palaeozoic sutures. Cenozoic shortening was controlled by tectonic inversion involving basement in the inner areas of the Coastal and Principal Cordilleras, while in the eastern Principal Cordillera and Precordillera a thin-skinned fold-and-thrust belt developed. The boundary between the Sierras Pampeanas and the Precordillera has been tested by deep-seismic reprocessing, and shows oblique discontinuities down to more than 30 km. The boundary between the Precordillera and Frontal Cordillera coincides with the slope of the early Palaeozoic continental margin of Gondwanaland. It is concluded that most of the present morphostructural boundaries match crustal discontinuities inherited from the earlier accretionary history.

1 Introduction

Research conducted in several areas of southern Gondwanaland in recent years has revealed the existence of suspect terranes throughout a fringe parallel to the present subduction zones. These terrane boundaries are major crustal discontinuities which exerted a strong influence in younger deformational episodes, by either extensional or compressional regimes. The aim of this chapter is to describe the major terranes of southern Gondwanaland in the Central Andes and discuss how they have influenced the final geometry during Mesozoic and Andean deformations.

In this chapter the term terrane implies a block which has travelled a great distance and has accreted to a plate boundary; thus, the term is not used in a merely descriptive way (for further discussion, see Dover 1990). Analyses are based mainly on extensive fieldwork carried out over recent years in this sector of the Andes and on new geophysical data available for some of the proposed sutures. The description and geological basis for the distinction of different terranes may be found in several previous papers (Ramos et al. 1984, 1986; Ramos 1988, 1989; Mpodozis and Ramos 1990); this chapter will concentrate on the influence that these previously recognized terranes have on the Meso-Cenozoic tectonic evolution of the region.

Correspondence to: V. A. Ramos, Departamento de Ciencias Geologícas, Universidad de Buenos Aires, Ciudad Universitaria - Pabellon H, (1428) Buenos Aires, Argentina

2 Terranes in Gondwanaland

Since the early concepts of Keidel (1921) and Du Toit (1927) it has been evident that an orogenic Gondwanides belt extended across different continents from the Precordillera in Argentina, through the Ventania System, Cape Fold Belt, Transantarctic Mountains, Tasman Fold Belt, as far as eastern Australia. This belt, known as the Samfrau geosyncline, has been attributed several tectonic interpretations (Keidel 1921; Borrello 1969, among others). Some authors interpreted the Ventania System and related regions either as an aulacogenic chain (Harrington 1970; Miller 1983) or as an intracratonic belt (Visser 1985). On the other hand, other authors, with a more mobilistic approach, have interpreted the Samfrau belt as the result of a backarc deformation related to a flat subduction from the Pacific side (Lock 1980; Forsythe 1982; Uliana et al. 1986; Dalziel et al. 1987), or of a collision between different blocks and the continental margin (Ramos 1984, 1986; De La Winter 1984).

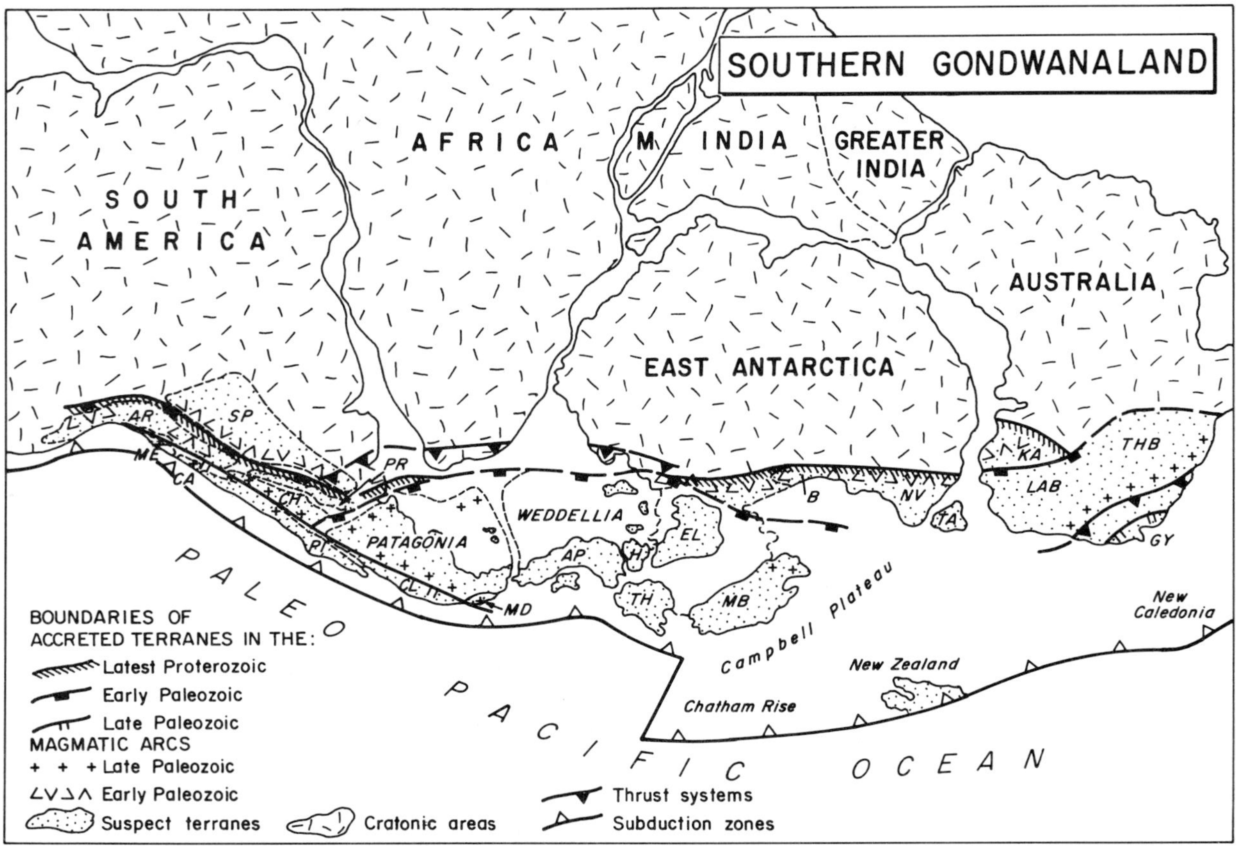

Fig. 1. Southern Gondwanaland and the accreted terranes. (After Ramos 1986, 1990; Powell 1990.) Terranes accreted in the latest Proterozoic: *SP* Sierras Pampeanas; *AR* Arequipa; *B* Beardmore; *KA* Kanmantoo belt. Terranes accreted by the late Ordovician: *PR* Precordillera; *NV* different terranes of Northern Victoria Land; *TA* Tasmania; *LAB* Lachlan belt; *THB* Thomson belt. Terranes accreted by the late Devonian-early Carboniferous: *CH* Chilenia; Patagonia and Weddellia: *AP* Antarctic Peninsula; *EL* Ellsworth; *H* Haack; *TH* Thurston; *MB* Marie Bird. Terranes accreted by the late Palaeozoic-Triassic: *ME* Mejillonia; *CA* Chañaral; *PI* Pichidangui; *CL* Chiloé; *MD* Madre de Dios; *GY* Gympie

The latter interpretation implies that most terranes located south of the Samfrau geosyncline are considered to be allochthonous or at least suspect terranes. This last hypothesis has gained support in recent years (Borg and De Paolo 1989; Ramos 1989; Storey 1989; Powell 1990; Rowell and Rees 1990). The still scarce palaeomagnetic evidence indicates the allochthony of part of these terranes, based on new palaeomagnetic poles that do not match the apparent polar wandering path of cratonic South America (Forsythe et al 1986; Rapalini and Vilas 1990; etc.).

The present state-of-the-art for the different terranes recognized in southern Gondwanaland is shown in Fig. 1. The following discussion will focus exclusively on the terranes of the Central Andes between 29° and 34°S lat.

3 Geological Provinces in the Central Andes

The present morphological and structural characteristics of the different cordilleras across the Central Andes show a strong involvement of the previous late Proterozoic-Palaeozoic terrane history on the superposed Andean deformation. The geological provinces that can be recognized between 30° and 33°S lat. are described below (see Fig. 2).

3.1 Sierras Pampeanas

This geological province located in central Argentina, is characterized by a series of crystalline basement blocks of Precambrian-early Palaeozoic age, uplifted

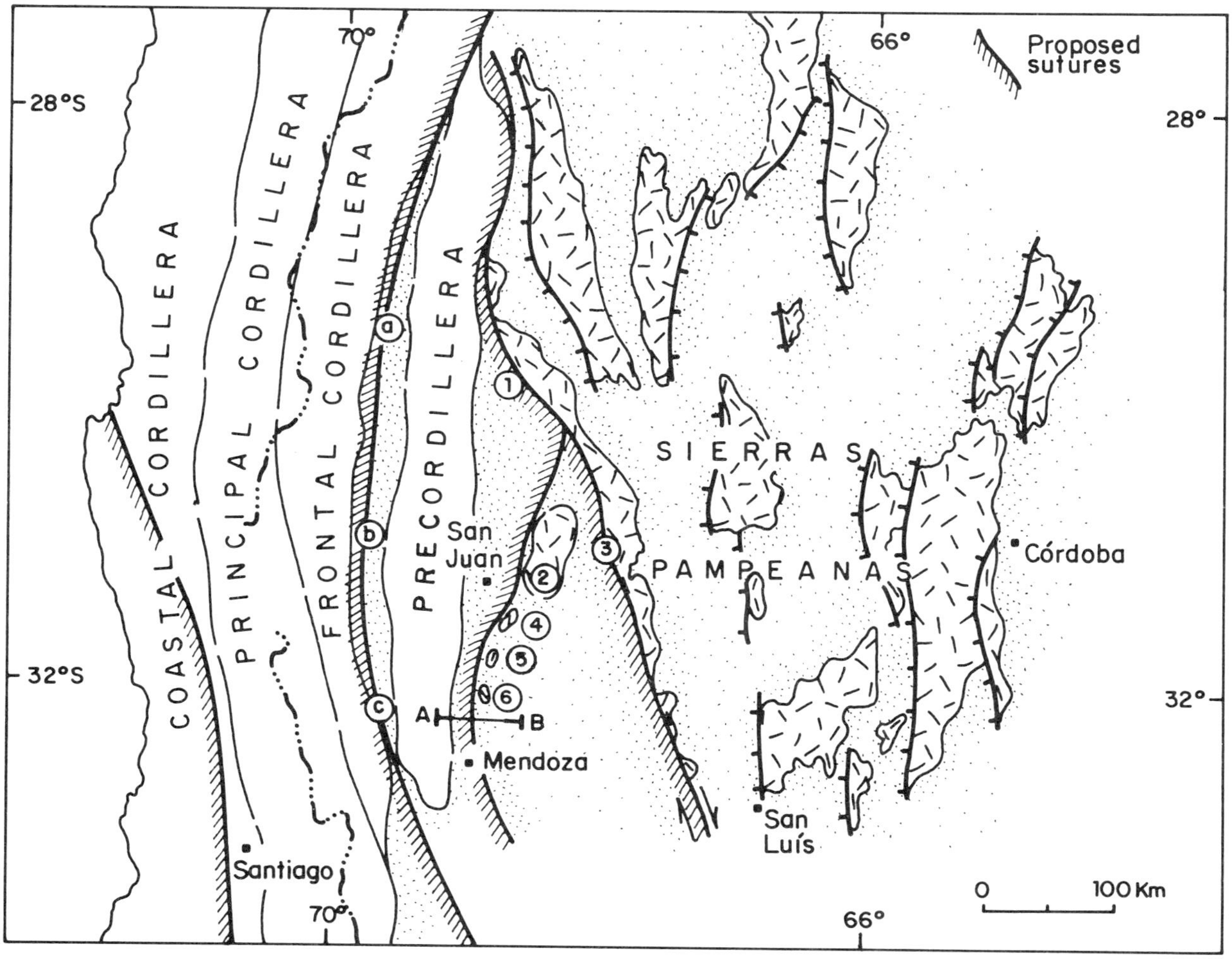

Fig. 2. Geological provinces of the Central Andes with indication of the proposed sutures. (After Ramos 1988b.) Localities cited: *1* Sierra de Valle Fértil; *2* Sierra de Pié de Palo; *3* Sierra de La Huerta; *4* Cerro Valdivia; *5* Cerro Barbosa; *6* Cerro Salinas. *a-c* Iglesia - Calingasta - Uspallata valley. *A-B* Section of Fig. 4. Dots indicate surrounding ranges and other topographical highs

and tilted during Tertiary Andean compression and associated with a flat subduction episode (Jordan et al. 1983). The resulting structure closely resembles that of the Laramide region of the United States (Jordan and Allmendinger 1986).

3.2 Precordillera

The Precordillera is an Andean thrust-and-fold belt that has developed on an early Palaeozoic carbonate platform with a typical thin-skinned structure. The western edge of the Precordillera coincides with a longitudinal depression known as the Iglesias-Calingasta-Uspallata valley. This present tectonic trough is similar to the Canadian Rocky Mountains trench (Price 1981). Both depressions have a strikingly similar present morphology controlled by the old Palaeozoic continental margin (Baldis et al. 1982). East of the trench the slope facies of the early Palaeozoic continental platform is preserved (Spalletti et al. 1989).

3.3 Frontal Cordillera

This unit was built during the Gondwanides orogeny in late Palaeozoic-early Mesozoic times. It is the result of an Andean-type subduction orogenesis followed by a generalized extension. Most of this province is composed of late Palaeozoic-Triassic magmatic rocks. During the Andean deformation the Frontal Cordillera behaved as a rigid block, and it is characterized by a few thick-skinned thrusts.

3.4 Principal Cordillera

The Principal Cordillera, or Main Andes, was the locus of the Andean orogeny during latest Mesozoic and Cenozoic times. The Jurassic and Cretaceous marine deposits were deformed in different ways based on the diverse participation of the basement in the deformation. Moscoso and Mpodozis (1988) described a thick-skinned tectonics for the northern sectors while the highest Andes are characterized by thin-skinned structures such as the Aconcagua fold-and-thrust belt (Ramos 1988a).

3.5 Coastal Cordillera

Along the present continental margin a series of late Palaeozoic metamorphic rocks have been preserved and these represent pieces of an accretionary prism which has developed since late Devonian times (Hervé 1988). This metamorphic basement is emplaced by a series of granitoids and mafic rocks of Jurassic and Cretaceous age. Recent palaeomagnetic studies indicate that most of this region is believed to have undergone significant latitudinal motions (Forsythe et al. 1986; García et al. 1988).

4 Terranes in the Central Andes

The proposed sutures among the different terranes are indicated in Fig. 2 and the tectonostratigraphic history represented in Fig. 3.

The Sierras Pampeanas terrane was accreted to the Río de La Plata craton during the late Proterozoic. A subduction zone, dipping away from the older craton, was responsible for the magmatic arc represented by a series of granitoids, gabbros and tonalites developed in the eastern Sierras Pampeanas during 640-570 Ma (Cingolani and Varela 1975). The metavolcanic rocks of this basement also show arc affinities (Delakowitz 1988). The final amalgamation and a 15-km uplift occurred during the latest Proterozoic-early Cambrian (Ramos 1988b, 1989). As a result of this the western Chaco-Paraná Basin was developed, and this shows a typical asymmetry characteristic of foreland peripheral basins (see Pezzi and Mozetic 1989, Figs. 1-2).

During the early Palaeozoic the western Sierras Pampeanas, were located more than 400 km away from the present trench but, even so, the magmatism has petrological and geochemical characteristics typical of an active continental margin (Lottner and Miller 1986; Aceñolaza and Toselli 1988; Pérez and Medina 1990). The ductile deformation of this basement which has been studied by many authors (Lottner and Miller 1986; Llambías et al. 1991), and the age of the magmatism indicate that this continental margin was active from the late Precambrian to the late Ordovician (González et al. 1985; Aceñolaza and Toselli 1988). The western border was characterized by a synthetic fold-and-thrust belt (sensu Roeder 1973) with a west vergence (Cominguez and Ramos 1991). As seen in the Sierra de Pie de Palo, two belts of different metamorphic grade are well exposed (Dalla Salda and Varela 1984). The eastern belt is represented by a middle to high metamorphic grade of amphibolite facies, while the western belt has an intense deformation with mylonites and cataclastic rocks in zeolite to green-schist facies. Based on such evidence Dalla Salda and Varela (1982) postulated a subduction zone in the western side of Sierras Pampeanas.

This zone of high deformation coincides with the present boundary between the Sierras Pampeanas and Precordillera. Different isolated patches of metamorphic basement are exposed in the small hills of Cerros Valdivia, Barbosa and Salinas (see Fig. 2). The metamorphic and structural analysis of this basement indicates a mylonitic deformation as suggested by Llano et al. (1984) and by Vaca and Rossa (1988).

Deformation is also present in the eastern Precordillera but is exposed at a higher structural level. The flysch deposits of the late Ordovician-Silurian Rinconada Formation are tightly folded with carbonate olistoliths of early Ordovician rocks preserved in Silurian beds (Peralta and Uliarte 1986). The combination of shear zones and intense folding has produced in the eastern border of the Precordillera a series of tectonic melanges as a result of several deformation phases in the late Ordovician during the Ocloyic movements (Furque 1972; Baldis et al. 1982). Most of this deformation took place close to the Ordovician-Silurian boundary, when the present Precordillera was finally amalgamated with the Sierras Pampeanas. Although the nature of this amalgamation is still a matter for debate (Baldis et al.

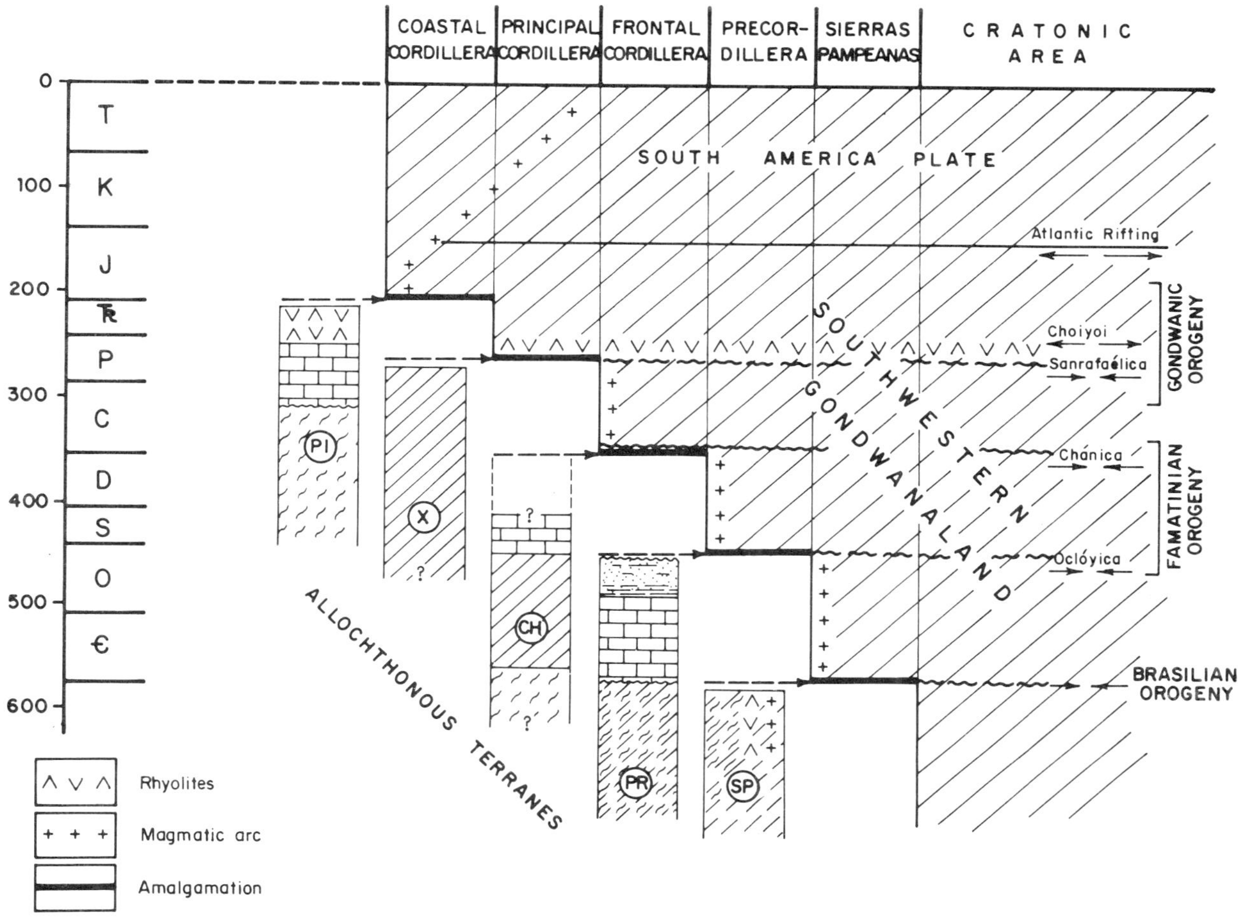

Fig. 3. Tectonostratigraphic diagram indicating the docking of the different allochthonous terranes and the successive margins of southwestern Gondwanaland. (After Forsythe et al. 1986; Ramos et al. 1986; Mpodozis and Kay 1990.) Terranes: *PI* Pichidangui; *X* Equis; *CH* Chilenia; *PR* Precordillera; *SP* Sierras Pampeanas

1984; Aceñolaza and Toselli 1988; Ramos et al. 1986), the striking effects of this accretion are detected in the Sierras Pampeanas, where subduction-related magmatism had ceased by the end of the Ordovician and was shifted westwards. During the Silurian-Devonian scarce magmatism is recorded in the Precordillera with the emplacement of the Cacheuta granodioritic stock (Caminos et al. 1982) and of several other plutons in the northern Precordillera, such as Cerro Imán (Furque 1968).

The western border of the Precordillera coincided with the continental margin of western Gondwanaland during most of the early Palaeozoic, as is recorded for Sierra de Tontal by Cingolani et al. (1989). The reconstruction of this continental margin is based on the sedimentary facies, and palaeontological evidence (Baldis et al. 1982; Spalletti et al. 1989), and geochemical characteristics of the ophiolitic assemblages which developed along the western border of the Precordillera (Haller and Ramos 1984; Kay et al. 1984). The imbricated structure of the Ordovician and Siluro-Devonian rocks indicates intense deformation, during the middle to late Devonian (Cucchi 1972) and lasting up to the early Carboniferous (Caminos et al. 1982). The thrusting has a west vergence as seen in the Cordones de Bonilla y Farallón (Keidel 1939) and in the Sierras de Cortaderas and Sandalio (Cortés 1989). This deformation, related to the Chanic movements, was interpreted as the result of a collision of the Chilenia terrane with the western Precordillera continental margin (for details see Ramos et al. 1986).

There is little information on the nature of the basement of the Chilenia terrane, known only from a few outcrops of metamorphic rocks in the Cordón del Plata (Caminos 1965), and by the La Pampa gneisses (Mpodozis and Cornejo 1988). However, the extended late Palaeozoic acidic magmatism of the Frontal Cordillera of Argentina and Chile, has been taken as indirect evidence of a sialic basement (Nasi et al. 1985).

The Carboniferous marine units are the first sedimentary record of the amalgamation of the previously described terranes. Continental deposits are widespread in the Sierras Pampeanas and eastern Precordillera, nearshore marine facies are widespread in the western Precordillera and estuarine to turbiditic facies are found along the Frontal and Principal Cordilleras. All these units, encompassing most of the Carboniferous up to the early Permian, were intensely folded and thrust during the Sanrafael orogenic phase (Ramos 1988a). Important subduction-related magmatic activity is recorded during the late Palaeozoic. When compared with the location of the early Palaeozoic magmatic belt, this late Palaeozoic belt was shifted more than 300 km towards the ocean (Ramos et al. 1986). Some authors have interpreted this shift as the result of a collision with an unidentified terrane associated with the compression of the Sanrafael orogenic phase ("X" terrane of Mpodozis and Kay 1990). According to Mpodozis and Kay (1990) this proposed terrane, west of the present location of Chilenia, has been displaced from its original location.

Soon after this deformation a general extension took place, while the Gondwana plate remained stationary; this is evidenced by the polar wandering path deduced from palaeomagnetic data (Ramos 1988a). The magmatism was thermally induced during this extensional period and, as a consequence, batholiths such as the Colangüil were emplaced between 264 and 247 Ma ago (Llambías and Sato 1990).

This plutonic activity was followed by a fringe of rhyolitic volcanism within the Palaeozoic accreted terranes, presently exposed along most of the Frontal Cordillera and along the western Sierras Pampeanas and Precordillera. This acidic volcanism, known as the Choiyoi province, is related to a generalized extension during late Permian-Triassic times (Zeil 1981; Kay et al. 1989).

At that time most of Gondwanaland, was part of the Pangaea supercontinent and only minor pieces were displaced along the continental margin, such as the Pichidangui terrane during the late Triassic (Forsythe et al. 1986). These displacements may have been responsible for the elimination of the "X" terrane from the Coastal Cordillera prior to the present emplacement there of the Pichidangui terrane, as has been proposed by Mpodozis and Kay (1990).

The accretion of the Pichidangui terrane marked the final amalgamation of western Gondwanaland at these latitudes. From that time onwards, an Andean subduction regime controlled the tectonic history of the Central Andes (Mpodozis and Ramos 1989).

5 Structure Between Sierras Pampeanas and Precordillera

Since the studies of Borrello (1969) it has been known that there was a zone of intense deformation between the Sierras Pampeanas and Precordillera. Recently, a deep sounding seismic profile was obtained from along the proposed suture to analyse the geometry and nature of this boundary. A mathematical retreatment of uncorrelated Vibroseis seismic data, obtained by YPF (the Argentine oil company), was performed in order to produce a longer seismic profile. An interpretation depth of about 40 km was achieved, and the resolution was good enough to determine deep crustal reflection events (Cominguez and Ramos 1991).

Based on this, it is possible to confirm that the proposed suture coincides with a major crustal boundary that extends down to almost the Moho (Fig. 4). Such boundaries have already been found in many other orogens related to the collision of continental blocks (for example, Ando et al. 1984, in the Appalaches; Meissner et al. 1987, in the Hercynian belts of Europe).

The late Proterozoic synthetic thrust-and-fold belt of Sierras Pampeanas overthrusts the Precordilleran basement, as evidenced by a series of east-dipping oblique reflectors. The interpretation is based on the correlation of Cerro Salinas late Proterozoic structures, intersected by drilling further south of Cerro Salinas, with the time interval recognized in the seismic lines (for details see Cominguez and Ramos 1991). These reflectors were interpreted as mylonitic zones based on a correlation with northern outcrops, as well as with the typical seismic pattern also observed by Allmendinger et al. (1987). These imbrications also indicate a crustal delamination of the lower crust of the Sierras Pampeanas. The structural grain of the suture zone, dominated by east-dipping ancient faults across the middle and upper crust, controls Andean deformation. Most of the

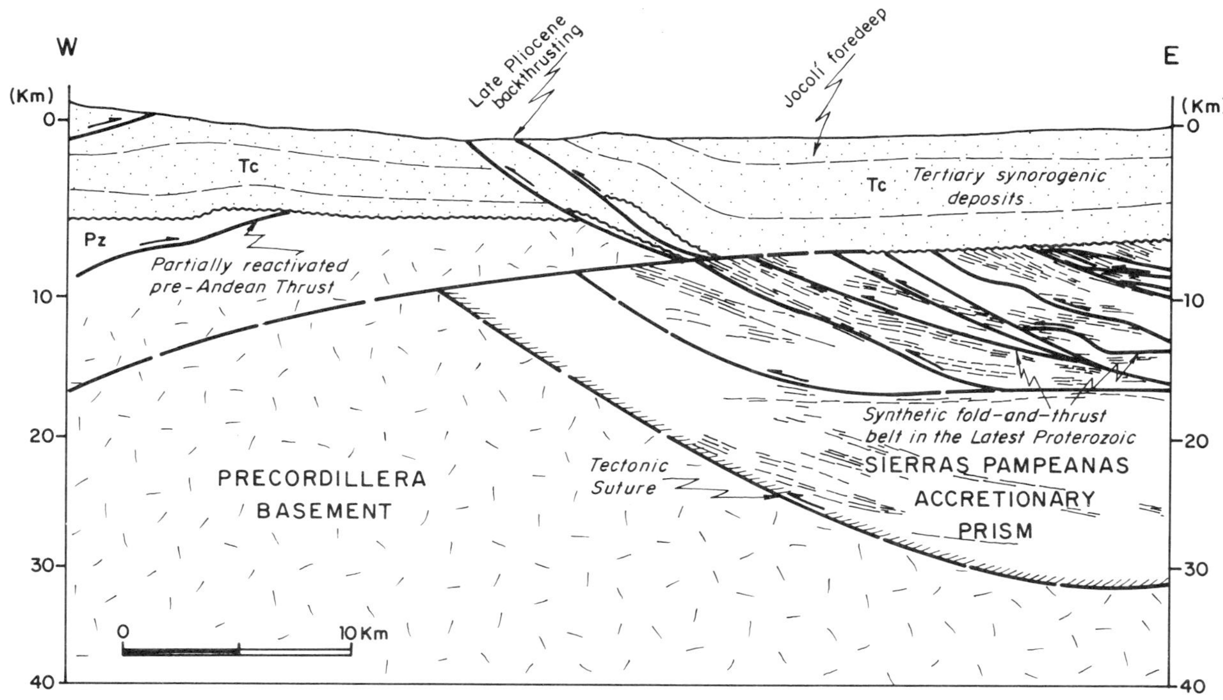

Fig. 4. Schematic section showing the deep structure between the Precordillera and Sierras Pampeanas suture based on the deep seismic sounding profile by Comínguez and Ramos (1990). *Pz* Palaeozoic deposits; *Tc* Late Tertiary deposits.

western Sierras Pampeanas of this segment, such as the Sierras de Valle Fértil, La Huerta, and the southern extension into the San Luis province (see Fig. 2), was uplifted by basement thrusts controlled by this structural grain. Even the eastern blocks of Precordillera, such as the Sierras de Villicum, Zonda and Pedernales, are split by east-dipping faults, in spite of the normal west-dipping thrusts that are dominant in the entire central and western Precordillera.

The present position of the active orogenic front of the Precordillera at Sierras de Zonda and Villicum is characterized by a series of emergent back thrusts (see Uliarte et al. 1987). These thrusts, and also the late Pliocene thrusts depicted in Fig. 4, are controlled by reactivation of the collision boundary.

6 Structure Between Precordillera and Frontal Cordillera

The present structure between these geological provinces is the result of a Mesozoic extension and an Andean compression, superposed with different intensities along the boundary.

6.1 Mesozoic Extension

After the final amalgamation of western Gondwanaland by the end of the late Palaeozoic, and after the formation of the Pangaea supercontinent, a generalized extensional episode occurred along the fringe of the Palaeozoic accreted terranes (Fig. 5) surrounding cratonic Gondwanaland (Uliana and Biddle 1988; Ramos and Kay 1991). The location and distribution of the Triassic rifting was closely controlled by the terrane boundaries. One of the best examples is the Cuyo Triassic rift, which developed entirely in the upper plate of the suture corresponding to the Precordillera terrane. No Triassic rift deposits are present on the Chilenia side (present Frontal Cordillera) of the suture (Fig. 6). In an analogous situation, the Marayes and Beazley Triassic rifts also developed in the upper plate of the Sierras Pampeanas-Precordillera suture. This situation is not unique since in many other parts of the world, the inception of rifting is heavily controlled by crustal weakness zones of previous sutures (e.g. Appalaches, Phinney 1986; Baikal rift, Logatchev 1984).

Another interesting feature is the asymmetry of these Triassic rift basins. Along the analysed segment

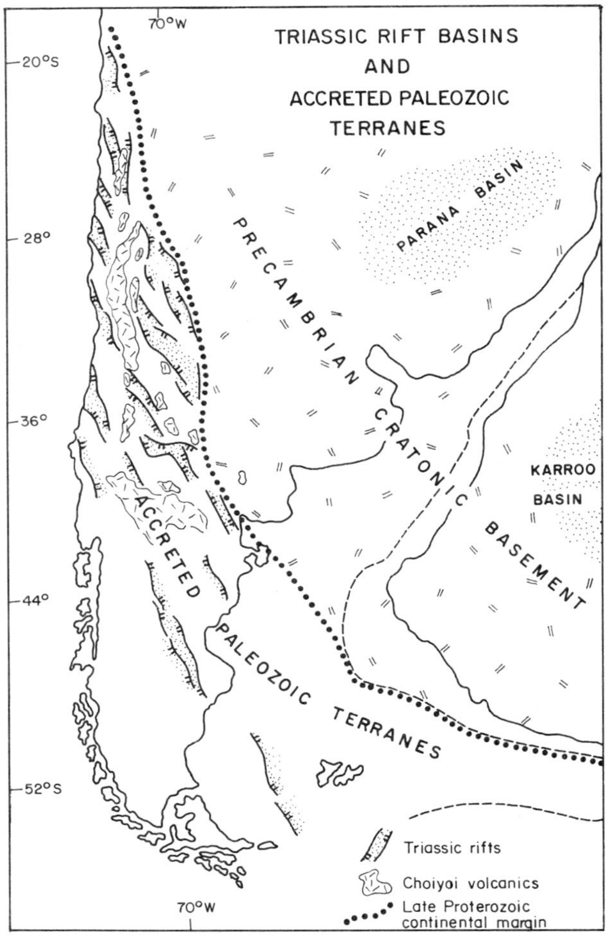

Fig. 5. Development of extension along a fringe of accreted terranes in western Gondwanaland. (After Uliana and Biddle 1988)

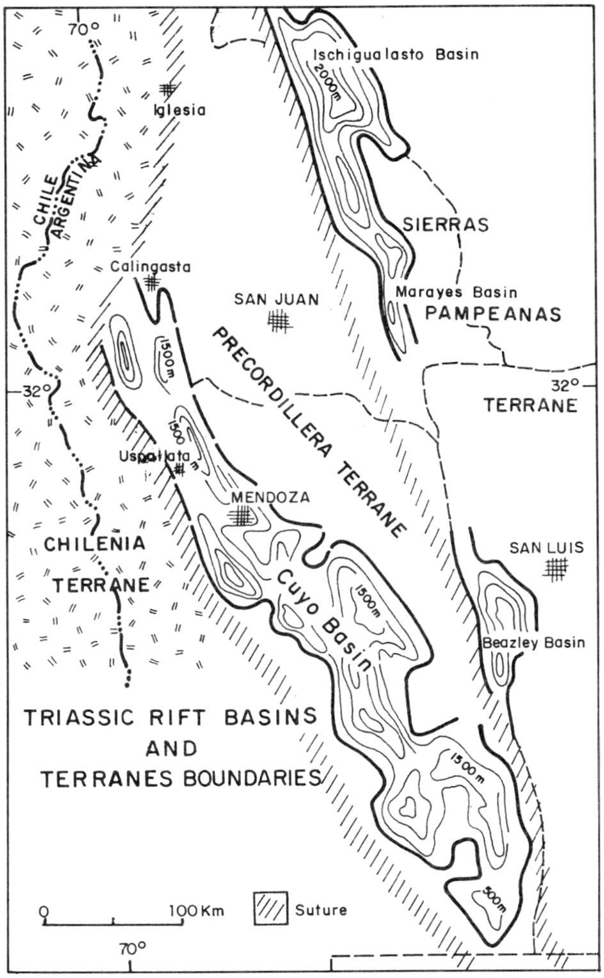

Fig. 6. Inception of the Triassic rifts in the upper plate of the sutured terranes. (After Ramos and Kay 1990)

of the Andes, the deeper parts of the basins are located on the eastern side of the rift. This asymmetry infers an east-dipping detachment that can be correlated with the previous suture (Fig. 7), as has been found in several rift systems (Bosworth et al. 1986; Rosendahl 1987). Similar asymmetries with the same polarity are observed in the Marayes rift and in the other Triassic basins developed along the western Sierras Pampeanas.

Tectonic inversion of these rifts produced the complex structure that characterizes the present Andean geometries.

6.2 Andean structure

The Andean boundary between the Frontal Cordillera and Precordillera corresponds to a tectonic longitudinal valley that coincides with the inception of the early Palaeozoic continental margin of Gondwana, as previously discussed. Although of a tectonic nature, this boundary has different structural styles from north to south. The geometry of the previous Triassic asymmetric rifts, where the location of the master detachment was mainly inferred from the relationship between the synrift and sag facies of the rift, seems to control the present vergence.

The northern segment around Iglesias represents a piggyback basin where a non-emergent thrust ramp

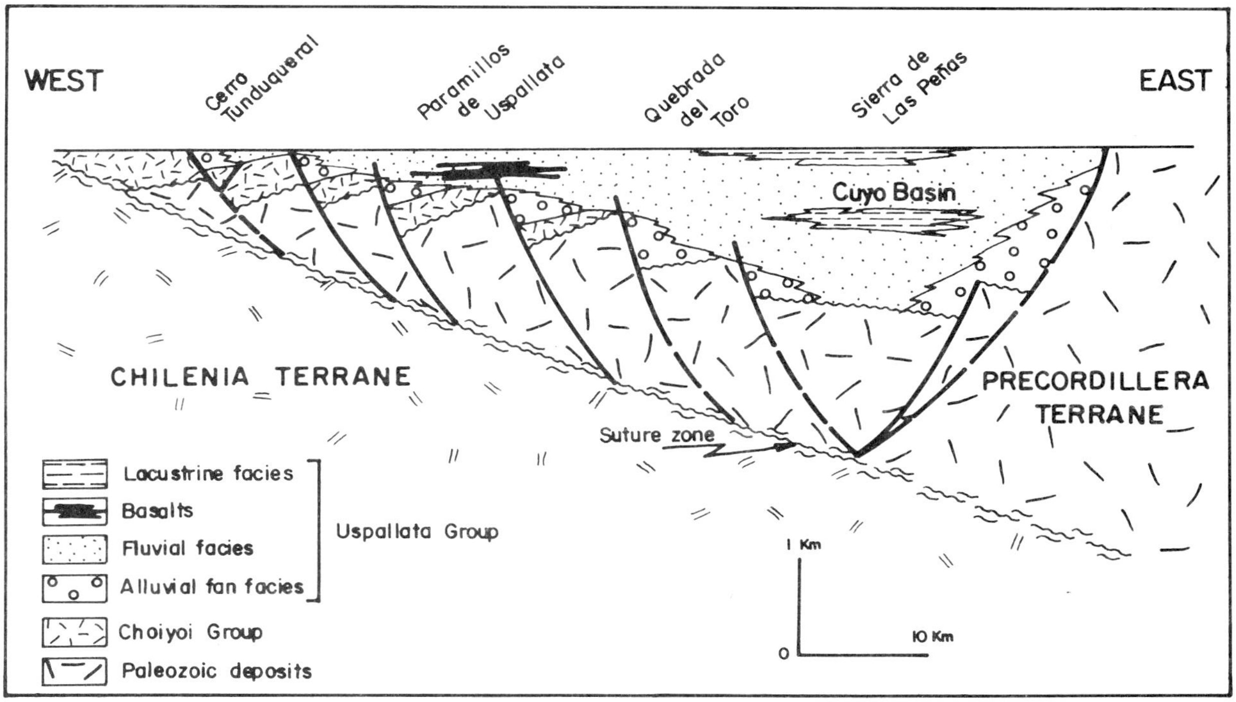

Fig. 7. Conceptual model indicating the asymmetry of a Triassic rift basin developed in the upper plate of the Precordillera terrane. (After Ramos and Kay 1990)

controls the uplift of the Frontal Cordillera (Beer and Allmendinger 1988).

The central segment, south of Calingasta, shows a tectonic inversion of the previous margin with a series of west dipping thrusts such as the ones described at Sierra de Tontal by Cingolani et al. (1989), or those in the eastern Frontal Cordilleran foothills. Present reactivation of these structures indicates strike-slip motion, as observed in the Tigre Fault. This interpretation was based on morphological evidence such as the present displacement of creeks and small valleys, and on kinematic indicators observed in trenches, especially those dug in the Quaternary fault scarps (Bastías and Uliarte 1987).

The southern segment, north of Uspallata, shows the eastern side of the valley to be controlled by a series of back thrusts as seen in Sierras de Cortaderas and Farallón (see Keidel 1939). At this latitude east-dipping thrusts imbricate the ophiolites and turbiditic deposits, in connection with the early Palaeozoic structure of the accretionary prism.

The diverse structural behaviours among the different segments seem to be controlled by the geometry of the previous extension. The northern segment did not have a significant Triassic extension; the eastern border of the Calingasta valley of the central segment corresponds with the inverted west-dipping eastern half of the rift, whereas the eastern border of the southern segment coincides with the western east-dipping side of the rift.

7 Structure of the Principal Cordillera

The incipient extension of Triassic rifting was controlled by the northwest trend of the sutures, and was restricted to the vicinity of the boundaries between the different terranes. Subsequent extension, during most of the Jurassic and up to the early Cretaceous, expanded to reach the axis of the Principal Cordillera (Fig. 8). Extension was again controlled by an east-dipping detachment, responsible for the asymmetry of the intra-arc basins, and for the differing intensity of magmatism across the magmatic arc. The best exposures of these detachments are in Quebrada Paipote, north of Copiapó in Chile, where

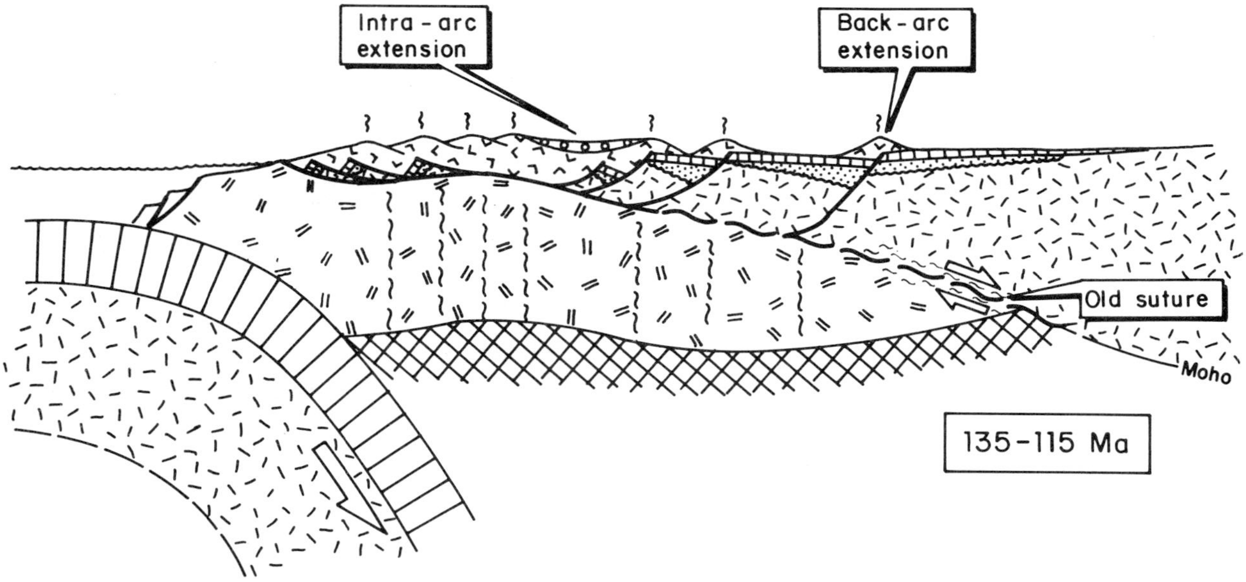

Fig. 8. Asymmetric extension of the intra-arc and backarc basins at the latitude of Aconcagua (32°30'S) during the early Cretaceous

several subhorizontal normal faults extend to the pre-Lower Cretaceous limestones (Mpodozis, in prep.).

The development of the main magmatic activity during the early Cretaceous occurred along the eastern slope of the Coastal Cordillera (Lo Prado, Veta Negra and equivalent units). At the same time, an outer arc occurred along the present international boundary in the Principal Cordillera, and is exemplified by the Pelambres, Juncal and equivalent units. Minor isolated patches representing backarc magmatism were generated further east (Fig. 7).

The Andean compression produced a tectonic inversion of the intra-arc basins, deformed in a thick-skinned style with the basement partially involved in the deformation. The eastern backarc platforms display a thin-skinned deformation, as observed in the Aconcagua fold-and-thrust belt.

8 Concluding remarks

The present morphostructural units of the Central Andes, such as the Sierras Pampeanas, Precordillera, and Frontal and Principal Cordilleras, match first-order crustal boundaries.

These crustal boundaries have had a complex history, beginning with collisions during the late Proterozoic and early Palaeozoic orogenies. During these periods thrust systems developed in the upper crustal levels and shear zones developed at depth.

The boundaries were weakness zones that controlled the inception of normal faults and rifting during Mesozoic times, produced in the upper plate of the sutured blocks. Subsequent thrusting and strike-slip displacement during Andean deformation produced the different structural geometries of the Andes. These structures vary from partially inverted tectonics involving basement at the inner regions to a more detached thin-skinned deformation in the eastern platform areas.

Most of the present and past tectonic activity is linked to old crustal boundaries that have been areas of intense compression during collisions, the site of rifting during extension, and affected by tectonic inversion during Andean times.

Acknowledgements. This study was supported by a grant from the Antorchas Foundation and by financial assistance from the UBACYT ex-139 and PIA CONICET.

References

Aceñolaza FG, Toselli A (1988) El Sistema de Famatina, Argentina: Su interpretación como orógeno de margen continental activo. V Congr Geol Chil, Santiago, Actas 1(A): 55-67

Allmendinger RW, Nelson KD, Potter JC, Barazangi M, Brown LD, Oliver JE, (1987) Deep seismic reflection characteristics of the continental crust. Geology 15: 304-310

Ando CJ, Czuchra BL, Klemperer SL, Brown LD, Cheadle MJ, Cook FA, Oliver JE, Kaufman S, Walsh T, Thompson Jr JB, Lyons JB, Rosenfeld JL (1984) Crustal profile of mountain belt: COCORP deep seismic reflection profiling in New England Appalachians and implications for architecture of convergent mountain chains. Am Assoc Pet Geol Bull 68(7): 819- 837.

Baldis B, Beresi MS, Bordonaro O, Vaca A (1982) Síntesis evolutiva de la Precordillera Argentina. V Congr Geol Latinoam Geol, Buenos Aires, Actas 4: 399-445

Baldis B, Beresi MS, Bordonaro O, Vaca A (1984) The Argentine Precordillera as a key to Andean structure. Episodes 7(3): 14-19

Bastías HE, Uliarte E (1987) Morfología de la falla rumbo-deslizante El Tigre, entre los ríos Jachal y San Juan. X Congr Geol Argent, Tucumán, Actas 1: 251-254

Beer JA, Allmendinger RW (1988) Structural development of the Central Andes constrained by seismic stratigraphy of the Iglesia basin, Argentina. Geol Soc Am, Abstr 20(A): 8

Borg SG, De Paolo DJ (1989) Continental lithospheric studies: Integration of isotopic and geophysical investigations. ANTALITH Meet, Washington, July 1989

Borrello AV (1969) Los geosinclinales de la Argentina. Direc Nac Geología y Minería, An 14: 1-136

Bosworth W, Lambiase J, Keiser R (1986). A new look at Gregory's rift: The structural style of continental rifting. EOS 67: 577-583

Caminos R (1965) Geología de la vertiente oriental del Cordón del Plata, Cordillera Frontal de Mendoza. Asoc Geol Argent Rev 20(3): 351-392

Caminos R, Cingolani CA, Hervé F, Linares E (1982) Geochronology of the Preandean metamorphism and magmatism in the Andean Cordillera between latitudes 30° and 36°S. Earth Sci Rev 18: 332-352

Cingolani C, Varela R (1975) Geocronología Rubidio-Estroncio de rocas ígneas y metamórficas de las Sierras Chica y Grande de Córdoba, República Argentina. II Congr Iberoam Geol Econ, Buenos Aires, Actas 1: 9-35

Cingolani C, Cuerda A, Varela R, Schauer O (1989) Geología de la Precordillera occidental en la comarca de la Sierra de Tontal, provincia de San Juan, República Argentina. Comunicaciones 40: 39-56

Cominguez AH, Ramos VA (1991). La estructura profunda entre la Precordillera y Sierras Pampeanas (Argentina): Evidencias de la sísmica de reflexión profunda. Rev Geol Chile 18(1): 3-14

Cortés JM (1989) Tectónica de colisión en el extremo noroccidental de la Precordillera Mendocina. Reunión sobre Transectas de América del Sur, Resúmenes 13, Mar del Plata.

Cucchi RJ (1972) Geología y estructura de la Sierra de Cortaderas, San Juan-Mendoza, República Argentina. Asoc Geol Argent Rev 27(2): 229-248

Dalla Salda L, Varela R (1982) La estructura del basamento del tercio sur de la Sierra de Pié de Palo, Provincia de San Juan, Argentina. V Congr Latinoam de Geol, Actas 1: 451-468

Dalla Salda L, Varela R (1984) El metamorfismo en el tercio sur de la Sierra de Pié de Palo, San Juan. Asoc Geol Argent Rev 39(1-2): 68-93

Dalziel IWD, Storey BC, Garrett SW, Grunow AM, Herrod LDB, Pankhurst RJ (1987) Extensional tectonics and the fragmentation of Gondwana. In: Coward MP et al (eds) Continental extensional tectonics, Geol Soc Spec Publ 28: 433-441

Delakowitz B (1988) Geologisch-geochemisch-lagerstättenkundliche Untersuchungen zur Genese von Wolfram-vorkommen in der Sierra del Morro-Oeste, Provinz San Luis, Argentinien. Münch Geowiss Abh B4: 1-107

De la Winter H (1984) Tectonostratigraphy, as applied to analysis of South African Phanerozoic basins. Trans geol Soc S Afr 87: 169-179

Dover JH (1990) Opinion: Problems of terrane terminology - causes and effects. Geology 18(6): 487-488

Du Toit AL (1927) A geological comparison of South America with South Africa. Publ Carnegie Inst 381:1-157

Forsythe R (1982) The late Paleozoic to early Mesozoic evolution of southern South America: A plate tectonic interpretation. J Geol Soc Lond 139: 671-682

Forsythe RD, Kent DV, Mpodozis C, Davidson J (1986) Paleomagnetism of Permian and Triassic rocks, central Chilean Andes. In: McKenzie GD (ed) Gondwana six: Structure, tectonics, and geophysics. American Geophysical Union, Geophys Monogr 40: 241-252

Furque G (1968) Descripción Geológica de la Hoja 17b Guandacol, provincias de La Rioja y San Juan. Direc Nac de Geología y Minería Bol 92: 1-104

Furque G (1972) Los movimientos caledónicos en la Argentina. Rev Mus La Plata 8: 129-136

García AR, Beck ME, Burmester RF, Munizaga F, Hervé F (1988) Paleomagnetic reconnaissance of the region de los Lagos, southern Chile, and its tectonic implications. Rev Geol Chile 15(1): 13-30

González R, Cabrera MA, Castellote P, Omil M, Bortolotti P, Moyano R and Ojeda J (1985) Esquematización de la ubicacion espacial y temporal de la eruptividad en Sierras Pampeanas Noroccidentales, República Argentina. IV Congr Geol Chil Antofagasta, Actas 3(4): 138-150

Haller M, Ramos VA (1984) Las ofiolitas famatinianas (Eopaleozoico) de las provincias de San Juan y Mendoza. IX Congr Geol Argent Buenos Aires, Actas 2: 66-83

Harrington HJ (1970) Las Sierras Australes de Buenos Aires, República Argentina: Cadena aulacogénica. Asoc Geol Argent Rev 25(2): 151-181

Hervé F (1988) late Paleozoic subduction and accretion in southern Chile. Episodes 11(3): 183-188

Jordan TE, Allmendinger RW (1986) The Sierras Pampeanas of Argentina: A modern analogue of Laramide deformation. Am J Sci 286: 737-764

Jordan TE, Isacks B, Ramos VA, Allmendinger RW (1983) Mountain building in the Central Andes. Episodes 6(3): 20-26

Kay SM, Ramos VA, Kay R (1984) Elementos mayoritarios y trazas de las vulcanitas ordovícicas en la Precordillera Occidental: Basaltos de rift oceánicos tempranos (?) próximos al margen continental. IX Congr Geol Argent Buenos Aires, Actas 2: 48-65

Kay SM, Ramos VA, Mpodozis C, Sruoga P (1989) Late Paleozoic to Jurassic silicic magmatism at the Gondwanaland margin: Analogy to the middle Proterozoic in North America? Geology 17(4):324-328

Keidel J (1921) Sobre la distribución de los depósitos glaciares del Pérmico conocidos en la Argentina y su significación para la estratigrafía de la serie del Gondwana y la paleogeografía del Hemisferio Austral. Acad Nac de Ciencias Bol 25: 239-368

Keidel J (1939) Las estructuras de corrimientos paleozoicos de la Sierra de Uspallata, provincia de Mendoza. Physis 14(46): 1-96

Llambías E and Sato AM (1990) El batolito de Colangüil (29-31°S), Cordillera Frontal de Argentina: Estructura y marco tectónico. Rev Geol Chile 17(1): 89-108

Llambías E, Cingolani C, Varela R, Prozzi C, Ortiz-Suárez A, Caminos R, Toselli A, Saavedra J (1991) Leucogranodioritas sincinemáticas ordovícicas en la Sierra de San Luis, República Argentina. VI Congr Geol Chileno Santiago, Resúmenes: 187-191

Llano JA, Esparza AM, Rossa N and Vaca A (1984) Geología y petrografía del cerro Salinas, provincia de San Juan. IX° Congreso Geológico Argentino Buenos Aires, Actas 1: 298-309

Lock BE (1980) Flat-plate subduction of the Cape Fold Belt of South Africa. Geology 8: 35-39

Logatchev NA (1984) The Baikal rift system. Episodes 7(1): 38-42

Lottner US, Miller H (1986) The Sierra de Ancasti as an example of the structurally controlled magmatic evolution in the Lower Paleozoic basement of the NW Argentine Andes. Zentralbl Geol Paläont (9/10): 1269-1281

Meissner R, Wever T, Flüch ER (1987) The Moho in Europe - implications for crustal development. Ann Geophys 5B(4): 357-364.

Miller H (1983) Der Antarktische Kontinent - Kernstück von Gondwana. Geol Rundsch 72: 101-103

Moscoso R, Mpodozis C (1988) Estilos estructurales en el Norte Chico de Chile (28-31°S), regiones de Atacama y Coquimbo. Rev Geol Chile 15(2): 151-166

Mpodozis C and Cornejo P (1988) Hoja Pisco Elqui. IV° Región de Coquimbo. Carta Geológica de Chile 68. Serv Nac Geología y Minería, Santiago, 160 pp

Mpodozis C, Kay SM (1990) Late Paleozoic to Triassic evolution of the Gondwana margin: Evidence from Chilean Frontal Cordillera batholiths (28°-33°S). Rev Geol Chile 17(2): 153-180

Mpodozis C, Ramos VA (1989) The Andes of Chile and Argentina. In: Ericksen GE, Cañas Pinochet MT, Reinemud JD (eds) Geology of the Andes and its relation to hydrocarbon and mineral resources, Circumpacific Council for Energy and Mineral Resources, Houston, Earth Sci Ser 11: 59-90

Nasi C, Mpodozis C, Cornejo P, Moscoso R, Maksaev V (1985) El batolito de Elqui-Limarí (Paleozoico superio - Triásico): Características petrográficas, geoquímicas y significado tectónico. Rev Geol Chile 25-26: 77-111

Peralta SH, Uliarte ER (1986) Estructura de la Fm. Rinconada (Eopaleozoico) en su localidad tipo, Precordillera de San Juan. I Jorn Geol Precordillera San Juan, Actas: 237-242

Pérez WA, Medina ME (1990) Aspectos geológicos y geoquímicos las rocas ígneas del norte de la Sierra de Paimán. XI Congr Geol Argent San Juan, Actas I: 109-112

Pezzi EE, Mozetic ME (1989) Cuencas sedimentarias de la región chacoparanense. In: Chebli G, Spalletti L (eds) Cuencas sedimentarias Argentinas. Ser Correlación Geol 6: 65-78

Phinney RA (1986) A seismic cross section of the New England Appalachians: The orogen exposed. In: Barazangi M, Brown L (eds) Reflection seismology: The continental crust, Geodyn Ser 14: 157-172

Powell CM (1990) Gondwanaland context of the Tasman fold belt. In: Gondwana terranes and resources. X Aust Geol Convent Abstr: 190-191

Price RA (1981) The Cordilleran foreland thrust and fold belt in the southern Canadian Rocky Mountains. In: McClay FR, Price NJ (eds) Nap and thrust tectonics. Geol Soc Lond Spec Publ 9: 427-448

Ramos VA (1984) Patagonia: ¿Un continente paleozoico a la deriva? IX Congr Geol Argent Buenos Aires, Actas 2: 311-325

Ramos VA (1986) Tectonostratigraphy, as applied to analysis of South African Phanerozoic basins by H. de la R. Winter, discussion. Trans Geol Soc S Afr 87(2): 169-179

Ramos VA (1988a) The tectonics of the Central Andes; 30° to 33°S latitude. In Clark S, Burchfiel D (eds) Processes in continental lithospheric deformation, Geol Soc Am Spec Pap 218: 31-54

Ramos VA (1988b) Tectonics of the late Proterozoic - early Paleozoic: A collisional history of southern South America. Episodes 11(3): 168-174

Ramos VA (1989) The birth of southern South America. Am Scientist 77(5): 444-450

Ramos VA (1990) Terrane history of southern South America and Antarctica. In: Gondwanaland context of the Tasman fold belt. X Aust Geol Convent: 191

Ramos VA, Kay SM (1991) Triassic rifting and basalts of the Cuyo basin, Central Argentina. In: Symp Andean Magmatism. Geol Soc Am Spec Pap 265: 79-91

Ramos VA, Jordan TE, Allmendinger RW, Mpodozis C, Kay SM, Cortés JM, Palma MA (1986) Paleozoic terranes of the Central Argentine-Chilean Andes. Tectonics 5(6):855-880

Rapalini AE and Vilas JF (1990) Preliminary paleomagnetic data from the Sierra Grande Formation: tectonic consequences of the first mid-Paleozoic paleopoles from Patagonia. J S Am Earth Sci

Roeder DH (1973) Subduction and orogeny. J Geophys Res 78: 5005-5024

Rosendahl BR (1987) Architecture of continental rifts with special reference to East Africa. Ann Rev Earth Planet Sci 15: 445-503

Rowell AJ, Rees MN (1990) Suspect terranes of the central Transantarctic Mountains. In: Gondwana Terranes and Resources, X Aust Geol Convent: 247-248

Spalletti L, Cingolani CA, Varela R, Cuerda AJ (1989) Sediment gravity flow deposits of an Ordovician deep-sea fan system (western Precordillera, Argentina). Sediment Geol 61: 345-369

Storey BC (1989) West Antarctic mosaic: Tectonic evolution. 28 Int Geol Congr Washington, Abstr 3: 187

Uliana MA, Biddle KT (1988) Mesozoic-Cenozoic paleogeographic and geodynamic evolution of southern South America. Rev Bras Geociencias 18(2): 172-190

Uliana M, Biddle K, Phelps VM, Gust DA (1986) Significado del vulcanismo y extensión mesojurásicas en el extremo meridional de Sudamérica. Asoc Geol Argent Revista 40(3-4): 23-1-253

Uliarte ER, Bastías HE, Ruzycki de Berenstein L (1987) Morfología y neotectónica en el cerro La Chilca, Pedernal, Provincia de San Juan, Argentina. X Congr Geol Argent Tucumán, Actas 1: 227-230

Vaca A, Rossa N (1988) Petrología y estructura del frente occidental de Sierras Pampeanas en San Juan, Argentina. V Congr Geol Chil Santiago, Actas 1(A): 189-202

Visser JNJ (1985) Tectonostratigraphy, as applied to analysis of South African Phanerozoic basins. Discussion. Trans geol Soc S Afr 88: 183-184

Zeil W (1981) Volcanism and geodynamics at the turn of the Paleozoic to the Mesozoic in the Central and southern Andes, Zentralbl Geol Paläont I(3/4): 298-318

Some Isotopic and Geochemical Constraints on the Origin and Evolution of the Central Andean Basement (19°- 24°S)

KLAUS-WERNER DAMM, RUSSELL S. HARMON and SHARI KELLEY

Abstract. The Central Andes of northern Chile and northwestern Argentina developed in a largely autochthonous, intracontinental setting during Proterozoic and Palaeozoic times through a recurrent sequence of extensional and compressional tectonic regimes. Exposed pre-Mesozoic rocks comprise an intricate collage of crustal complexes consisting of metamorphosed basement, intrusive rocks, and weakly metamorphosed volcanic and sedimentary strata that rest upon and are intruded by various plutonic lithologies. U-Pb Proterozoic protolith ages range from 1460 to 1210 Ma and are broadly supported by Nd- and Sr- isotope model ages. Three Preandean orogenic cycles are recognized: (1) from c. 560 to 520 Ma, (2) from c. 505 to 405 Ma, and (3) from c. 350 to 240 Ma. Apatite and zircon fission track ages indicate the earliest crustal thickening and slowest cooling within the Precordillera south of c. 23°S, in the Limon Verde complex, and in the Cordon de Lila and Sierra Almeida intrusives. Further to the north, fission track ages for zircon reflect partial annealing during the Eocene magmatic event in this region, whereas apatite ages range from 40 to 25 Ma throughout the entire Precordillera. Preandean intrusive suites vary from dioritic to granitic in bulk composition. A combination of structural and field evidence, age relations, geochemical data, and isotopic data permit discrimination between anorogenic (A-type) and synorogenic (S-type) intrusives, and point to a predominantly crustal origin for all parental magmas. No I-type granitoids are observed in the Preandean basement. Pb-Pb and Sr-Nd isotopic systematics suggest a petrogenesis for the plutonic rocks involving variable extents of mixing between lower crustal sources and lower/upper crustal assimilants. During Phanerozoic orogenic cycles, anorogenic intrusions and related continental tholeiites, generated during extensional episodes, were followed by synorogenic rocks produced when the tectonic regime changed to one of compression and crustal thickening.

1 Introduction

A study involving the coordinated application of field mapping, structural relationships, regional geochronology, stable and radiogenic isotope ratio determinations, and geochemistry has been undertaken to improve the general understanding of the timing and nature of the magmatism and metamorphism that has contributed to the development and consolidation of the Central Andean basement. Additionally, the uplift and unroofing history of the basement rocks studied was constrained through zircon and apatite fission track analysis. Much of the detailed geological background information and geochronological data have been discussed by Damm et al. (1990, 1991). Here we present a brief overview of these observations and offer some thoughts on their geodynamic significance.

The geodynamic development of the western margin of the South American continent has been a topic of active discussion for the past decade, and the issue is by no means resolved. Because most studies to date have focussed on the post-Palaeozoic crustal history, i.e. the "Andean Cycle" sensu strictu, little is known about the older basement, its role in the development of the Andean arc, its contribution to crustal thickening of the Andes over the past 15-20 Ma, or its influence on Andean petrogenesis. Although it is widely accepted that subduction and convergent plate margin processes have played an important role in the geodynamic development of the continental margin during the last 700 Ma, it is a matter of continuing debate about (1) how and when the crustal nucleus of the Andean Cordillera was generated, (2) what sort of geodynamic and tectonic processes were responsible for the generation of the

Correspondence to: Russell S. Harmon, Isotopic Geosciences Laboratory, British Geological Survey, Keyworth, Nottingham NG12 5GG, UK

Subandean basement, and (3) the Preandean history of this basement.

Current hypotheses about these important geological questions are of three basic types:

1. A *continuous subduction model*, which presumes that the present Andean situation of eastward subduction against a stable continental margin has occurred more or less uninterrupted, yet with varying plate geometries, since Precambrian time (see e.g. Coira et al. 1982; Willner et al. 1985; Pichowiak this Vol.).

2. An *exotic terrane model*, which postulates that extensive crustal addition and consolidation occurred during pre-Mesozoic time through subduction of oceanic crust, crustal shortening, and the docking and accretion of many different kinds of allochthonous terranes to the stable margins of the Brazilian Shield (see e.g. Nur and Ben-Avraham 1977, 1982, 1983; Dalziel and Forsythe 1985; Ramos 1988).

3. An *ensialic margin model* which considers that the development of the continental margin area of the western Brazilian craton was largely complete by the late Proterozoic and, subsequently, this area has remained autochthonous and predominantly passive tectonically, with its late Precambrian to Palaeozoic history characterized by repeated opening and closure of intracontinental basins and with the current situation of Andean subduction only initiated in the Triassic with the break-up of Pangaea (see e.g. Dalziel and Forsythe 1985; Damm et al. 1990).

Most workers engaged in Preandean studies continue to favor the continuous subduction model and, as result, have produced a multitude of different, and often contradictory, late Proterozoic-Palaeozoic plate geometries. With the advent of the terrane concept in north America during the last two decades, it became increasingly fashionable to ascribe poorly understood tectonic and geological features of the Preandean basement to "exotic blocks", derived from unknown regions to the west and subsequently docked and accreted to the western South American continental margin during late Proterozoic to Mesozoic times. In order to evaluate these three essentially different ideas, we have attempted to compare the diagnostic features predicted by each model with the characteristics observed for the exposed basement. For example, subduction-related magmatic activity along narrow, well-defined belts that are parallel to an old, yet active, continental basement would be expected for the continuous subduction situation, whereas exotic basement blocks that differ significantly in chronological history, structural evolution, and geochemical signature of magmatic rocks would be manifestations of an exotic terrane situation. By contrast, an ensialic margin situation would be recognized by coherent patterns of magmatic, metamorphic, and structural evolution of regional extent related to repeated successions of crustal extension and compression throughout the history of the continental margin.

2 Geological Framework

Selected basement outcrops in north Chile and northwest Argentina between 18° and 24°S have been studied. Aspects of the field and structural data, metamorphic record, metamorphic P-T paths, and radiogenic isotope ages have been presented elsewhere (Damm et al. 1990, 1991). This chapter focuses on aspects of the the isotopic and geochemical character and cooling history of the various intrusive units of the Preandean basement in the study area, in an attempt to further discriminate among the three geodynamic models described above.

Between 18°S and 24°S, the basement of the Central Andes is composed of several discrete magmato-metamorphic segments that have been formed into a complex collage within the physiography of the present Andean Cordillera as a result of Mesozoic-Cenozoic mountain building. The most important of these basement elements (Fig.1) have been included in this study. From north to south within the Precordillera are the *Belén-Chapiquina* area (18°20'-18°40'S/69°35'-69°25'W), the *Quebrada Choja* and *Quebrada Chara* regions (21°00'-21°10'S/68°55'-68°45'W), and the *Sierra de Limon Verde* area (22°35'-22°45'S/69°00'-68°50'W). Within the Coast Range are the north-central regions of the *Mejillones Peninsula* (23°00-23°20'S/70°40'-70°25'W) and an area along the northwestern shore of the *Salar de Navidad* (23°35'-23°40'S/70°10'-70°05'W). Farther east in the Preandean Depression in the vicinity of 23°35'-24°20'S/68°30'-68°15'W are the *Cordon de Lila*, *Cordon de Tucucaro*, and *Cordon de Chinquilchoro* areas, the *Sierra de Pingo Pingo*, *Sierra de Alto del Inca*, and *Sierra Choschas*. The *Sierra de Macon* (23°35-24°20'S/68°30'-68°15'W) crops out in the Western Cordillera/Altiplano and the *Quebrada Tajamar, Ochaqui*, *Salar de Diabillos* areas (23°35'-24°20'S)/67°30'-66°20'W) and the *Santa Rose de Tastil* (24°00-24°30'S/65°00'-65°40'W) are located in the Eastern Cordilla and eastern Puna.

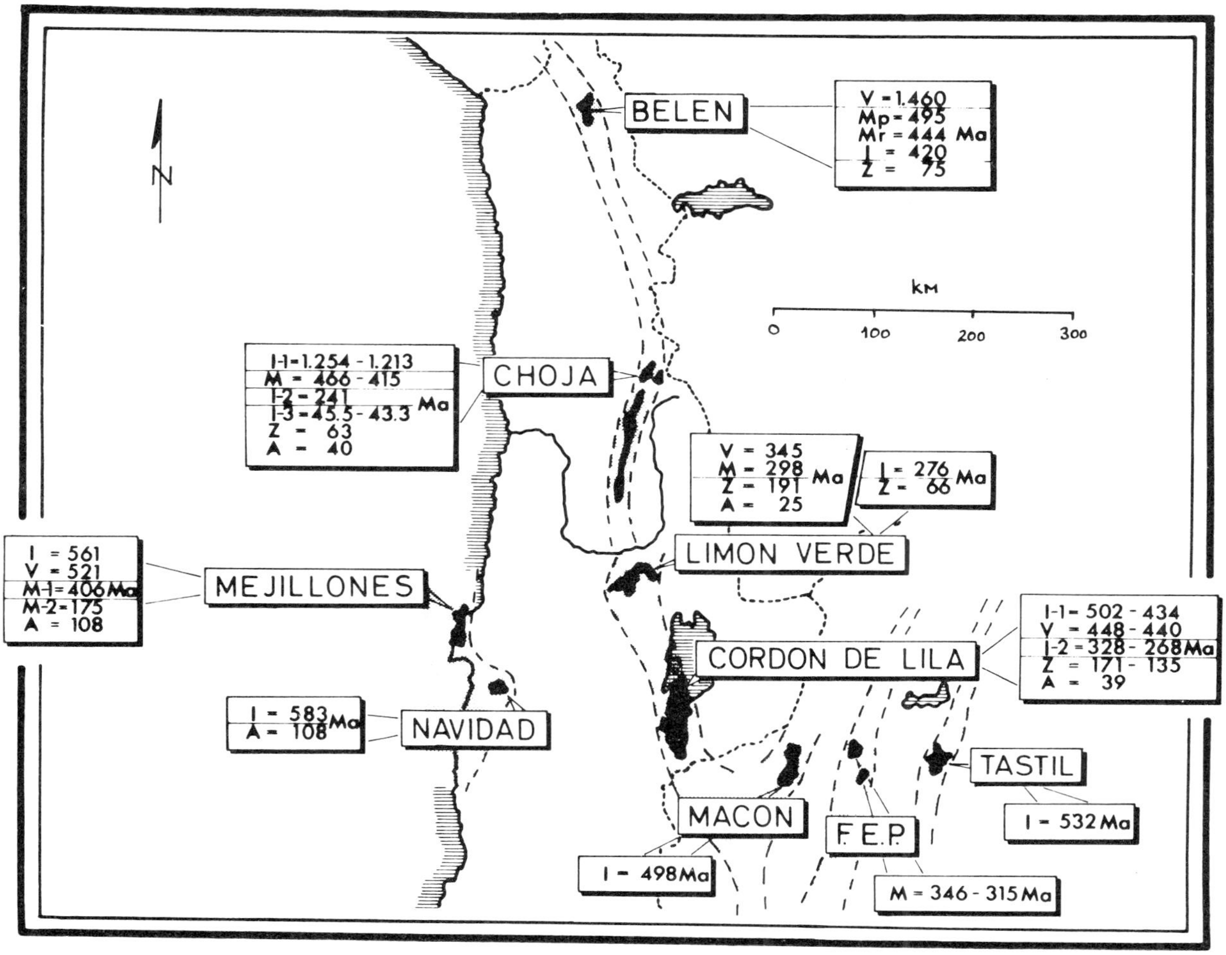

Fig. 1. Basement areas in Chile and Argentina investigated in this study. Radiogenic ages are indicated in the *box* next to each locality. *V*: volcanic intrusion; *M* metamorphic overprint *p* denotes peak metamorphism *r* retrograde metamorphism; *I* intrusive event; *Z* zircon fission track age; *A* apatite fission track age. Age determinations, other than for Z and A, were carried out using U-Pb, Rb-Sr, Sm-Nd, and K-Ar techniques. For details and interpretation see text and Damm et al. (1990, 1991). *Dashed lines* and *hatched areas* denote zones of basement exposed through Andean Cycle rocks and do not therefore necessarily signify Preandean geodynamic relationships

Geochronological data (discussed in detail by Damm et al. 1990, 1991) are summarized in Fig. 1, which presents magmatic, metamorphic, and uplift ages for these 16 basement areas. The oldest protolith ages are seen in the Belén amphibolites and Quebrada Choja orthogneiss and migmatite complexes, whose origin dates to middle Proterozoic time. However, abundant hints of pre-Phanerozoic magmatic events throughout the Central Andean basement are provided by inherited U-Pb ages for zircon populations from granitoids of almost all localities studied (Damm et al. 1990). In addition, both Nd- and Pb- model ages >1200 Ma support the idea of pre-Phanerozoic crustal generation. Products of the earliest Phanerozoic orogeny are preserved in the Coastal Ranges of Chile around Antofagasta and along a NNE-SSW trending zone extending from Tarija in Bolivia as far south as Cachi in Argentina (Fig.1). Palaeozoic igneous or metamorphic ages characterize all of the other basement outcrops studied. Uplift occurred predominantly in the Mesozoic, although evidence of late Palaeozoic uplift is documented in the Mejillones, Limon Verde, and Cordon de Lila areas. The following paragraphs summarize the age information in Fig. 1 into a geological history for the Central Andean basement between 18° and 24°S, as

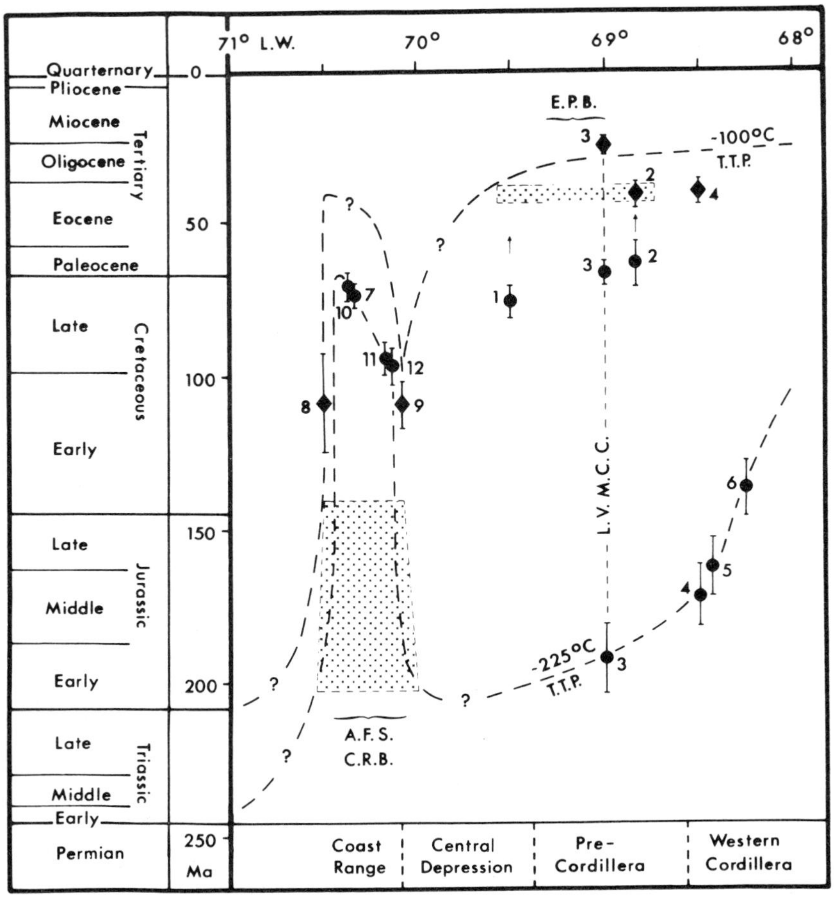

Fig. 2. Andean Cycle time-temperature paths (dashed lines labelled TTP) for the uplift and cooling history of the Coastal Range, Central Depression, Pre-cordillera and Western Cordillera. Fission-track ages for zircon are interpreted as denoting cooling through c. 225°C and fission-track ages for apatite indicate cooling through c. 100°C. Stippled areas mark U-Pb zircon ages of related intrusives of the Coast Range and Pre-Cordillera. *AFS* Atacama Fault System; *CRB* Coast Range Batholith; *EPB* Eocene Plutonic Bodies; *LVMCC* Limon Verde. Apatite fission-track ages are denoted by the *solid diamonds* and zircon fission-track ages indicated by the *solid circles*; *bars* 2σ uncertainty in age determinations. Numbers for this and subsequent figures identify intrusions as follows: *1* Bélen; *2* Choja; *3* Limon Verde; *4* Mejillones; *5* Navidad; *6* Chinquilchoro; *7* Tucucaro; *8* Lila; *9* Pingo Pingo *10* = Monturaqui; *11* Macon; *12* Tajamar *13* Ochaqui; *14* Diabillos *15* Tastil; *17* Arita

presently understood from the 16 localities studied (here ordered from north to south, with numbers in parentheses corresponding to those used in the figues to identify the individual intrusive centres).

Belén (1). Here a series of middle Proterozoic mica schists, gneisses, amphibolites, and strongly serpentinized ultramafic stocks and associated gabros were metamorphosed and folded between 495 and 445 Ma. Subsequently, the intrustion by granodiorites to monzogranites occurred at around 420 Ma and cooling of zircon through 225°C took place at c. 75 Ma (Fig.2).

Choja (2). This is a NNE-SSW trending sequence of volcaniclastic sediments, volcanics, and mid-Proterozoic instrusives that crop out in the Quebrada Choja and Quebrada Chara areas. These sequences reached peak metamorphic conditions between 465 and 415 Ma, which resulted in crustal anatexis and in situ emplacement of syenogranites. A second intrusive cycle, consisting of granodiorites to granites, occurred at about 240 Ma. A third and final magmatic cycle occurred at around 45 Ma and resulted in structurally controlled emplacement of monzodiorites, quartz diorites, tonalites, and tronhjemites as well as in the associated effusion of andesites. Uplift, as recorded by the cooling of apatite through 100°C, occurred at around 40 Ma, resulting in the partial reannealing of fission tracks in basement zircon at 63 Ma (Fig.2).

Sierra Limon Verde (3). This area forms a possible metamorphic core complex, as defined by Coney (1980), in the 18-24°S region of the Central Andes. It consists of both orthogneisses and paragneisses that were intruded by a range of Variscan granites, granodioritic, and more mafic plutons at c. 275 Ma. Uplift and cooling of zircons, through 225°C occurred at about 65 Ma. The roof zone consists of almost monomineralic ortho-amphibolites and intercalated garnet-mica gneisses. Deposition of the basaltic protolith occurred during the Carboniferous at c. 375 Ma (see also Rogers 1985). Metamorphism reached peak conditions at about 300 Ma, and was followed by severe retrogradation during uplift, which commenced at about 190 Ma as a result of extension that produced a marginal basin. Cooling of apatite to 100°C occurred at about 25 Ma (Fig.2).

Peninsula de Mejillones (4). In this area of the Atacama Fault Zone, outcrop a sequence of strongly folded mica schists, orthoamphibolites, and gneisses associated with granitoids emplaced at c. 560 Ma. A compositionally similar, yet undeformed granitoid emplaced at c. 585 Ma is present along the NNW shore of *Salar de Navidad (5).* This intrusive activity was followed by tholeiitic basalt volcanism during the late Cambrian at about 520 Ma and by regional metamorphism at around 405 Ma. The emplacement of gabbroic intrusions and related mafic/ultramafic stocks was associated with effusion of the widespread lavas of the La Negra Formation. This activity was associated with an increased rate of subduction and changes in plate geometry, and resulted in uplift of the western flank of the Jurassic marginal basin in the region, and in strong horizontal and vertical shearing. Because large amounts of magma were generated at this time (Pichowiak this Vol.) and were able to coalesce and rise along the Atacama Fault Zone, strong ductile deformation prevails up to very shallow levels in the crust (Scheuber and Andriessen 1990). Along the eastern and western flanks of the fault zone, cooling of apatite through 100°C took place at around 110 Ma (Fig. 2), whereas temperatures of >225°C were sustained in zircons of various Jurassic intrusives until between 95-70 Ma (see also Andriessen and Reutter this Vol.).

Cordon de Lila. This is a broad region in easternmost Chile, described by Damm et al. (1991), which trends SSE from the Salar de Atacama. Included in this area are: the *Cordon de Lila Complex (8)* and the *Chinquilchoro (6)*, the *Tucucaro (7)*, the *Pingo Pingo (9)*, the *Monturaqui (10)*, and *Arita (17)* plutons. Also present are a variety of intrusive bodies and a very weakly metamorphosed series of komatiitic tholeiite to plagidacite lavas and intercalated clastic sediments. Two major intrusive cycles are recognized, an early Palaeozoic one at c. 500-435 Ma and a late Palaeozoic one at c. 330-270 Ma. Initial uplift, as recorded in the cooling of zircons through the 225°C isograd occurred during the interval c. 170-135 Ma, with apatites yielding Eocene ages of 39 Ma (Fig.2).

Sierra de Macon (11). This c. 500 Ma granitoid forms a roughly N-S trending, elongate body to the west of Salar de Arizaro. It is comparable in both mineralogy and texture to the Choschas and Alto del Inca intrusives of the Cordon de Lila region to the west, with which it is contemporaneous (Fig.1).

Faja Eruptiva de La Puna. This is a segmented and tilted block consisting of a series of metamorphosed volcano-sedimentary units and numerous intrusive bodies including the *Quebrada Tajamar (12)* granodiorite, the *Diabillos (14)* granite, and the *Ochaqui (13)* granite. The metamorphic grade increases with crustal depth towards the south. U-Pb zircon ages of about 345-315 Ma are similar to Variscan ages documented for other intrusive bodies of the Argentinian Puna (Coira et al. 1982; Omarini et al. 1984).

Santa Rosa de Tastil (15). This is a granodiorite-granite intrusion in the Cordillera oriental that was emplaced at a high crustal level into folded sedimentary strata of the Puncoviscana Formation. Like other comparable plutons of the region (Bachmann et al. 1986, 1987), it has an early Palaeozoic age of c. 530 Ma (Fig.1).

3 Isotopic Characteristics

O-isotope ratios, as permil ($^{o}/_{oo}$) $\delta^{18}O$ values relative to the SMOW standard, and radiogenic isotope (Sr, Nd, & Pb) compositions have been measured for selected samples from all dated intrusive complexes. The isotopic data are summarized in Table 1. In order to use whole-rock $^{18}O/^{16}O$ ratios to study granitoid petrogenesis (e.g. as a meaningful guide to the O-isotopic composition of the source region from which the parental magma was derived), it is necessary to demonstrate that the rocks have not had their ^{18}O character modified by secondary, sub-solidus hydrothermal processes, which commonly affect plutonic rocks that crystallize and cool at high levels in the Earth's crust. Figure 3 presents mineral O-isotope relationships for the Preandean plutonic rocks examined in this study as δ-δ plots. On such diagrams, the $\delta^{18}O$ values for two minerals crystallized in O-isotope equilibrium at a particular temperature will define an array of 45° slope that is displaced away from the D = 0 line by the permil value of the ^{18}O fractionation between the two minerals at that temperature. The three diagrams (Fig. 3a-c) indicate O-isotope equilibrium between (1) plagioclase-quartz, (2) alkali feldspar-quartz, and (3) alkali feldspar-plagioclase at temperatures of 700-900°C, with the exception of only two specific samples (from intrusions #1 and #15, as noted by the dashed field boundaries in Fig. 3). These two samples have been affected by subsolidus interaction with a hydrothermal fluid such that plagioclase $\delta^{18}O$ values have been lowered with respect to coexisting quartz and alkali feldspar in the same rocks.

As a group, the intrusive rocks of the Preandean basement between 18° and 24°S exhibit a wide range

Table 1. Selected isotopic data for pre-Mesozic intrusive rocks of the Central Andes between 18° and 24° S

Intrusive	rock type	δ^{18} Owr (°/oo SMOW)	(^{87}Sr / ^{86}Sr)i	(^{143}Nd / ^{144}Nd)i	$^{206}Pb/^{204}Pb$	$^{207}Pb/^{204}Pb$	$^{208}Pb/^{204}Pb$
Belen	granodiorite, granite	8.5-10.5	0.71403 0.71809	0.51167- 0.51154	16.79- 17.17	15.54- 15.55	37.62- 38.35
Choja	granite	11.5-11.8	0.71817- 0.72021	0.51186- 0.51087	18.09- 19.37	15.65- 15.68	38.25- 38.41
Limon Verde	granodiorite, granite	6.8-9.0	0.70905- 0.71001	0.51217- 0.51210	17.36- 18.12	15.59- 15.63	38.06- 38.13
Mejillones	granodiorite, granite	10.4-11.9	0.71077- 0.71333	0.51190- 0.51176	18.05- 18.13	15.68- 15.75	38.59- 38.61
Navidad	granodiorite,granite	11.2-11.4	0.71043- 0.71084	0.51173- 0.51172	17.92- 17.83	15.73- 15.78	37.87- 37.92
Chinquilchoro	diorite,granodiorite	6.4-7.3	0.71533- 0.70522	0.51211- 0.51208	18.40- 18.50	15.58- 15.65	38.22- 38.32
Tucucaro	granite	7.6-8.5	0.70613- 0.70665	0.51179- 0.51174	17.68- 17.85	15.51- 15.61	37.57- 37.71
Lila	diorite	6.2-6.7	0.70395- 0.70426	0.51194- 0.51190	18.00- 18.05	15.44- 15.47	38.02- 38.13
Pingo Pingo	granodiorite, granite	7.4-8.3	0.70520- 0.70547	0.51206- 0.51196	18.06- 18.12	15.48- 15.61	38.01- 38.06
Monturaqui	granodiorite, granite	6.3-9.1	0.70545- 0.70566	0.51170- 0.51159	18.37- 18.48	15.58- 15.68	38.20- 38.30
Macon	granodiorite	6.1-9.0	0.70705- 0.70916	0.51189- 0.51181	18.28- 18.35	15.73- 15.82	38.23- 38.34
Tajamar	granite	10.1	0.71292	0.51168	17.01	15.39	37.43
Ochaqui	granite	10.4-11.3	0.71433- 0.71560	0.51160- 0.51159	16.82- 16.95	15.76- 15.83	36.58- 36.60
Diabillos	granite Diabillos	12.6-12.7	0.71189- 0.71212	0.51190- 0.51180	16.59- 16.61	15.40- 15.41	36.39- 36.40
Tastil	granite	11.2-13.1	0.71155- 0.71732	0.51175- 0.51160	17.29- 17.58	15.43- 15.55	37.55- 38.10
Arita	granodiorite, granite	9.6-10.9	0.71144- 0.71314	0.51202- 0.51201	17.76- 17.82	15.29- 15.33	37.77- 37.83

of $\delta^{18}O$ values from +6.1 to +13.1°/$_{oo}$. In particular, the O-isotope compositions are strongly biased toward higher $\delta^{18}O$ values than are characteristic of volcanic or plutonic rocks of the Andean Cycle (Longstaffe et al. 1983; Harmon et al.1984). The 7°/$_{oo}$ range in the $^{18}O/^{16}O$ ratio and the preponderance of $\delta^{18}O$ values in excess of +10°/$_{oo}$ (Table 1) indicates a largely crustal derivation for these Preandean intrusives, with varying contributions from metasedimentary and metaigneous protoliths.

Radiogenic isotope systematics for the intrusive rocks are similarly variable. Respective $^{87}Sr/^{86}Sr$ and $^{143}Nd/^{144}Nd$ initial ratios, calculated from zircon U-Pb emplacement ages, range from c.0.704 to 0.722 and from c.0.512 to 0.5114, and, like $^{18}O/^{16}O$ ratios, are displaced well away from the values expected for mantle-derived magmas (Fig. 4) This supports the idea of an origin for these plutonic suites predominantly from crustal sources. Pb-isotope ratios vary from unradiogenic to quite radiogenic values with $^{206}Pb/^{204}Pb$ ratios ranging from c. 16.2 - 19.3, $^{207}Pb/^{204}Pb$ ratios ranging from c.15.2 to 15.8, and $^{208}Pb/^{204}Pb$ ratios ranging from c.36.6 to 38.4 .

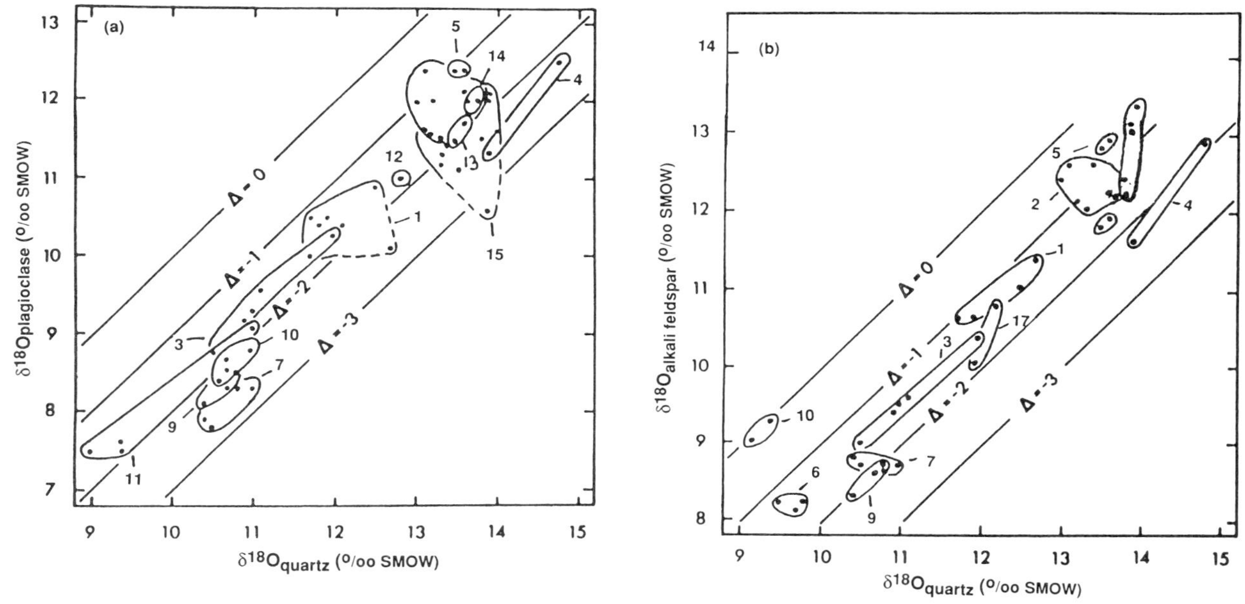

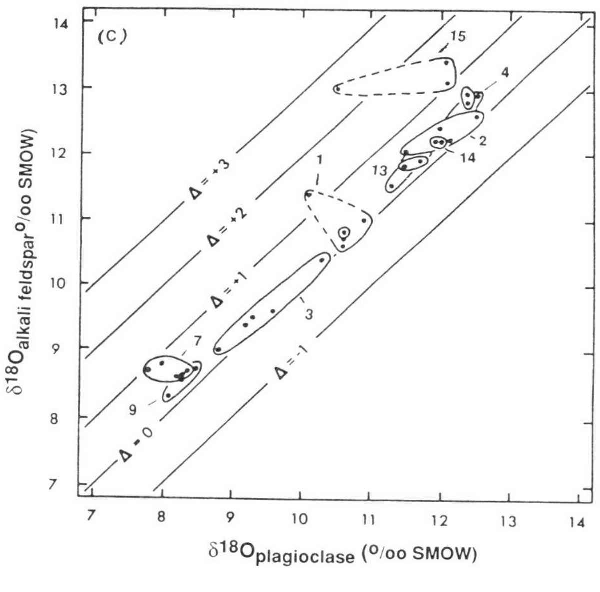

Fig. 3 a-c. Plots of **a** $\delta^{18}O$ plagioclase versus $\delta^{18}O$ quartz, **b** $\delta^{18}O$ alkali feldspar versus $\delta^{18}O$ quartz, and **c** $\delta^{18}O$ alkali feldspar versus $\delta^{18}O$ plagioclase for the pre-Mesozoic intrusive rocks of the Central Andes. Diagonal lines denote permil differences between the $\delta^{18}O$ values for the two minerals. According to thermodynamic theory, ^{18}O fractionations set at a particular temperature for different bulk rock ^{18}O compositions should be characterized by arrays of unit slope. Arrays which cut across the D=0 line at steep positive slope are indicative of isotopic disequilibrium. Locality identifiers as in Fig. 2. See text for discussion

4 Discussion

There is clear isotopic evidence that the intrusive rocks of the Preandean basement are largely of crustal origin. To further constrain the details of their origin, it is necessary to have a tectonic framework in which to consider isotopic and geochemical data. Approaches to this problem vary from the traditional classifications, like that of Peacock (1931), which in certain instances, can be interpreted in terms of a general tectonic setting, to the more recent chemical discrimination approaches of Chappell and White (1974) and Pearce et al. (1984). The latter approaches use a variety of major and trace element characteristics to discriminate between magmas generated from different sources or in different tectonic settings. For example, Chappell and White (1974) distinguish granitoids derived from a metasedimentary source from those generated from a metaigneous protolith on the basis of Na content, normative mineralogy, Sr-isotope ratios, and peraluminous character, whereas Pearce et al. (1984) use the trace element systematics of Rb-Y-Nb and Rb-Yb-Ta to discriminate between ocean-ridge

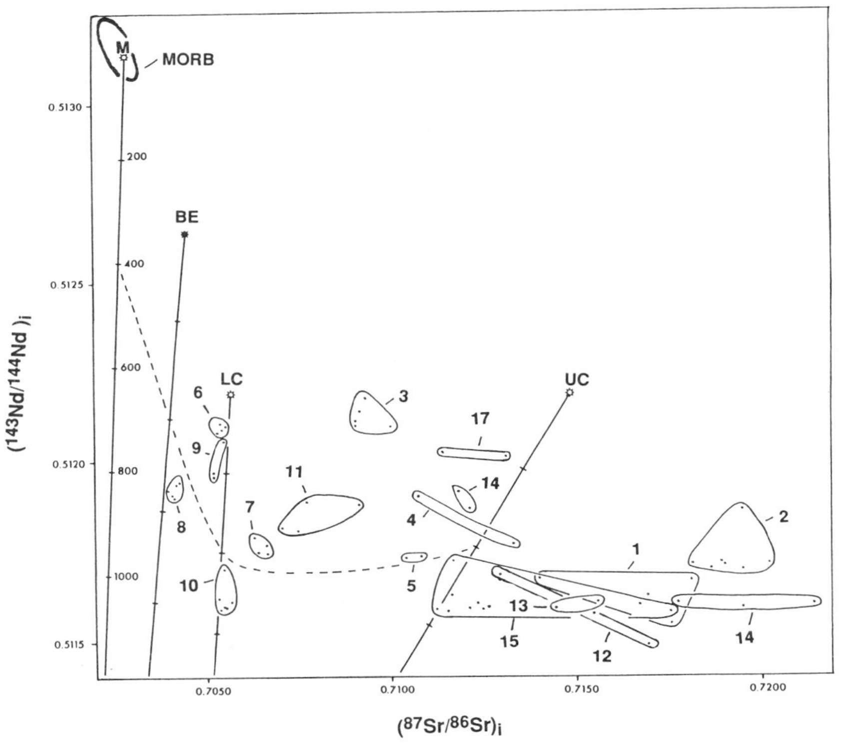

Fig. 4. Initial $^{143}Nd/^{144}Nd$ ratios versus initial $^{87}Sr/^{86}Sr$ ratios for the pre-Mesozoic intrusives of the Central Andes. Indicated on the diagram are model reservoir evolution curves at 200 Ma increments for: *BE* Bulk Earth; *LC* average Lower Crust; *UC* average Upper Crust; *M* Mantle. From Zartman and Haines (1988). The dashed line indicates a mixing trajectory between MORB-source mantle and average upper crust at 400 Ma. Locality identifiers as in Fig. 2. See text for discussion

granites, volcanic-arc granites, within-plate granites, and collision-zone granites. We have not, however, been able to employ this latter approach effectively because the pre-Mesozoic intrusions of the Central Andes fall into the ternary boundary region of the two discrimination diagrams, overlapping the fields for volcanic-arc granites, within-plate granites, and collision-zone granites. Instead, we have found the approach pioneered by Chappell and White (1974) to be more useful; they first defined the distinction between peraluminous granitoids of metasedimentary derivation (S-type) and metaluminous granitoids of metaigneous derivation within the Berridale batholith in southeastern Australia. As pointed out subsequently by O'Neil and Chappell (1977) and O'Neil et al. (1977), granitoids derived from sedimentary and igneous protoliths can also be distinguished on the basis of their O-isotope ratios. This is because sedimentary rocks consist of materials that have become enriched in ^{18}O as a result of having been through a cycle of residence at the Earth's surface involving low-temperature weathering, chemical sedimentation, and diagenesis. Recognizing the conditions of hydrous versus anhydrous melting in a granitoid source region can be more difficult, but Collins et al. (1982) have argued that the Ga/Al ratio is a reliable discriminant. This transpires because Ga is more mobile than Al in dry, fluorine-rich conditions of anhydrous melting and, therefore, melts derived from a dry source region (A-type) are characterized by higher Ga/Al ratios than are melts generated from a normal metaigneous protolith. Thus, it has been argued that, on this basis, it is possible to distinguish between a granitoid generated from a residual granulite protolith that had been subjected to a previous episode of partial melting and melt extraction and one produced from a typical tonalitic or granodioritic protolith (Collins et al. 1982).

By combining these two approaches, it is possible to gain further insight into the origin of the intrusive rocks of the Central Andean basement. In Fig. 5, in which $\delta^{18}O$ values are plotted against Ga/Al ratios, clear distinctions are observed between Andean intrusions of Eocene age and older basement intrusives, as well as among the basement intrusives themselves. In this diagram, $^{18}O/^{16}O$ ratios differentiate between magmas generated from metasedimentary and metaigneous protoliths, whereas Ga/Al ratios distinguish anhydrous from hydrous melting conditions. Such discrimination has geodynamic significance, because the three models described above are each expected to produce a particular kind of granitoid melt: I-type during subduction, S-type during collision, and A-type during extension and crustal thinning.

Notable from all three diagrams (Fig. 5a-c) is the absence of Proterozoic and Palaeozoic plutonic rocks

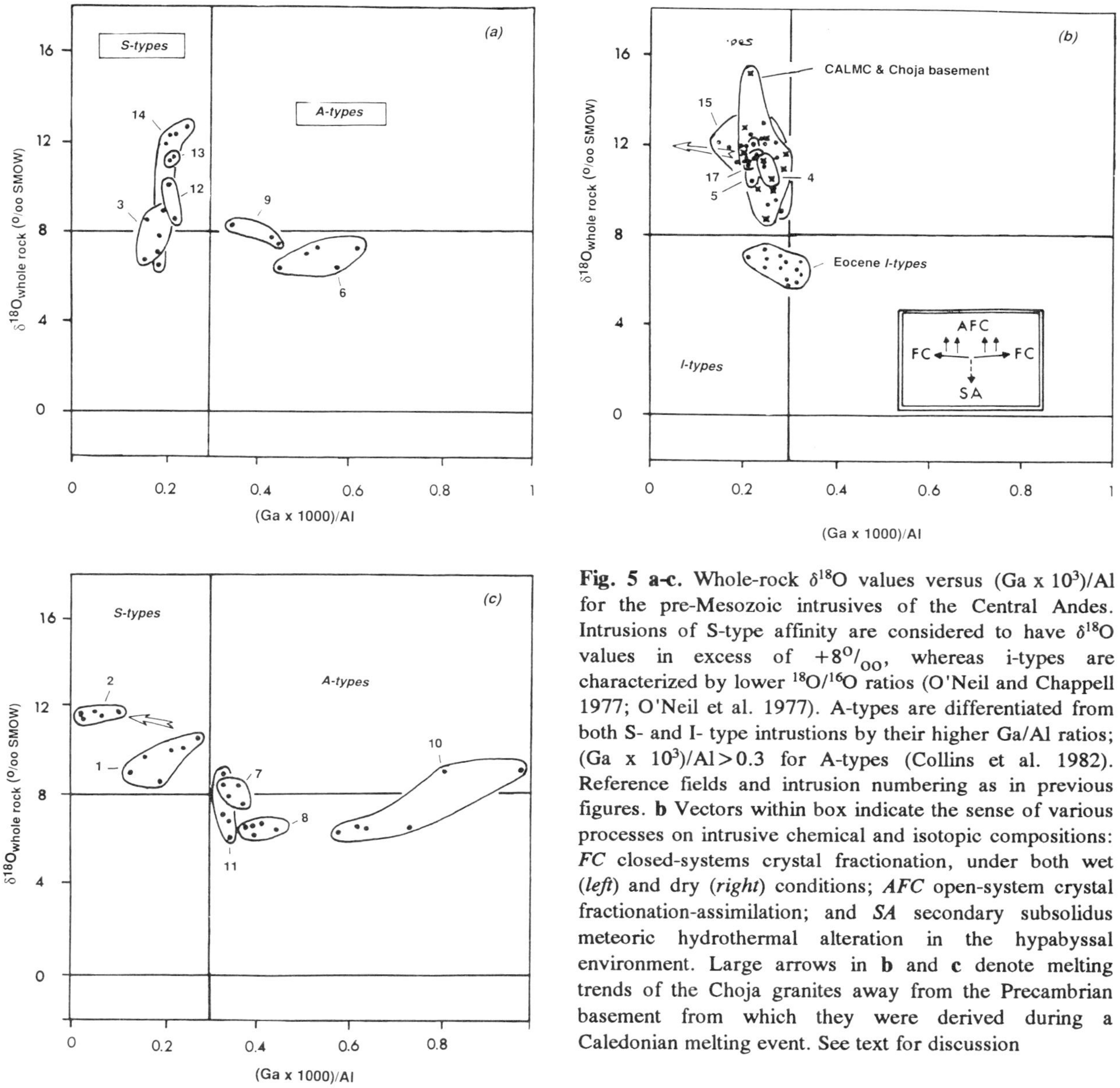

Fig. 5 a-c. Whole-rock $\delta^{18}O$ values versus (Ga x 10^3)/Al for the pre-Mesozoic intrusives of the Central Andes. Intrusions of S-type affinity are considered to have $\delta^{18}O$ values in excess of $+8^o/_{oo}$, whereas i-types are characterized by lower $^{18}O/^{16}O$ ratios (O'Neil and Chappell 1977; O'Neil et al. 1977). A-types are differentiated from both S- and I- type intrustions by their higher Ga/Al ratios; (Ga x 10^3)/Al>0.3 for A-types (Collins et al. 1982). Reference fields and intrusion numbering as in previous figures. **b** Vectors within box indicate the sense of various processes on intrusive chemical and isotopic compositions: *FC* closed-systems crystal fractionation, under both wet (*left*) and dry (*right*) conditions; *AFC* open-system crystal fractionation-assimilation; and *SA* secondary subsolidus meteoric hydrothermal alteration in the hypabyssal environment. Large arrows in **b** and **c** denote melting trends of the Choja granites away from the Precambrian basement from which they were derived during a Caledonian melting event. See text for discussion

with the I-type features that characterize the Eocene intrusive rocks of the Andean Cycle. The 560-520 Ma suite (Fig. 6b) is entirely of S-type affinity. Rocks of similar S-type character are present in both the 505-405 Ma (Caledonian) and 350-240 Ma (Variscan) suites (Fig. 6a), which also include A-type compositions. In addition, a progressive shift from early A-type to later S-type magmatism is observed within these two orogenic cycles when the data are viewed in a chronological context (e.g. intrusives #10 and #11 in Fig. 5). The field for the metamorphic basement of this region of the Central Andes largely coincides with the fields for the S-type granitoids (Fig. 5b), supporting the idea that these intrusives were generated within the upper crust as a result of anatectic melting of a metasedimentary protolith. The tendency for the A-type rocks to have lower $\delta^{18}O$ values than the S-type rocks suggests that they were not derived from the same sources.

This distinction between S-type and A-type melts is also observed in Fig. 6, the ilmenite-magnetite log fO_2-temperature diagram, which gives us additional confidence in the $\delta^{18}O$-Ga/Al discrimination. The anorogenic A-type melts appear to have crystallized

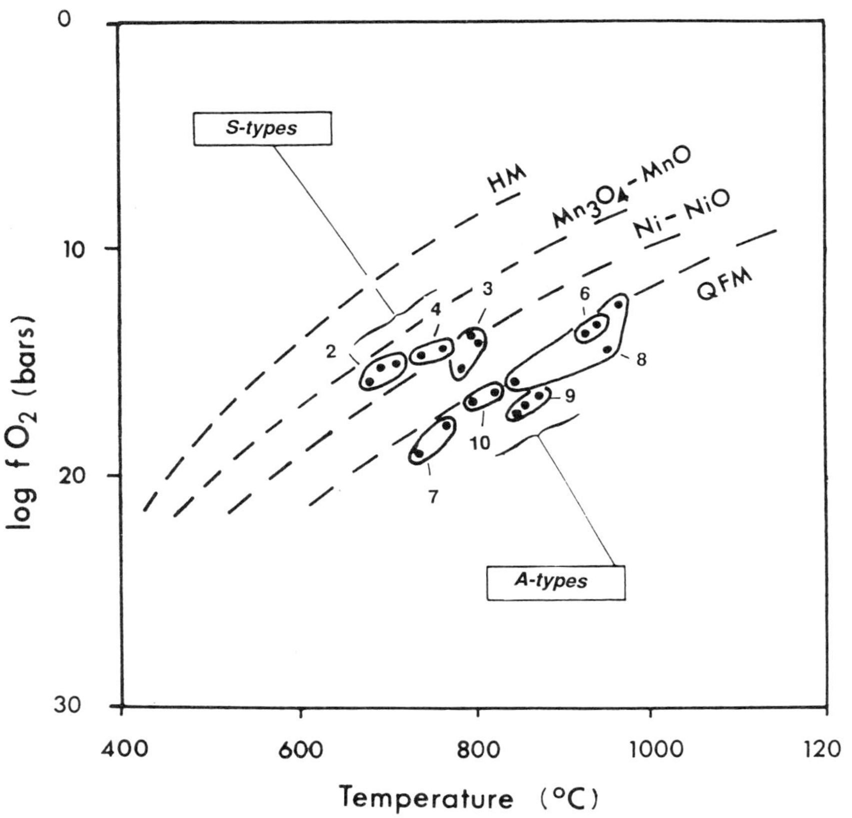

Fig. 6. Plot of log fO_2 versus temperature for the pre-Mesozoic intrusive for the Central Andes. Buffers: *HM* Fe_2O_3-Fe_3O_4; Mn_3O_4-MnO; Ni-NiO; *QFM* SiO_2-Fe_2SiO_4-Fe_3O_4. Locality identifiers as in Fig. 2. See text for discussion.

between 960 and 730°C, whereas the synorogenic S-type melts crystallized between 800 and 630°C. The logfO_2/T paths for the two groups of intrusives are also distinctly different: the anorogenic series following the QFM buffer and the synorogenic series following a path between the Ni-NiO and Mn_3O_4 buffers. Thus oxygen fugacities were higher for the synorogenic magmas than for the anorogenic ones.

The radiogenic isotope data shown in Fig. 4 also fall into two broad groups, with S-types having higher 87Sr/86Sr ratios and lower ^{143}Nd/^{144}Nd ratios than I- or S-types. The A-type intrusives (#6-10 in Figs 3-6), which are confined to the Cordon de Lila and Sierra de Macon areas (Fig. 1), plot with Sr-isotope ratios close to the lower crustal model evolution curve of Zartman and Haines (1988). In contrast, the S-type rocks with higher Sr-isotope ratios plot in a broad field about, and displaced to the right of, the lower crustal model evolution curve. These features, together with the distinct lack of Nd-isotope ratios less radiogenic than "Bulk Earth" (Fig. 4), indicate that all these Central Andean intrusive suites were largely, if not entirely, of crustal derivation. The S-types were predominantly the product of anatectic melting of middle-upper crustal sources that had a relatively restricted range of Nd-isotope variation, but which were extremely heterogeneous in terms of their Sr-isotope compositions. This suggests protoliths of approximately the same age (i.e. mid-late Proterozoic) having a variety of Rb-Sr ratios and supports the ideas of Goldstein (1988) about decoupled Rb-Sr and Sm-Nd systems in the continental crust. In some cases, there is an indication that the mixing of melts derived from multiple crustal sources played an important role in granitoid petrogenesis. In contrast, the A-type magmas were generated largely within a mafic lower crust, with perhaps minor crustal inputs in individual instances (e.g. Sierra de Macon - intrusion #11 in Fig. 4).

Feldspar Pb-isotope variations for the intrusive suites are shown in Fig. 7. Unlike the subduction-generated rocks of the Andean Cycle, which plot in a restricted field along the uppermost portion of the Stacey and Kramers (1975) average crustal Pb-evolution curve, the Proterozoic-Palaeozoic intrusives are characterized by their large range of Pb-isotope variation and, in some instances, their retarded Pb-isotope evolution. In both the ^{207}Pb/^{204}Pb vs. ^{206}Pb/^{204}Pb (Fig. 7a) and ^{208}Pb/^{204}Pb vs. ^{206}Pb/^{204}Pb (Fig.7b) diagrams, the Preandean rocks plot between the Lower Crustal and Upper Crustal model evolution curves of Zartman and Haines (1988). The A-types tend to cluster about the Stacey-Kramers average crustal Pb-evolution curve, with a slope that is sub-parallel to that of the Geochron. The exception is the Sierra de Macon rocks which, as noted above, exhibit ^{87}Sr/^{86}Sr-^{143}Nd/^{144}Nd systematics indicative of a complex petrogenesis involving both lower crustal

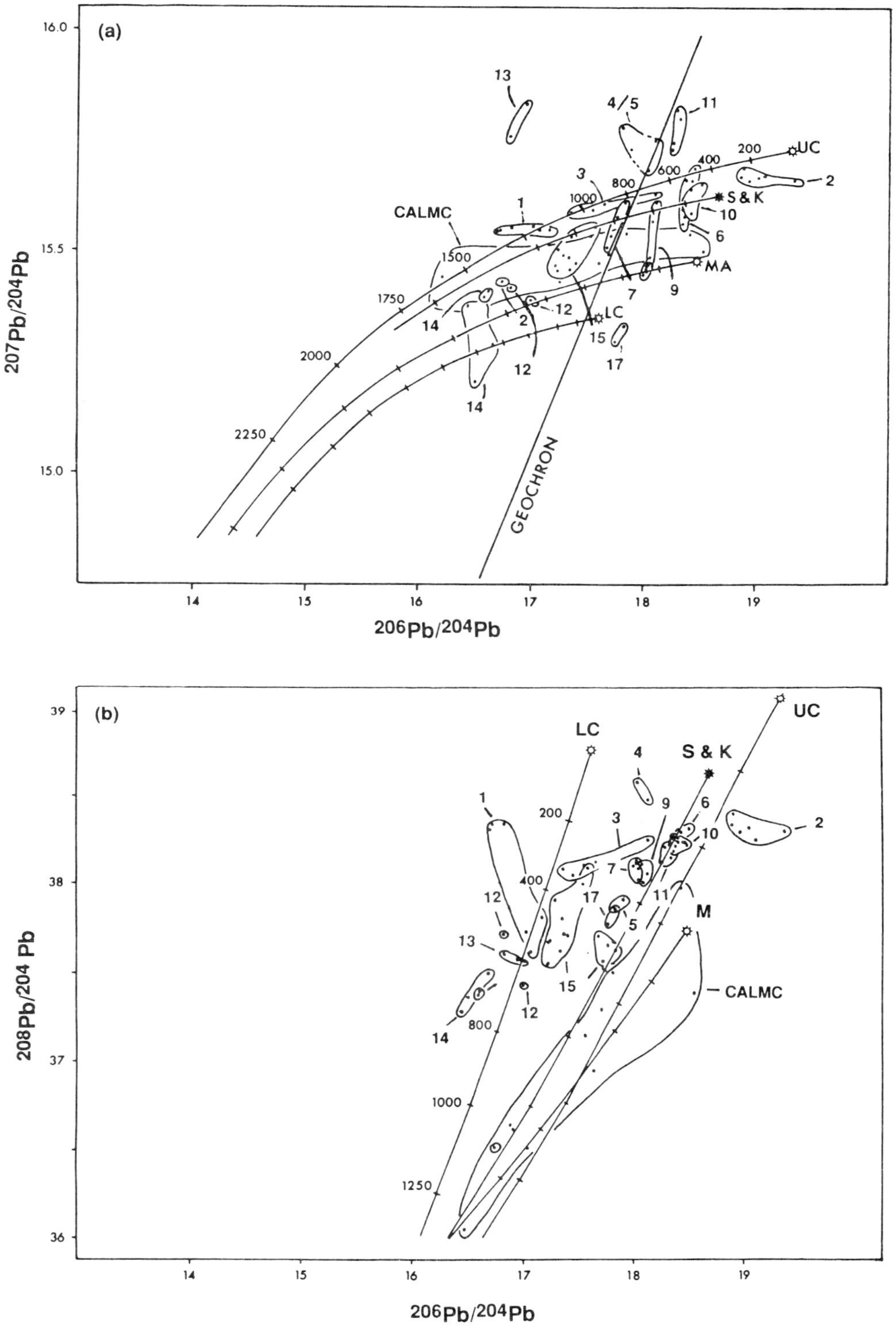

Fig. 7 a and b. Lead isotope data and evolution curves in the system **a** $^{207}Pb/^{204}Pb$ versus $^{206}Pb/^{204}Pb$ and **b** $^{208}Pb/^{204}Pb$ versus $^{206}Pb/^{204}Pb$. Lead isotope evolution curves are for designated model reservoirs from Stacey and Kramers (1975) and Zartman and Haines (1988) are as follows: *S & K* = Stacey and Kramers (1975) two-stage average crustal lead model evolution; *LC* lower crust; *UC* upper crust; *M* mantle. Marks along curves indicate 200 Ma increments. Locality identifiers as in Fig. 2. See text for discussion

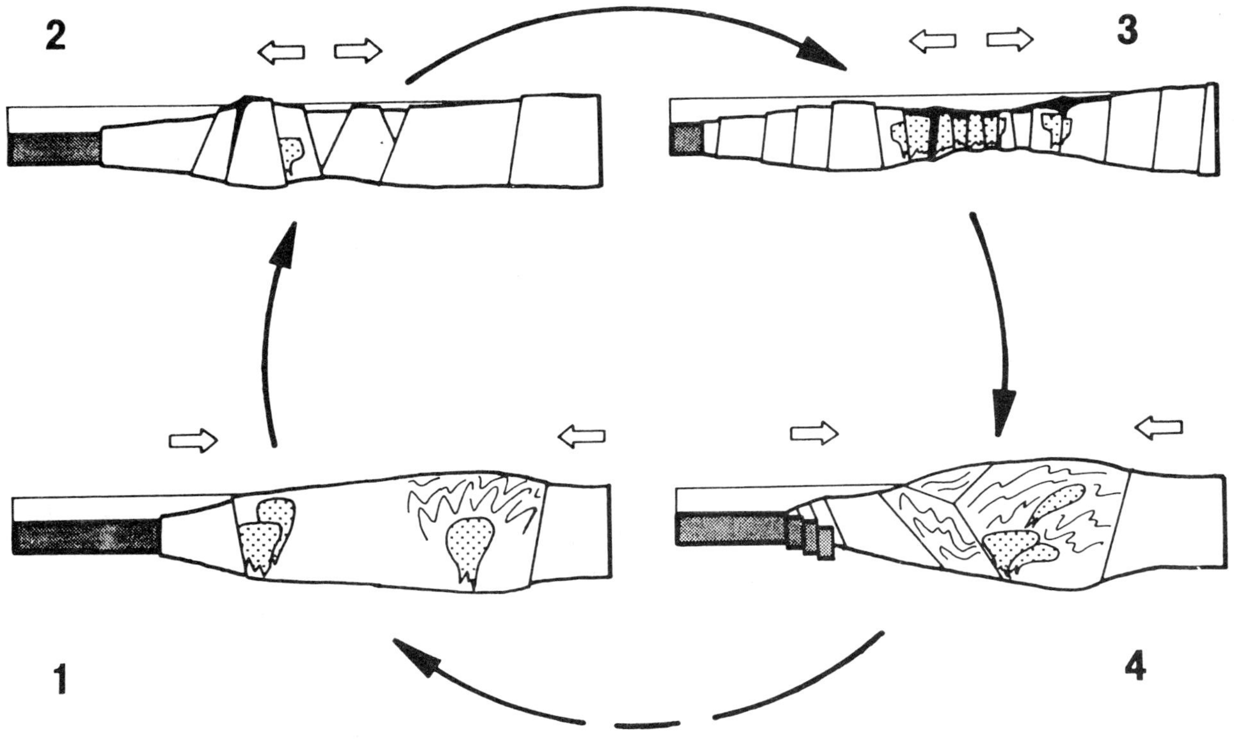

Fig. 8. Geodynamic model for the Preandean evolution of the Central Andes between 18° and 24°S. From our results we envisage a cyclicity of geodynamic events, exemplified here for the time span from the latest Precambrian to early Silurian. *1* **Early Phanerozoic consolidation** and crustal thickining along the western foreland of the Brazilian shield, folding of Pincoviscana Formation and intrusion of the Mejillones and Salar de Navidad plutons and the chain of plutons ranging from Canani to Cachi. *2* **Late Cambrian:** Crustal thinning triggers marine ingressions and early A-type plutonism (e.g. Choschas-Montiraqui plutons and Mejillones basaltic volcanism). *3* **Ordovician:** Increasing crustal thinning, with resulting widening of marine environment and associated A-type plutonism and volcanic activity (e.g. Cordon de Lila lavas). *4* **Latest Ordovician/early Silurian:** Geodynamic reversal causes crustal thickening, synorogenic magmatism, and further crustal accumulation and consolidation - i.e. crustal growth along the margin of the old cratonic terrane. See text for discussion

and upper crustal components. The tendency toward low Pb-isotope ratios in many of the intrusive suites probably reflects the influence of unradiogenic Pb derived from the western margin of the Precambrian craton (Brazilian Shield), or possibly from the southerly extension of the Arequippa Massif which underlies at least the northern portion of the study area (Dalziel and Forsythe 1985). The metamorphic rocks of the region, indicated by the area labelled CALMC, cover a large area on both diagrams, and also show a tendency toward unradiogenic Pb-isotopic compositions, typical of granulitic crustal terranes that have suffered a depletion in U during dehydration or partial melting. This displacement of the Central Andean basement intrusive suites above the lower crustal evolution curve in both diagrams, but particularly in Fig. 7a, indicates the involvement of a more radiogenic source in the origin of these rocks or the mixing of Pb from both upper and lower crustal sources.

5 Summary and Conclusions

The described geochronological and isotopic characteristics of the Pre-Mesozoic crystalline basement of the Central Andes are a direct manifestation of the late Proterozoic-Palaeozoic history of the region and can therefore be used to interpret this history in the context of the three tectonic models for the evolution of the western margin of central South America. By inspection of the data, shown in Fig. 5 it is clear that there is no firm evidence that the pre-Mesozoic history of the Central Andes was characterized, for any substantial period of time, by a destructive plate margin tectonic regime involving eastward subduction of oceanic lithosphere beneath the western margin of the South American continent, as is the case at present and has been at least since the Cretaceous. The older intrusives studied do not have any of the features of the I-type magmas that are characteristic of such a magmato-tectonic regime, and which are observed in both the plutonic and volcanic rocks of the Andean cycle. The situation of uninterrupted eastward subduction against a stable continental margin throughout the Palaeozoic ought to have produced magmatic rocks with stable and radiogenic isotopic compositions similar to those in Andean Cycle rocks generated in the same areas over the past 100 Ma. Instead, plutonic rocks of essentially identical S-type and I-type character were intruded into the upper crust across a wide (>1000 km) transect perpendicular to the continental margin at c. 23°-24°S during three separate pre-Mesozoic orogenic cycles. Thus, the continuous subduction model can be ruled out.

The exotic terrane model postulates that extensive crustal addition and consolidation occurred through subduction of oceanic crust, crustal shortening, and the accretion and amalgamation of different kinds of terranes to the stable cratonic margins of the Brazilian Shield. The diversity and complexity of the Central Andean basement outcrops investigated is not inconsistent with the idea of suspect terranes (Dalziel and Forsythe 1985; Ramos 1988), but careful and detailed field observations between 18° and 24°S in the Central Andes has not not yielded any evidence for bona fide allochthonous terranes of Phanerozoic age. Also, the striking similarity in petrographic, structural, geochemical, and isotopic character of the basement remnants studied contrasts sharply with the extremely heterogeneous nature of allocthonous terranes found in North America and in other portions of the Circum-Pacific mobile belt. However, the exotic terrane model cannot be entirely eliminated because of the scarcity of reliable data for pre-Phanerozoic times, when such a tectonic process may have contributed substantially to the formation of ancient continental crust along the margins of the Archean cratonic nucleii.

Based upon the data discussed here, we favour an ensialic margin model for the pre-Mesozoic evolution of the western margin of central South America. Such a model proposes that the development of the continental margin area was largely complete by late Proterozoic time and, subsequently, this area remained autochthonous and mainly passive, with its late Precambrian and Palaeozoic history characterized by repeated opening and closure of intracontiental basins. We see the current situation of Andean subduction initiated only in the Triassic with the break-up of Pangaea. These ideas are supported by two important observations: (1) the similarity in the structural, geochronological, and petrographic records among the various basement remnants, and (2) the cyclicity of anorogenic versus synorogenic magmatic activity and the strong crustal signature of all magmatic products studied.

Thus, the overall picture we envisage for the Central Andean tectonic setting from late Proterozoic through Palaeozoic times is one of a generally passive continental margin, within which extensional basins periodically opened and closed in response to global plate interactions (Fig. 8). This concept is supported by both the similarity of the structural, petrographic, and chronological records for the various basement rtemnants studied as well as the cyclicity of anorogenic and synorogenic magmatic activity and the strong signature of continental crust in all of the basement intrusives. The absence of I-type magmatic rocks in the region implies that subduction was not an important feature of this continental margin until the current Andean Cycle commenced in the early Mesozoic with the break-up of Pangaea. We therefore postulate two major geodynamic changes during the Palaeozoic. An early phase of accretion along the Archean cratonic nucleus was superceded by intracontinental consolidation which, in turn, was followed by the present compressional margin regime. With forced subduction of oceanic crust along a pre-consolidated margin during the Jurassic, crustal accumulation due to forearc accretion and voluminous magma generation resulted in a termination of earlier extensional backarc basins and marine environments and led directly to parallel uplift and unroofing along the present N-S trending Coast Range and Precordillera. Cooling of basement rocks below 100°C occurred during Palaeogene time.

Acknowledgements. Over the years this project has become an example of a truly multidisciplinary and international cooperation. Conceived and initiated by W. Zeil (Technische Universität Berlin), its success was due to the enthusiasm of many contributors, particularly those South American colleagues who assisted with fieldwork. However, little would have been accomplished without the support of the German Science Foundation (Deutsche Forschungsgemeinschaft) and the Alexander von Humboldt Foundation. We wish to record our special thanks to the following individuals for their support in many different ways: S. Pichowiak and L.J.A. Schermerhorn (Freie Universität Berlin); A. W. Hofmann, W. Todt, and I. Raczekj (Max-Planck-Institut für Chemie); M.J. Holdaway (Southern Methodist University); W.I. Mantion and R.J. Stern (University of Texas at Dallas); H. Niemeyer (Universidad del Norte); R. Omarini and J. Viramonte (Universidad Nacional de Salta); and, particularly to the late J. Borthwick, who not only made the stable isotope work possible but whose friendship and professional expertise are very much missed. NERC Isotope Geosciences Laboratory Publication No. 53.

References

Bachmann G, Grauert B, Kramm U, Lork A, Miller H (1986) Oberkambrischer Magmatismus im Grundgebirge Nordwest Argentiniens: Isotopengeologische Untersuchungen an Granitoiden der Intrusivkomplexe von Santa Rosa de Tastil und Cañani. Berl Geowis Abh A, Sonderbd Geowiss Lateinamerika Kolloq: 111-112

Bachmann G, Grauert B, Kramm U, Lork A, Miller H (1987) El magmatismo del cambrio Medio/Cambrico Superior en el basamento del Noroeste Argentino: investigaciones isotopicas y geochronologicas sobre los granitoides de los complejos intrusivos de Santa Rose de Tastil y Cañani. X Congr Geol Argent Actas 4: 125-127

Chappell BW, White AJR (1974) Two contrasting granite types. Pac Geol 8: 173-174

Coira B, Davidson J, Mpodozis C, Ramos V (1982) Tectonic and magmatic evolution of the Andes in northern Argentina and Chile. Earth Sci Rev, 18: 303-332

Collins WJ, Beams SD, White AJR, Chappell BW (1982) Nature and origin of A-type granites with particular reference to southeastern Australia. Contrib Mineral Petrol 80: 189-200

Coney P (1980) Cordilleran metamorphic core complexes: an overview. Geol Soc Am Mem 153: 7-31

Dalziel IWD, Forsythe RS (1985) Andean evolution and the terrane concept. In: Howell DG (ed) Tectonostratigraphic Terranes of the Circum-Pacific Region. Earth Sciences Series, Circum-Pacific Council for Energy and Mineral Resources 1: 565-581

Damm K-W, Pichowiak S, Harmon RS, Todt W, Omarini R, Niemeyer H (1990) Pre-Mesozoic Evolution of the Central Andes - the basement revisited. Geol Soc Am Spec Pap 241: 101-126

Damm, K-W, Pichowiak S, Harmon RS, Todt W, Breitkreuz C, Buchelt M (1991) The Cordon de Lila Complex, Central Andes, N-Chile: An Ordovician Continental Volcanic Province Geol. So Am Spec Paper 265:.179-188

Goldstein SL (1988) Decoupled evolution of Nd and Sr isotopes in the continental crust and the mantle. Nature 336: 733-738

Harmon RS, Barreiro BA, Moorbath S, Hoefs J, Francis PW, Thorpe RS, Deruelle B, McHugh J, Viglino JA (1984) Regional O-, Sr-, and Pb-isotope relationships in late Cenozoic calc-alkaline lavas of the Andean Cordillera. J Geol Soc 141: 803-822

Longstaffe FJ, Clarks AH, McNutt RH, Zentilli M (1983) Oxygen isotopic compositions of Central Andean plutonic and volcanic rocks. Earth Planet Sci Lett 64: 9-18

Nur A, Ben Avraham Z (1977) Lost Pacifica continent. Nature 270: 41-43

Nur A, Ben Avraham Z (1982) Oceanic plateaus, the fragmentation of continents and mountain building. J Geophys Res 87: 3644-3661

Nur A, Ben Avraham Z (1983) Displaced terranes and mountain building. In: Hsu KJ (ed) Mountain building processes. Academic Press, London, pp 73-84

Omarini RH, Cordani UG, Viramonte JG, Salfity JA, Kawashita K (1984) Estudio geochronologico Rb-Sr de Faja Eruptiva de la Puna en el sector de San Antonio de los Cobres, provincia de Salta. 9 Congr Geol Argent Actas 3: 146-158

O'Neil JR, Chappell BW (1977) Oxygen and hydrogen isotope relations in the Berridale batholith. J Geol Soc Lond 133: 559-571

O'Neil JR, Shaw SE, Flood RH (1977) Oxygen and hydrogen isotope compositions as indicators of granite genesis in the New England batholith, Australia. Contrib Mineral Petrol 62: 313-328

Peacock MA (1931) Classification of igneous rock series. J Geol 39: 65-67

Pearce JA, Harris NBW, Tindle AG (1984) Trace element discrimination diagrams for the tectonic interpretation of granitic rocks. J Petrol 25: 956-983

Pitcher WS, Atherton MP, Cobbing EJ, Beckinsale RD (1985) Magmatism at a plate edge - the Peruvian Andes. Blackie-Halsted Press, 328 pp

Ramos VA (1988) Late Proterozoic-Early Paleozoic of South America - a collisional history. Episodes 11: 168-173

Rogers G (1985) A geochemical traverse across the north Chilean Andes. PhD Thesis, The Open University Milton Keynes, 333 pp (unpubl)

Scheuber E, Andriessen PAM (1990) The kinematic and geodynamic significance of the Atacama Fault Zone, northern Chile. J Struct Geol 12: 243-257

Stacey JS, Kramers JD (1975) Approximation to terrestrial lead isotope evolution by a two-stage model. Earth Planet Sci Lett 26: 207-221

Wilner AP, Miller H, Lottner US (1985) The evolution of the Andean convergent plate margin in the early Paleozoic between latitudes 15°S and 34°S. Communicaciones 35: 257-259

Zartman RE, Haines SM (1988) The plumbotectonic model for Pb isotopic systematics among major terrestrial reservoirs - a case for bidirectional transport. Geochim Cosmochim Acta 52: 1327-1339

The Late Carboniferous to Triassic Volcanic Belt in Northern Chile

CHRISTOPH BREITKREUZ and WERNER ZEIL

Abstract. A prominent feature of the pre-Jurassic Andean basement are outcrops of late Palaeozoic to Triassic, mainly siliceous volcanoplutonic complexes. Published models assume a magmatic trend to have developed from a "normal" subduction setting with calc-alkaline magmatism during Carboniferous times towards an ensialic marginal rift during the Permo-Triassic, which gave rise to (calc-)alkaline and peralkaline magmas. In the Precordillera and Cordillera Occidental of northern Chile (20°-25°S), widespread outcrops of late Carboniferous to Triassic volcanosedimentary successions occur (Peine Group). Caldera eruption seems to have been important in this region. The volcanic rocks are of calc-alkaline to slightly alkaline, predominantly siliceous composition (SiO_2 range 50-80%). The geochemical features of the volcanic rocks display a clear volcanic arc affiliation. In parts, a diagenesis/anchimetamorphism of the volcanic rocks caused considerable variations of mobile elements (e.g. K, Na, Rb, Sr, Cu, Si). HFS and RE elements display patterns of a non-tensional subduction-controlled setting of the magmatism at a continental margin of moderate thickness. In the Salar de Atacama area, a conspicuous alluvial-limnic volcanosedimentary intercalation occurs. The terrestrial basin which accommodated this sequence formed during the latest Carboniferous-?early Permian on top of the deposits of the pre-existing late Carboniferous composite volcanism. The lower part of the intercalation consists of green limnic and multicoloured alluvial fan deposits and the upper part is made up of red alluvial sedimentary rocks. Throughout the sequence, minor basic volcanic rocks occur. The considerable size of the basin, the limnic-alluvial fan facies association and the accompanying intrabasinal basic volcanism indicate a basin formation controlled by extensional arc tectonics. With regard to the north Chilean Peine Group, this arc-graben represents the only support to the aforementioned model of a tensional regime during the Permo-Triassic.

1 Introduction

Late Carboniferous to middle Triassic volcanosedimentary successions and associated intrusions crop out extensively in the north Chilean Precordillera and Cordillera Occidental (Fig. 1a; Ramirez and Gardeweg 1982; Marinovic and Lahsen 1984; Vergara and Thomas 1984; Naranjo and Puig 1984; Bogdanic 1990). Coeval intrusions are recorded from the Cordillera de la Costa (Fig. 1a; Skarmeta and Marinovic 1981; Chavez 1983; Naranjo and Puig 1985). Associated volcanic successions in the coastal area are scarce and stratigraphically poorly defined (Breitkreuz 1986; Scheuber and Andriessen 1990). In spite of the wide distribution of the volcanosedimentary successions, few detailed studies have so far been published (e.g. Davidson et al. 1985; Osorio and Rivano 1985).

Correspondence to: C. Breitkreuz, Institut für Geologie und Paläontologie, Technische Universität Berlin, Ernst-Reuter-Platz 1, D-1000 Berlin 10

The north Chilean volcanosedimentary successions represent the products of a magmatic zone at the Pacific margin of Gondwana, which can be traced from the Peruvian Andes to the southern Andes and the Antarctic Peninsula (Zeil 1981; Mpodozis et al. 1985; Mahlburg Kay et al. 1989; Breitkreuz 1990). Palaeomagnetic data from the Quebrada Pajonales and Sierra de Argomedo areas (Fig. 1a) in the north Chilean Precordillera document a concordant position of the magmatic zone with respect to South America (Jesinskey et al. 1987). Thus, there is no reason to suspect any complications in palaeogeography due to terrane accretion.

This chapter presents the results of investigations on the late Carboniferous to Triassic volcanosedimentary successions in the north Chilean Precordillera and Cordillera Occidental, grouped under the informal name "*Peine Group*" by Bahlburg and Breitkreuz (1991). The results will be discussed in the light of the geodynamic models that exist for the Gondwana margin evolution during late Palaeozoic-Triassic times.

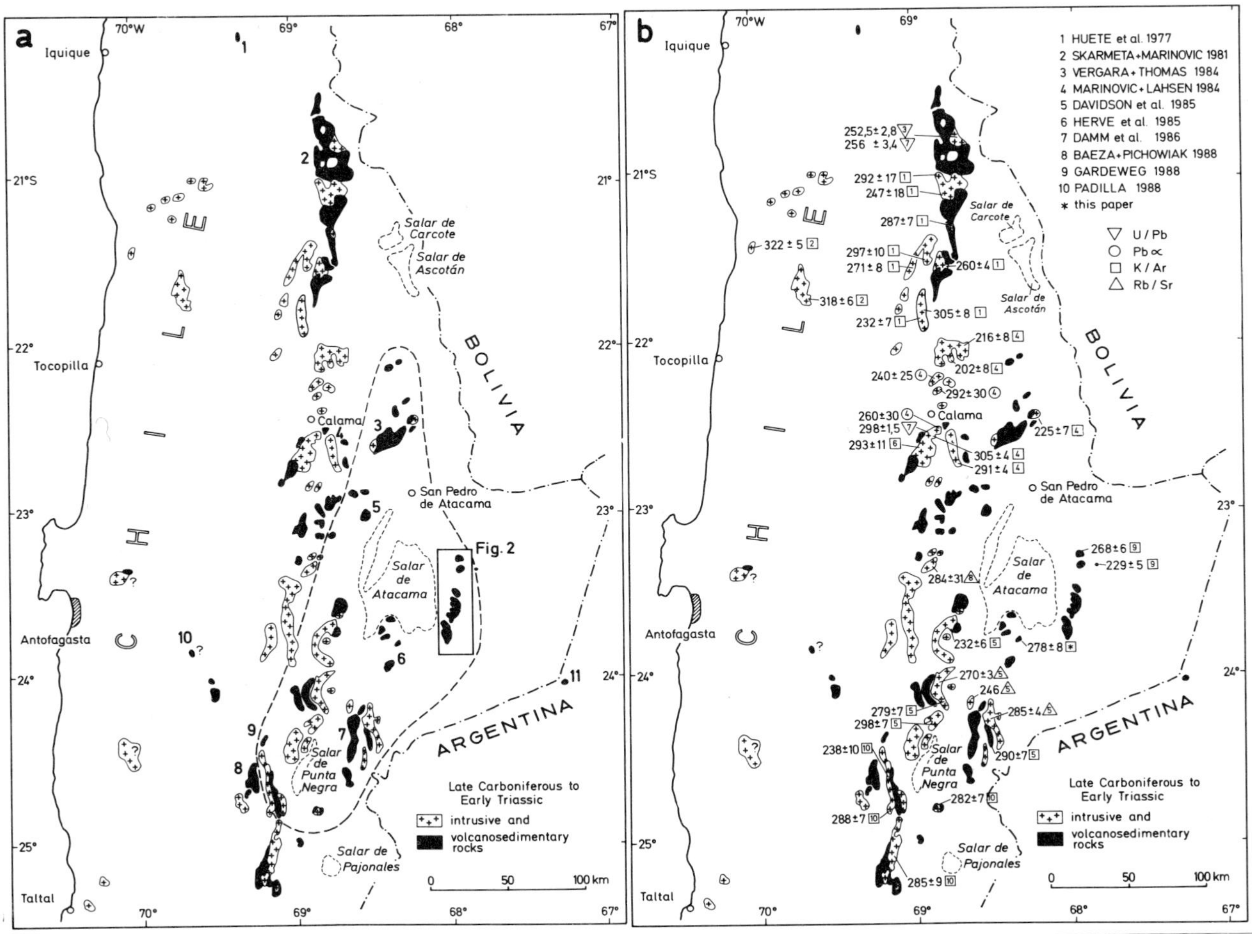

Fig. 1. a Outcrops of late Carboniferous to Triassic volcanic and plutonic rocks in northern Chile and northwest Argentina. For references see Sect. 1. *Heavy dashed line* indicates minimum extension of the latest Carboniferous-?early Permian limnic facies of the Miembro Medio (see Sect. 3.2); Localities: *1* Juan de Morales; *2* Collahuasi; *3* Tuina; *4* Cerro Moctezuma; *5* El Bordo; *6* Cordón de Lila; *7* Quebrada Pajonales; *8* Sierra de Argomedo; *9* Cerro Bayo; *10* Augusta Victoria; *11* Salar de Arizaro. **b** Radiometric ages of the late Carboniferous to Triassic volcanic and plutonic rocks of northern Chile

2 Stratigraphy

2.1 Age Constraints and Field Relations of the Peine Group

Many radiometric ages of the associated intrusive rocks and of thermal events within metamorphic series from the area have been published; these ages range from 305 to 202 Ma (i.e. Moscovian to Sinemurian; Fig. 1b; Huete et al. 1977; Skarmeta and Marinovic 1981; Marinovic and Lahsen 1984; Vergara and Thomas 1984; Davidson et al. 1985; Hervé et al. 1985; Damm et al. 1986; Baeza and Pichowiak 1988; Padilla 1988). However, only four ages, ranging from 290 to 229 Ma, are available from volcanic rocks of the Peine Group (Davidson et al. 1985; Gardeweg 1988; Cordón de Lila area: see below). Thus, in north Chile, volcanism started somewhat earlier than in the southern Andes, where the first volcanic deposits of the Choiyoi Group are of Permian age (Caminos 1979; Zeil 1981). As a general rule, the associated plutonic rocks intruded the Peine Group or older rocks (e.g. Quebrada Pajonales, Fig. 2i, Davidson et al. 1981, 1985). The Cerro Moctezuma area south of Calama is the only place in which a late Carboniferous-Permian pluton (Limon Verde pluton) is unconformably overlain by the sedimentary and volcanic rocks of the Peine

Group (Fm. Agua Dulce, Fig. 2c; Marinovic and Lahsen 1984).

In the Sierra de Almeida, the Peine Group is underlain by the marine and continental clastic rocks of the Devonian to early Carboniferous Zorritas Fm. (Fig. 2h and i; Davidson et al. 1981; Breitkreuz 1986). An onset of volcanic activity as early as in the late Carboniferous is evident from the intercalation of alluvial-limnic sediments ("Miembro Medio", Fig. 3) containing late Carboniferous ostracods (Osorio and Rivano 1985; Breitkreuz et al. 1992). Correlation between the units of the Peine Group is limited, due to the pronounced lateral facies variations and the lack of radiometric and palaeontological data. Only the "Miembro Medio" (see below) helps to distinguish "pre-Miembro Medio" from "post-Miembro Medio" volcanic units in several outcrops in the vicinity of the Salar de Atacama (Figs. 2 and 3). Assuming that the respective Miembro Medio units are broadly synchronous, we propose a division of the successions into three parts, the Miembro Inferior (and older units), Miembro Medio, and Miembro Superior (and younger units) for every outcrop where the sedimentary unit is under- and overlain by volcanic units (following the division of the Peine Fm. given by Ramirez and Gardeweg 1982, and of the Tuina Fm. given by Marinovic and Lahsen 1984). This assumption is supported by the fact that the respective Miembro Medios comprise similar successions of lithotypes, as well as comparable biostratigraphic and palaeoclimatic features (see below and Breitkreuz 1991; Breitkreuz et al. 1992).

Volcanic activity terminated in the Precordillera during the middle Triassic, as is indicated by the transition from continental volcanosedimentary to non-volcanic marine facies (Fig. 2k: Sierra de Argomedo; Chong and von Hillebrandt 1985; Breitkreuz 1986).

In general, the successions of the Peine Group are only gently tilted and faulted; in some places large open folds occur (e.g. Fig. 3b). Davidson et al. (1985) suggested syndepositional tilting and folding caused by caldera subsidence. In fact, many intraformational unconformities can be observed. However, the Peine Group was also affected by late Cretaceous and Cenozoic compressional tectonics (Reutter et al. 1988).

2.2 The Precordillera north of Calama

The northernmost occurrence of late Palaeozoic volcanic rocks in Chile is reported by Galli (1968) from the Cerro Juan de Morales south of 20°S (Fig. 1a). Here, strongly altered massive rhyolites (Fm. Quipisca; Galli 1968) are overlain by continental red beds and shallow marine limestones (Fm. Juan de Morales; Galli 1968) which yield Permian brachiopods (Zeil 1964).

Vergara and Thomas (1984) described widespread outcrops of thick continental volcanosedimentary successions (Fm. Collahuasi) in the Precordillera around 21°S, which are intruded by Permian plutons (Fig. 1a and b). The volcanic rocks are andesitic to rhyolitic in composition. Limestone intercalations, quoted as of lacustrian origin by Vergara and Thomas (1984), have been classified by Breitkreuz (1986) as shallow marine deposits, based on the occurrence of sea-urchin spines.

Bogdanic (1990) suggested that some of the widespread outcrops of continental volcanosedimentary successions that occur in the Precordillera between 21° and 22°S, and which have been described as Cretaceous units by previous authors, should be correlated with the Fm. Collahuasi (Fig. 1).

2.3 Tuina area

Marinovic and Lahsen (1984) divided the thick volcanosedimentary successions of the Fm. Tuina into three "Miembros" (Fig. 2a and b). The Fm. Tuina is intruded by a Triassic porphyric dacite (Fig. 1b; Lahsen and Marinovic 1984). Thick siliceous ignimbrites and devitrified obsidian flows dominate the lithology of the "Miembro Inferior". At the Morro del Inca (for any geographical locality not depicted in the figures, please refer to the geological maps of the Chilean Geological Survey (Sernageomin) cited in the references) the Miembro Inferior is overlain by a c. 150-m-thick sedimentary unit comprising green alluvial-limnic clastic and calcareous sediments in the lower part and red alluvial clastic sediments in the upper part (Miembro Medio). The transition from green to red sediments is marked by isolated basic lavas, tuffs and peperites. The upper part of the red beds also contains basic lavas (Fig. 2b).

Only 2 km to the north, near the S. José mine, the green alluvial-limnic beds of the Miembro Medio are overlain by a c. 2-km-thick succession of siliceous ignimbrites and subordinate basic lavas with some intercalations of red volcanigenic sandstones and conglomerates ("Miembro Superior", Fig. 2a). Thus, the red beds, which comprise the upper part of the

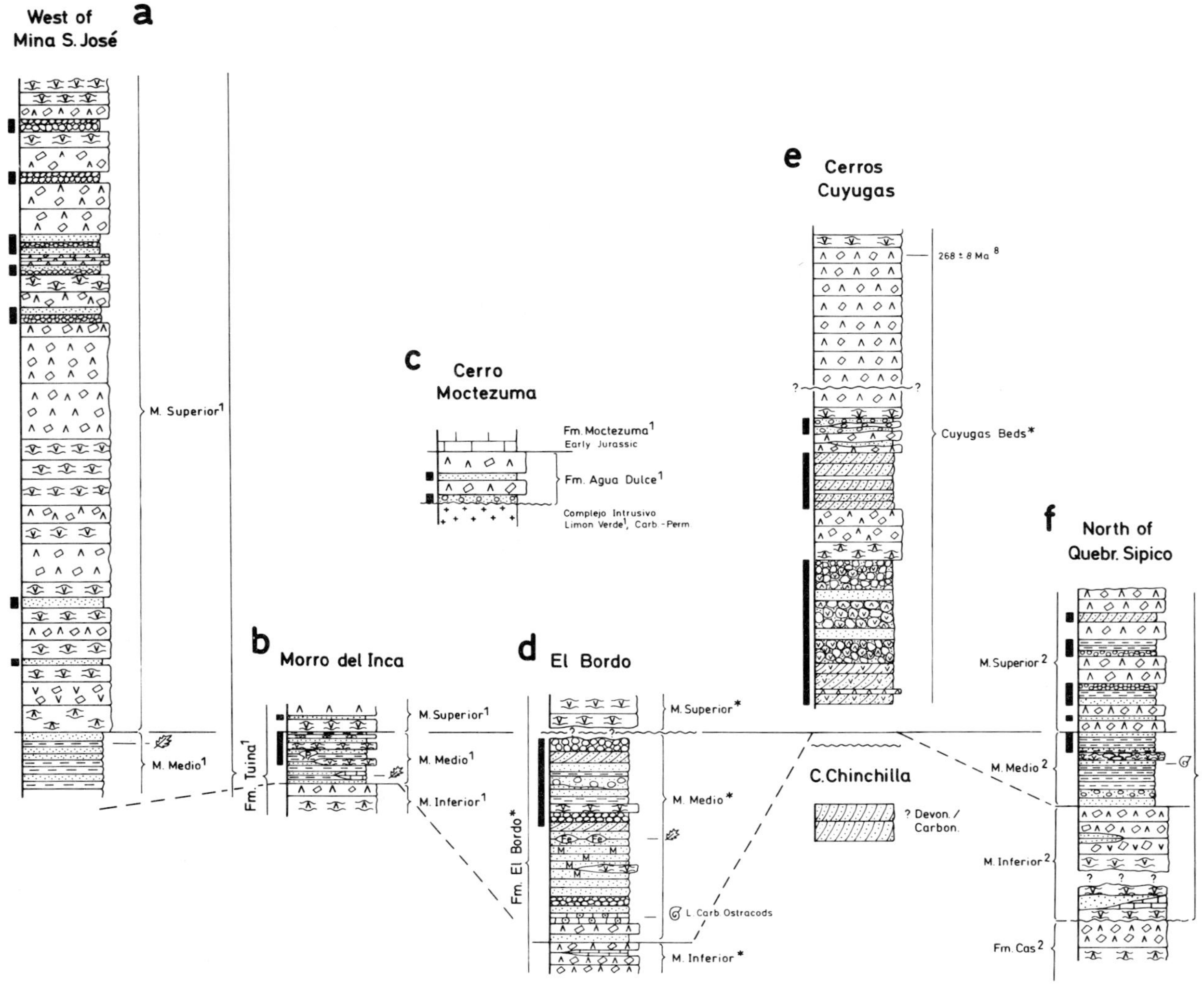

Fig. 2a-k. Schematic lithological columns of sections taken by the authors in the volcanosedimentary successions of the Peine Group. Bed thicknesses are not to scale; *horizontal correlation lines* indicate the Miembro Medio-Superior boundary, but they are not necessarily synchronous. The same applies to the *dashed correlation lines*, which indicate the Miembro

Miembro Medio at the Morro del Inca (Fig. 2b) had either been eroded or were never deposited in this area. Lahsen and Marinovic (1984) and Breitkreuz et al. (1992) describe estherians and late Carboniferous plant fossils from the alluvial-limnic beds of the Miembro Medio.

2.4 El Bordo Area

South of Calama, extended outcrops of volcanic and sedimentary rocks are summarized as Fm. Agua Dulce (Triassic to early Jurassic) by Jensen and Quinzio (1979), Ramirez and Gardeweg (1982) and Lahsen and Marinovic (1984)(Fig. 1a). With the

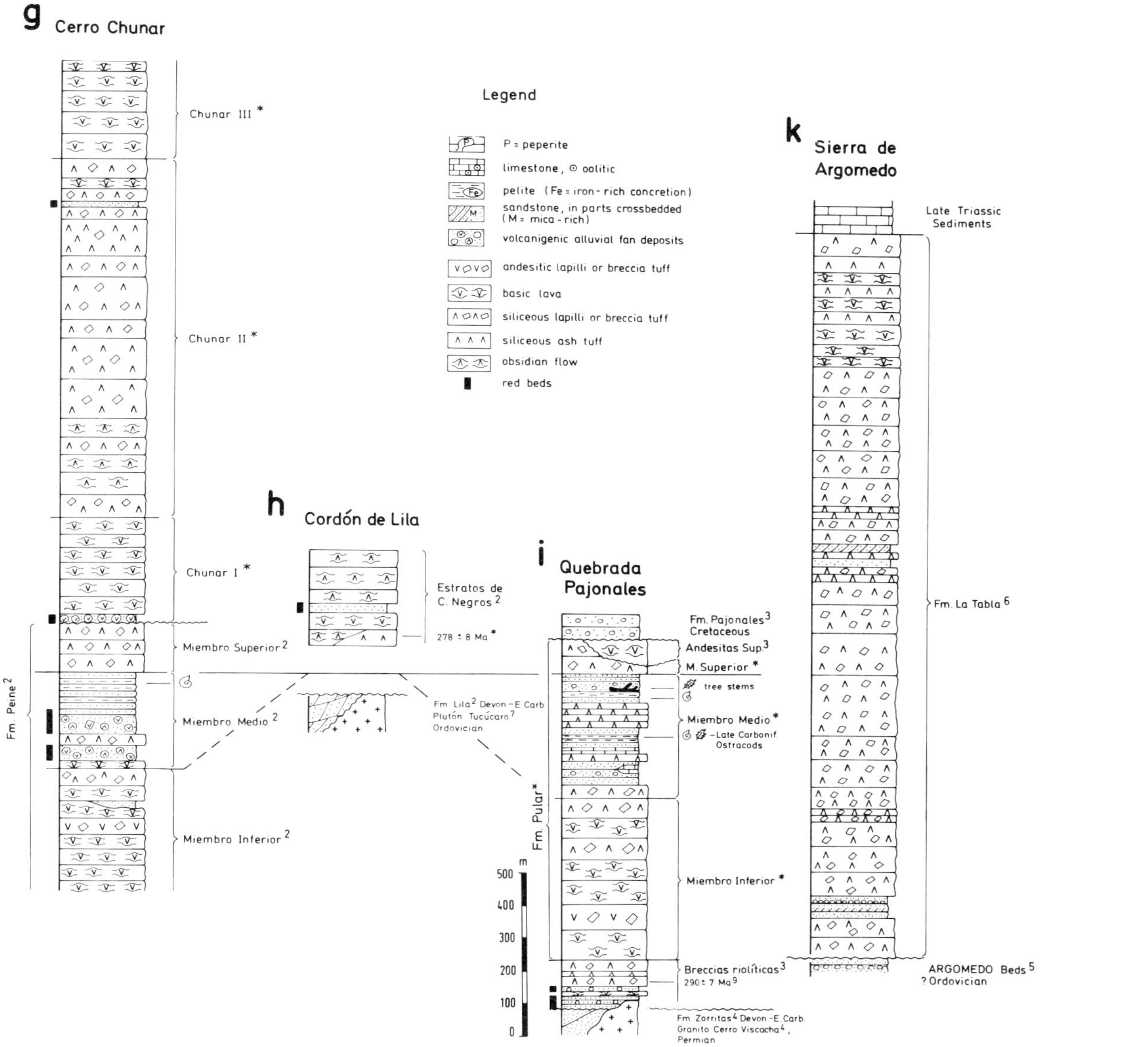

Inferior-Medio transition. [1] Marinovic and Lahsen (1984); [2] Ramirez and Gardeweg (1982); [3] Osorio and Rivano (1985); [4] Davidson et al. (1981); [5] Breitkreuz (1985); [6] Naranjo and Puig (1984); [7] Mpodozis et al. (1983); [8] Gardeweg (1988); [9] Davidson et al. (1985); *this chapter

exception of the rocks exposed in the Cerro Moctezuma area (see Sect. 2.1 and paragraph below), the only age constraint for these successions, is that they underlie early Jurassic marine sediments. With the reservation that further investigations may suggest otherwise, we include the Agua Dulce Fm. in the Peine Group.

The outcrop at Cerro Moctezuma was the only one to be visited by the authors. Here, a 150-m-thick succession of red volcanigenic sandstones, conglomerates and siliceous ignimbrites overlie the Permo-Carboniferous "complejo intrusivo Limon Verde" with an erosional unconformity (Fig. 2c; Marinovic and Lahsen 1984). The succession is

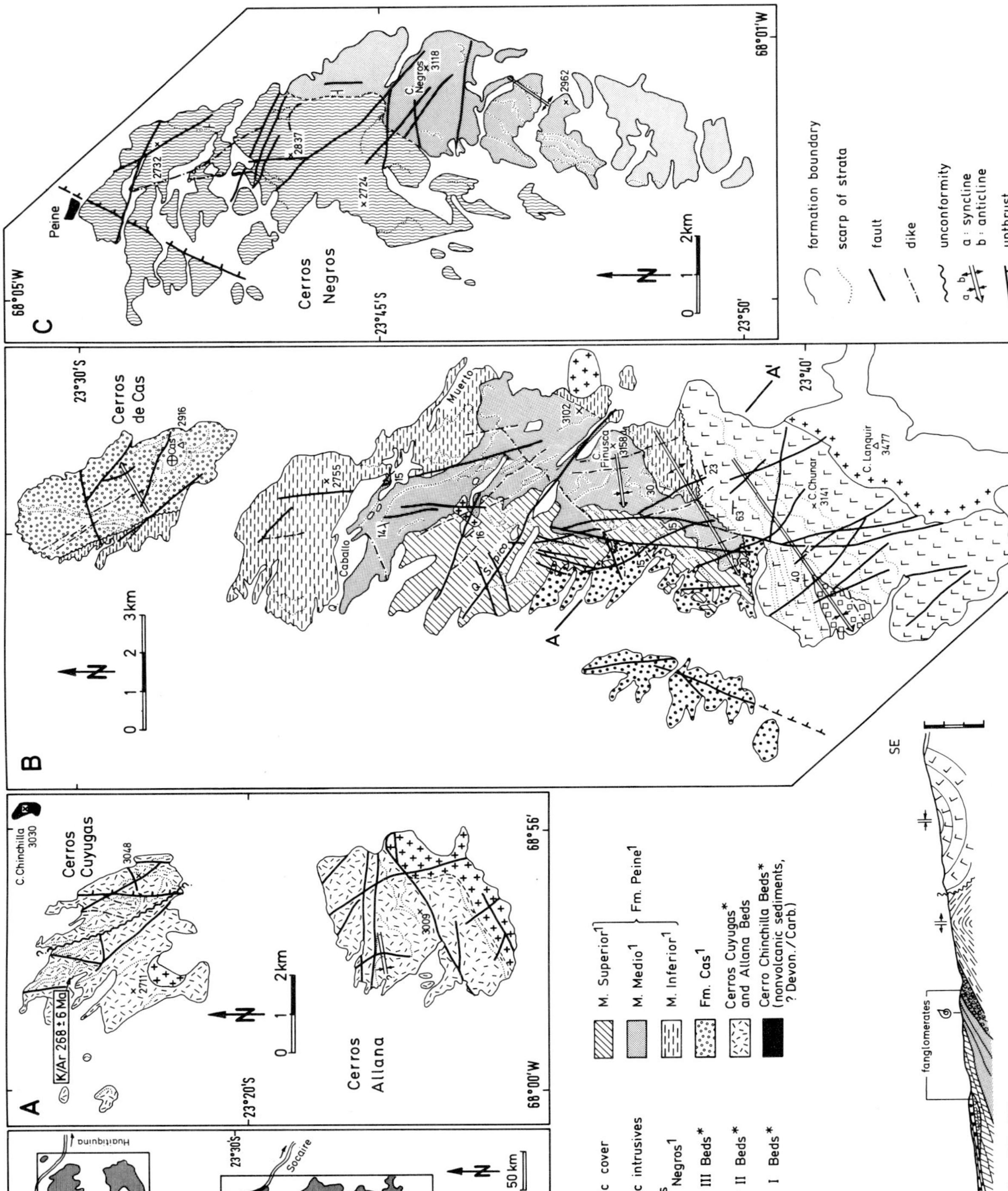

Fig. 3a-c. Geological map of the area east of Salar de Atacama. For location see Fig. 1a. K-Ar age, Gardeweg (1988); sample locality, Gardeweg (pers. comm.); [1] Ramirez and Gardeweg (1982); *this chapter overlain by the limestones of the early Jurassic Fm. Moctezuma (Marinovic and Lahsen op.cit.).

From the El Bordo area (Fig. 1a), Ramirez and Gardeweg (1982) reported outcrops of continental fossiliferous sediments and volcanic rocks of possible Carboniferous to Triassic age ("Estratos El Bordo"; Ramirez and Gardeweg 1982). They indicate that the Estratos El Bordo are overlain by volcanic rocks of the Fm. Agua Dulce. The results of our field study of the El Bordo area favour a modification of the division made by Ramirez and Gardeweg. The sediments of the El Bordo area are under- and overlain by volcanic successions. Therefore, we propose the name Fm. El Bordo and a subdivison into three members following the denominations given by Ramirez and Gardeweg (1982) and Lahsen and Marinovic (1984) for the Fms. Peine and Tuina, respectively (Fig. 2a, b, f and g). The oldest strata comprise siliceous ignimbrites (Miembro Inferior; Fig. 2d). They are concordantly overlain by a 600-m-thick sedimentary succession (Miembro Medio) which starts with partly phosphatic and oolitic limestones containing abundant late Carboniferous ostracods (Breitkreuz et al. 1992). The limestones grade into a succession of grey to green, fine- and coarse-grained clastic sediments and rare basic lavas. Apart from volcanigenic detritus, a high mica content and iron-rich concretions are characteristic of this unit. The upper 250 m of the Miembro Medio are made up of red sandstones and coarse-grained alluvial fan deposits. The Miembro Medio of the Fm. El Bordo is unconformably overlain by a c. 100-m-thick unit of basic lavas (Miembro Superior).

2.5 The Area East of the Salar de Atacama

At the eastern border of the Salar de Atacama, a long chain of excellent outcrops of volcanosedimentary units of the Peine Group is exposed (Figs. 1 and 2). In a modification of the results published by Ramirez and Gardeweg (1982), a succession of seven units, which have a cumulative thickness of nearly 3 km, has been established by the authors for the central area, which includes the Cerros Cas, Pinusca, Chunar and Lanquir (Figs. 2f, 2g and 3b). These units are, from the base to the top: the Fm. Cas, unconformably overlain by the Fm. Peine (M. Inferior, M. Medio and M. Superior), and the Chunar I, II, III Beds which overlie unconformably the Fm. Peine.

We summarized several volcanosedimentary units occurring in the Cerros Cuyugas and Allana (Fig. 3a) under the name "Cerro Cuyugas and Allana Beds"; the relationship of these to those of the Cerros de Cas-C. Lanquir area is not clear. The K-Ar dating of 268 ± 6 Ma (Fig. 3a; Gardeweg 1988), obtained from an ignimbrite of the Cuyugas area, is the only indication of a "post-Miembro Medio" age of the Cuyugas Beds (Fig. 2e). Similarly, the only information as to the stratigraphic position of the Estratos Cerros Negros (Fig. 3c) is the fact that they are underlain by the sedimentary unit of the Miembro Medio.

Formación Cas

In the southwestern part of the Cerros de Cas area, a NNW-SSE trending, nearly vertical, angular unconformity is exposed (see also Moraga et al. 1974); this interpreted as a fault by Ramirez and Gardeweg (1982). Below the unconformity, the beds of the Fm. Cas form a gentle, NE-dipping syncline (Fig. 2b) made up mainly of siliceous ash-, lapilli tuffs and flows (Fig. 2f).

Formación Peine

On the WSW side of the unconformity exposed in the Cerros de Cas, the base of the lower volcanic unit of the Fm. Peine (*Miembro Inferior*; Ramirez and Gardeweg 1982) is exposed. The first beds on top of the unconformity are in a near vertical position. Bedding dip shallows progressively towards the WSW. Here, the lithology comprises basic lavas intercalated by volcanigenic sandstones and ostracod-bearing stromatolitic limestones (Fig. 2f). The upper part of the Miembro Inferior crops out in the Quebrada Caballo Muerto (Fig. 2f) and south of Cerro Pinusca (Fig. 2g), where the Miembro Inferior resembles the core of a SW-dipping anticline (Fig. 3b, section A-A'). Basic aphanitic lavas predominate in the the Miembro Inferior. Towards the top, andesitic and rhyolitic lapilli tuffs, rich in phenocrysts and volcanigenic xenoliths, occur.

Figure 3b and c depicts the extended outcrops of the sedimentary *Miembro Medio* in the area east of the Salar de Atacama. In the Quebrada Caballo Muerto and south of Cerro Pinusca, the base of this miembro is exposed. It lies conformably on top of the Miembro Inferior volcanic beds. In the Quebrada Caballo Muerto, the unit starts with green parallel- and convolute-laminated, fine-grained beds rich in volcanigenic detritus. North of Quebrada Sipico, estherians occur in green pelites (Ramirez and

Gardeweg 1982). This limnic environment received abundant silica-pour hydroclastic detritus (indicated by quenched plagioclase crystals), which is assumed to have originated from intralacustrian volcanic activity (Breitkreuz 1991). North of Quebrada Sipico the transition from green limnic to red alluvial clastic sediments is marked locally by intercalations of basic lavas (partly in the form of pillow lavas) and tuffs (Fig. 2f), as is the case in the Miembro Medio of the Fm. Tuina (Fig. 2b and Sect. 2.3).

In contrast to the limnic facies at the base of the Miembro Medio between Quebrada Caballo Muerto and Sipico SW of C. Pinusca, the Miembro Medio starts with red volcanigenic alluvial fan deposits (Fig. 2g). These subaerial deposits, up to 150 m thick, are overlain by green fine-grained limnic clastic sediments containing estherians. Here, below the volcanic units of the Miembro Superior, no upper red beds occur, as is case for the area north of Quebrada Sipico; this indicates differential deposition and/or erosion during "Miembro Medio" time. The red beds north of Quebrada Sipico are characterized by fine-grained flood plain or playa deposits, displaying desiccation cracks, wave ripples and reworked pelitic layers.

The *Miembro Superior* of the Fm. Peine also shows lateral lithofacial variations (Fig. 2f and g). North of Quebrada Sipico, a succession of silica-rich ash tuffs and non-welded ignimbrites of a minimum thickness of 400 m is exposed. It contains volcanigenic sediment intercalations, the facies of which grades from predominantly flood plain deposits at the base to channel deposits near the top of the exposed section. In contrast, west of Cerro Pinusca, a 150 m thick succession of massive silica-rich welded ignimbrites without sedimentary intercalations occurs. The cumulative thickness of the Fm. Peine is c. 1000 m.

Chunar Beds

Ramirez and Gardeweg (1982) correlated the succession of basic lavas which overlie the Miembro Superior west of Cerro Pinusca (Fig. 3b) with the composite volcanic successions cropping out south of Peine ("Estratos Cerros Negros", Fig. 3c). Furthermore, Ramirez and Gardeweg (1982) compared the thick ignimbrite beds that occur in the area of the Cerro Chunar with the Fm. Cas, suggesting a formation prior to the deposition of the Fm. Peine. However, the contact between the respective miembros of the Fm. Peine and the alleged Fm. Cas is depositional and not faulted, representing a steeply dipping angular unconformity similar to the one described from the western Cerros de Cas (Fig. 3b, section A-A'). Accordingly, the volcanic units of the Cerro Chunar area are younger than the Fm. Peine. Therefore, we propose the informal names Chunar I Beds for the basic volcanic successions overlying the Miembro Superior in the Cerro Pinusca area and Chunar II Beds for the thick ignimbritic succession which comprise the SW-dipping syncline in the Cerro Chunar area (Fig. 3b, section A-A'). In the SW, the core of the syncline is occupied by another succession of basic lavas, named Chunar III Beds. The three Chunar units have a cumulative thickness of c. 1700 m.

The *Chunar I Beds* are up to 320 m thick and are composed of porphyritic andesitic lavas and sand- to gravel-sized volcanigenic red beds (Fig. 2g). The *Chunar II Beds* (1100 m) start with thick dacitic tuffs and lavas give way to deposits of rhyolitic composition towards the top. Large amounts of cogenetic crystals, mantled by devitrified glass, are a characteristic feature of the Chunar II ignimbrites. Volcanic xenoliths occur subordinately only in the upper part. Furthermore, the upper part of the Chunar II Beds displays some intercalations of red volcanigenic channel sediments. The *Chunar III Beds* (c. 300 m) consist of well-bedded andesitic, partly porphyritic lavas.

Cerros Cuyugas and Allana Beds

In Fig. 3a we summarized the volcanosedimentary units cropping out in the Cerros Cuyugas and Allana under the name "Cerros Cuyugas and Allana Beds". However, apart from a certain lithological similarity, no indications of the stratigraphic relation between the units of the two areas have yet been found. Ramirez and Gardeweg (1982) correlated the volcanic successions of both areas with the Fm. Cas, which represents the oldest volcanic unit of the region. For the reasons mentioned above, at least the successions at Cerro Cuyugas seem to be younger, presumably "post-Miembro Medio".

No detailed section was made in the Cerro Allana area, because its andesitic lavas and rhyolitic lapilli and breccia tuffs were subject to intense hydrothermal activity (see also Diaz 1978).

In the Cerros Cuyugas, a thick westward-dipping volcanosedimentary succession (minimum thickness 1400 m) is exposed (Fig. 3a). At its base, violet, well-sorted and crossbedded sandstones occur;

Ramirez and Gardeweg (1982) correlated these with the sandstones cropping out at Cerro Chinchilla to the NE (Fig. 3a). However, the Cerro Chinchilla Beds are quartz-rich mature sandstones, whereas the basal violet clastic beds in the eastern part of Cerros de Cuyugas contain only volcanigenic detritus, and therefore belong to the overlying volcanosedimentary successions (Breitkreuz 1986). An abundant clast type is a post-depositionally deformed silica-rich ash tuff. Some thin ash tuff layers, lithologically similar to the tuff clasts, are intercalated in the violet sandstone unit.

As shown in Fig. 2e, the lower part of the section is characterized by andesitic lavas and rhyolitic welded ignimbrites intercalated with volcanigenic alluvial fan and fluvial sediments. Intraformational erosional unconformities are a prominent feature. The upper 500 m of the section are built up of monotonous, thick rhyolitic ignimbrite sheets which frequently display vitrophyric basal layers. The top of the section is made up of an andesitic flow. However, the succession continues in the western part of Cerros Cuyugas where it is partly covered by young aeolian deposits (Fig. 3a).

Estratos Cerros Negros

No detailed section was made in the volcanic successions and sediments cropping out to the south of Peine (Estratos Cerros Negros, Miembro Medio)(Fig. 3c). Ramirez and Gardeweg (1982) describe the Estratos Cerros Negros as porphyritic andesitic lavas with some sedimentary intercalations suggesting a Triassic to Jurassic age.

2.6 Cordón de Lila Area

Ramirez and Gardeweg (1982) and Niemeyer et al. (1985) reported volcanosedimentary units from the Cordón de Lila area, south of the Salar de Atacama (Fig. 1a), which unconformably overlie early Palaeozoic and Devonian-early Carboniferous rocks. On the basis of lithological similarities, Ramirez and Gardeweg (1982) correlated these beds with the units cropping out in the Cerros Negros area (Estratos Cerros Negros, Fig. 3c).

The 300-m-thick succession is composed of andesitic flows and volcanigenic red beds at the base, grading into flows of dacitic composition towards the top (Fig. 2h). Biotite separated from a crystal-rich tuff, which occurs at a hill north of Cerro Cornisa, yielded the following values (for location refer to Fig. 16 and unit a_{20} of Fig. 17 in Breitkreuz 1986):

K_2O: 7.271 $\pm$ 0.012 wt %,
atm ^{40}Ar: 28.54 $\pm$ 1.01%,
age 278 $\pm$ 8 Ma.

This age indicates an early Permian volcanic activity in the Cordón de Lila area.

2.7 Quebrada Pajonales, Sierra de Almeida

From the Sierra de Almeida, several outcrops of the Peine Group are known (Fig. 1a; Davidson et al. 1981; Niemeyer et al. 1985; Breitkreuz 1986). Figure 2i shows a section taken in the Quebrada Pajonales, located in the western central part of the Sierra de Almeida. As already reported by Osorio and Rivano (1985), the Pajonales section contains a sedimentary intercalation which can be correlated with the other Miembro Medio outcrops that surround the Salar de Atacama. In order to standardize (see above), we propose a modification of the division published by Osorio and Rivano (1985) as depicted in Fig. 2i.

The c. 1150-m-thick succession starts with coarse-grained volcanigenic red beds, andesitic lavas, thick ignimbrites and breccia tuffs, lying unconformably on sandstones of the Devonian-early Carboniferous Fm. Zorritas (Fig. 2i). This unit, which Osorio and Rivano (1985) called "Brechas Riolíticas", is intruded by the Permian "Granito Cerro Viscacha" (Davidson et al. 1985). Davidson et al. (1985) published a K-Ar age of 290 $\pm$ 7 Ma for an ignimbrite of this unit. The section continues with a c. 950-m-thick succession for which we propose the name Fm. Pular. It starts with a c. 500 m volcanic succession of aphanitic andesitic lavas, andesitic breccia tuffs and silica-rich ignimbrites (Miembro Inferior). It is overlain by a c. 350-m-thick succession of alluvial and limnic sediments which contain thin-bedded ash tuff intercalations (Miembro Medio). The volcanigenic alluvial fan deposits (partly debris flows) contain tree trunks and pieces of bark. The fine-grained, partly calcareous limnic beds yield abundant estherians and ostracods. From the latter, a late Carboniferous age (?Westphalian) could be obtained (Osorio and Rivano 1985; Breitkreuz et al. 1992). On top of the Miembro Medio, c. 100 m of massive pink ignimbrites occur (Miembro Superior). These rocks are overlain unconformably by another c. 100-m-thick unit of andesitic lavas ("Andesitas Superiores"; Osorio and Rivano 1985). Both units are overlain unconformably by Cretaceous red beds (Fm. Pajonales; Osorio and Rivano 1985).

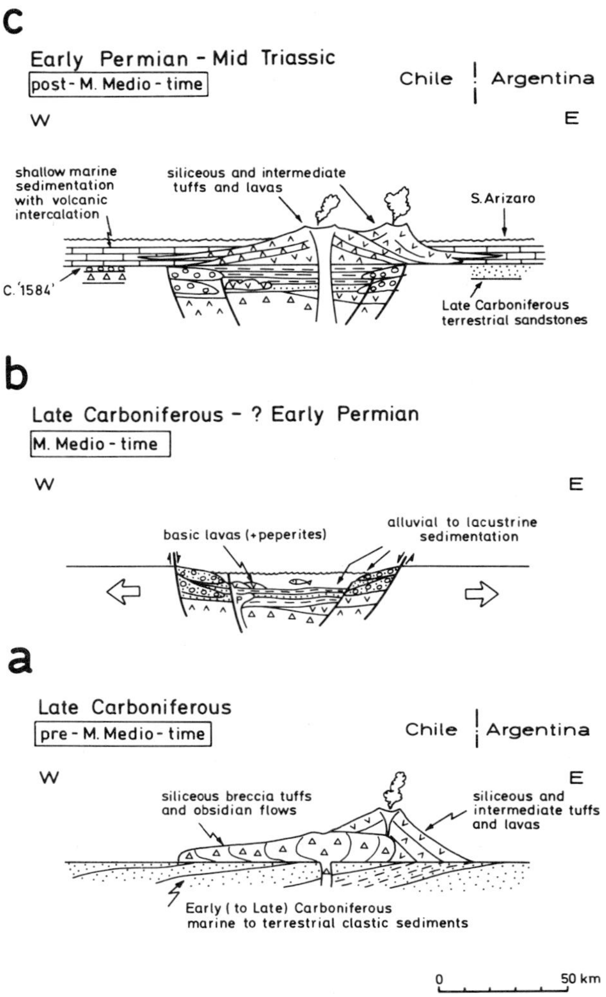

Fig. 4. Volcanism and sedimentation *a* before, *b* during and *c* after "Miembro Medio" time, shown for W-E sketches at 24°S

2.8 Cerro Bayo and Sierra de Argomedo

Widespread outcrops of the Peine Group are also known from the Chilean Precordillera to the west of the Salar Punta Negra (Davidson et al. 1985; Naranjo and Puig 1984; Padilla 1988). In the Cerros Bayos area (Fig. 1a), a transition is documented from a c. 3000 m thick succession of continental predominantly silica-rich volcanic and sedimentary deposits to late Triassic marine sediments (Chong and von Hillebrandt 1985; detailed section in Breitkreuz 1986).

In the northern Sierra de Argomedo (Fig. 1a), about 40 km south of the Cerro Bayo area, late Triassic marine sediments overlie a c. 2250 m thick volcanic succession which contain only a few sedimentary intercalations (Fm. La Tabla; Fig. 2k). Its lower part, which rests unconformably on marine clastic sediments of presumably Ordovician age (Argomedo Beds, Breitkreuz 1985), is built up of a monotonous succession of massive silica-rich ignimbrites. Intercalations of fluvial volcanigenic sediments are sporadically observed. The upper 400 m of the section contain andesitic lavas and rhyolitic ignimbrites (Fig. 2k).

Padilla (1988) reported ostracod-bearing sediments from the Sierra de Varas (c. 20 km to the east of the Sierra de Argomedo), which he correlated with the Fm. Pular (see Sect. 2.7). According to Padilla, the sediments overlie the Fm. La Tabla.

2.9 Augusta Victoria and Salar de Arizaro Areas

West of the Chilean Precordillera, in the Augusta Victoria area, early Permian limestones are underlain by volcanic rocks (Fig. 1a; Niemeyer et al. 1985). These are porphyric subvolcanic stocks of rhyolitic composition. Breitkreuz (1986) emphasized the fact that in neighbouring outcrops (Cerro 1584, Cerro Palestina), the early Permian marine strata also contain volcanigenic sediments and intercalations of basic lavas and silica-rich tuffs.

Volcanic intercalations have also been reported by Donato and Vergani (1985) from early Permian marine sediments of the Fm. Arizaro at Salar de Arizaro (Argentine Puna; Fig. 1a).

3 Volcanosedimentary Facies and Depositional Environment of the Peine Group

The volcanosedimentary units of the Peine Group are mainly continental deposits (exceptions are mentioned in Sect. 2.1 and 2.9). Most volcanic rocks of the Peine Group were deposited subaerially, apart from some basic volcanic intercalations in the limnic deposits of the Miembro Medio. The predominant rock type of the volcanic units is a thick, massive sheet-like ignimbrite. Basic lavas and tuffs occur subordinately within the measured sections; this is also true of ash tuffs and silica-rich lavas.

Caldera eruption seems to have been an important volcanic activity as indicated by the predominance of thick sheet-like ignimbrite units (e.g. Cerros Cuyugas

Beds, Chunar II, Fm. La Tabla; Fig. 2e, g and k), by frequent intraformational unconformities, which were caused by tilting and erosion, and in places by the circular outcrop arrangement of volcanic and cogenetic plutonic rocks (Fig. 1; Cas and Wright 1987; see also Davidson et al. 1985).

The detritus of the Peine Group sediments consists almost entirely of volcanigenic clasts. Only a few non-volcanigenic clasts were recognized as xenoliths in the volcanic rocks and as accessory clasts within the volcanigenic sediments. Thus, the Peine Group itself provided most of its detrital material.

3.1 "Pre-Miembro Medio" Time

Silica-rich lavas have been observed only in pre-Miembro Medio units (e.g. Miembro Inferior of the Fm. Tuina and Fm. Cas)(Fig. 4a). The volcanigenic sedimentary intercalations display proximal alluvial facies. Only the stromatolitic limestone intercalation in the Miembro Inferior of the Fm. Peine indicates temporary limnic conditions.

The occurrence of red beds (e.g. intercalated with the "Brechias Riolíticas", Fig. 2i) indicate a warm climate. For southern Bolivia, Sempere (1987) assumed a climatic change from cold to warm-humid conditions in the late Moscovian. We assume that in northern Chile the climate was not very different from that in southern Bolivia. This would imply that volcanic activity did not start much earlier than the late Moscovian in northern Chile. This coincides with the radiometric data (Fig. 1b).

3.2. "Miembro Medio" Time

During "Miembro Medio" time, silica-rich volcanism was less important. Subsidence led to the formation of a continental basin, whose depocentre accommodated a lake (or several lakes) of 100 x 300 km minimum extension (Figs. 1 and 4b)(Breitkreuz 1991). Limnic and alluvial fan facies are in close vertical and horizontal relation in the lower part of the Miembro Medio (see also Fig. 2). Where the clastic influx was low, stromatolitic and oolitic limestones were able to develop. Basic, in parts intralacustrian volcanism occurred. The depositional environment of the Miembro Medio changed from alluvial and limnic conditions in the lower part to alluvial in the upper part.

Plant fossils and ostracods from the lower part of the Miembro Medio of the Fms. Tuina, El Bordo and Pular yielded a latest Carboniferous-?early Permian age (Breitkreuz et al. 1992). The alluvial red beds and the flora indicate a warm-humid climate, while the growth rings of the tree trunks prove seasonal climatic variations (Breitkreuz et al. 1992). Deposition of the Miembro Medio possibly lasted until the early Permian.

3.3 "Post-Miembro Medio" Time

During the early Permian-middle Triassic, silica-rich volcanism prevailed. In the continental environment, sedimentary intercalations consisted exclusively of alluvial red beds. As indicated by the radiometric ages and the field relations, most of the cogenetic intrusions took place during this time. At least during the early Permian, the continental volcanic zone represented a peninsula which extended northwards into a shallow marine sea (Breitkreuz et al. 1988). This is indicated by the early Permian marine limestones occurring to the west ("Cerro 1584"; Breitkreuz 1986), north (Juan de Morales; Zeil 1964) east (Fm. Arizaro; Donato and Vergani 1985) of the continental volcanosedimentary successions (Fig. 4c).

4 Geochemistry of the Volcanic Rocks of the Peine Group

The volcanic rocks of the Peine Group show a variable diagenetic and/or anchimetamorphic alteration. The volcanosedimentary successions, which are in a close outcrop relation to the cogenetic intrusive rocks, show (hydro-)thermal alteration. Some rocks show only devitrification of glassy material and incipient hydration of mafic phenocrysts and feldspar. Progressively altered rocks show a late diagenetic/anchimetamorphic quartz-epidote-albite paragenesis. Carbonatization is also frequent. For geochemical analysis, only xenolith-free samples with "minor" diagenetic alterations were selected.

Figure 5a shows the alkali-silica variation of 53 samples and includes the data published by Breitkreuz et al. (1989). For reference, data of fresh plutonic rocks given, by Baeza and Pichowiak (1988), are shown. Most samples follow a calc-alkaline trend from basalts to rhyolites (AFM-triangle not presented here). However, some lavas of low to intermediate silica content are rich in alkali elements. Most of these originate from the volcanic intercalations of the Miembro Medio, thus giving further support to a tensional tectonic regime during that time. However,

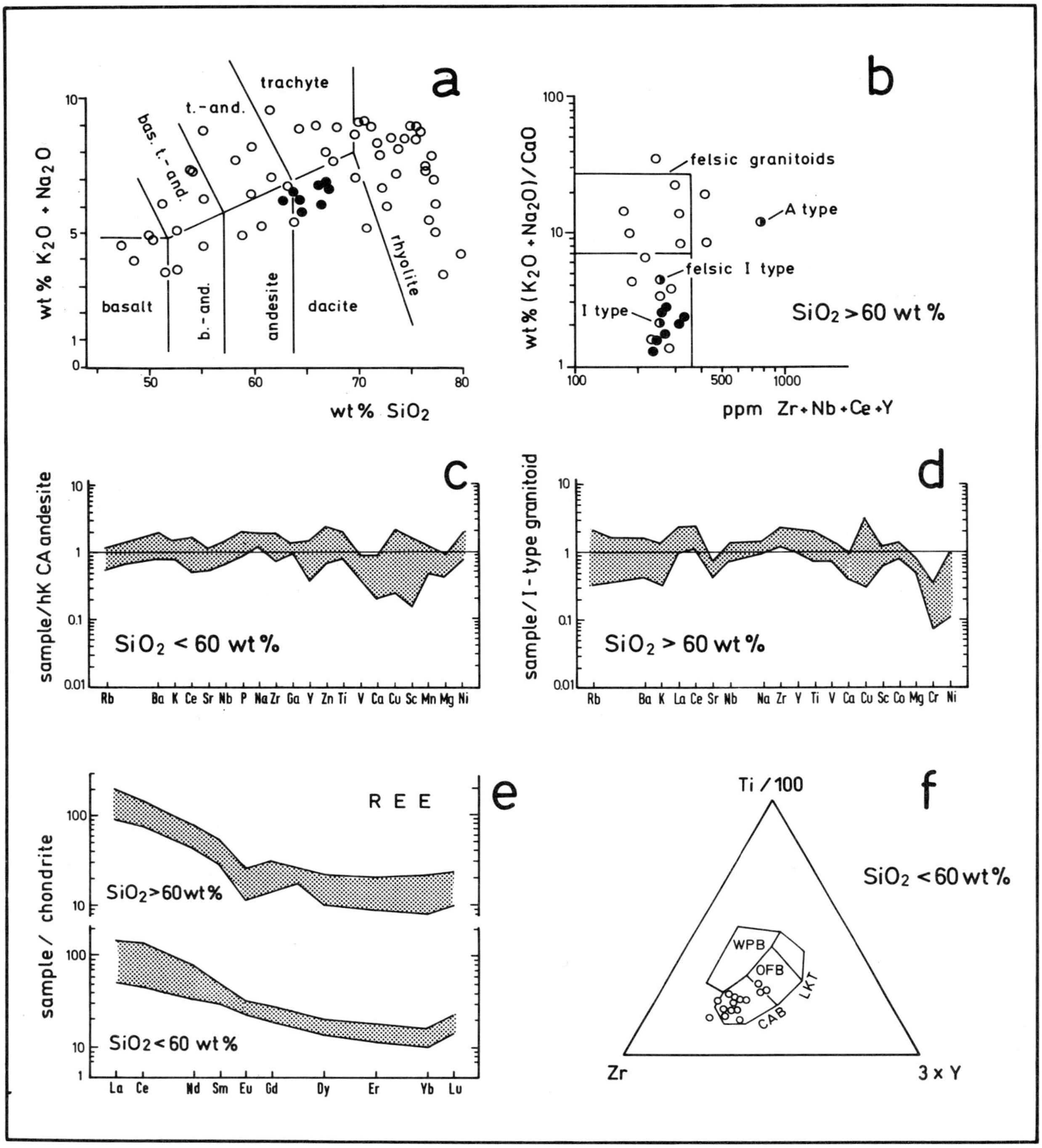

Fig. 5. a alkali-silica plot of the volcanic rocks of the Peine Group (*open circles*). *Filled circles* Permian plutonic rocks from the area SW of El Bordo (Baeza and Pichowiak 1988); classification grid according to Le Bas et al. 1986. **b** Discrimination diagram for granitoid rocks. After Whalen et al. (1987); *open circles* silica-rich volcanic rocks of the Peine Group; *half filled circles* averages of Whalen et al.; *filled circles* see Fig. 5a. **c** Multi element range (*shaded area*) of six basic volcanic rocks of the Peine Group, normalized to the average hK CA andesite of Gill (1981). **d** Multi element range of six silica-rich volcanic rocks of the Peine Group, normalized to the average I-type granitoid of Whalen et al. (1987). **e** ranges of REE patterns of six basic and six silica-rich volcanic rocks of the Peine Group (normalized to chondrite; Nakamura 1974). **f** Discrimination triangle for basic rocks. After Pearce and Cann (1973); *WPB* within-plate basalt; *LKT* low-potassium tholeiite; *OFB* ocean floor basalt; *CAB* calc-alkaline basalt; *open circles* basic volcanic rocks of the Peine Group

because these partly intralacustrian lavas are altered, post-depositional alkali enrichment cannot be ruled out. Samples with SiO_2 >75 wt% show an alkali depletion. We assume that the silica enrichment and the alkali depletion were caused by hydration reactions (Surdam and Boles 1979).

Figure 5c-e gives a survey of the chemistry of selected basic and siliceous volcanic rocks. The basic compositions show a fairly good correlation with potassium-rich calc-alkaline andesites of modern subduction-controlled continental margin arcs (Fig. 5c). Chalcophile elements (e.g. Cu, Sc) suffered considerable variation during diagenesis, whereas LIL elements (Rb, Ba, K, Sr, Na) show only moderate depletions or enrichments. Furthermore, the range of the REE patterns of basic rocks, with more or less flat HREE and a moderate to strong LREE enrichment, are typical of evolved andesites of subduction controlled continental margin arcs (see e.g. Thorpe et al. 1976; Cullers and Graf 1984) (Fig. 5e).

Six samples of silica-rich rocks of the Peine Group are compared with the average I-type granitoid given by Whalen et al. (1987) (Fig. 5d). Here, not only Cu but also Rb, Ba and K show strong variation, presumably caused by diagenetic alterations. Figure 5e depicts the range of six REE patterns of silica-rich samples which display flat HREE patterns, a negative Eu-anomaly and strongly enriched LREE. The flat HREE patterns of both the silica-rich and silica-poor compositions exclude garnet fractionation, which is characteristic of a magmatism active in rifting regimes or in thick continental crust (Francis et al. 1989, Kontak et al. 1990). This implies that the Peine Group magmatism took place in a non-tensional subduction setting with a moderate crustal thickness.

The late Carboniferous ("pre-Miembro Medio") volcanic rocks show no systematic differences compared with the Permo-Triassic ("post-Miembro") volcanic deposits. Another point should be made with respect to the geotectonic models published (see Sect. 5). Comparision of element patterns of the basic and silica-rich rocks of the Peine Group with those of average continental andesites (e.g. Gill 1981) or those of average A-type granitoids (e.g. Whalen et al. 1987), respectively, show no clear within-plate affinity. This is also evident from the respective discrimination diagrams of Pearce and Cann (1973) and Whalen et al. (1987) (Fig. 5b and f). Variation of the alkali content, in part due to alteration, is the only "trend" visible in Fig. 5b. The HFS elements (Fig. 5b and f) do not seem to have been affected by diagenetic alteration, as they display no, or only moderate, systematic variation in Fig. 5c and d.

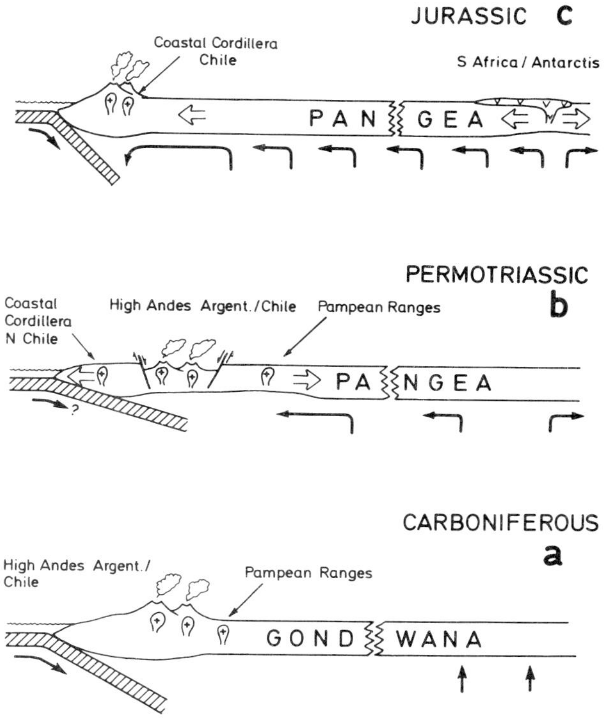

Fig. 6. Plate tectonic sketches of the southern Central Andes and of the continental area to the east, depicting *a* mantle convection, *b* continental extension and *c* magmatism from the Carboniferous to the Jurassic

5 Geodynamic Evolution

For the southern Andes, Mahlburg Kay et al. (1989) presume an evolution of the magmatic arc from a subduction controlled setting during the Carboniferous to a tensional setting during the Permo-Triassic (see also Mpodozis et al. 1985; Parada 1988; Mpodozis and Mahlburg Kay 1990). In the Peruvian Cordillera Oriental, orogenic plutonism took place at the Devonian-Carboniferous boundary (Carlier et al. 1982) and, after a magmatic lull, subduction magmatism controlled by a rift (Mitu Group) occurred during the late Permian-early Jurassic (Kontak et al. 1985, 1990; Clark et al. 1990).

This shift from subduction control to a tensional regime has been quoted by Davidson et al. (1985) also for the plutonism in northern Chile. However, on the grounds of our lithological and geochemical data, we cannot distinguish a Carboniferous subduction-related volcanic suite from a Permo-Triassic tensional-related volcanic suite within the Peine Group. Instead, the data of the Permo-Triassic volcanic rocks are also consistent with a non-tensional subduction-controlled setting. The only indication of a tensional regime in northern Chile, at least for the latest Carboniferous to ?early Permian, can be inferred from the volcanosedimentary facies association of the Miembro Medio. The association of volcanigenic limnic and alluvial fan deposits intercalated into volcanic-arc successions is characteristic of an intra-arc basin. Together with the considerable size of the basin and the syndepositional basic volcanism, this favours the model of an arc-graben for the Miembro Medio basin (Breitkreuz 1991; see also Busby-Spera 1988).

From the late Proterozoic to the Carboniferous, the Pacific margin of South America was of an accretionary type, growing by "normal" arc magmatism and terrane collision (see e.g. Ramos et al. 1986; Breitkreuz et al. 1989; Bahlburg and Breitkreuz 1991) (Fig. 6a). When Pangaea was assembled in the Permian, continental insolation led to an enhanced radial mantle convection away from the centre of the supercontinent (Anderson 1982) (Fig. 6b). The mantle convection induced a tensional regime in the overlying continental plate (Gurnis 1988). This continent-wide tension first provoked extension in the Andean continental margin due to the fact that, in contrast to the inner parts of Pangaea, this marginal area was structurally weakened and heated by the previous Carboniferous arc magmatism. In the southern Andes, melting of recently accreted crust led to significant siliceous magmatic activity in a broad zone from the Coastal Cordillera to the Patagonian platform (Mahlburg Kay et al. 1989). In the Central Andes of Peru and northern Chile, extension seems to have been focussed more in the High and Eastern Cordillera areas, where extensional arc-grabens developed; in the Eastern Cordillera these were associated with pronounced rift volcanism (Kontak et al. 1985; Clark et al. 1990) (Fig. 6b).

Siliceous volcanism continued until the Jurassic on the Patagonian platform (Rapela and Kay 1988) (Fig. 6c). The abrupt westward shift of the main magmatic zone from the High Andes to the Coastal Cordillera at the beginning of the Jurassic remains a puzzling feature of Andean evolution. Mpodozis and Mahlburg Kay (1990) assumed that this shift, which also represents a shift from silica-rich crustal-contaminated compositions to basic mantle-controlled magmas, was provoked by the collision of a postulated terrane with the margin of Gondwanaland during the middle Permian so that, consequently, there was a westward flip of the subduction zone. However, Breitkreuz (1990) argued that enhanced radial mantle convection, which, in the Jurassic, was strong enough to have induced basaltic rift volcanism in the inner part of Pangaea (South Africa, Antarctica, Fig. 6c), bent the subducting oceanic slab westwards to a steeper dip. This might have caused the westward shift of the magmatic arc to the Coastal Cordillera and to the basic magmatism of that area (see also Pankhurst el al. 1988).

The break up of Pangaea accelerated convergence at the Pacific plate margin of South America. As a consequence, a destructive continental margin with subduction erosion was established in the Andes.

Acknowledgements. This study was funded by the Deutsche Forschungsgemeinschaft (ref. no. Gi-31/51-3+5). R. Harmon kindly facilitated the K-Ar dating which was carried out by G. Wörner and H. Appel at the NERC Isotope Geology Centre, London. F. Jurtan checked the English text. Reviews by H. Bahlburg, A. Clark, and C. Mpodozis improved the manuscript.

References

Anderson DC (1982) Hotspots, polar wander, Mesozoic convection and the geoid. Nature 297:347-355

Baeza L, Pichowiak S (1988) Complejos plutónicos controlados por estructuras en la Precordillera del Norte de Chile - geoquimica y geocronología de Limón Verde y Catorce de Febrero. Actas V Congr Geol Chile, Santiago 3:I/91-108

Bahlburg H, Breitkreuz C (1991) The Paleozoic evolution of active margin basins in the southern Central Andes (NW Argentina, northern Chile). J S Am Earth Sci 4: 171-188

Bogdanic T (1990) Kontinentale Sedimentation der Kreide und des Alttertiärs im Umfeld des subduktionsbedingten Magmatismus in der chilenischen Präkordillere (21°-23°S). Berl Geowiss Abh A 123: 117 pp

Breitkreuz C (1985) Presentation of a marine volcano-sedimentary sequence of presumably Pre-Devonian age in the Sierra de Argomedo (24°45'S-69°22'W). Actas IV Congr Geol Chile, Antofagasta 1:1/76-88

Breitkreuz C (1986) Das Paläozoikum in den Kordilleren Nordchiles (21°-25°S). Geotekt Forsch 70:1-88

Breitkreuz C (1990) Late Carboniferous to Triassic magmatism in the Central and southern Andes: the change from an accretionary to an erosive plate margin mirrors the Pangea history. Symp int Géodyn Andin. ORSTOM Coll Sèmin, Paris, pp 359-362

Breitkreuz C (1991) Alluvio-limnic sedimentation and volcanism in a late Carboniferous arc-graben, northern Chile. Sediment Geol 74:173-188

Breitkreuz C, Bahlburg H, Zeil W (1988) The middle to late Paleozoic evolution of north Chile: geotectonic implications. Lect Notes Earth Sci 17:87-101

Breitkreuz C, Bahlburg H, Delakowitz B, Pichowiak S (1989) Volcanic events in the Paleozoic Central Andes. J S Am Earth Sci 2:171-189

Breitkreuz C, Helmdach F, Kohring R, Mosbrugger V (1992) Late Carboniferous intra-arc sedimentary rocks in the north Chilean Andes: Stratigraphy, paleogeography and paleoclimate. Facies 26: 67-80

Busby-Spera C (1988) Speculative tectonic model for the early Mesozoic arc of southwest Cordilleran United States. Geology 16:1121-1125

Caminos R (1979) Cordillera Frontal. In: Geología regional Argentina I, Acad Nac Sci, Córdoba, pp 397-453

Carlier G, Grandin G, Laubacher G, Marocco R, Megard F (1982) Present knowledge of the magmatic evolution of the Eastern Cordillera of Peru. Earth Sci Rev 18:253-283

Cas R, Wright JV (1987) Volcanic successions: Modern and ancient. Allen & Unwin, London

Chavez W (1983) The geologic setting of disseminated copper sulfide mineralization of the Mantos Blancos copper-silver district, Antofagasta Province, Chile. AIME, ann Meet Geol Sect, Preprint, Atlanta 193:1-200

Chong G, von Hillebrandt A (1985) El Triásico preandino de Chile entre los 23°30' y 26°00' de Lat. Sur. Actas IV Congr Geol Chile, Antofagasta 1:1/162-210

Clark AH, Kontak DJ, Farrar E 1990) The San Judas Tadeo W (Mo, Au) deposit: Permian lithophile mineralization in southeastern Peru. Econ Geol 85: 1651-1668

Cullers RL, Graf JL (1984) Rare earth elements in igneous rocks of the continental crust: Intermediate and silicic rocks - Ore petrogenesis. In: Henderson P (ed) Rare earth element geochemistry. Developments in Geochemistry 2. Elsevier, Amsterdam, pp 275-316

Damm K-W, Pichowiak S, Todt W (1986) Geochemie, Petrologie und Geochronologie der Plutonite und des metamorphen Grundgebirges in Nordchile. Berl Geowiss Abh A 66:73-146

Davidson J, Mpodozis C, Rivano S (1981) El Paleozoico de Sierra de Almeida, al oeste de Monturaqui, Alta Cordillera de Antofagasta, Chile. Rev Geol Chile 12:3-23

Davidson J, Ramirez CF, Gardeweg M, Herve M, Brook M, Pankhurst R (1985) Calderas del Paleozoico Superior-Triásico Superior y mineralización asociada en la Cordillera de Domeyko, Norte de Chile. Comun Univ Chile, Santiago 35:53-57

Diaz F (1978) Estudio geológico-económico y geoquímico del área de los Cerros de Allana (Chámar). Inst Invest Geol Chile (unpubl)

Donato EO, Vergani G (1985) Geología del Devónico y Neopaleozoico de la zona del Cerro Rincon, Provincia de Salta, Argentina. Actas IV Congr Geol Chil, Antofagasta 1:1/262-283

Francis PW, Sparks RSJ, Hawkesworth CJ, Thorpe RS, Pyle DM, Tait SR, Montovani MS, McDermott F (1989) Petrology and geochemistry of volcanic rocks of the Cerro Galan caldera, northwest Argentina. Geol Mag 126:515-547

Galli C (1968) Cuadrángulo Juan de Morales, Provincia de Tarapacá, Carta Geológica de Chile, 1:50 000. Inst Invest Geol Chile 18

Gardeweg M (1988) Petrografía y geoquimica del complejo volcanico Tumisa, Altiplano de Antofagasta, Andes del Norte de Chile. Actas V Congr Geol Chile, Santiago 3:I183-208

Gill J (1981) Orogenic andesites and plate tectonics. Springer, Berlin Heidelberg New York

Gurnis M (1988) Large-scale mantle convection and the aggregation and dispersal of supercontinents. Nature 332:695-699

Hervé F, Munizaga F, Marinovic N, Hervé M, Kawashita K, Brook M, Snelling N (1985) Geocronología Rb-Sr y K-Ar del basamento cristalino de Sierra Limon Verde. Actas IV Congr Geol Chile, Antofagasta 3:4/235-253

Huete C, Maksaev V, Moscoso R, Ulriksen C, Vergara H (1977) Antecedentes geocronológicos de rocas intrusivas y volcanicas en la cordillera de los Andes comprendida entre la Sierra Moreno y el Río Loa, y los 21° y 22° Lat. Sur, II. Región, Chile. Rev Geol Chile 4:35-41

Jensen P, Quinzio L (1979) Geología del área de Pampa Elvira y contribución al conocimiento del Jurásico marino entre los 23°00' y 23°30' Lat. sur y los 68°45' y 69°03' Long. oeste, II Región, Chile. Mem Título Univ Chile, Dep Geol, Santiago (unpubl)

Jesinskey C, Forsythe RD, Mpodozis C, Davidson J (1987) Concordant late Paleozoic paleomagnetizations from the Atacama Desert: Implications for tectonic models of the Chilean Andes. Earth Planet Sci Lett 85:461-472

Kontak DJ, Clark AH, Farrar E, Strong DF (1985) The rift-associated Permo-triassic magmatism of the Eastern Cordillera: A precursor to the Andean orogeny. In: Pitcher WS, Atherton MP, Cobbing EJ, Beckinsale RD (eds) Magmatism at a plate edge - the Peruvian Andes, Wiley, New York, pp 36-44

Kontak DJ, Clark AH, Farrar E, Archibald DA, Baadsgaard H (1990) Late Paleozoic-early Mesozoic magmatism in the Cordillera de Carabaya, Puno, SE Peru: geochronology and petrochemistry. J S Am Earth Sci 3:213-230

Le Bas MJ, Le Maitre RW, Streckeisen A, Zanettin B (1986) A chemical classification of volcanic rocks based on total alkali-silica diagram. J Petrol 27:745-750

Mahlburg Kay S, Ramos VA, Mpodozis C, Sruoga P (1989) Late Paleozoic to Jurassic silicic magmatism at the Gondwana margin: analogy to the middle Proterozoic in North America? Geology 17:324-328

Marinovic N, Lahsen A (1984) Hoja Calama, Región de Antofagasta, Carta Geológica de Chile, 1:250 000. Serv Nac Geol Miner Chile 58

Moraga A, Chong G, Fortt MA, Henriquez H (1974) Estudio geologico del Salar de Atacama, Provincia de Antofagasta. IIG (Chile), Bol 29

Mpodozis C, Mahlburg Kay S (1990) Provincias magmáticas acidas, "Singularidad Triásica" y ruptura de Gondwana: La Superunidad Ingaguás y la Provincia Choiyoi (Cordillera Frontal Chilena, 28°-31°S). Rev Geol Chile 17: 153-180

Mpodozis C, Hervé F, Davidson J, Rivano S (1983) Los granitoides de Cerros de Lila, manifestaciones de un episodio intrusivo y termal del Paleozoico inferior en los Andes del Norte de Chile. Rev Geol Chile 18:3-14

Mpodozis C, Nasi C, Moscoso R, Cornejo P, Maksaev V, Parada MA (1985) El cinturon magmatico del Paleozoico superior-Triásico de la Cordillera Frontal Chilena entre los 28°-31°S: "Estratigrafía" ígnea y marco tectonico. Comun Univ Chile, Santiago 35:161-165

Nakamura N (1974) Determination of REE, Ba, Fe, Mg, Na and K in carbonaceous and ordinary chondrites. Geochim Cosmochim Acta 38:757-775

Naranjo JA, Puig A (1984) Hojas Taltal y Chañaral, Carta Geológica de Chile, 1:250 000. Serv Nacion Geol Miner Chile 62, 63

Niemeyer H, Urzúa F, Aceñolaza G, Gonzalez R (1985) Progresos recientes en el conocimiento del Paleozoico de la región de Antofagasta. Actas IV Congr Geol Chile, Antofagasta 1:1/410-438

Osorio R, Rivano S (1985) Paraparchitidae (Ostracoda) del Paleozoico Superior en la Formación Pular (Harrington, 1961), Quebrada de Pajonales, vertiente occidental de la Sierra de Almeida, Antofagasta. Actas IV Congr Geol Chile, Antofagasta 1:1/439-457

Padilla H (1988) Eventos intrusivos y deformaciones en la Cordillera de Domeyko a la latitud del Salar de Punta Negra - antecedentes geocronológicos K-Ar. Actas V Congr Geol Chile, Santiago 3:I229-243

Pankhurst RJ, Hole MJ, Brook M (1988) Isotope evidence for the origin of Andean granites. Trans R Soc Edinb Earth Sci 79:123-133

Parada M (1988) Pre-Andean peraluminous and metaaluminous leucogranitoid suites in the High Andes of central Chile. J S Am Earth Sci 1:211-221

Pearce JA, Cann JR (1973) Tectonic setting of basic volcanic rocks determined using trace element analyses. Earth Planet Sci Lett 19:290-300

Ramirez R, Gardeweg M (1982) Hoja Toconao, Región de Antofagasta, Carta Geológica de Chile, 1:250 000. Serv Nac Geol Miner Chile 54

Ramos VA, Jordan TE, Allmendinger RW, Mpodozis C, Kay SM, Cortés JM, Palma M (1986) Paleozoic terranes of the central Argentine-Chilean Andes. Tectonics 5:855-880

Rapela CW, Kay SM (1988) Late Paleozoic to recent magmatic evolution of northern Patagonia. Episodes 11:175-182

Reutter K-J, Giese P, Götze H-J, Scheuber E, Schwab K, Schwarz G, Wigger P (1988) Structures and crustal development of the Central Andes between 21° and 25°S. Lect Notes Earth Sci 17:231-261

Scheuber E, Andriessen PAM (1990) The kinematic and geodynamic significance of the Atacama Fault Zone, northern Chile. J Struct Geol 12:243-257

Sempere T (1987) Caracteres geodinamicas generales del Paleozoico Superior de Bolivia. Ann Meet work group, IGCP 211: Late Paleozoic of South America, Santa Cruz, Bolivia, pp 9-19

Skarmeta J, Marinovic N (1981) Hoja Quillagua, Región de Antofagasta, 1:250 000, Carta geológica de Chile. Inst Invest Geol Chile 51

Surdam RC, Boles JR (1979) Diagenesis of volcanic sandstones. Soc Econ Paleontol Mineral Spec Publ 26:227-242

Thorpe RS, Potts PJ, Francis PW (1976) Rare earth data and petrogenesis of andesites from the north Chilean Andes. Contrib Mineral Petrol 54:65-78

Vergara H, Thomas A (1984) Hoja Collacagua, Región de Tarapacá, Carta Geológica de Chile, 1:250 000. Serv Nac Geol Miner Chile 59

Whalen JB, Currie KL, Chappell BW (1987) A-type granites: geochemical characteristics, discrimination and petrogenesis. Contrib Mineral Petrol 95:407-419

Zeil W (1964) Geologie von Chile. Borntraeger, Berlin

Zeil W (1981) Vulkanismus und Geodynamik an der Wende Paläozoikum/Mesozoikum in den zentralen und südlichen Anden (Chile - Argentinien). Zentralbl Geol Paläontol I 1981:298-318

Geodynamic Evolution of the Early Palaeozoic Continental Margin of Gondwana in the Southern Central Andes of Northwestern Argentina and Northern Chile

HEINRICH BAHLBURG, M. CHRISTINA MOYA and WERNER ZEIL

Abstract. The oldest non-metamorphic rocks of the southern Central Andes (NW Argentina and N Chile) are represented by the quartz-sandstone succession of the late Cambrian Mesón Group exposed in the Eastern Cordillera. The Mesón Group formed on a shallow marine platform deepening towards the NW. This platform developed presumably as an extensional structure on Pampean basement. Above an erosional unconformity of late Cambrian age, the shallow marine sandstones and shales of the Santa Victoria group (latest Cambrian-early Llanvirn) were deposited in a shelf basin located in the Eastern Cordillera. Farther west in the Puna volcanosedimentary successions (VS, c. 3500 m thick) derived from a magmatic arc located in northern Chile, are connected to eastward subduction and active volcanism during the middle and late Arenig. During the middle Ordovician, the Puna Basin deepened significantly leading to the deposition of the Puna turbidite complex (PTC, c. 3500 m thick). High subsidence and sedimentation rates were caused by the onset of collision of the para-autochthonous Arequipa Massif terrane (AMT), transforming the Arenig ensialic backarc basin into a middle Ordovician marine foreland basin. In response to thrusting in the west, a flexural bulge formed in the east, resulting in the emergence of the Eastern Cordillera shelf (Guandacol event). During the Ashgill, the basin fill was folded in the Oclóyic orogeny, resulting in the formation of the positive area of the Arco Puneño, which since suffered only marginal transgressions in the Llandovery and the late Palaeozoic. The folded Ordovician sedimentary rocks were post-tectonically intruded by the presumably Silurian, partly sheared granitoids of the "Faja Eruptiva de la Puna Oriental" (FE). The shear zones in the FE granitoids document sinistral strike-slip movements of possible Silurian age. Magmatism and the strike-slip system were connected to the probably oblique, SE directed collision of the AMT.

1 Introduction

In the Andes of NW Argentina and N Chile, late Cambrian and Ordovician clastic sedimentary and volcanogenic rocks are well exposed in the Argentinian Cordillera Oriental and to the west in the adjacent Puna highlands (Figs. 1 and 2). The highlands are subdivided into the northern and southern Puna by the NW-SE trending Calama-Olacapato-El Toro Fault (Fig. 1). This fault has been active repeatedly since the Precambrian (Salfity 1985) and is seen as an important factor controlling the differing structural histories of the southern and northern Puna (Alonso et al. 1984). In the Cordillera Oriental and the Puna, the late Cambrian and Ordovician units unconformably overlie the metaturbidites of the late Proterozoic-early Cambrian Puncoviscana Formation; these were folded and metamorphosed during the mid-Cambrian Pampean orogeny (Fig. 2; Willner et al. 1987). This region forms part of the continental margin of South America and is considered to have been an active one since at least the Ordovician (e.g. Coira et al. 1982; Hervé et al. 1987).

Correspondence to: H. Bahlburg, Geologisch-Paläontologisches Institut, Im Neuenheimer Feld 234, D-6900 Heidelberg

2 Geodynamic Concepts

While the late Cambrian basin in NW Argentina is commonly interpreted in terms of an extensional, graben-like structure on Pampean basement (Salfity et al. 1975; Aceñolaza et al. 1982), the geodynamic evolution of the Ordovician basin has been the subject of intense debate. The debate originated mainly from (1) the scarcity of basin-analytical studies of the Ordovician basin in the Cordillera Oriental and the Puna, and (2) from mutually exclusive interpretations of the origin and geotectonic significance of the magmatic rocks of the FE (Fig. 1).

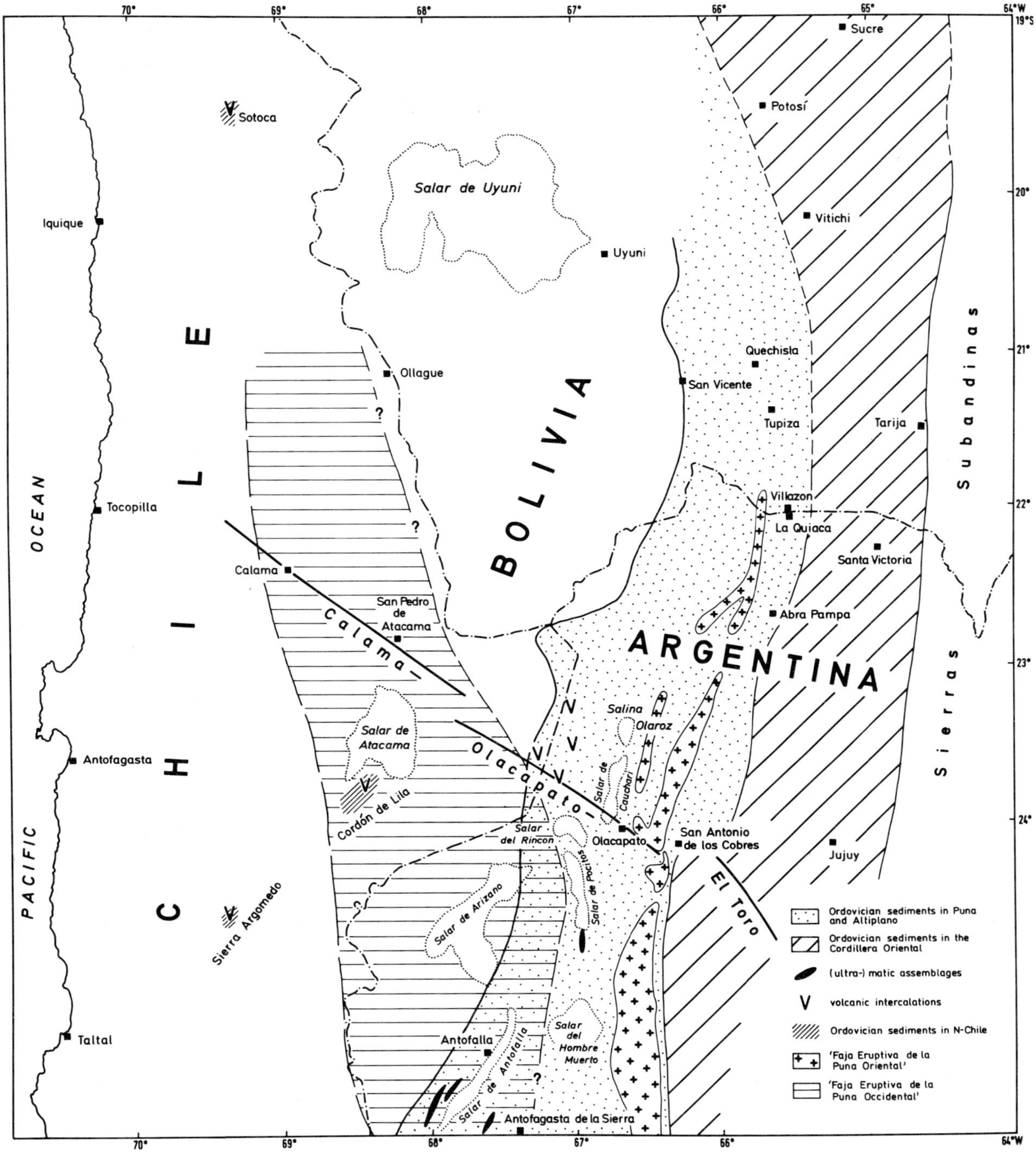

Fig. 1. Distribution of Ordovician rocks and Lower Palaeozoic intrusives in the southern Central Andes of southern Bolivia, northwestern Argentina and northern Chile

Méndez et al. (1973) were the first to map the magmatic rocks of the FE in the northern and southern Puna. They described the generally intrusive nature of the rocks and defined their age as postdating the folding of the Ordovician sedimentary country rocks in the Oclóyic orogeny (Turner and Méndez

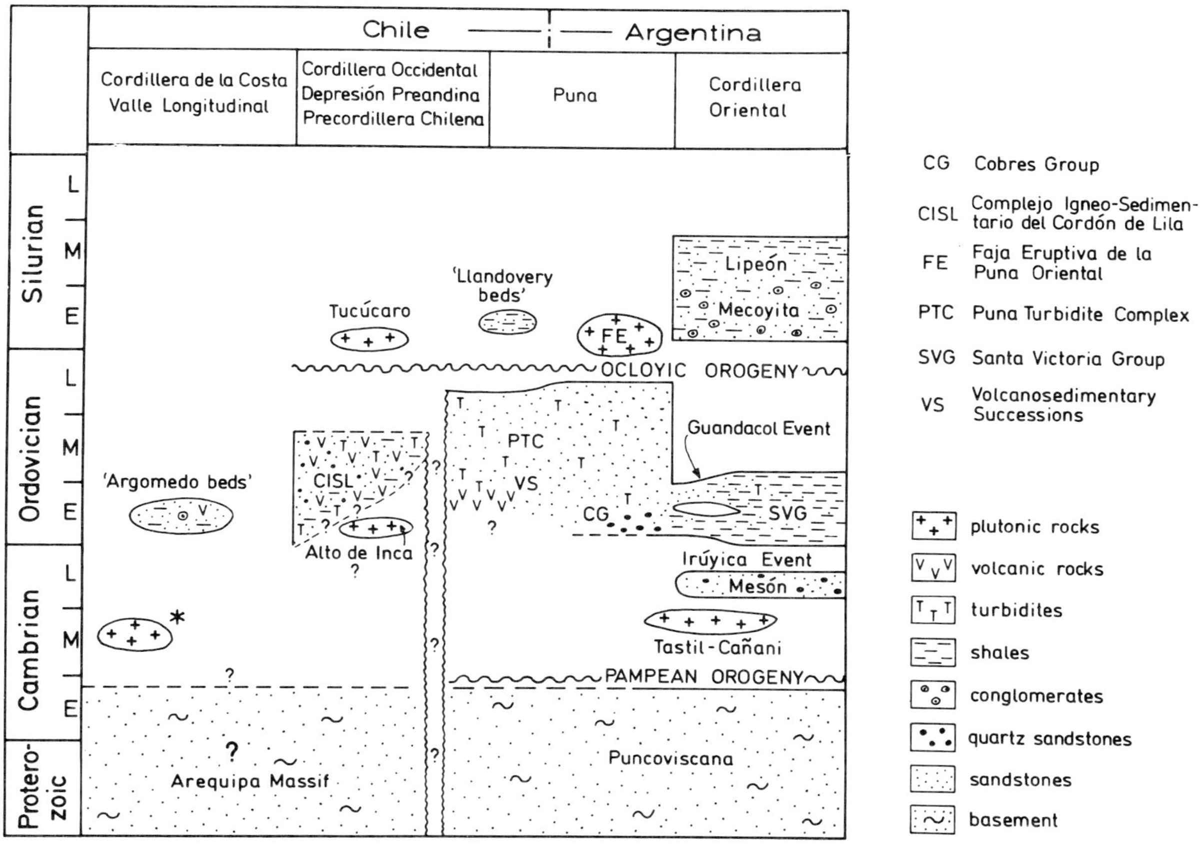

Fig. 2. Time-stratigraphic diagram of the early Palaeozoic rock units exposed in the NW Argentine Cordillera Oriental and Puna, and in the N Chilean Cordillera Occidental, Depresión Preandina, Precordillera, Valle Longitudinal and Cordillera de la Costa

1979) which took place during the Ashgill (Monaldi and Boso 1987). This was confirmed by Schwab (1973) who studied the area to the NW of San Antonio de los Cobres. However, Coira (1973, 1979) and Koukharsky and Mirre (1974) interpreted similar rocks exposed to the west of Abra Pampa and to the west of San Antonio de los Cobres (Fig. 1) as rhyolitic submarine ignimbrites and dacites intercalated with the Ordovician sedimentary rocks. Coira and Koukharsky and Mirre correlated these alleged volcanic rocks with early Ordovician lavas and volcaniclastic rocks occurring in the western part of the northern Puna (Fig. 1; Schwab 1973).

Radiometric dating of granitoids of the FE to the west of San Antonio de los Cobres by Omarini et al. (1984) points to magmatic activity during the middle Ordovician (Rb-Sr: 471 ±12 Ma). Further dates obtained for granitoids to the northwest of Abra Pampa (Rb-Sr: 374 ±7 Ma; Omarini et al. 1979) are seen by Coira et al. (1982) as representing a thermal overprint of late Devonian age. Coira and Ramos (in Coira et al. 1982) as well as Allmendinger et al. (1983) and Grissom et al. (1991) interpreted the FE as the magmatic arc of an east-dipping Ordovician subduction zone, the activity of which ended with the collision of the para-autochthonous AMT in the Oclóyic orogeny (Shackleton et al. 1979; Ramos 1988; Forsythe et al. in prep.). Accordingly, the sedimentary successions of the northern and southern Puna are described as forearc deposits. Isolated outcrops of (ultra)basic rock associations in the southern Puna, south of 25°S (Argañaraz et al. 1973; Allmendinger et al. 1983), are correlated with similar rocks occurring in the Argentine Precordillera at approx. 31°S by Allmendinger et al. (1983) and Ramos et al. (1986). In both papers, these associations are interpreted as ophiolites of an Ordovician subduction zone. In contrast, Davidson and Mpodozis (in Coira et al. 1982) and Aceñolaza and Toselli (1984) ascribe the magmatism of the FE to an ensialic extensional regime. Accordingly, the Ordovician Puna Basin is looked upon as an ensialic marginal basin between a southern prolongation of the Arequipa Massif in the west and of the Paraguay craton in the east. A further hypothesis is offered by Dalziel and Forsythe (1985). They classify the Puna Basin as oceanic. It was closed during the Oclóyic orogeny as a result of the collision of the AMT with the Paraguay craton. Lithospheric shortening is compensated for by two subduction zones of opposed polarity. The magmatic arc of the eastern, east-dipping subduction zone allegedly constitutes the FE. A model with two east-dipping subduction zones was proposed by Ramos (1988). The western subduction zone is thought to have been located in the Cordón de Lila to the west of the "Faja Eruptiva de la Puna Occidental" (FW; Palma et al. 1986; Fig. 1); the eastern one would have been positioned at the eastern margin of the Puna basin. The magmatic belts of both the FE and the FW would thus represent the respective magmatic arcs.

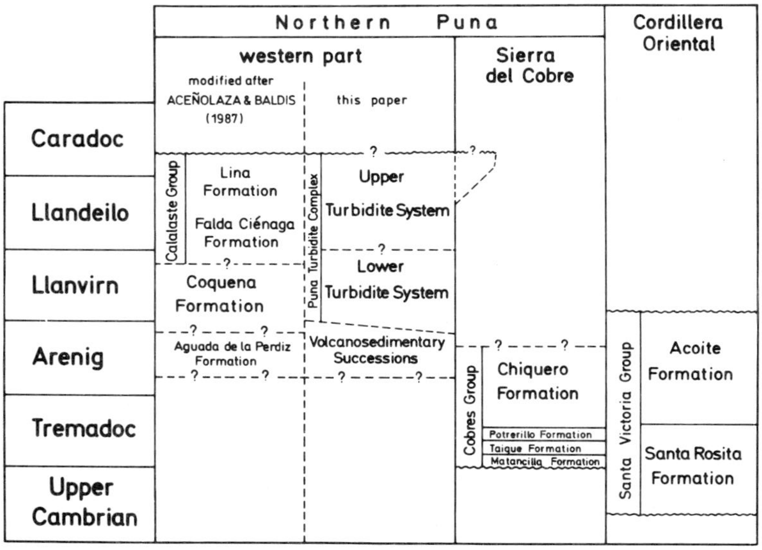

Fig. 3. Stratigraphy of the Ordovician in the NW Argentine Cordillera Oriental and the Puna (from Bahlburg et al. 1990)

This chapter presents a summary of the early Palaeozoic stratigraphic record and of the sedimentological features especially of the Ordovician units of the northern Puna. It is a resumé of our recent studies (Bahlburg et al. 1988, 1990; Moya 1988; Bahlburg 1990, 1991) and will elucidate both the development of the early stages of the active continental margin of Gondwana in the region of the southern Central Andes, and the relationship between the geodynamic evolution of the Ordovician sedimentary basin and the magmatic rocks of the FE.

3 Regional Stratigraphic Development

3.1 Cambrian

The oldest non-metamorphic sedimentary rocks are represented by the c. 3000-m-thick succession of quartz-sandstones and shales of the Mesón Group which are exposed in the northern Argentine Cordillera Oriental (Figs. 1 and 2; Turner 1960; Kumpa and Sanchez 1988). An angular unconformity separates the Mesón Group from the underlying metaturbidites of the Puncoviscana Formation, which were folded in the mid-Cambrian Pampean orogeny (Fig. 2; Aceñolaza et al. 1988), whereas it rests with erosional contact on the mid-Cambrian Cañani and Santa Rosa de Tastil plutons (Fig. 2; Turner 1960).

3.2 Ordovician

Above an erosional unconformity (Irúyica event: Turner and Méndez 1979; Fig. 2), deposition of the shallow marine sandstones and shales of the Santa Victoria Group, containing abundant trilobites and graptolites, began in the Cordillera Oriental during the latest Cambrian and ended in the early Llanvirn (Figs. 2 and 3; Harrington and Leanza 1957; Turner 1960; Moya 1988). At the eastern margin of the westwardly adjacent Puna highland, transgression started in the early Tremadoc with deposition of thick quartz-sandstone successions which grade into turbidites containing intercalations of pebbly mudstones (Schwab 1973; Cobres Group: Aceñolaza and Baldis 1987; Figs. 1-3). In the eastern Puna, sedimentation is thought to have ended in the Arenig (Schwab 1973; Aceñolaza and Baldis 1987) but may have continued into the late Ordovician (Bahlburg et al. 1990).

Stratigraphic results based on a new graptolite collection in the northern Puna led to the proposition by Bahlburg et al. (1990; Fig. 3) of a modified stratigraphic subdivision of the Ordovician units of this area. Reference will also be made to the stratigraphic scheme of Aceñolaza and Baldis (1987; Fig. 3).

In the western Puna, the stratigraphic record is established during the middle Arenig with the c. 3500-m-thick Volcanosedimentary Successions (VS, Figs. 1-3; middle-late Arenig), which encompass the north Chilean Aguada de la Perdiz Formation and its southerly continuations in Argentina, as well as the lower part of the Coquena Formation (Figs. 1 and 3; Garcia et al. 1962; Schwab 1973; Breitkreuz 1986; Aceñolaza and Baldis 1987; Koukharsky et al. 1988; Bahlburg et al. 1990). In their lower part, the VS consist of vesicular basic lavas, hydroclastic rocks and volcaniclastic debris flow deposits which are overlain in the upper part by siliceous ash tuffs and volcaniclastic turbidites (Breitkreuz et al. 1989; Bahlburg 1990). Geochemical data characterize the lavas as products of a volcanic arc (Koukharsky et al. 1988; Breitkreuz et al. 1989; Bahlburg 1990). Ordovician volcanic rocks younger than Arenig are not recorded. The vesicular lavas together with intercalated stromatolites (Coira and Barber 1987) document a very shallow marine environment in the lower part of the VS, whereas the turbidites in the upper part indicate a marked deepening of the depositional site of the VS. The VS represent deposits of a volcaniclastic apron (White and Busby-Spera 1987). During the latest Arenig, the VS graded into the volcaniclastic PTC (c. 3500 m; Figs. 2 and 3) which consists of the Lower and Upper Turbidite Systems (LTS and UTS, respectively; Fig. 3). The PTC encompasses the upper part of the Coquena Formation as well as the Calalaste Group (Fig. 3; Ramos 1972; Schwab 1973; Aceñolaza and Baldis 1987; Bahlburg et al. 1990). Deposition of the PTC ended presumably at the Llandeilo-Caradoc transition (Bahlburg 1990; Bahlburg et al. 1990). The turbidites of the PTC were deposited by axial, N-directed palaeocurrents. However, there is no regional proximal-distal trend (Macdonald 1986). The PTC is composed almost exclusively of erosional debris from the westward lying volcanic arc line source (Figs. 4 and 5), the volcanic activity of which was extinct after the Arenig. During the Llanvirn and Llandeilo increasing, although minor, amounts of detritus of (meta)sedimentary rocks were shed into the basin in addition to the volcanogenic components (Fig. 4).

In the southern Puna west of the Salar de Antofalla, ophiolitic associations of mafic lavas and turbidites, including cherts, are exposed in a number of outcrops (Fig. 1). No index fossils or geochronological data are known from these rocks. Due to the presence of similar ophiolite assemblages of determined Ordovician age in the Argentine Precordillera to the south, an Ordovician age is also assumed for the ophiolitic rocks in the southern Puna (Ramos et al. 1986). However, the ophiolitic rocks in this area display tectonic features attributed to polyphase deformation typical of Precambrian rocks in this region of the Andes (Mon et al. 1988). In contrast, according to these authors, the Ordovician sedimentary rocks in the Puna were subjected to less intense deformation which occurred in a single event during the Oclóyic orogeny (Fig. 2). Consequently, Mon et al. (1988) infer a late Proterozoic age for the ophiolitic rocks in the southern Puna.

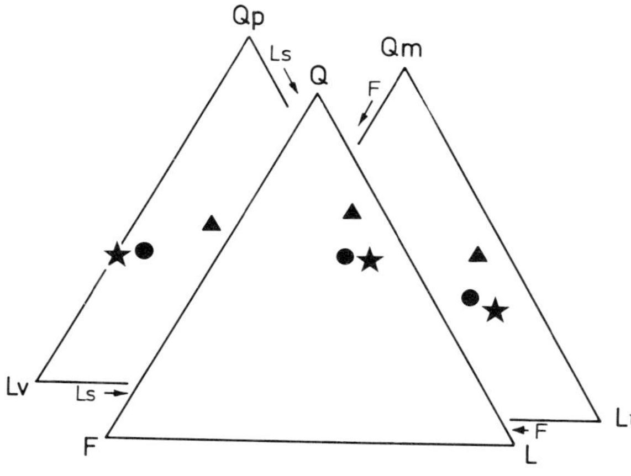

Fig. 4 Averages of framework modes of the volcanosedimentary successions (*stars* n=21), the Lower Turbidite System (*dots* n=17) and Upper Turbidite System (*triangles* n=9). After Dickinson (1985) and Bahlburg (1991). *Q* total quartz; *F* total feldspar; *L* total rock fragments; *Qm* monocrystalline quartz; *Lt* total rock fragments plus polycrystalline quartz; *Qp* polycrystalline quartz; *Lv* (meta)volcanic rock fragments; *Ls* (meta)sedimentary rock fragments; *P* plagioclase; *K* K-feldspar

3.3 Tectonic Features

In the Ashgill, the strata in the Cordillera Oriental and Puna were folded into NNW-SSE striking symmetrical folds during the Oclóyic orogeny (Turner and Méndez 1979; Mon and Hongn 1987; Monaldi and Boso 1987; Bahlburg 1990). The strata in the southern Puna, however, were strongly deformed into uniformly west verging folds (Allmendinger et al. 1983; Mon and Hongn 1987). In the northern Puna, locally developed east and west vergences are the product of post-Palaeozoic tectonics (Schwab 1973; Mon and Hongn 1987). After the Oclóyic orogeny, the Puna became a positive area (Arco Puneño: Padula et al. 1967).

3.4 Silurian

Silurian beds are exposed in the western Puna in the Salar del Rincón area and in the Cordillera Oriental (Figs. 1 and 2). At a locality a few kilometres SW of the Salar del Rincón, shallow marine clastic rocks overlie folded Ordovician strata with an angular unconformity. They were assigned to the early Devonian Salar del Rincón Formation by Aceñolaza et al. (1972). According to Isaacson et al. (1976), brachiopods from the same locality rather point to an early Llandovery age. In a recent visit to this outcrop, the brachiopod fauna we collected from several beds is similar to the one presented by Isaacson et al. It includes the genus *Cryptothyrella*, which confirms the early Llandovery age of these strata (Isaacson et al. 1976; A.J. Boucot, pers. comm.). With the reservation that a later publication may suggest otherwise, we refer to these rocks as the "Llandovery beds" (Fig. 2; Bahlburg and Breitkreuz 1991). They represent the western reaches of a short-lived transgression which originated in the Bolivian and Argentine Cordillera Oriental (Isaacson et al. 1976).

In the Cordillera Oriental, the glacio-marine sedimentary rocks of the Mecoyita and Zapla Formations (late Ashgill-?Wenlock; Turner 1960; Méndez et al. 1979; Monaldi and Boso 1987; Fig. 2) were also deposited above folded Ordovician strata. The Mecoyita Formation grades into the fine-grained, shallow marine clastic rocks of the Lipeón Formation (Wenlock-Ludlow, Méndez et al. 1979; Malanca and Monaldi 1987).

4 Faja Eruptiva de la Puna Oriental (FE)

The folded Ordovician strata were post-tectonically intruded by the granitoids of the FE (Figs. 1 and 2; Méndez et al. 1973; Salfity et al. 1975; Bahlburg 1990). Intrusive contacts with the Ordovician sedimentary rocks are well exposed in many localities between La Quiaca and San Antonio de los Cobres (Fig. 1). The magmatic rocks consist of porphyritic and equigranular, in parts hypabyssal, granitoids (Méndez et al. 1973) which are rich in sedimentary xenoliths. The porphyritic varieties contain macroscopically blue, resorption-embayed quartz and plagioclase phenocrysts in a fine-grained groundmass. They probably crystallized in a subvolcanic intrusive level. The equigranular rocks consist of quartz-plagioclase-K-feldspar aggregates; biotite is the only mafic mineral present. K-feldspar crystals of up to 15 cm diameter are common to both varieties. They are a characteristic feature of the magmatic rocks and do not occur in the country rocks. The large K-feldspars crystallized around primary minerals and partly replaced them. They presumably formed during late magmatic-metasomatic processes (Emmermann 1969). The granitoids are calc-alkaline, peraluminous and have a dacitic to rhyolitic composition (Bahlburg 1990). In contrast to the descriptions of, for example Coira (1973) and Coira et al. (1982), no volcanic rocks were observed in the FE north of 24°S (Bahlburg 1990).

The occurrence of *Dicellograptus* sp. in strata exposed to the west of Abra Pampa implies that the folded country rocks of the FE in this region were deposited in the Llandeilo or even late Ordovician (Fig. 1; Bahlburg et al. 1990). Thus, a very late Ordovician to post-Ordovician age has to be inferred for the FE. This constitutes a pronounced discrepancy from the 471 Ma age (Llanvirn) of the granitoids west of San Antonio de los Cobres (Fig. 1; Omarini et al. 1984). However, late Ordovician-early Silurian magmatic activity, possibly connected with the Oclóyic orogeny is also recorded to the west of the FE in northern Chile in the magmatic belt of the FW (Tucúcaro pluton, Fig. 2; Mpodozis et al. 1983). In view of the varying interpretations of the age of the FE, we give preference to field observations and stratigraphic data indicating a very late to post-Ordovician age for the FE. Thus, the FE does not constitute the magmatic arc of the Ordovician subduction zone in this area, as assumed by Coira et al. (1982). Accordingly, the interpretation of the Ordovician Puna Basin as a fore-arc basin (Coira et al. 1982; Hervé et al. 1987) can no longer be upheld.

5 Geodynamic Evolution

5.1 Cambrian

The late Cambrian basin of the Mesón Group developed as an extensional graben-like structure on Pampean basement (Salfity et al. 1975). The quartz-

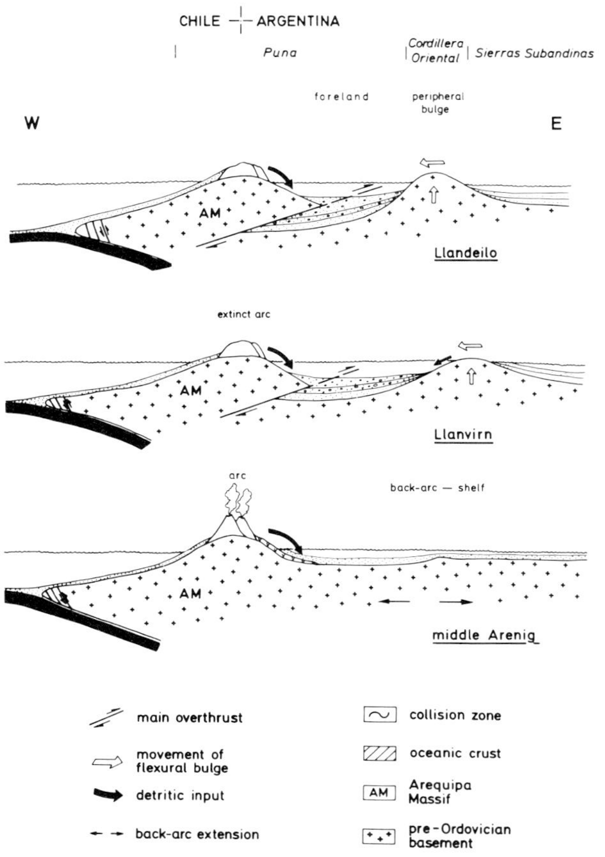

Fig. 5. Hypothetical cross sections (at approx. 23°30') depicting the geodynamic evolution of the Gondwana margin in the southern Central Andes during the Ordovician. After Bahlburg (1990) with palaeomagnetic data by Forsythe et al. (in prep.)

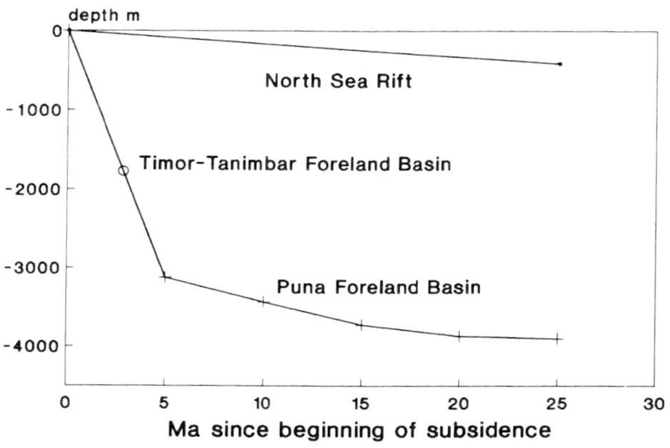

Fig. 6. Tectonic subsidence curve for the western Puna Basin. From Bahlburg (1990); calculated according to Steckler and Watts (1978) and Sclater and Christie (1980). Comparison with the tectonic subsidence curves of the North Sea graben during the late Cretaceous-Tertiary (Sclater and Christie 1980), and the sub-Recent Timor-Tanimbar Foreland Basin (preliminary calculation after Audley-Charles 1986)

sandstones and shales of the Mesón Group were deposited on a north-dipping, shallow marine platform which was connected to the open sea towards the NW (Aceñolaza et al. 1972; Kumpa and Sanchez 1988). The erosional unconformity of the Irúyica event was potentially linked to further extension of the basin, but could also have been effected by the global regression during the latest Cambrian-Ordovician transition (Erdtmann 1986).

5.2 Ordovician

Above the Irúyica unconformity, transgression reached the Cordillera Oriental in the latest Cambrian when c. 4500 m of sandstones and shales were deposited on a shallow marine shelf (Turner 1960; Moya 1988). This shelf extended into the eastern Puna (Cobres Group, Figs. 2 and 3) and deepened westward. However, a western border to the basin is not recorded until the Arenig, when the magmatic arc of an east-dipping subduction zone was active in northern Chile (Breitkreuz et al. 1989; Niemeyer 1989). The volcanic products of this arc are represented by the VS in the western Puna close to the Chilean-Argentine border (Figs. 1 and 2; Koukharsky et al. 1988; Breitkreuz et al. 1989; Bahlburg 1990). The arc was probably connected to the magmatic belt of the FW. Thus, during the Arenig, the Ordovician basin in Puna and Cordillera Oriental was in a backarc position (Fig. 5). According to Forsythe et al. (in prep.), the magmatic arc of the FW formed part of the para-autochthonous AMT. The Puna extensional backarc basin formed when the AMT moved away from the continent by clockwise rotation, the rotational pole being located to the north in Peru or Bolivia.

The ophiolitic associations of assumed Ordovician age in the southern Puna are correlated by Ramos et al. (1986) with Ordovician ophiolites exposed in the Argentine Precordillera and are similarily interpreted as ocean floor remnants (Fig. 1; Allmendinger et al. 1983; Ramos 1988). Accordingly, the backarc basin is presumed to have been floored by oceanic crust in the southern Puna south of c. 24°S (Ramos 1988), whereas it was ensialic in the northern Puna and in southern Bolivia (Sempéré 1989; Bahlburg 1990). However, no geochemical and geochronological data are available for the (ultra)basic rocks of the southern Puna.

The ophiolites of the southern Puna are situated to the east of the presumed arc of the FW, and those of the Precordillera are located to the west of an along-strike prolongation of this magmatic belt. There is is therefore doubt as to whether the (ultra)basic rocks of the southern Puna are in fact connected to those of the Precordillera and whether it is justified to apply the interpretation of the geochemical features of the Precordillera ophiolites to the respective complexes of the southern Puna.

Consequently, it seems prudent to interpret only the Puna Basin as an early Ordovician backarc basin of the FW magmatic arc complex. In this context the "ophiolites" of the southern Puna may document a higher degree of extension of the backarc basin in this area, potentially leading to the generation of oceanic crust, than in the northern Puna (Forsythe et al. in prep.). However, this interpretation rests on the assumption of an Ordovician age for the ophiolites. In opposition to this, based on structural data, Mon et al. (1988) infer a Proterozoic age for these rocks.

During the late Arenig pronounced (decompacted) subsidence of up to 1100 m Ma^{-1}, corresponding to a tectonic subsidence of c. 620 m Ma^{-1}, was initiated in the western reaches of the Puna Basin and continued until the end of the Llandeilo (Fig. 6; Bahlburg 1991). This was accompanied by the deposition of c. 7000 m of lavas and tuffs grading into very poorly sorted and immature volcaniclastic turbidites (VS, PTC: Figs. 2 and 3; Bahlburg 1990). Subsidence and sedimentation rates of this order of magnitude are typical features of marine foreland basins, where a thrust load is emplaced on attenuated crust (Beaumont et al. 1982; Allen et al. 1986). A recent example represents the modern Timor-Tanimbar foreland basin off northern Australia; this basin subsided during the last 2.3 Ma at similar rates (Audley-Charles 1986; Fig. 6).

In the case of the Puna basin, the thrust load is interpreted as representing the Arenig magmatic arc connected to the AMT. The arc was thrust eastwards upon its backarc basin at the Arenig-Llanvirn boundary when the drift of the AMT had been reversed by anticlockwise rotation. Therefore, this thrusting event is interpreted as the beginning of the collision between the AMT and the Gondwana margin in this area (Ramos 1988; Bahlburg and Breitkreuz 1991; Forsythe et al. in prep.). This collision transformed the Arenig backarc basin into a middle Ordovician foreland successor basin (Fig. 5; Bahlburg 1991). As a response to thrusting in the west, a flexural bulge formed contemporaneously further east in the Cordillera Oriental. Its formation led to uplift and emergence during the Guandacol diastrophic phase (Figs. 2 and 5; Salfity et al. 1984; Bahlburg 1990). This tectonically active regime controlled sedimentary patterns in the basin until the end of the stratigraphic record at the Llandeilo-Caradoc transition (Bahlburg et al. 1990; Bahlburg 1991). In this scenario, the north Chilean CISL and Argomedo beds (Figs. 1 and 2; Breitkreuz et al. 1989; Niemeyer 1989) form part of the Arequipa Massif. However, due to the scarcity of index fossils, a stratigraphic correlation with the Ordovician units of NW Argentina is not possible at present (Bahlburg et al. 1988). The collision of the AMT culminated in the Oclóyic orogeny (Figs. 2,5) and led to the folding of the basin fill and to the formation of a positive area in this region (Arco Puneño: Padula et al. 1967), which suffered only marginal transgressions in the Llandovery (Fig. 2) and late Palaeozoic times.

5.3 Silurian

Post-tectonically, the peraluminous, calc-alkaline granitoids of the FE intruded into the folded sedimentary rocks, presumably during the early Silurian (Figs. 1 and 2; Méndez et al. 1973; Bahlburg 1990). The rocks are characterized by N-S striking, subvertical shear zones which accommodated sinistral strike-slip movements. The extent of lateral displacement is still unclear. Similar strike-slip faults are also present farther west in the Ordovician CISL rocks in the north Chilean Cordón de Lila (Niemeyer 1989). Magma evolution and granitoid emplacement in the FE may have been connected with a transtensional-transpressive strike-slip cycle which may have developed in response to the SE directed, oblique collision of the AMT (Bahlburg 1990). A comparable interpretation is given by Scheuber and Andriessen (1990) for late Jurassic-early Cretaceous strike-slip movements along the north Chilean Atacama Fault, and associated mechanisms of granitoid emplacement (Pichowiak et al. 1990).

Acknowledgements. This chapter is a contribution to the IGCP project 279 "Terranes of Latin America". It benefited from reviews by C.M. Bell, Cheltenham; H. Miller, München; and R.M. Shackleton, Milton Keynes. Our study was carried out in cooperation with J.A. Salfity, Universidad Nacional de Salta, Argentina. F. Jurtan, Berlin, checked the English text.

References

Aceñolaza FG, Baldis B (1987) The Ordovician system of South America. Correlation chart and explanatory notes. IUGS Publ 22: 68 pp

Aceñolaza FG, Toselli AJ (1984) Lower Ordovician volcanism in northwest Argentina. In: Bruton DL (ed) Aspects of the Ordovician System. Paleontol Contrib Univ Oslo 295: 203-209

Aceñolaza FG, Benedetto JL, Koukharsky M, Salfity JA, Viera O (1972) Presencia de sedimentitas devónicas y neopaleozóicas en la Puna de Atacama, Provincia de Salta, Argentina. Rev Asoc Geol Argent 27:345-346

Aceñolaza FG, Fernández R, Manca N (1982) Carácteres bioestratigráficos y paleoambientales del Grupo Mesón (Cámbrico medio-superior), centro-oeste de América del Sur. Estud Geol 38:385-392

Aceñolaza FG, Miller H, Toselli AJ (1988) The Puncoviscana Formation (late Precambrian-early Cambrian). Sedimentology, tectonometamorphic history and age of the oldest rocks of NW Argentina. In: Bahlburg H, Breitkreuz C, Giese P (eds) The southern Central Andes: contributions to structure and evolution of an active continental margin. Lecture Notes in Earth Science 17, Springer, Berlin Heidelberg New York, pp 25-38

Allen PA, Homewood P, Williams GD (1986) Foreland basins: An introduction. In: Allen PA, Homewood P (eds) Foreland Basins. Spec Publ Int Assoc Sediment 8: 3-12

Allmendinger RW, Ramos VA, Jordan TE, Palma M, Isacks BL (1983) Paleogeography and Andean structural geometry, northwest Argentina. Tectonics 2: 1-16

Alonso R, Viramonte J, Gutierrez R (1984) Puna Austral - bases para el subprovincialismo geológico de la Puna Argentina. IX Congr Geol Argent Actas 1: 43-63

Argañaraz R, Viramonte J, Salazar L (1973) Sobre el hallazgo de serpentinitas en la Puna Argentina. V Congr Geol Argent Actas 1: 23-32

Audley-Charles MG (1986) Timor-Tanimbar Trough: the foreland basin of the evolving Banda orogen. In: Allen PA, Homewood P (eds) Foreland Basins. Spec Publ Int Assoc Sediment 8: 91-102

Bahlburg H (1990) The Ordovician basin in the Puna of NW Argentina and N Chile: geodynamic evolution from back-arc basin to foreland basin. Geotekt Forsch 75: 1-107

Bahlburg H (1991) The Ordovician back-arc to foreland successor basin in the Argentine-Chilean Puna: Tectonosedimentary trends and sea-level changes. In: Macdonald DIW (ed) Sedimentation, tectonics, and eustasy. Spec Publ Int Assoc Sediment 12

Bahlburg H, Breitkreuz C (1991) The evolution of marginal basins in the southern Central Andes of Argentina and Chile during the Paleozoic. Jour South Amer Earth Sci 4: 171-188

Bahlburg H, Breitkreuz C, Zeil W (1988) Geology of the Coquena Formation (Arenigian-Llanvirnian) in the NW Argentine Puna: constraints on geodynamic interpretation. In: Bahlburg H, Breitkreuz C, Giese P (eds) The southern Central Andes: contributions to structure and evolution of an active continental margin. Lect Notes Earth Science 17, Springer, Berlin Heidelberg New York, pp 71-86

Bahlburg H, Breitkreuz C, Maletz J, Moya MC, Salfity JA (1990) The Ordovician sedimentary rocks in the northern Puna of Argentina and Chile: new stratigraphical data based on graptolites. Newsl Stratigr 23: 69-89

Beaumont C, Keen CE, Boutilier R (1982) A comparison of foreland and rift margin sedimentary basins. Philos Trans R Soc Lond A 305: 295-317

Breitkreuz C (1986) Das Paläozoikum in den Kordilleren Nordchiles (21°-25°S). Geotekt Forsch 70: 1-88

Breitkreuz C, Bahlburg H, Delakowitz B, Pichowiak S (1989) Volcanic events in the Paleozoic central Andes. J S Am Earth Sci 2: 171-189

Coira B (1973) Resultados preliminares sobre la petrología del ciclo eruptivo concomitante con la sedimentación de la Formación Acoite en la zona de Abra Pampa, provincia de Jujuy. Rev Asoc Geol Argent 23: 85-88

Coira B (1979) Descripción geológica de la Hoja 3c, Abra Pampa; Carta geol.-econ. República Argentina, 1:200 000. Inst Nac Geol Min Bol 170: 90 pp

Coira B, Barber E (1987) Vulcanismo submarino Ordovícico (Arenigiano-Llanvirniano) del Rio Huaytiquina, Provincia de Salta, Argentina. X Congr Geol Argent Actas 4: 305-307

Coira B, Davidson J, Mpodozis C, Ramos V (1982) Tectonic and magmatic evolution of the Andes of northern Argentina and Chile. Earth Sci Rev 18: 303-332

Dalziel IWD, Forsythe RD (1985) Andean evolution and the terrane concept. In: Howell DG (ed) Tectonostratigraphic terranes of the Circum-Pacific region. Circum-Pacific-Council energy mineral resources Earth Sci Ser 1: 565-581

Dickinson WR (1985) Interpreting provenance relations from detrital modes of sandstones. In: Zuffa GG (ed) Provenance of arenites. Nato ASI Ser C 148: 333-361

Emmermann R (1969) Genetic relations between two generations of K-feldspar in a granite pluton. Neues Jahrb Min Abh 111: 289-313

Erdtmann B-D (1986) Early Ordovician eustatic cycles and their bearing on punctuations in early nemtophorid (planctic) graptolite evolution. In: Walliser OH (ed) Global bio-events. A critical approach. Lect Notes Earth Sci 8: 139-152

Forsythe RD, Davidson J, Mpodozis, C, Jesinkey C Lower Paleozoic relative motion of the Arequipa block and Gondwana; paleomagnetic evidence from Sierra Almeida of northern Chile. Tectonics (in prep).

García AF, Pérez D'Angelo E, Ceballos SE (1962) El Ordovícico de Aguada de la Perdiz, Puna de Atacama, provincia de Antofagasta. Rev Mineral 77: 52-61

Grissom GC, De Bari SM, Page R, Page S, Villar L, Coleman RG, Ramirez MV de (1991) The deep crust of an early Paleozoic arc; the Sierra de Fiambalá, northwestern Argentina. Geol Soc Amer Spec Pap 265: 189-200

Harrington H, Leanza AF (1957) Ordovician trilobites of Argentina. Univ Kansas Press Spec Publ 1: 1-276

Hervé F, Godoy E, Parada MA, Ramos VA, Rapela C, Mpodozis C, Davidson J (1987) A general view on the Chilean-Argentine Andes, with emphasis on their early history. Geodyn Ser 18: 97-113

Isaacson PE, Antelo B, Boucot AJ (1976) Implications of a Llandovery (early Silurian) brachiopod fauna from Salta province, Argentina. J Paleontol 50: 1103-1112

Koukharsky M, Mirre JC (1974) Nuevas evidencias del vulcanismo Ordovícico en la Puna. Rev Asoc Geol Argent 26: 128-134

Koukharsky M, Coira B, Barber E, Hanning M (1988) Geoquímica de vulcanitas ordovícicas de la Puna (Argentina) y sus implicancias tectónicas. V Congr Geol Chil 3: I137-158

Kumpa M, Sánchez MC (1988) Geology and sedimentology of the Cambrian Grupo Mesón (NW Argentina). In: Bahlburg H, Breitkreuz C, Giese P (eds) The southern Central Andes: Contributions to structure and evolution of an active continental margin. Lecture Notes in Earth Science 17, Springer, Berlin Heidelberg New York, pp 39-54

Macdonald DIM (1986) Proximal to distal sedimentological variation in a linear turbidite trough: Implications for the fan model. Sedimentology 33: 243-259

Malanca S, Monaldi CR (1987) Lichiidae de la Formación Lipeón (Silurico), Sierra de Zapla, Jujuy, Argentina. IV Congr Latinoam Paleontol 1: 141-147

Méndez V, Navarini A, Plaza D, Viera V (1973) Faja Eruptiva de la Puna Oriental. V Congr Geol Argent Actas 4: 89-100

Mendez V, Turner JCM, Navarini A, Amengual R, Viera V (1979) Geología de la región noroeste, Provincias Salta y Jujuy, Republica Argentina. Dir Gral Fabr Milit: 1-118

Mon R, Hongn F (1987) Estructura del Ordovícico de la Puna. Rev Asoc Geol Argent 42: 31-38

Mon R, Hongn F, Omarini R (1988) Estructura del basamento andino entre los paralelos 24° y 28° latitud Sur. V Congr Geol Chil 1: A19-36

Monaldi CR, Boso MA (1987) Dalmatina (Dalmatina) subandina nov. sp. (Trilobita) en la Formación Zapla del norte argentino. IV Congr Latinoam Paleontol 1: 149-157

Moya MC (1988) Lower Ordovician in the southern part of the Argentine Eastern Cordillera. In: Bahlburg H, Breitkreuz C, Giese P (eds) The southern Central Andes: Contributions to structure and evolution of an active continental margin. Lecture Notes in Earth Science 17, Springer, Berlin Heidelberg New York, pp 55-70

Mpodozis C, Hervé F, Davidson J, Rivano S (1983) Los granitóides de Cerros de Lila, manifestaciones de un episodio intrusivo y termal del Paleozóico inferior en los Andes del Norte de Chile. Rev Geol Chile 18: 3-14

Niemeyer H (1989) El complejo ígneo-sedimentario del Cordón de Lila, región de Antofagasta: significado tectónico. Rev Geol Chile 16: 163-181

Omarini R, Cordani U, Viramonte J, Salfity J, Kawashita K (1979) Estudio isotópico de la 'Faja Eruptiva de la Puna' a los 22°35'LS, Argentina. II Congr Geol Chil Actas E: 258-269

Omarini R, Viramonte J, Cordani U, Salfity J, Kawashita K (1984) Estudio geochronológico Rb-Sr de la Faja Eruptiva de la Puna en el sector de San Antonio de los Cobres, Provincia de Salta. IX Congr Geol Argent Actas 3: 146-158

Padula E, Rolleri EO, Mingramm ARG, Criado Roque P, Flores MA, Baldis BA (1967) Devonian of Argentina. Int Symp Devonian Syst 2: 165-199

Palma MA, Parica PD, Ramos VA (1986) El granito de Archibarca: su edad y significado tectónico, provincia de Catamarca. Rev Asoc Geol Argent 41: 414-419

Pichowiak S, Buchelt M, Damm K-W (1990) Magmatic activity and tectonic setting of the early stages of the Andean Cycle in northern Chile. In: Kay SM, Rapela CW (eds) Plutonism from Antarctica to Alaska. Geol Soc Am Spec Pap 241: 127-144

Ramos VA (1972) El Ordovícico fosilífero de la Sierra de Lina, departamento de Susques, provincia de Jujuy, República Argentina. Rev Asoc Geol Argent 27: 84-94

Ramos VA (1988) Late Proterozoic-early Paleozoic of South America - A collisional history. Episodes 11: 168-174

Ramos VA, Jordan TE, Allmendinger RW, Mpodozis C, Kay SM, Cortés JM, Palma M (1986) Paleozoic terranes of the central Argentine-Chilean Andes. Tectonics 5: 855-880

Salfity JA (1985) Lineamentos transversales al rumbo andino en el noroeste argentino. IV Congr Geol Chil Actas 2: 2/119-137

Salfity JA, Omarini RH, Baldis B, Gutiérrez WJ (1975) Consideraciones sobre la evolución geológica del Precámbrico y Paleozóico del norte Argentino. II Congr Iberoam Geol Econ 4: 341-361

Salfity JA, Malanca S, Brandan ME, Monaldi CR, Moya C (1984) La Fase Guandacol en el norte de la Argentina. IX Congr Geol Argent Actas 1: 555-567

Scheuber E, Andriessen P (1990) Kinematic and geodynamic significance of the Atacama fault zone, northern Chile. J Struct Geol 12: 243-257

Schwab K (1973) Die Stratigraphie in der Umgebung des Salar de Cauchari (NW Argentinien). Ein Beitrag zur erdgeschichtlichen Entwicklung der Puna. Geotekt Forsch 43: 1-168

Sclater JG, Christie PAF (1980) Continental stretching: An explanation of the post-mid Cretaceous subsidence of the central North Sea Basin. Jour Geophys Res 85: 3711-3739

Sempéré T (1989) Paleozoic evolution of Central Andes. 28th Int Geol Cong Abstr 3: 73

Shackleton RM, Ries AC, Coward MP, Cobbold PR (1979) Structure, metamorphism and geochronology of the Arequipa Massif of coastal Peru. J Geol Soc 136: 195-214

Steckler MS, Watts AB (1978) Subsidence of the Atlantic-type continental margin off New York. Earth Planet Sci Lett 41: 1-13

Turner JCM (1960) Estratigrafía de la Sierra de Santa Victoria y adyaciencias. Bol Acad Nac Cienc 41: 163-196

Turner JCM, Méndez V (1979) Puna. II Simp Geol Reg Argent 1: 13-56

White JDL, Busby-Spera C (1987) Deep marine arc apron deposits and syndepositional magmatism in the Alisitos Group at Punta Cono, Baja California, Mexico. Sedimentology 34: 911-927

Willner A, Lottner U, Miller H (1987) Early Paleozoic structural development in the NW Argentine basement of the Andes and its implications for geodynamic reconstructions. Geophys Monogr 40: 229-239

The Nitrate Deposits of Chile

Guillermo Chong Diaz

Abstract. The geology of the unique nitrate deposits of Chile is described. The preliminary results of a current multidisciplinary research programme are included. Sedimentary, "in rocks", and miscellaneous deposits have been defined and their geographical, geomorphological, geological and, metallogenic characteristics described. From the geological viewpoint, different lithostratigraphic and intrusive units have been recognized. Alternative theories of ore genesis are discussed and it is argued that original brines, enriched in nitrates and associated salts, are of magmatic origin. These brines were emplaced by means of geothermal mechanisms, either through isolated geothermal fields or through regional fault systems, and the spatial relationship with the emplacement of precious and basic metal ores is considered. Finally, the post-emplacement geological evolution of the deposits is discussed.

1 Introduction

The Chilean nitrate deposits are non-metallic ores from which sodium and potassium nitrates are obtained, along with iodine as a coproduct, and sodium sulphate and borates as byproducts. The ore is known as "caliche" and the final product is called "salitre". The mining establishments are called "oficina" or "campamentos" and, in former times, they were grouped in "cantones". The ores have been exploited for more than 150 years in northern Chile, despite severe economic and technical problems at various times.

These deposits have attracted the attention of geologists for more than a century, largely because of two important facts. First, they are composed of an assemblage of water-soluble saline minerals that is only rarely found in nature. Second, although variable amounts of nitrates are are found in deserts around the world, in northern Chile they are unique in forming deposits of hundreds of million tons. In addition, the genesis of these ores has been discussed by geologists for more than a century without producing of a satisfactory genetic theory. The deposits are so complicated that they pose more questions than answers, as was concluded by Ericksen (1983) who stated: "In fact, they are so extraordinary that, were it not for their existence, geologists could easily conclude that such deposits could not form in nature". Curiously, and notwithstanding their economic importance and scientific interest, systematic studies have not been made. Between the work of Whitehead (1920) and that of Ericksen (1963, 1979, 1981, 1983) there is a substantial information gap. Most of the technical literature during this time refers to specific aspects, principally to the genesis of the nitrates. On the other hand, the different companies that exploited these ores, have assumed that reserves exist and have shown no interest in their geology. Indeed, the industry has had a very irregular development and for more than 60 years its only goal was survival. This chapter, which is a preliminary report, summarizes the previous known facts, abridges the results of a research programme headed by the author and includes new information.

For the first time, a multidisciplinary research programme is in progress in this field. Geoscientists from Chile (Universidad Católica del Norte), Spain (Universidades de Barcelona and Complutense of Madrid) and England (Birmingham University) are participating. The research includes techniques (fluid inclusions, thin sections of saline minerals, stable isotopy and SEM/TEM) which have never before been applied to the deposits and also makes use of traditional chemical analysis.

2. Location

For a better understanding of these deposits we define their location in geographical, geomorphological, geological, and metallogenic terms (Figs. 1 and 2). The "oficinas salitreras" are located along a N-S trend of about 700 km between Pampa Tana (19°30` S) and

Correspondence to: G. Chong Diaz, Depto. de Sciencias Geologicas, Universidad Católica del Norte, Av. Angamos 0610, Casilla 1280, Antofagasta-Chile

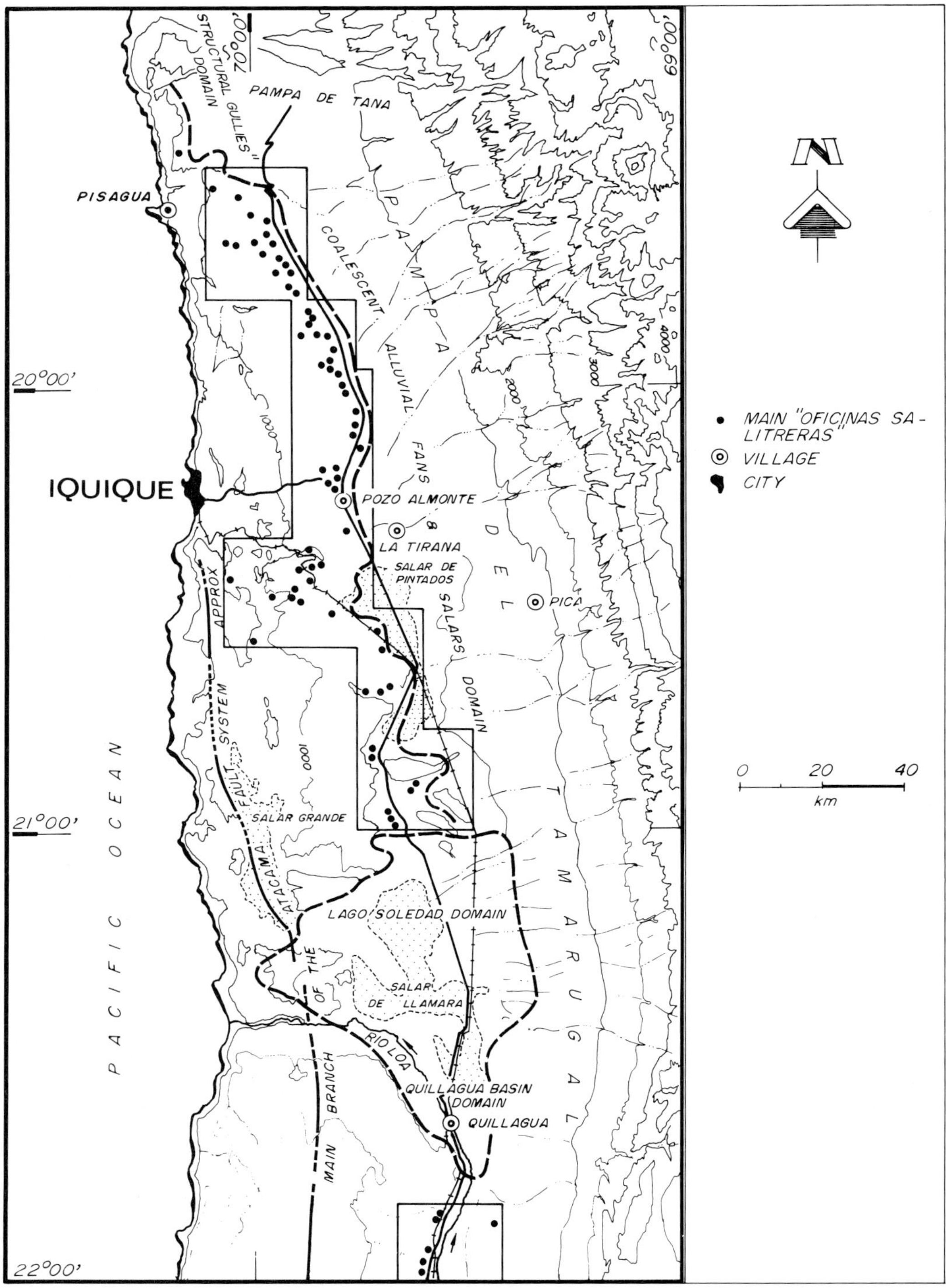

Fig. 1. *top* Index map of northern Chile showing the location of nitrate deposits between 19°30'and 22°00' S lat

Fig. 2. *right* Index map of northern Chile showing the ▶ location of nitrate deposits between 22°00' and 25°30' S lat

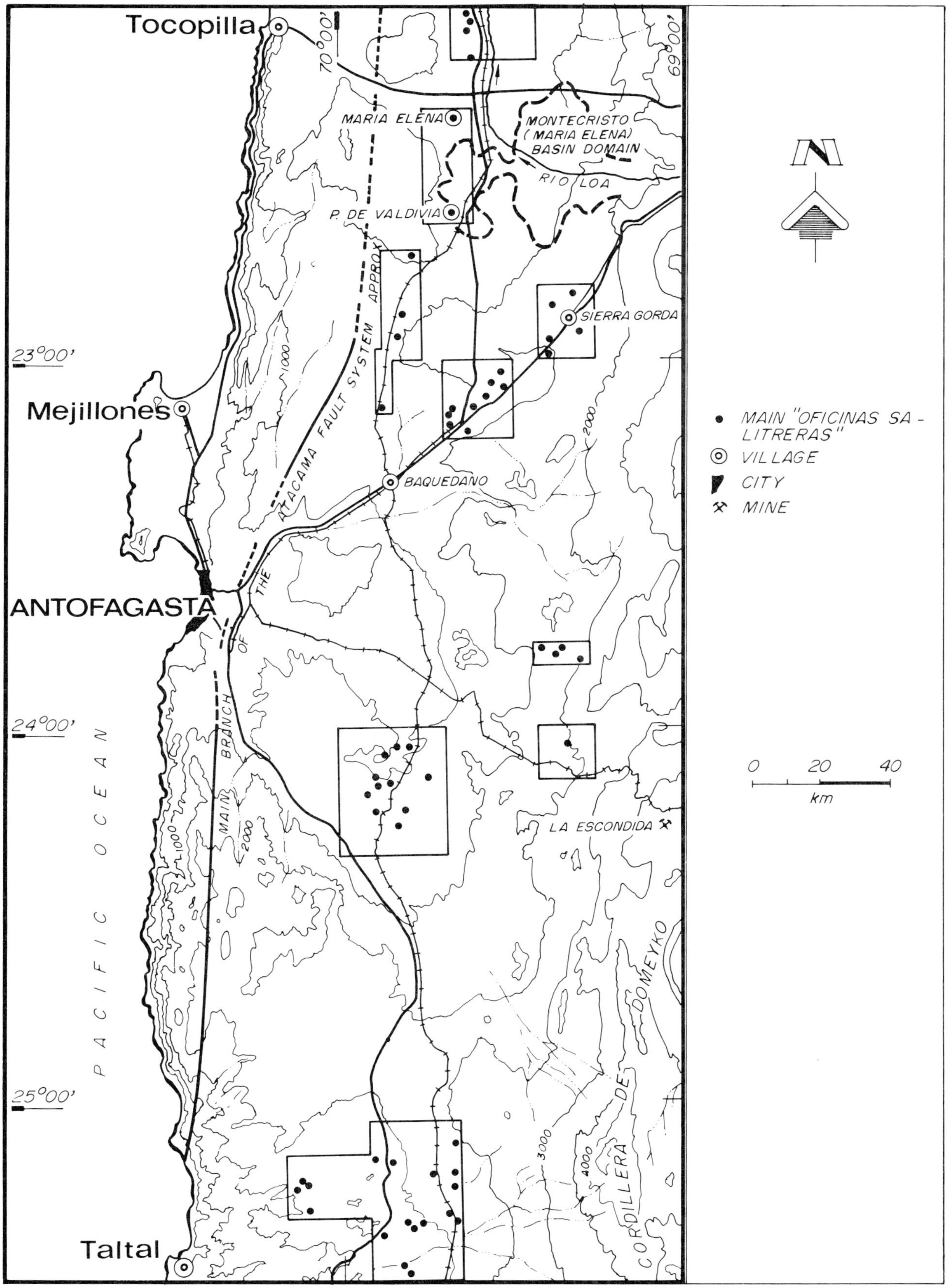

Tocopilla
70°00'
69°00'
MARIA ELENA
MONTECRISTO
(MARIA ELENA)
BASIN DOMAIN
RIO LOA
P. DE VALDIVIA
SIERRA GORDA
23°00'
Mejillones
ATACAMA FAULT SYSTEM APPROX.
1000
2000
BAQUEDANO
ANTOFAGASTA
MAIN BRANCH OF THE
24°00'
PACIFIC OCEAN
LA ESCONDIDA
25°00'
3000
4000
CORDILLERA DE DOMEYKO
Taltal
N
MAIN "OFICINAS SALITRERAS"
VILLAGE
CITY
MINE
0 20 40
km

Taltal (25°30'S). This zone is more or less regular from its northern extreme to, approximately, Baquedano (23°49'S/69°49'W). Here we find an irregular 60 km long strip of ores distributed in a general NE trend between Baquedano and Sierra Gorda (22°53'S/69°19'W). Then, to the SW, the trend resumes a regular N-S strip of deposits from the locality of Aguas Blancas (24°10'S/69°50'W) to Taltal. The western boundary of the ore zone is roughly located along the 70°10'W longitude, (Oficinas Trinidad and La Gloria). Far to the east the boundary is very irregular, with isolated groups of "oficinas" such as Savona, Pissis, Cochrane, and Domeyko (23°48'S/69°20'W); Augusta Victoria (24°03'S/69°20'W) and Dominador (24°23'S/69°30'W). The reasons for this irregular distribution involve geological and geomorphological factors but, also, the distribution of nitrate ores is based on the location of known exploited deposits. This means that low grade and unexploited ores are not represented. Nevertheless, the rough limits of the ore distributions seem to be well represented on this basis.

The geomorphology in northern Chile corresponds to that defined by a tectonically active desert. Major elements are controlled by regional systems of faults, with general N-S and E-W trends (Atacama and West Fissure Systems respectively; "alignments" of Zapiga, Olacapato-Toro, La Escondida, and Salar de Punta Negra). The main geomorphological units are from W to E: the Cordillera de la Costa (Coastal Range); the Depresión Central (Central Valley); the Preandean Ranges (i.e. Cordillera de Domeyko); the Preandean Basins (Atacama-Punta Negra Salars) and the Altiplano (=Puna=Alta Cordillera). As far as the geomorphological location of the nitrate deposits is concerned they are distributed in the Coastal Range and the Central Valley. Some ores are far to the west in the Coastal Range (i.e. La Gloria at 20°34'S/70°00'W). To the east they occasionally reach the western border of the Preandean Ranges. Their altitude varies between 700 and nearly 2500 m. Isolated deposits in the High Andes (Volcán Maricunga of Quaternary age) have been described (Ericksen 1981) and their presence has been documented in salt flats in the Altiplano. However, these deposits are not related with those of Lower Tertiary age described in this chapter. Previous papers do not describe a geological setting for the ores, largely because they are found in rocks of different lithology and age (Ericksen, 1981). We propose a "geological location" with the following main aspects:

- The ores are not Quaternary, as is indicated in most of the descriptions. We think they are of Lower Tertiary age (Eocene-Oligocene). This is documented by the close relationship in time and space with rocks of that age.
- They are located between the main N-S trending fault systems of the region, bounded to the west by the Atacama Fault System and to the east by the "West Fissure" (Fig. 1).
- They have been affected by the reactivation of some E-W structures.
- Lacustrine and fluvial activity, plus huge coalescent alluvial fans mainly deposited since the Upper Tertiary, have eroded or covered the nitrate deposits.

Finally, with reference to the relationship between nitrate and metallic ores, we postulate a metallogenic belt of Lower Tertiary age along the Central Valley; this belt includes epithermal gold-silver ores and "atypical" porphyry-copper deposits. The nitrate deposits appear to have a geological relationship with these metallic ores.

3 The Deposits

The nitrate deposits consist of an aggregation of saline minerals associated with detrital materials. The common major components are sulphates, chlorides and subsidiary borates and carbonates. What makes these ores distinctive is that the rest of the saline compounds are salts that seldom occur in nature. They correspond to highly water-soluble salts such as nitrates, iodates, chromates, dichromates, seleniates and perchlorates. A diversity of silicates, like clay minerals and zeolites are also present. The main cations are Na^+, K^+, Ca^{2+}, Mg^{2+} and a wide variety of trace elements.

The resulting mineralogy is complex and poorly documented. In the last 20 years a number of new minerals, like Humberstonite ($Na_7K_3Mg_2[SO_4]_6[NO_3]_2 \cdot 6H_2O$) and Brüggenite ($Ca[JO_3]_2 \cdot H_2O$) (Ericksen et al., 1974, 1986, 1989; Mrose et al. 1970), have been described. Our own studies clearly show that there are still other saline minerals which have probably not yet been described. Furthermore, silicates (mainly zeolites and clay minerals) and sulphides are present. We do not know which minerals contain perchlorates and seleniates.

3.1 Types of Ores

There is no universally accepted classification of the nitrate ores. Whitehead (1920) divides them into old, mature and young, but this classification has not been useful. Ericksen (1963) recognized alluvial deposits, deposits in rocks and deposits in salars. Rivera and Stephens (1988) separated primary from secondary deposits, but these terms cannot be used because of an erroneous appreciation of the regional geological setting. Today, geologists of the nitrate industry use, in an informal way, the term "caliche origen" to describe primary ores. In this chapter we use the classification of Ericksen (1963), but substitute the term "sedimentary" for alluvial deposits. We also change the term "deposits" in salars to "miscellaneous deposits". Use of this classification is temporary because we believe that a new classification should be produced, based on a more complete knowledge of the different deposits.

3.2 Sedimentary Deposits

These deposits are defined as those in which the economically important salts are found in clastic rocks deposited in volcanic-desertic (?) environments, mainly in intramontane basins. They show different stages of diagenesis, and the normal lithology includes breccias, coarse sandstones and, occasionally conglomerates and fine grained sediments. The ore forms a cavity filling matrix, veinlets, veins, irregular bodies of high grade nitrate and the cement of the detrital material. The economically important horizon, called "caliche", is enclosed in a horizontal sequence that has been described systematically in the geological and technical literature. The "caliche" has many different local names, according to its physical appearance; however, this nomenclature is not of practical use, with the probable exception of the terms "caliche blanco" and "caliche negro". The former is restricted to high-grade ore in which almost pure salts (nitrates and halite) can be macroscopically recognized and is present in all types of deposits. "Caliche negro" corresponds to a type of well cemented sedimentary ore, which is light to dark tan. The sequence incorporating the "caliche" is described in detail in Fig. 3 and, from top to bottom, the units "losa", "chuca", "costra", "caliche", and "conjelo/coba" are recognized. In the present mining operations they are described as "superficie" (including "losa" and "chuca"), "sobrecarga" (="costra"); "explotable" (="caliche"), and "asiento" or "subyacente" (="conjelo/coba"). Most of the descriptions assume a unique sedimentary affinity with lithological and saline changes related to the genesis and evolution of the deposits (Ericksen 1981; Chong 1984; van Moort 1985).

We have no definitive arguments with which to establish a stratigraphic type column of a sedimentary deposit but, after studying many sections (e.g. Fig. 4), we can present some preliminary conclusions:

- The saline-clastic sequence of the sedimentary deposits has been interpreted systematically as of Quaternary or, in some cases, as of Upper Tertiary age. We think that it should be older and partly of sedimentary origin with Lower Tertiary volcanic rocks.
- The associated volcanism was due to the development of a volcanic arc of Upper Cretaceous-Eocene age (Reutter et al. 1988; Scheuber et al this Vol.). We propose that the sediments were deposited during the upper part of the Eocene or during the Lower Oligocene, linked to a development to a more humid climate. This is documented by the thick continental sedimentary sequences of this age (Chong 1977).
- A regional palaeoslope to the east is assumed.
- It is believed that there exists a main phase of mineralization represented by the "caliche" horizon, but that subordinate mineralized levels are also present in the sequence. Thus, the process of mineralization was recurrent.
- Saline-clastic dykes (Fig. 5), termed sand dykes by Ericksen, (1981) are related to the sedimentary sequence. These dykes are of different ages, and sometimes are assumed to be "feeders" of the economically important horizon. In some cases they have a close relationship with conduits of palaeogeothermal fields. There we can see that they were injected with a high hydraulic pressure, splitting the host rock or cutting across strata. In some other cases, a recurrent passive injection, documented by fibrous salts, is evident.
- We believe that there are other sedimentary cycles of post nitrate deposition which are not genetically related to the nitrates. These cycles correspond to a series of clastic and saline strata with coarse sandstones, fine grained breccias and, rarely, fine-grained sediments such as siltstones or claystones. In turn, the saline compounds are gypsum, anhydrite, halite and other salts with a detritic fine grained fraction. There is a special saline clastic unit known as "panqueque". It is formed by fine grained clastic material plus anhydrite, gypsum and halite. The "panqueque" can be very well

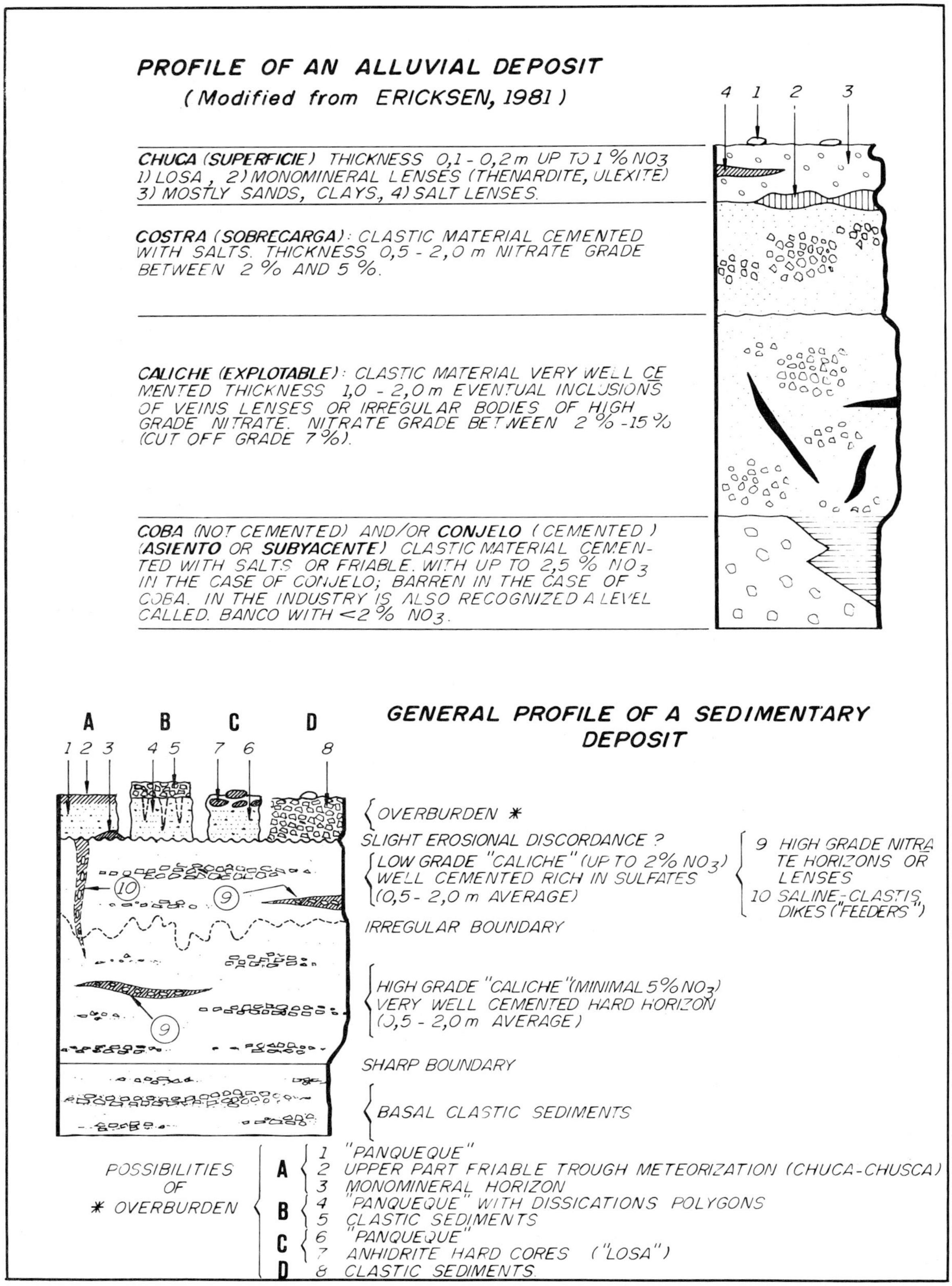

Fig. 3. Theoretical sections of sedimentary deposits

Fig. 4. "Atypical" section of a sedimentary deposit. Oficina California; Primera Region de Tarapaca

Fig. 5. "Sand Dyke" in a sedimentary deposit. Oficina Celia; Segunda Region de Antofagasta

cemented and may include monomineral horizons in its lowest division (tenardite, ulexite). The thickness of the whole "panqueque" can reach 1.5 m, though commonly it varies from millimetres to 60-80 cm. It is pink to yellow and can be found at different levels with the youngest deposits overlying almost the whole desert with no relation to topography. In contrast, the oldest deposits show very deep desiccation polygons and, in many cases they directly overlie the nitrate-rich levels. They are interpreted as "B" horizons of desert saline soils, as former mud flows sedimented in salt flats and even locally as related to deposition from marine fogs.

- We assume that the units traditionally called "costra" and "caliche" actually form one horizon of differing vertical composition due to vertical leaching. The upper "costra" is enriched in sulphates (low-grade ore), whilst the underlying "caliche" is enriched with highly soluble compounds like nitrates and chlorides (high grade ore).
- We have not distinguished between the units "coba" or "conjelo". We have observed sediments without ore subjacent to the "caliche", and these we interpret as a physical boundary for the original fluids due to a lack of permeability.

3.3 Deposits in Rocks

These are high grade nitrate deposits in which the ore is located in rocks as veins, veinlets, irregular massive bodies or fine disseminations (Fig. 6). The

Fig. 6. Underground nitrate mine. Oficina San Antonio de Zapiga. I Region de Tarapaca. An almost pure nitrate "pillar" in a rough "room and pillar" exploitation can be observed

host rocks are of different age and lithology, and include marine, sedimentary, volcanic and intrusive rocks. The dimensions of the ore bodies vary from centimetres to more than one metre in thickness; their length can reach hundreds of metres.

These deposits have been exploited by means of underground mining, either using shafts to reach the nitrate bodies and then following them, or through traditional underground mining methods (Fig. 6). In some places (i.e. Sierra Borda Mining District) these massive nitrate bodies have a close relationship with metallic mineralization (Fig. 1). Here, supergene copper oxides appear in the same assemblages as nitrate bodies.

With the exception of veinlets and disseminations in ignimbritic rhyolitic tuffs, these ores show mineralization through conduits. The original enriched solutions circulated through faults and fractures and were later deposited in bedding planes and open fractures. The relationship between "feeders", (i.e. nitrate-enriched "dykes" with sediments and salts or sand dykes) can be directly observed. Sometimes these features were used as surface markers to locate the underground bodies.

3.4 Miscellaneous Deposits

This group includes some salars with exploitable nitrates in their crusts (e.g. Salar del Carmen) (Fig. 2). Ericksen (1981) argued that the nitrates were carried to the basin by underground water which had leached former ores; once there, deposition was achieved by capillarity. Such a process should occur periodically and the ore could be "harvested" after periods of redeposition. We have not studied this model exhaustively. A second type of ore considered in this group is formed directly from deposition in the domain of palaeogeothermal fields. We have found palaeodeposits of this type with saline terraces of high grade iodine ore (more than 1300 ppm/ton) and nitrates (up to 25%) (Figs. 7 and 8).

Fig. 7. Section of a Lower Tertiary palaeogeothermal field. II. Region de Antofagasta. From *top to bottom*: basalt lava flow; travertine lense; lacustrine-volcanic sediments including low temperature silica and manganese oxides. Palaeoconduits with salts relicts

4 Geological Setting

With the exception of Whitehead (1920), the literature does not attempt to relate the nitrate deposits to a specific geological setting. Our working hypothesis is that this geological setting exists and, moreover, is closely related to the genesis of the ores. Obviously, subsequent geological events may affect and overprint this setting. In applying this hypothesis four main geological units are distinguished: Three are directly related to the genesis of the deposits and their subsequent development, while the fourth plays a secondary, though also important role. Here, these units are called basement, Meso-Cenozoic volcanic arc, Neogene sedimentary sequences and Miocene-Pliocene volcanic arc.

4.1 Basement

This unit corresponds to Palaeozoic-Mesozoic rocks outcropping in the uplifted Coastal Range and Preandean Ranges and sporadically as structurally isolated rocks in the Central Valley. Its main characteristics are:

- Palaeozoic: Marine and continental sequences locally affected by low grade metamorphism. Palaeontological data are rather scarce and the sequences are described as of Devonian, Carboniferous and Permian age (Chong, 1977).
- Triassic: Marine and continental sequences interfingered with important volcanic series.
- Jurassic: Marine, continental, evaporitic sequences and thick series of volcanites of intermediate composition. Some intrusions are intermediate to basic in composition. Copious fossils demonstrate that the whole system is present with the predominance of Dogger and Malm (Chong, 1973, 1977).
- Lower Cretaceous: Represented by volcanic and continental rocks and some sparse outcrops of marine sequences. Intrusives are of intermediate composition.

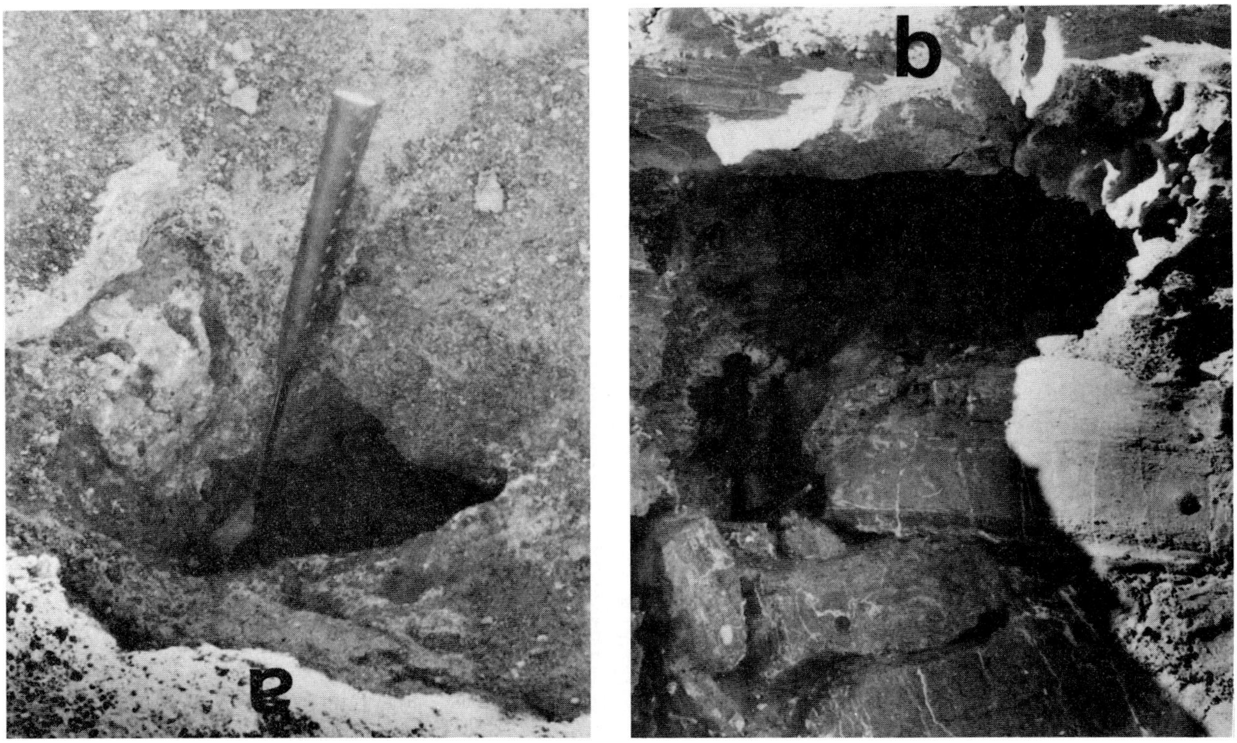

Fig. 8. *a* Vertical palaeoconduit in basalts. *b* Horizontal palaeoconduit in lacustrine-volcanic sediments

- Upper Cretaceous: Thick volcanic series with less continental sediments. Volcanic rocks are of acidic to intermediate composition but include some rare basic rocks. Hypabyssal intrusives of similar composition are present.

4.2 Meso-Cenozoic Volcanic Arcs

Reutter et al. (1988) described an Upper Cretaceous-Eocene volcanic arc emplaced along the present Preandean Range. We propose that it was also partly emplaced in both the present Central Valley and along sectors of the Coastal Range and that its terminal activity possibly extended to Lowest Oligocene times. The lithology corresponds to rhyolites, dacites with subordinate andesites and some basalts and volcanic glasses. These rocks were deposited as lava and ignimbritic flows and as pyroclastics associated with hypabyssal intrusives of the same composition intruded as dykes, sills, domes, plugs and quartz-feldspar stocks. We have also found extensive outcrops with copious evidence of palaeogeothermal fields. These include lacustrine deposits with abundant manganese oxides, low temperature quartz, travertine and sinter. Continental-related sediments are rather scarce because they were deposited far to the east, along the western border of the Preandean Ranges. With respect to this chapter we must consider the following:

- This volcanic arc was partly emplaced during Lower Tertiary times and is related to the genesis of epithermal gold/silver deposits and perhaps of some porphyry copper deposits.
- There is a relationship between the spatial distribution of these metallic ores and the nitrate deposits. In some cases these volcanic rocks are the sole geological setting of the nitrates.
- This volcanism was emplaced along N-S and E-W trends. It cuts across the basement that provides the many different elements now present in the nitrate deposits (i.e. the thick Jurassic sequences were able to provide important amounts of iodine, halite, gypsum and carbonates.

4.3 Neogene Sedimentary Sequences

This unit includes lacustrine, fluviatile and alluvial strata as well as interbedded volcanic rocks deposited in the internal basin of the Central Valley. The relationship with nitrate deposits is as follows:

- At the latitude of Pisagua (Fig. 1) some lacustrine deposits are uplifted in the eastern border of the Central Valley. They are similar to those of the "pampitas" area (Chong, 1984, 1988) and of other locations to the south (i.e. Cerros de la Joya, Lagunas). These sediments overlie the nitrate deposits and may have partly destroyed them by erosion. This is evidenced by the isolated nitrate ores surrounded by these lacustrine sediments.
- Salt flats and alluvial sediments of the "bajadas" of coalescent alluvial fans, with their terminal part to the west cut and "seal" the nitrate deposits along the eastern sector of the Coastal Range (i.e. Salares de Bellavista and Pintados).
- There are some basins, like Montecristo-Maria Elena, Quillagua, and Lago Soledad, linked to the development of the Loa River. Their sediments are younger than the nitrate deposits.

4.4 Miocene-Recent Volcanic Arc

This unit corresponds to the youngest, still active volcanism located in the Altiplano along the Chilean-Bolivian-Argentine frontier. It consists of stratovolcanoes of Upper Tertiary and Recent age and a volcaniclastic plateau; both overlying a Palaeozoic-Mesozoic basement. This system, being younger, is not directly related to the nitrate deposits, but it may have provided some elements which have been deposited in the nitrate basins, either directly or through underground waters.

5 Discussion on the Genesis of the Nitrate Deposits

Whitehead (1920) stated: "Many quite divergent hypotheses on the origin of the deposits have been suggested with the universal acceptance of none". This statement still reflects the opinion of many geologists, since Darwin (in Ericksen, 1981) published his theory one century ago. In this section we do not propose to discuss the many theories put forward in the past (see Ericksen 1981).

Nitrate genesis has been attributed to sources such as marine seaweeds, vegetation in saline lakes or marine water, leaching of ammoniacal material derived from the guano of sea birds or from bacterial decay, from volcanic activity or mudflows enriched in salts, from atmospheric nitrogen, from marine fogs, or from volcanic or marine sedimentary rocks.

Nitrate genesis cannot be addressed as a single question. Several aspects must be considered: Where did the original nitrates come from? In what chemical form were the saline compounds precipitated? Which other salts were present in the original assemblage? Which other elements were later incorporated? What are the natural sources of some compounds even more exotic than nitrates (e.g. iodates, chromates, dichromates, seleniates and perchlorates)? What constituted the original mineral assemblage and the saline paragenesis? How many types of deposit exist? How useful are terms like primary, secondary and even tertiary ores? Was the redox potential stabilizing all these labile compounds the same during and after the emplacement of the ores? What are the real geomorphological factors? How were the ores preserved? Were there enrichment processes? What are the main hydrochemical and geochemical aspects of the sedimentary deposits? What was the palaeogeography when the ores were emplaced? What is the geological setting of the ores? Answers to these and many other questions require an interdisciplinary scientific research and the development of certain new laboratory techniques.

5.1 Some Arguments and Facts Used to Establish a Genetic Theory

The basic tenet proposed herein is that the genesis of the nitrates is related to the magmatism of the Upper Cretaceous-Lower Tertiary volcanic arc. This idea of relating nitrate genesis with volcanic activity is not new. Moreover, considering the geological history of the region this possibility is one of the more obvious either in a direct way or through secondary activities (De Kalb, Fiestas and Steinmann in Ericksen, 1981).

Whitehead (1920) and Ericksen (1981) stated similar possibilities when they considered Jurassic and Upper Tertiary volcanic rocks, respectively. In Araya and Toro (1983) and Chong (1984) a relationship with hydrothermal alteration activity is discussed. Rivera and Stephens (1988) attempt to explain an association between nitrate deposits and geothermal activity. Other pertinent points include:

1. Whitehead (1920) mentioned a relationship between Jurassic tuffs and nitrate deposits in the southern part of the Antofagasta province. However, due to the lack of data at that time, he

mistook Tertiary volcanic rocks for Jurassic volcanic rocks but, nevertheless, he did attempt to test the relationship between igneous activity and nitrate genesis.

2. Chong (1973, 1977) described acidic and intermediate volcanic rocks of the Lower Tertiary in the southern part of the Central Valley (Chile-Alemania Formation; Chong 1973). Formerly these rocks had been assumed to be of Upper Tertiary age by correlation with the Riolítica Formation of Brüggen (1950). Later, they became the target of exploration in the search for silver/gold epithermal deposits. During this study some saline-metallic ore relations were observed.
3. In metallic ore deposits, as in hydrothermal alteration zones present in these volcanic rocks, the presence of saline assemblages is common. The saline compounds appear to be spatially related to supergene enrichment. Macroscopically, one can recognize gypsum, anhydrite, halite and tenardite. Other saline minerals have not yet been investigated but it seems that carbonates and nitrates are not uncommon.
4. Nitrate ore, in the form of veinlets or disseminations, found exclusively in rhyolitic tuffs and without a relationship with other rocks, is common in some nitrate deposits (e.g. Taltal District). Also of note is the presence of nitrate ore disseminated in sediments of intermontane basins bounded exclusively by these rhyolitic tuffs.
5. Silica, carbonate and nitrate veinlets are related to gold/silver anomalies in volcanic rocks and to hypabyssal intrusives of acidic composition. Suites of halide minerals in the oxidation zones of silver deposits, and of iodates in the mineralogy of some porphyry copper deposits, can be observed.
6. SEM and thin section studies showed stibnite to be included in nitrate and nitrate "intruding" zones along cleavage planes (J.J.Pueyo pers. commun.).
7. We have observed some alteration boundaries between nitrate veins and the host rocks. We interpret these as the effects of hydrothermal fluids.
8. We postulate the presence of a metallogenic belt along the Central Depression and part of the Coastal Range between latitudes 20°00'S and 24°30'S. This belt includes the nitrate deposits.
9. Finally, recent studies (Pueyo and Chong, in prep.) showed an extensive palaeogeothermal area emplaced in Lower Tertiary rocks. The outcrops reveal the presence of travertine, sinter, lacustrine deposits with copious manganese oxides, and palaeoconduits with abundant salt relicts (Figs. 7 and 8). Some samples collected from the area contain nitrates (up to 6%) and iodine (up to 1400 ppm). We interpret this as the result of nitrate- and iodine-enriched brines that circulated in this geothermal field as original hydrothermal solutions.

5.2 Working hypothesis for ore genesis

We assume that the original nitrogen was derived from magmatic sources during the Lower Tertiary volcanic arc evolution. Nitrogen in the form of oxides, ammonium compounds and even elemental nitrogen is abundant in volcanic activity (Rankama and Sahama 1962; Babor and Ibarz 1973). There are neither easy nor obvious answers to explain how this nitrogen was fixed as nitrates. We are working with the idea that this type of magma, related to metallic compounds, includes elements that would act as catalyzers. In this respect we can expect deposition of nitrates from ammonium compounds in the presence of sulphides as catalyzers. At the same time, abundant oxygen can be expected from the reaction between water and nitrogen, thus contributing to the production of a high redox potential. Other sources of abundant water are the ignimbritic eruptions. However, we think that most of the meteoric water to provide the brines and later distribution of the saline compounds was produced during climatic changes in Lower Oligocene time. In short, we propose the following:

- The original nitrate and most of the associated saline components had their source in magmatic fluids produced during magmatic activity in the Lower Tertiary.
- In Oligocene times the climate changes provided plenty of water that promoted an intense geothermal activity during the terminal phases of volcanism. This geothermal activity took place in local fields or along regional faults. Meteoric water was always the main transport mechanism.
- Enriched brines, which circulated near the surface, were concentrated in ponds or produced mudflows. In the case of "Deposits in Rocks", the brines used favourable conduits (faults, joints, bedding) for their migration. In sedimentary deposits, the permeability of certain horizons was an important factor.
- Fractional crystallization occurred due to leaching; capillarity and leaching of elements from the host rocks were later enrichment processes.
- The increasing amounts of water during the Oligocene may have been responsible for the

oxidation of metallic ores, forming all the halides and sulphates that we now find in the oxidation zones.
- It is estimated that the huge regional fault systems prevailing during Oligocene times (the Atacama and West Fissure Systems) would provide a geomorphological "trap" to locate the ores. It is also assumed that there was a palaeoregional slope from west to east (Mortimer 1973).
- The late geological evolution resulted in partial dissolution and redeposition, supplying other saline compounds to the basins and leaching elements from the host rocks.

6 Subsequent Geological Evolution of the Deposits

Accepting that the ores were emplaced in Lower Tertiary time, their geological evolution can be summarized as follows: In the Lower Oligocene the uplifting of the Andes started together with intensive erosion. The reactivation of the regional faults produced the setting of the present main geomorphological units. During this time the climate wass more humid, and pediments of Choja (Galli 1967) or Tarapaca (Mortimer and Saric, 1972) are defined together with the deposition of thick sedimentary sequences (i.e. Atacama Gravels, Sillitoe et al. 1968; Azapa Formation, Salas et al. 1966; Pampa de Mulas Formation, Chong 1977). Volcanism was still active but was less intense than in Paleocene-Eocene times. During this period the biggest Chilean porphyry copper deposits were emplaced. In the Oligocene the watershed was shifted estward (Mortimer 1973) and we assume that some lacustrine systems appeared (e.g. the Salar de Atacama and Punta Negra systems). At the same time, lacustrine conditions developed in northernmost Chile, between Arica and Iquique (Fig. 1) with the consequent destruction of some nitrate deposits. Nevertheless most of the nitrates were protected from erosion because they were uplifted or "sealed" with later saline blankets.

In the Middle and Upper Miocene the formation of the Altiplano began and the Central Valley became an endorrheic basin with continuous sedimentation. Lacustrine systems were formed (e.g. María Elena, Lago Soledad and Quillagua Basins) and were linked through the Loa River development. All these fluviolacustrine systems eroded nitrate deposits in the Central Valley.

At the Miocene-Pliocene boundary a new contractional tectonic phase took place (Diaguita phase; Maksaev 1979). Some saline basins began to form in the Altiplano. Major structural valleys were cut south of Arica, causing the drainage of lakes, and the Loa River started to incise to the sea. In Pliocene times the climate changed to hyperarid because of the uplifting of the Andes (Naranjo and Paskoff 1980) and the initiation of the Humboldt Current. The High Andes basins evolved to evaporitic basins and the Loa River completed the drainage of the Soledad Lake. Some isolated bodies of water became salt flats (Lomas de la Sal, Cerro Soledad, South Lagunas) (Fig. 1). The marine deposits in the Coastal Range were completed; in some parts the coastal uplift ceased and abrasion terraces were formed. The reactivation of faults caused some lacustrine deposits to be uplifted in the eastern part of the Coastal Range and some valleys became "hanging" valleys at its western border. Volcanic activity to the east was intense and some materials were transported westwards to the nitrate basins.

In Upper Pliocene times the current geomorphological/climatic model was established. The regional slope was to the west and large coalescent alluvial fans were formed. In their distal parts the deposition of salars and mud flows destroyed some nitrate deposits. From the volcanic activity more materials were introduced into the basins via groundwater. The extreme aridity produced desert soils with the formation of the sulphate horizons known today as "panqueques". Their evolution generated the "losa" and "chuca" levels. Capillarity and fogs played a secondary role in the evolution of the nitrate ores through the solution and redeposition of surface salts.

Acknowledgements. This work has the financial support of the Fondo Nacional de Desarrollo Científico y Tecnológico de Chile [FONDECYT]) as project number 90/105.

References

Araya H, Toro JC (1983) Geología de los yacimientos de Pedro de Valdivia y María Elena. II Región, Chile. Thesis, Dep Geociencias, Universidad del Norte, Antofagasta, 287 pp (unpubl)

Babor J, Ibarz J (1973) Química general moderna. Marín, Barcelona, 1144 pp

Brüggen J (1950) Fundamentos de Geología de Chile. Inst. Geogr. Militar. 374 p. Santiago de Chile.

Chong G (1973) Reconocimiento geológico del área Catalina-Sierra de Varas y estratigrafía del Jurásico del Profeta. Memoria para optar al Título de Geólogo. Departamento de Geología, Antofagasta

Chong G (1977) Contribution to the knowledge of the Domeyko Range in the Andes of northern Chile. Geol Rundsch 66(2): 374-404

Chong G (1984) Die Salare in Nordchile. Geologie, Struktur und Geochemie. Geotek Forsch 67: 146 pp

Chong G (1988) The Cenozoic saline deposits of the Chilean Andes between 18°00' and 27°00' South latitude. In: Bahlburg H; Breitkreuz C, Giese P (eds): The southern Central Andes. Lecture Notes in Earth Sciences 17: Springer, Berlin Heidelberg New York, pp 137-151

Ericksen GE (1963) Geology of the salt deposits and the salt industry of northern Chile. US Geol Surv Open File Rep, Washington DC, 164 pp

Ericksen GE (1979) Origin of the nitrate deposits of northern Chile. Actas II Congr Geol Chil Arica, 2: 181-205

Ericksen GE (1981) Geology and origin of the chilean nitrate deposits. US Geol Surv Prof Pap 1188: 37 pp

Ericksen GE (1983) The Chilean nitrate deposits. Am Sci 71: 366-374

Ericksen GE, Mrose M, Marinenko JJ (1974) Mineralogical studies of the nitrate deposits of Chile. IV. Bruggenite, $Ca(JO_3)_2.H_2O$. A new saline mineral. J Res US Geol Surv 2(4): 471-478

Ericksen GE, Mrose M, Marinenko J, McGee J (1986) Mineralogical studies of the nitrate deposits of Chile. V. Iquiqueite, $Na_4K_3(CrO_4)B_{24}O_{39}(OH)\cdot 12H_2O$. A new saline mineral. Am Mineral 71: 830-836

Ericksen GE, Evans H, Mrose M, McGee J, Marinenko J,, Konnert J (1989) Mineralogical studies of the nitrate deposits of Chile. VI. Hectorfloresite, $Na_9(JO_3)_4$. A new saline mineral. Am Mineral 74: 1207-1214

Galli C (1967) Pediplain in northern Chile and the Andean uplift. Science 158: 653-655

Maksaev V (1979) Las fases tectónicas Incaica y Quechua en la Cordillera de los Andes del Norte Grande de Chile. Actas II Congr Geol Chil Arica 1: B63-67

Mortimer C (1974) The Cenozoic history of the southern Atacama Desert. J Geol Soc Lond 129: 505-526

Mortimer C, Saric N (1972) Landform evolution in the coastal region of Tarapaca Province, Chile. Rev Geol Din 21: 162-170

Mrose M, Fahey I, Ericksen GE (1970) Mineralogical studies of the nitrate deposits of Chile. III. Humberstonite, $K_3Na_7Mg_2(SO_4)_6(NO_3)_2.6H_2O$. A new saline mineral. Amer. Sci., 55: 1518-1533. New Haven.

Naranjo J, Paskoff R (1980) Evolución geomorfológica del Desierto de Atacama entre los 26° y 33° lat. sur. Revisión cronológica. Rev Geol Chile 10: 85-89

Rankama K, Sahama THG (1962) Geoquímica. Aguilar, Málaga, 862 pp

Reutter K-J, Giese P, Götze H-J, Scheuber E, Schwab K, Schwarz G, Wigger P (1988) Structural and crustal development of the Central Andes between 21° and 25°S. In: Bahlburg H; Breitkreuz C, Giese P (eds): The southern Central Andes. Lecture Notes in Earth Sciences 17: Springer, Berlin Heidelberg New York, pp 231-261

Rivera S, Stephens AJ (1988) Cuerpos geotermales fósiles de edad Terciario Inferior y mineralización asociada en la Región de Antofagasta. Actas V Congr Geol Chil Santiago 1: B-39-64

Salas R, Kast R, Montecinos F, Salas I (1966) Geología y recursos minerales del Departamento de Arica, Provincia de Tarapacá. Inst Invest Geol Bol 21. 113 pp

Sillitoe RH, Mortimer C, Clark AH (1968) A chronology of landform evolution and supergene mineral alteration, southern Atacama Desert, Chile. Earth Sci Trans Sec B, Inst Min Met 77: 66-99

van Moort JC (1985) Natural enrichment processes of nitrate, surphate, chloride, iodate, borate, perchlorate and chromate in the caliches of northern Chile. Actas IV Congr Geol Chil Antofagasta IV: 3-674-3-702

Whitehead WL (1920) The Chilean nitrate deposits. Econ Geol 15: 187-224

Petrochemical Factors Governing the Metallogeny of the Bolivian Tin Belt

Bernd Lehmann

Abstract. Geochemical data both from tin-bearing peraluminous granitic plutons of Triassic and Middle Tertiary age in northern Bolivia and from the peraluminous ash-flow tuffs of late Tertiary age from Macusani, southern Peru, and the Morococala and Los Frailes volcanic fields in the central part of the Bolivian tin belt define systematic tin enrichment trends which are consistent with a magmatic evolution controlled by fractional crystallization at a bulk tin distribution coefficient D_{Sn}(xtls/melt) <1. Locally similarly fractionated plutonic and volcanic rocks from outside the tin belt (Western Cordillera and NW Argentina) show no tin enrichment, i.e. $D_{Sn} \approx 1$. The metasedimentary source of the igneous systems of the Bolivian tin belt was probably not anomalous in tin. However, it provided the reduced environment which gave rise to the ilmenite-series character of the tin-bearing rocks. This contrasts with the magnetite-series affiliation of the non-tin granites and porphyries of the central Andes. A high degree of fractionation together with a low oxidation state of the igneous system are considered to have been the major controls on the origin of the Bolivian tin belt.

1 Geological Setting of the Tin Belt

The geological framework of the Bolivian tin belt consists of a more than 10 000-m-thick intracratonic pile of clastic marine sedimentary rocks of Ordovician to Devonian age. The Lower Palaeozoic sequence is structurally deformed and mostly low-grade metamorphosed (Hercynian orogeny), and is locally overlain by inliers of Upper Mesozoic red beds and Lower Tertiary continental sedimentary and volcanic rocks. The stratified rock sequence is intruded by several large Permo-Triassic granitic plutons in southern Peru and northern Bolivia which form the NW-trending chain of glacial peaks of the Cordillera Real (Eastern Cordillera). Oligocene to Miocene granitic plutons are exposed further south (Cordillera Quimsa Cruz) and are coeval with widespread and generally intensely altered rhyodacitic to quartzlatitic porphyry stocks which occur mainly in central and southern Bolivia, but extend also into southern Peru and northernmost Argentina. Extensive Miocene ash-flow tuff sheets in association with caldera centres cover part of the tin belt, of which the Los Frailes (8500 km^2) and Morococala (1500 km^2) volcanic fields predominate in central Bolivia (Ericksen et al. 1985). The northern end of the tin belt in Peru is outlined by the 2500-km^2 Miocene-Pliocene Macusani volcanic field (Noble et al. 1984).

The polymetallic hydrothermal systems of the Bolivian tin belt are associated with both the Permo-Triassic tin granites and the Middle Tertiary tin granites and porphyries (Evernden et al. 1977; Grant et al. 1979; Clark et al. 1983; McBride et al. 1983). Mineralization accompanies late-phase intrusive and hydrothermal activity characterized by satellite subintrusions, subvolcanic domes and pipes, hydrothermal breccias, and pervasive alteration (Sillitoe et al. 1975; Grant et al. 1977; Francis et al. 1983). The two most important ore deposits of the tin belt are the Cerro Rico de Potosi, which ranks historically as the largest silver producer in the world and which is currently one of the major tin mines of Bolivia, and Llallagua, which is the world's largest hard-rock tin deposit (Fig. 1).

Both plutonic and volcanic rocks in the tin belt are mostly peraluminous, have high levels of lithophile trace elements such as Rb, Li, Cs, B and Sn, and belong to the ilmenite-series rock group (Carlier et al. 1982; Halls and Schneider 1988; Pichavant et al. 1987; Lehmann et al. 1990). Hydrothermal alteration consists chiefly of the mineral association quartz-sericite-tourmaline. Boron enrichment is a characteristic and very widespread feature of the hydrothermal tin ore systems of both Triassic and Tertiary age. Boron and lithium are also concentrated through the action of thermal springs on volcanic

Correspondence to: B. Lehmann, Fachbereich Lagerstättenforschung, TU Clausthal, Adolph-Roemer-Straße 2a, D-3392 Clausthal-Zellerfeld

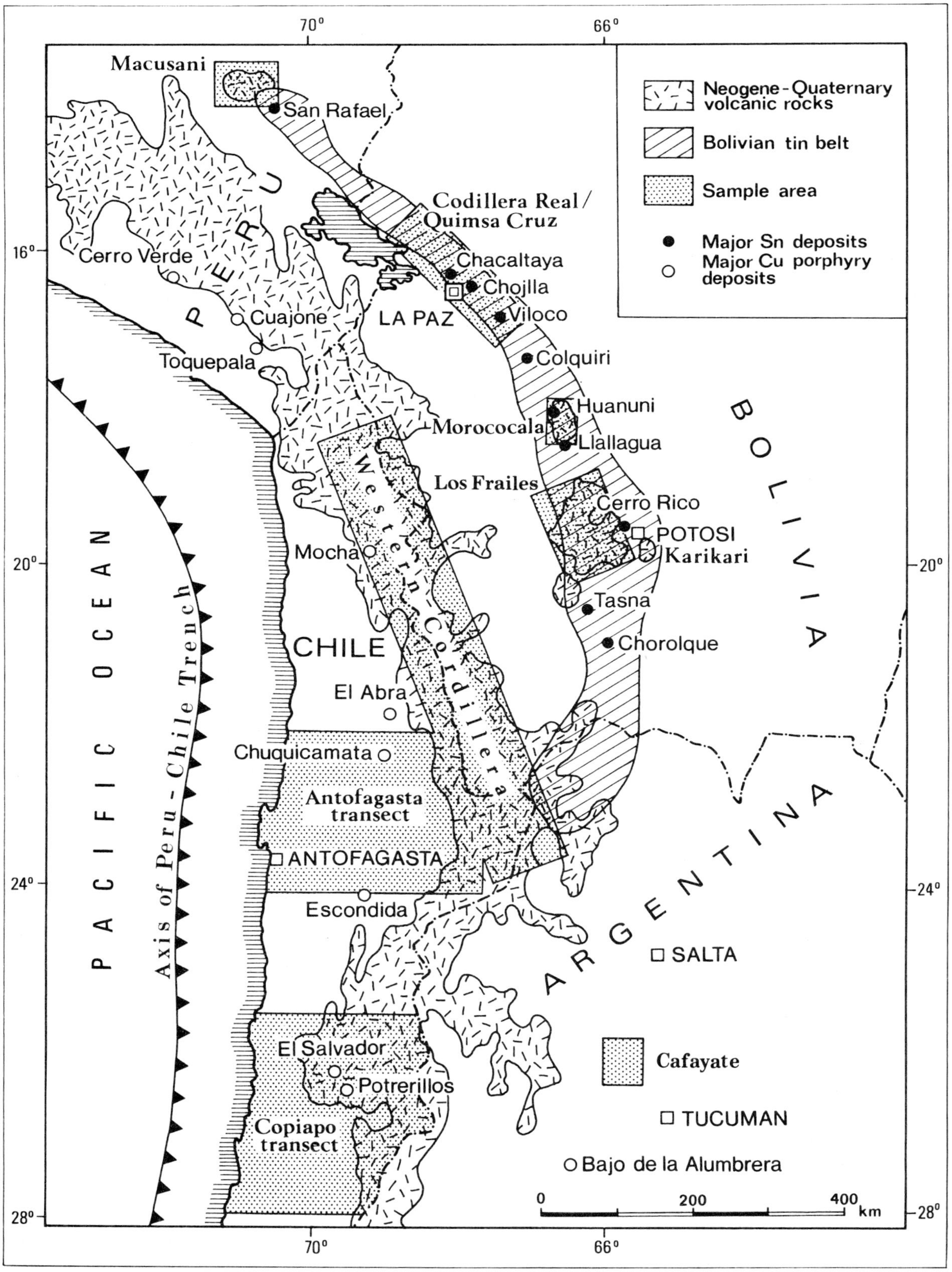

Fig. 1. Location of sample groups discussed in text, and of major copper-molybdenum porphyry and tin ore deposits in the Central Andes. Distribution of Neogene-Quaternary volcanic rocks from Ericksen et al. (1987)

rocks in the salars on the Altiplano. The hydrothermal mineral association of the ore deposits in the tin belt contains cassiterite as the dominant tin mineral, and a variety of sulphides of Ag, Sb, As, Cu, Pb, Zn, and Bi, with locally elevated Au contents. Detailed accounts of the Bolivian ore systems are given by Ahlfeld and Schneider-Scherbina (1964), Grant et al. (1977, 1980), Kelly and Turneaure (1970) and Sillitoe et al. (1975).

2 Geochemical Data Base and Petrogenetic Background

The following discussion is based on geochemical data from plutonic and volcanic rocks of the Bolivian tin belt and from some complementary tin-barren rock suites from northern Chile and NW Argentina (Fig. 1). Emphasis is on the tin granites of the Cordilleras Real and Quimsa Cruz which have been described in detail by Lehmann (1979), Miller (1988), Tistl (1985) and Winkelmann (1983). Additional data on the Los Frailes and Morococala volcanic fields of the central Bolivian tin belt are from Ericksen et al. (1985, 1990) and from H. Michel (unpublished). The geochemical data on the Macusani volcanic field in southern Peru are from Noble et al. (1984) and Pichavant et al. (1988).

Both the Permo-Triassic and the Tertiary silicic igneous rocks of the tin belt are dominantly of crustal origin. This is deduced from isotope data on the Triassic granite plutons of southern Peru and northern Bolivia, which give a range in $^{87}Sr/^{86}Sr_i$ of 0.708-0.717 (McNutt and Clark 1983; Kontak et al. 1984), and a late Proterozoic Nd model age (Miller and Harris 1989). Rocks from the Cenozoic volcanic fields of central and southern Bolivia have initial Sr ratios of 0.707-0.713 and ε_{Nd} values around -8 (Klerkx et al. 1977; Schneider 1985), whereas the Macusani volcanics have initial $^{87}Sr/^{86}Sr$ ratios between 0.721 and 0.726, and ε_{Nd} values around -9 (Noble et al. 1984; Pichavant et al. 1988). The strongly peraluminous character of the Macusani volcanics in southern Peru and of the Los Frailes, Morococala and Karikari volcanics in central Bolivia (phenocryst mineralogy consisting of Al-rich phases such as garnet, biotite, cordierite, andalusite; normative corundum >2 wt%; molecular Al_2O_3 / $[Na_2O+K_2O+CaO]$ >1.2) suggests a meta-sedimentary pelitic source for these rocks (Halls and Schneider 1988; Pichavant et al. 1988; Ericksen et al. 1990).

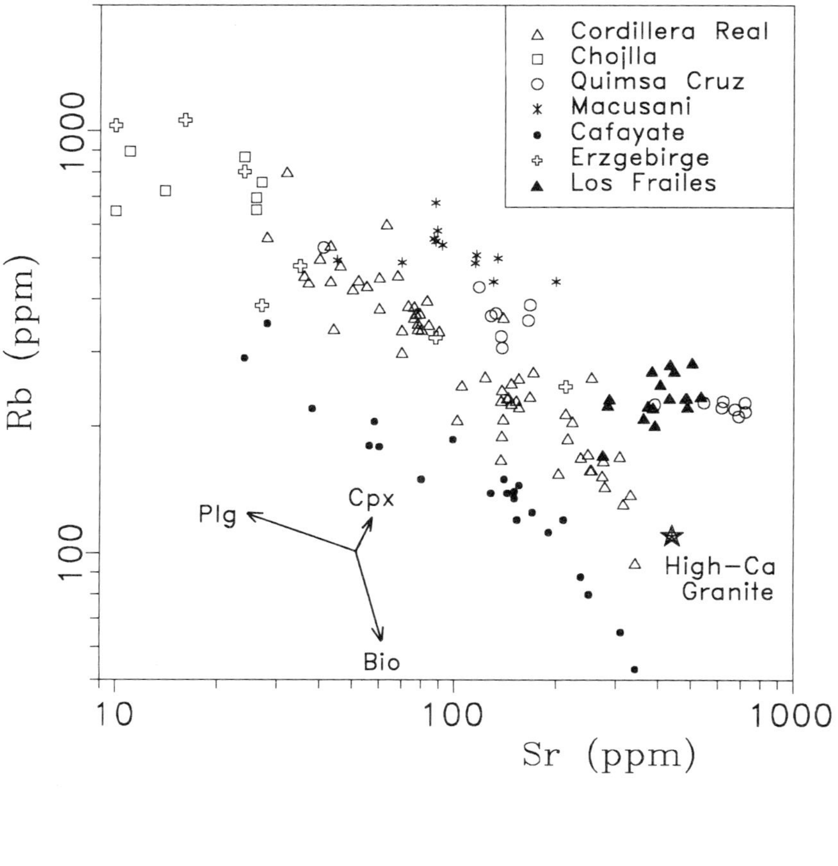

Fig. 2. Rb-Sr variation diagram for granitic rocks from the northern Bolivian tin belt (Cordillera Real granites, Chojlla mine granite, Cordillera Quimsa Cruz granites), from the Cafayate granite suite of northwestern Argentina, and for the Macusani (southern Peru) and Los Frailes (central Bolivia) volcanic rocks. Data from Lehmann (1979), Michel (unpubl.), Miller (1988), Noble et al. (1984), Pichavant et al. (1988), Rapela and Shaw (1979), Tistl (1985), Winkelmann (1983). Reference data from the Erzgebirge granite suite are from Tischendorf (1989). The vectors indicate the calculated change in melt composition as a result of fractional crystallization with F (degree of fractionation) 0.8 for plagioclase (*Plg*), biotite (*Bio*) and clinopyroxene (*Cpx*). Mineral/melt partition coefficients from Nabelek (1986). Composition of average high-Ca granite from Turekian and Wedepohl (1961)

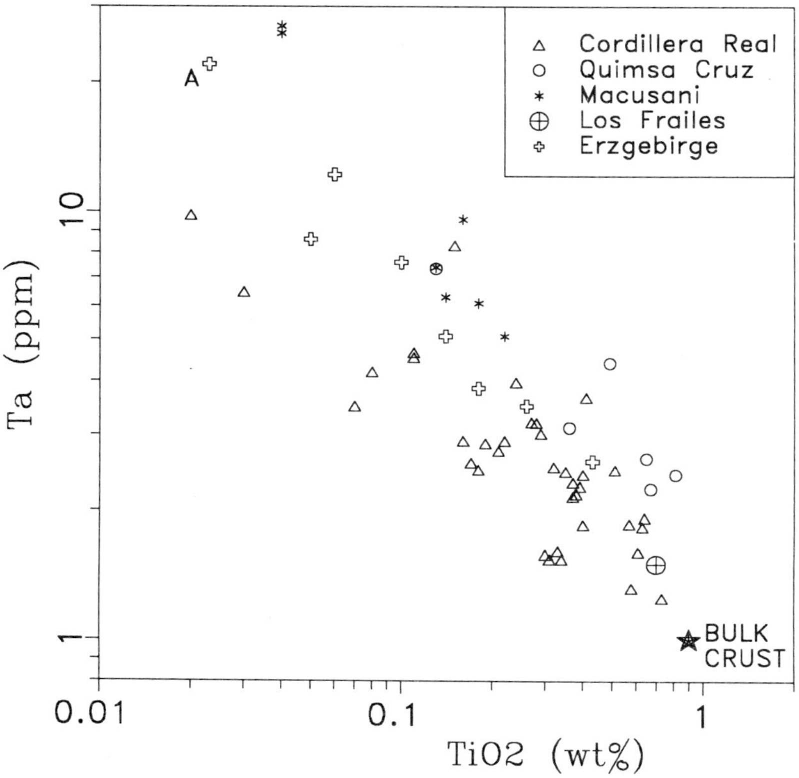

Fig. 3. Ta-TiO_2 variation diagram for granitic rocks from the northern Bolivian tin belt (Cordilleras Real and Quimsa Cruz) and for silicic volcanic rocks from Macusani (southern Peru) and Los Frailes (central Bolivia). Data from Ericksen et al. (1985), Lehmann (1979), Michel (unpubl.), Miller (1988), Noble et al. (1984), Pichavant et al. (1988). The reference data for the Erzgebirge granite suite are arithmetic means for major granite units as compiled in Tischendorf (1989). The *open cross marked A* locates the greisenized Altenberg granite stock (Just et al. 1987). Composition of bulk crust from Taylor and McLennan (1985)

This situation contrasts with the petrogenesis of the Cenozoic volcanic rocks of the Western Cordillera (border region of Chile and Bolivia), which is generally understood as a result of mixing between upper mantle and crustal material combined with intracrustal fractionation processes (Klerkx et al. 1977; Hawkesworth et al. 1982; Hildreth and Moorbath 1988). Tin data on the Western Cordillera rocks are from Gardeweg et al. (1984) and Lehmann and Pichler (1980). These sample suites are from ignimbritic sheets and dacitic to rhyolitic stocks of magnetite-series affiliation.

The samples from northern Chile are from Mesozoic-Cenozoic plutonic and subvolcanic rocks in the Antofagasta and Copiapó transects, and have been described by Ishihara et al. (1984). They belong to the calc-alkaline, mostly metaluminous suite of quartz-monzodioritic to granodioritic composition, with very important copper-molybdenum mineralizations associated with deeply eroded stratovolcano complexes. The porphyry-copper related rocks are known, so far, to have low initial $^{87}Sr/^{86}Sr$ ratios around 0.704 (Gustafson and Hunt 1975; Halpern 1979; Shibata et al. 1984). The Mesozoic-Cenozoic granitic rocks are generally of the magnetite-series, are hornblende-bearing and correspond to the I-type group of the classification scheme of Chappell and White (1974).

Another sample suite of tin-barren granitic rocks is from the Cafayate area in the Pampean Ranges of NW Argentina and has been described by Rapela and Shaw (1979) and Rapela et al. (1982). The Lower Palaeozoic Cafayate plutons have a compositional spectrum from biotite tonalite to biotite-muscovite granite, peraluminous composition, and initial $^{87}Sr/^{86}Sr$ ratios in between 0.703 and 0.707 (Rapela et al. 1982; Saavedra et al. 1987). Their chemical and mineralogical characteristics do not fit the typical features of either I- or S-type granites; the occurrence of accessory magnetite suggests a magnetite-series affiliation.

3 Geochemical Evolution of Tin

The magmatic evolution of both the Cordillera Real tin granites and the Pampean Ranges (Cafayate) non-tin granites is largely controlled by fractional crystallization. This has been shown by trace-element modeling, REE distribution patterns, and a general trend towards low-pressure thermal-minimum composition combined with a multiple intrusion history (Lehmann 1979; Rapela and Shaw 1979; Saavedra et al. 1987).

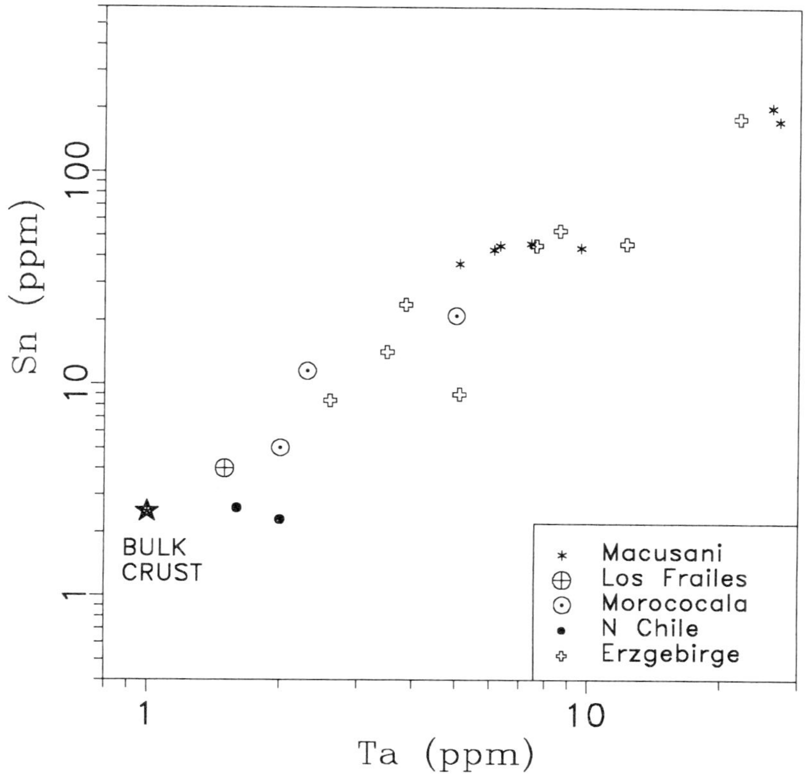

Fig. 4. Sn-Ta variation diagram for silicic volcanic rocks from northern Chile and the Macusani, Los Frailes and Morococala volcanic fields. Data from Ericksen et al. (1985, 1990). The reference data for the Erzgebirge granite suite are from Tischendorf (1989)

The systematic trends in the Rb-Sr variation diagram of Fig. 2 confirm this earlier interpretation. The linear correlation trends in log-log space point to substantial feldspar and/or biotite fractionation in both the Cordillera Real and Cafayate rock suites. The Chojlla samples come from an underground exposure of a granitic stock in the Chojlla tin-tungsten mine, and plot in the geochemically most evolved portion of the Cordillera Real trend. However, these samples are affected by strong subsolidus muscovite blastesis (with K-feldspar and oligoclase stable), and have therefore a magmatic element pattern modified by hydrothermal overprint.

The Cordillera Real samples plot very closely to the distribution trend defined by the Erzgebirge granite sequence. The Erzgebirge reference data are from the compilation of Tischendorf (1989) and represent arithmetic means for the intrusion sequence from earliest and least evolved to latest and most evolved granite phases. Tin mineralization in the Erzgebirge (and elsewhere) is associated with the latest and most evolved granite subintrusions. The samples from the Los Frailes volcanic field in central Bolivia show little variation in their Rb and Sr contents. This is either due to little variation in the degree of fractionation or reflects a fractionation process which does not involve silicic minerals.

The Ta-TiO_2 variation diagram of Fig. 3 further emphasizes the role of fractional crystallization for those sample suites for which Ta data are available. The low hydrothermal solubility of Ta and Ti makes these two elements particularly suitable for monitoring of magmatic processes in granitic rocks, which are invariably affected by hydrothermal overprint. Again, the Cordillera Real tin granites give a fractionation trend similar to the Erzgebirge granites. The Macusani samples plot in the most evolved portion of the general fractionation pattern. Assuming a perfectly incompatible behaviour of Ta (D_{Ta}xtls/melt = 0), the degree of fractionation F is in the order of 0.05-0.1 in most evolved rock portions of both the Erzgebirge and central Andean rock suites. The Ta-Ti correlation trend points back to least evolved material of bulk crustal composition.

Fig. 4 shows the distribution of tin as a function of tantalum, which is taken as a measure of the degree of magmatic fractionation of a given rock sample. Tin displays a highly incompatible behaviour which is broadly similar to tantalum. The data for the volcanic rocks from the Bolivian tin belt (Los Frailes, Morococala, Macusani) define a linear correlation trend which is identical to the Erzgebirge fractionation pattern. The general fractionation trend leads back to bulk-crust composition. Additional Sn-Ta data in Fig. 5 show much larger scatter for the northern Bolivian tin granites and the subvolcanic Karikari dome complex of central Bolivia. This may be due to hydrothermal redistribution of tin which is

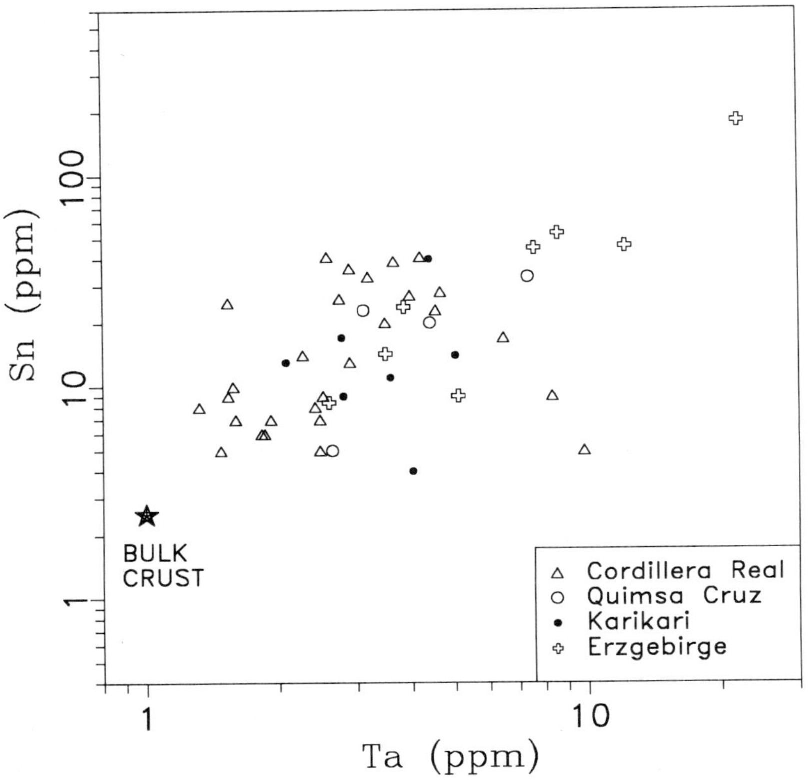

Fig. 5. Sn-Ta variation diagram for granitic rocks from the Cordilleras Real and Quimsa Cruz, and the Karikari dome complex. Data from Lehmann (1979), Miller (1988), Schneider (1985). The reference data for the Erzgebirge granite suite are from Tischendorf (1989)

an integral part of the process of hydrothermal tin ore formation (Lehmann 1990).

Figures 6 and 7 give the distribution pattern of tin as a function of Rb/Sr and TiO_2, respectively. The samples from the Bolivian tin belt define a general tin enrichment trend with increasing Rb/Sr and decreasing TiO_2. This trend broadly coincides with the pattern of the reference data from the Erzgebirge granite suite. Again, the samples from the Chojlla tin granite and the Macusani volcanics plot in the most evolved portions of the diagrams. The general tin enrichment trend has considerable scatter which is probably a result of subsolidus hydrothermal overprint.

The plutonic rock samples from tin-barren rock suites of northern Chile and NW Argentina (Cafayate suite) as well as the volcanic rocks of the Western Cordillera do not give systematic tin enrichment patterns. Instead, these samples plot chiefly at low tin levels of <3 ppm in spite of their partly evolved degree of fractionation. All 36 samples of the Cafayate fractionation suite and the great majority of the western Cordillera rhyolite samples plot below our XRFA detection limit of ≤3 ppm Sn, and 34 samples from the granitic rocks of northern Chile give a mean of 1.6 $\pm$0.9 (1σ) (analyses by AAS in Ishihara et al. 1984). Such a distribution pattern implies a compatible behavior of tin in these rocks, i.e. a bulk tin distribution coefficient D_{Sn}(xtls/melt) $\approx$1.

The difference in bulk distribution coefficient of tin in the granitic sample populations from inside and outside the Bolivian tin belt is interpreted as a consequence of their ilmenite- and magnetite-series affiliation, respectively. Critical factors controlling D_{Sn} in silicic melts are modal composition and oxidation state of the system (Lehmann 1990). Melt structure in terms of aluminium excess is probably a less critical parameter in view of the existence of non-peraluminous tin granites in the Southeast Asian tin belt (Lehmann and Harmanto 1990). The petrological classification concept of ilmenite- versus magnetite-series granitic rocks (Ishihara 1977) combines both modal composition and oxidation state of a melt in terms of its Fe-oxide mineralogy. According to the equilibrium

$$\text{titanite} + 2/3\ \text{magnetite} + \text{quartz} = \text{hedenbergite} + \text{ilmenite} + 1/3\ O_2$$

relatively oxidized silicic systems with the accessory mineral assemblage magnetite + titanite can be distinguished from those which are more reduced and contain ilmenite + clinopyroxene (or amphibole). The above mineral equilibrium is located roughly in the middle, between the hematite-magnetite and quartz-fayalite-magnetite buffers (Wones 1989).

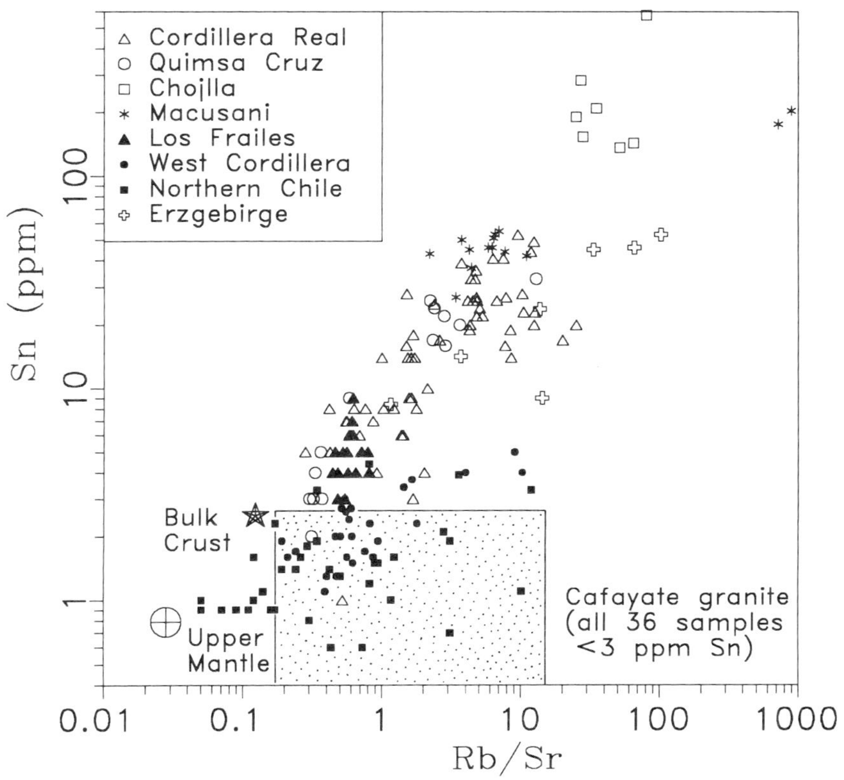

Fig. 6. Sn-Rb/Sr variation diagram for granitic rocks from the northern Bolivian tin belt (Cordillera Real, Chojlla Mine, Quimsa Cruz), northern Chile, and for felsic volcanic rocks from the western Cordillera and the Macusani (southern Peru) and Los Frailes (central Bolivia) ignimbrite suites. Data from Gardeweg et al. (1984), Ishihara et al. (1984), Klerkx et al. (1977), Lehmann (1979), Lehmann and Pichler (1980), Michel (unpubl.), Miller (1988), Noble et al. (1984), Pichler and Zeil (1969, 1972), Pichavant et al. (1988), Rapela and Shaw (1979), Tistl (1985), Winkelmann (1983). Reference data for the Erzgebirge granite suite are from Tischendorf (1989). All granite samples from the Cafayate area in northwestern Argentina plot below the analytical detection limit of 3 ppm Sn (*stippled area*). Compositions of upper mantle (*crossed circle*) and bulk crust (*heavy star*) are from Anderson (1983) and Taylor and McLennan (1985)

The oxidation state of a melt system will also control the speciation of its redox-dependent trace elements. According to the equilibrium

$$O^{2-} + Sn^{4+} = Sn^{2+} + 1/2\ O_2$$

the tetravalent tin species will be favoured in relatively oxidized melts and its crystal-chemical properties similar to Ti^{4+} and Fe^{3+} will establish a compatible behaviour, whereas divalent tin in reduced environments can behave incompatibly (Lehmann 1990).

Ilmenite-series granitic rocks are not only less oxidized rocks, but they have also a lower content of opaque minerals than magnetite-series rocks (Ishihara 1981). The lower modal content of opaques, which are the main carriers of tin in silicic rocks, together with a higher Sn^{2+}/Sn^{4+} ratio in reduced melts, explains the contrasting behaviour of tin which is incompatible in ilmenite-series rocks and compatible in magnetite-series rocks. This behaviour is opposite to the one of molybdenum which behaves more compatibly with decreasing oxidation state of a melt (Tacker and Candela 1987). In fact, molybdenum data from the central Andes give a countercurrent distribution pattern compared to tin (Ericksen et al. 1990: Fig. 7). This situation is also reflected in the metallogenic polarity of tin (Eastern Cordillera) and molybdenum (Western Cordillera).

4 Conclusions

The tin distribution in plutonic and volcanic silicic rock suites from the central Andes follows two different patterns, i.e. systematic tin enrichment trends during magmatic evolution in rocks from the Bolivian tin belt, and no tin enrichment in rock suites from outside the tin belt. The magmatic evolution in both general rock groups is, at least in some better defined sample populations, controlled by fractional crystallization. The reason for the contrasting behaviour of tin is seen in a variable bulk tin distribution coefficient D_{Sn}(xtls/melt) which must be ≈1 in non-tin granites and <1 in tin granites (Lehmann 1982). An explanation for such a behaviour is provided by the oxidation state of an igneous system which is reflected in its opaque mineralogy. The ilmenite

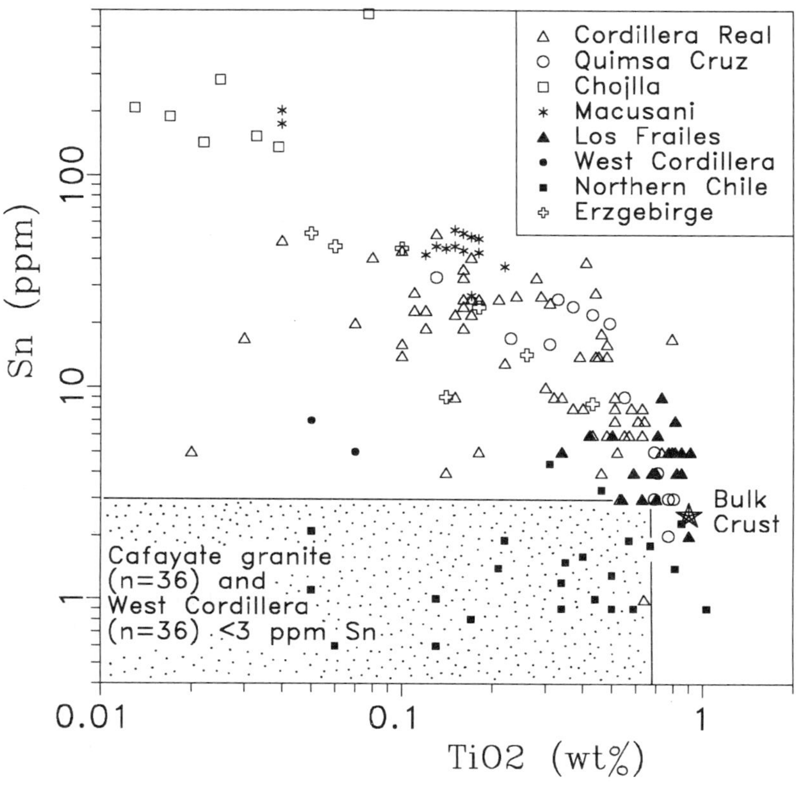

Fig. 7. Sn-TiO_2 variation diagram for the same sample populations as plotted in Fig. 6

-series rock suites of the tin belt indicate reduced melt conditions and low opaque content favourable for magmatic tin enrichment, whereas the more oxidized magnetite-series igneous systems outside the tin belt allow no tin enrichment.

In their least evolved portions the geochemical evolution trends for the igneous rocks of the tin belt approach bulk crustal composition. A geochemical tin anomaly in the crustal source rocks of the tin belt is therefore not likely. However, pelitic source rocks with their typically elevated carbon and boron levels could explain the peraluminous character and low oxidation state of the igneous systems of the tin belt, as well as their extended late magmatic history in which high boron levels are instrumental in depressing the solidus temperature. The boron content of average shale (around 100 ppm) is distinctly different from the crustal average (10 ppm) (Taylor and McLennan 1985). Tin granites from Bolivia, Cornwall and Portugal display boron enrichment trends with boron levels in least differentiated rock samples near average shale composition (Lehmann 1982).

The establishment of a tin ore system is related to large-scale hydrothermal redistribution which acts upon fractionated granitic phases magmatically pre-enriched in tin (Lehmann and Mahawat 1989). Such a tin redistribution process is again dependent on the oxidation state of an igneous system, because the hydrothermal solubility of tin is orders of magnitude greater near the quartz-fayalite-magnetite buffer as compared with the hematite-magnetite buffer (Eugster 1986). The paragenetic study of Kelly and Turneaure (1970) indicates conditions of oxygen fugacity below the pyrite-magnetite-pyrrhotite buffer during the tin-tungsten ore stage in the Bolivian ore deposits. This is a situation typical of tin ore deposits worldwide, and is different from the much more oxidized copper and molybdenum porphyry ore environment (Burnham and Ohmoto 1980). The low oxidation state of the ilmenite-series silicic rocks of the Bolivian tin belt will therefore not only favour the magmatic enrichment of tin, but also provide conditions for an effective hydrothermal redistribution process of tin. In contrast, the high oxidation state of non-tin bearing magnetite-series rocks is a prerequisite for effective magmatic enrichment and hydrothermal mobility of molybdenum.

References

Ahlfeld F, Schneider-Scherbina A (1964) Los yacimientos minerales y de hidrocarburos de Bolivia. Bol Dept Nac Geol La Paz 5: 1-388

Anderson DL (1983) Chemical composition of the mantle. Proc 14th Lunar Planet Sci Conf, part 1, J Geophys Res 88 (Suppl): B41-B52

Burnham CW, Ohmoto H (1980) Late-stage processes of felsic magmatism. Min Geol, Spec Issue 8: 1-11

Carlier G, Grandin G, Laubacher G, Marocco R, Mégard F (1982) Present knowledge of the magmatic evolution of the eastern Cordillera of Peru. Earth Sci Rev 18: 253-283

Chappell BW, White AJR (1974) Two contrasting granite types. Pac Geol 8: 173-174

Clark AH, Palma VV, Archibald DA, Farrar E, Arenas F. MJ, Roberstson RCR (1983) Occurrence and age of tin mineralization in the Cordillera Oriental, southern Peru. Econ Geol 78: 514-520

Ericksen GE, Smith RL, Luedke RG, Flores M, Espinosa A, Urquidi BF, Saravia F (1985) Preliminary geochemical study of ash-flow tuffs in the Morococala and Los Frailes volcanic fields, central Bolivian tin belt. US Geol Surv, Open-File Rep 85-258: 1-8

Ericksen GE, Eyzaguirre VR, Urquidi BF, Salas OR (1987) Neogene-Quaternary volcanism and mineralization in the central Andes. US Geol Surv, Open-File Rep 87-634: 1-35

Ericksen GE, Luedke RG, Smith RL, Koeppen RP, Urquidi B F (1990) Peraluminous igneous rocks of the Bolivian tin belt. Geology 13: 3-8

Eugster HP (1986) Minerals in hot water. Am Mineral 71: 655-673

Evernden JF, Kriz SJ, Cherroni C (1977) Potassium-argon ages of some Bolivian rocks. Econ Geol 72: 1042-1061

Francis PW, Halls C, Baker MCW (1983) Relationships between mineralization and silicic volcanism in the Central Andes. J Volcanol Geotherm Res 18: 165-190

Gardeweg M, Ishihara S, Matsuhisa Y, Shibata K, Terashima S (1984) Geochemical studies of Upper Cenozoic igneous rocks from the Altiplano of Antofagasta, Chile. Bull Geol Surv Jpn 35: 547-563

Grant JN, Halls C, Avila W, Avila G (1977) Igneous geology and the evolution of hydrothermal systems in some sub-volcanic tin deposits of Bolivia. Geol Soc Lond, Spec Publ 7: 117-126

Grant JN, Halls C, Avila W, Snelling NJ (1979) K-Ar ages of igneous rocks and mineralization in part of the Bolivian tin belt. Econ Geol 74: 838-851

Grant JN, Halls C, Sheppard SMF, Avila W (1980) Evolution of the porphyry tin deposits of Bolivia. Min Geol, Spec Issue 8: 151-173

Gustafson LB, Hunt JP (1975) The porphyry copper deposit at El Salvador, Chile. Econ Geol 70: 857-912

Halls C, Schneider A (1988) Comentarios sobre la génesis de los yacimientos del cinturón estañífero boliviano. Rev Geol Chile 15: 41-56

Halpern M (1979) Strontium isotope composition of rocks from the Disputada copper mine, Chile. Econ Geol 74: 129-130

Hawkesworth CJ, Hammill M, Gledhill AR, van Calsteren P, Rogers G (1982) Isotope and trace element evidence for late-stage intra-crustal melting in the High Andes. Earth Planet Sci Lett 58: 240-254

Hildreth W, Moorbath S (1988) Crustal contributions to arc magmatism in the Andes of central Chile. Contrib Mineral Petrol 98: 455-489

Ishihara S (1977) The magnetite-series and ilmenite-series granitic rocks. Min Geol 27: 293-305

Ishihara S (1981) The granitoid series and mineralization. Econ Geol, 75th Anniv Vol, pp 458-484

Ishihara S, Ulriksen CE, Sato K, Terashima S, Sato T, Endo Y (1984) Plutonic rocks of north-central Chile. Geol Surv Jpn Bull 35: 503-536

Just G, Schilka W, Seltmann R (1987) INAA investigations in tin-bearing granites of the Altenberg and Sadisdorf ore deposits. 4th Meet on Nuclear Analytical Methods, Dresden May 4-8, 1987, Proc 1: pp 242-251

Kelly WC, Turneaure FS (1970) Mineralogy, paragenesis and geothermometry of the tin and tungsten deposits of the eastern Andes, Bolivia. Econ Geol 65: 609-680

Klerkx J, Deutsch S, Pichler H, Zeil W (1977) Strontium isotopic composition and trace element data bearing on the origin of Cenozoic volcanic rocks of the central and southern Andes. J Volcanol Geotherm Res 2: 49-71

Kontak DJ, Clark AH, Farrar E (1984) The magmatic evolution of the Cordillera Oriental, southeastern Peru. In: Harmon RS, Barreiro BA (eds) Andean magmatism: Chemical and isotopic constraints. Shiva, Nantwich, pp 203-219

Lehmann B (1979) Schichtgebundene Sn-Lagerstätten in der Cordillera Real/Bolivien. Berl Geowiss Abh A 14: 1-135

Lehmann B (1982) Metallogeny of tin: Magmatic differentiation versus geochemical heritage. Econ Geol 77: 50-59

Lehmann B (1990) Metallogeny of tin. Springer, Berlin Heidelberg New York, 212 pp

Lehmann B, Harmanto (1990) Large-scale tin depletion in the Tanjungpandan tin granite, Belitung Island, Indonesia. Econ Geol 85: 99-111

Lehmann B, Mahawat C (1989) Metallogeny of tin in central Thailand: A genetic concept. Geology 17: 426-429

Lehmann B, Pichler H (1980) Tin distribution in mid-Andean volcanic rocks. Mineral Deposita 15: 35-39

Lehmann B, Ishihara S, Michel H, Miller J, Rapela C, Sanchez A, Tistl M, Winkelmann L (1990) The Bolivian tin province and regional tin distribution in the Central Andes: A reassessment. Econ Geol 85: 1044-1058

McBride SL, Robertson RCR, Clark AH, Farrar E (1983) Magmatic and metallogenetic episodes in the northern tin belt, Cordillera Real, Bolivia. Geol Rundsch 72: 685-713

McNutt RH, Clark AH (1983) Implications of the initial strontium isotope ratios of Central Andean, Triassic-to-Quaternary, igneous rocks in N Chile, S Peru and NW Bolivia. EOS (Transact Am Geophys Union) 64: p 329

Miller JF (1988) Granite petrogenesis in the Cordillera Real, Bolivia and crustal evolution in the Central Andes. PhD Thesis, Open Univ Milton Keynes, 198 pp (unpubl)

Miller JF, Harris NBW (1989) Evolution of continental crust in the Central Andes; constraints from Nd isotope systematics. Geology 17: 615-617

Nabelek PI (1986) Trace-element modeling of the petrogenesis of granophyres and aplites in the Notch Peak granite stock, Utah. Am Mineral 71: 460-471

Noble DC, Vogel TA, Peterson PS, Landis GP, Grant NK, Jezek PA, McKee EH (1984) Rare-element enriched, S-type ash-flow tuffs containing phenocrysts of muscovite, andalusite, and sillimanite, southeastern Peru. Geology 12: 35-39

Pichavant M, Valencia HJ, Boulmier S, Briqueu L, Joron JL, Juteau M, Marin L, Michard A, Sheppard SMF, Treuil M, Vernet M (1987) The Macusani glasses, SE Peru: Evidence of chemical fractionation in peraluminous magmas. In: Mysen BO (ed) Magmatic processes: Physicochemical principles. Geochem Soc, Spec Publ 1: 359-373

Pichavant M, Kontak DJ, Briqueu L, Valencia HJ, Clark AH (1988) The Miocene-Pliocene Macusani volcanics, SE Peru. II. Geochemistry and origin of a felsic peraluminous magma. Contrib Mineral Petrol 100: 325-338

Pichler H, Zeil W (1969) Die quartäre "Andesit"-Formation in der Hochkordillere Nord-Chiles. Geol Rundsch 58: 866-903

Pichler H, Zeil W (1972) Paleozoic and Mesozoic ignimbrites of northern Chile. Neues Jahrb Miner Abh 116: 196-207

Rapela CW, Shaw DM (1979) Trace and major element models of granitoid genesis in the Pampean Ranges, Argentina. Geochim Cosmochim Acta 43: 1117-1129

Rapela CW, Heaman LM, McNutt RH (1982) Rb-Sr geochronology of granitoid rocks from the Pampean Ranges, Argentina. J Geol 90: 574-582

Saavedra J, Toselli AJ, Rossi de Toselli JN, Rapela CW (1987) Role of tectonism and fractional crystallization in the origin of Lower Paleozoic epidote-bearing granitoids, northwestern Argentina. Geology 15: 709-713

Schneider A (1985) Eruptive processes, mineralization and isotopic evolution of the Los Frailes-Karikari region/Bolivia. PhD Thesis, Univ London, London, 280 pp

Shibata K, Ishihara S, Ulriksen E (1984) Rb-Sr ages and initial $^{87}Sr/^{86}Sr$ ratios of late Paleozoic granitic rocks from northern Chile. Bull Geol Surv Jpn 35: 537-545

Sillitoe RH, Halls C, Grant JN (1975) Porphyry tin deposits in Bolivia. Econ Geol 70: 913-927

Tacker RC, Candela PA (1987) Partitioning of molybdenum between magnetite and melt: A preliminary experimental study of partitioning of ore metals between silicic magmas and crystalline phases. Econ Geol 82: 1827-1838

Taylor SR, McLennan SM (1985) The continental crust: Its composition and evolution. Blackwell, Oxford, 312 pp

Tischendorf G (1989) Silicic magmatism and metallogenesis of the Erzgebirge (compiled by G. Tischendorf). Veröff Zentralinst Physik Erde Potsdam 107: 1-316

Tistl M (1985) Die Goldlagerstätten der nördlichen Cordillera Real/Bolivien und ihr geologischer Rahmen. Berl Geowiss Abh A 65: 1-93

Turekian KK, Wedepohl KH (1961) Distribution of the elements in some major units of the Earth's crust. Bull Geol Soc Am 72: 175-191

Winkelmann L (1983) Geologie und Lagerstätten im Bereich Palca (Mururata) und die Geochemie der Silursedimentite in der Cordillera La Paz/Bolivien. Berl Geowiss Abh A 51: 1-110

Wones DR (1989) Significance of the assemblage titanite + magnetite + quartz in granitic rocks. Am Mineral 74: 744-749

MAPS OF THE CENTRAL ANDEAN SEGMENT BETWEEN 20° AND 26°S

Comments on the Geological and Geophysical Maps

KLAUS-J. REUTTER and HANS-J. GÖTZE

1 Introduction

The book on hand contains the results of a symposium (workshop) held in Berlin, May 23-25, 1990, and is dedicated to geoscientific problems of the Central Andes. This symposium, to which scientists from Argentina, Bolivia, Chile, France, Great Britain and the United States had been invited, was the final meeting of a research programme ("Mobility of Active Continental Margins") of geoscientists from the universities of Berlin (Freie Universität and Technische Universität). As their research work was concentrated on the segment of the Southern Central Andes between the parallels 20° and 26°S, most of the contributions of the symposium also treated this region. Therefore the editors were of the opinion that a set of maps accompanying this book would not only provide a better understanding of those chapters dealing with the segment mentiond but also bridge the gap between surface geology and deeper crustal structures evidenced by geophysics.

At an earlier stage of the research activities of the Berlin research group a first geological map had been compiled at a scale of 1:1 000 000 for internal use. This map played an important role in the interdisciplinary interpretation when the first results of geophysical fieldwork became available. Maps and cross sections containing both geological and geophysical data were drawn to common scales so that the results of different disciplines within the research group could be directly compared.

For publication with this book the internal geological map was updated and maps of gravity and seismology were compiled on the same cartographical base to ease the task of direct correlations and interdisciplinary interpretation among geologists and geophysicists. The selected scale of 1:1 000 000 provides both a general view to surface geology and tectonic units as well as sufficient insight into details in order to recognize important local features at a handy map size. Also, the gravity maps fit this scale, providing information on both regional trends and local anomalies of the earth's gravity field from some 4600 gravity stations. For map presentation the Lambert Conical Projection was selected. It has its reference meridian at 67°W and the two standard parallels at 21°S and 25°S. Morphological and geographical data of the geological map were taken from the Operational Aviation Chart at the same scale. The sets of digitized original and gridded data of gravity maps are available on request from H.-J. Götze.

As mentioned in the Preface to this book, Transect 6 of the Global Geoscience Transect Programme ("Central Andean Transect, Nazca Plate to Chaco Plains, southwestern Pacific Ocean, northern Chile and northern Argentina") passes through the area of the maps. This transect, which was compiled by geoscientists from the Universidad Nacional de Salta (Argentina), the Universidad Católica del Norte in Antofagasta (Chile) and by members of the universities of Berlin provides more detailed information on the geology and geophysics of this part of the Andes (Omarini and Götze 1991). Also, a digitized version of the transect is being prepared (Schmidt and Götze, in prep.).

2 Geological Map of the Central Andes between 20°S and 26°S

The only available geological map that extending beyond the international boundaries in this area of the triple junction of the frontiers between Argentina, Bolivia and Chile is at a scale of 1:5 000 000 (de Almeida 1978), all other geological maps are limited to the national territories of these countries. Therefore, already in 1986, at a very early stage of the research programme of the Freie Universität and the Technische Universität Berlin, a map of the Andean segment between 20°S and 26°S was compiled by R. Döbel (one of the authors of Map 1 of this Vol.) on the basis of the existing national maps at scales between 1:400 000 (Argentina: Amengual et al. 1979) and 1:1 000 000 (Bolivia:

Correspondence to: Klaus-J. Reutter, Institut für Geologie und Geophysik, Freie Universität Berlin, Malteserstr. 74-100, D 1000 Berlin 46

Yacimientos Petrolíferos Fiscales Bolivianos and Servicio Geológico de Bolivia 1978; Chile: Servicio Nacional de Geología y Minería 1982). Since in this preliminary compilation maps at scales greater than 1:400 000 had not been considered and, in the meantime, new maps had been published or unpublished maps were made available to the research group, a completely new map had to be drawn for publication with this volume. Thus, the geological data base of the present "Geological Map of the Central Andes between 20°S and 26°S" (enclosed Map 1) is wider and more current than that of its unpublished predecessor.

For the Chilean part of the present geological map the basis information originated mainly from the geological maps at a scale of 1:250 000 published by the Servico Nacional de Geología y Minería, Santiago (Fig. 1). For great parts of the Bolivian sector we obtained unpublished maps at the scale of 1:250 000 from Yacimientos Petrolíferos Fiscales Bolivianos, and for the westernmost parts, the recently published "Geological map of the Altiplano and Cordillera Occidental, Bolivia, 1:500 000" (Marsh et al. 1992) could be used. The geology of the Argentine territory of the geological map was adopted from the "Mapa del Noroeste Argentino 1:500 000" (Yacimientos Petrolíferos Fiscales 1984). The maps (1:1 000 000) of Bolivia and Chile were used only for those parts not covered by the more detailed maps mentioned.

The maps cited in Fig. 1 form the geological basis. In addition, some other published (see references) or unpublished data were incorporated by the authors (Reutter, Dõbel, Bogdanic, and Kley) and the contributors (R. Omarini, J. Salfity, J. Viramonte from Argentina; O. Aranibar, M. Cirbian, J. Jarandilla, H. Perez, D. Tufiño from Bolivia; G. Chong, A. Jensen, C. Mpodozis, H. Niemeyer from Chile). However, this was rather sporadic and, hence, the map does not contain all the geological information available and representable at this scale and with the adopted legend.

The Mesozoic-Cainozoic geology of Chile is characterized by magmatic rocks extruded and intruded in magmatic arc systems. These rocks differ fundamentally from the mostly sedimentary rocks of Bolivia and Argentina, where also Palaeozoic sedimentary and Precambrian basement rocks are broadly distributed and structures are quite different. It was therefore a rather difficult task to find a legend which, for economic reasons, could not be very detailed, but had to reflect the lithological and, to some degree, structural differences between the major tectonic units. These are mainly the arc, forearc and backarc areas which existed during the Mesozoic and Cainozoic in this part of the Andes. Compromises had to be made and, thus, many important geological data were lost, especially regarding the Palaeozoic. We did this hoping that the synoptic representation of this Andean segment from the Pacific coast to the Chaco plains would offer new insight into the problems of this orogen.

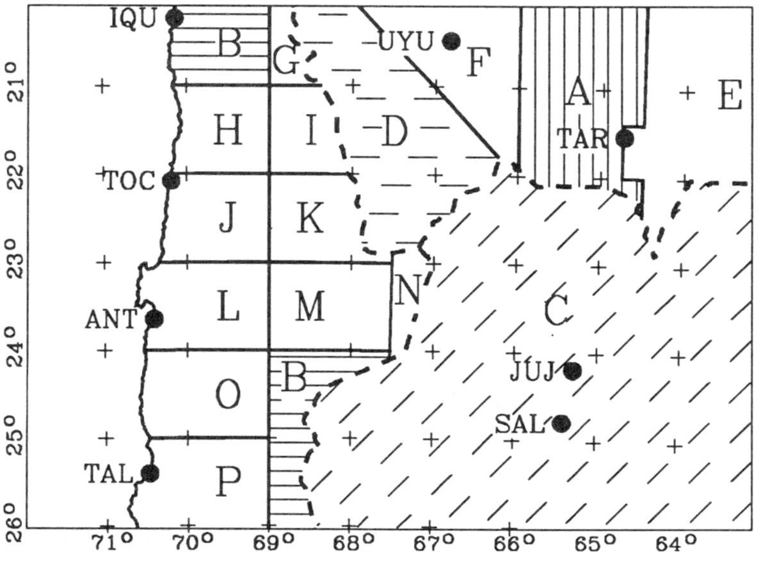

Fig. 1. Distribution of geological maps which served as the main basis for the compilation of the "Geological map of the Central Andes between 20°S and 26°S" (Map 1, enclosed in this Vol.). *A-B* 1:1 000 000: *A* Yacimientos Petrolíferos and Servicio Geológico de Bolivia (1978); *B* Servicio Nacional de Geología y Minería (1982); *C-D* 1:500 000: *C* Yacimientos Petrolíferos Fiscales (1984); *D* Marsh et al. (1992); *E-Q* 1:250 000: *E* Yacimientos Petrolíferos Fiscales Bolivianos (1988); *F* Yacimientos Petrolíferos Fiscales Bolivianos (1989); *G* Vergara and Thomas (1984); *H* Skarmeta and Marinovic (1981); *I* Ramirez and Huete (1981); *J* Ferraris (1978); *K* Marinovic and Lahsen (1984); *M* Ramirez and Gardeweg (1982); *N* Gardeweg and Ramirez (1985); *O* Mpodozis et al. in prep., *P* Naranjo and Puig (1984)

3 Gravity Maps (Bouguer Anomaly and Isostatic Anomaly) of the Central Andes between 20°S and 26°S

3.1 Data Base and Corrections

Within the activities of the working group "Mobility of Active Continental Margins" of the Berlin universities, a project was started in 1982 in close cooperation with the Universidad Católica del Norte (G. Chong, Antofagasta, Chile), Universidad de Chile (M. Araneda, Santiago, Chile) and Universidad Nacional de Salta (J. Viramonte and R. Omarini, Salta, Argentina) on the following subject: Construction of a new base of gravity data in the Andean geotraverse in N-Chile and NW-Argentina by collecting and updating already existing measurements and completing them by additional field surveys (Götze et al. 1988, 1990, 1991a). Here, the new data base is presented in the form of maps of the Bouguer anomaly (Map 2) and the isostatic anomaly (Map 3; both maps enclosed in this Vol.).

From 1982 to 1986 2568 gravity measurements were made in an Andean geotraverse covering N-Chile and NW-Argentina between 64°-71°W and 20°-26°S. Including 2100 reprocessed, older data, there is now a data base of about 4600 gravity values available. The spacing of stations was approximately 5 km along all passable tracks, although some local areas had a higher station density. To complete this data file 344 additional gravity sites were used in the offshore area of the Pacific Ocean, released by the BGI (Bureau Gravimétrique International, Toulouse), 462 in Bolivia from the IGM (Instituto Geográfico Militar, La Paz), 53 from the Geodetic Institute of the University of Buenos Aires, 311 from YPF (Buenos Aires, National Oil Company of Argentina) and 873 from YPFB (Santa Cruz, National Oil Company of Bolivia).

All measurements are related to the IGSN71 gravity datum. The two base stations in Oran/Argentina (IGSN71 40334K) and Iquique/Chile (IGSN71 40400K) were used as well as the following reference stations: Jujuy/Argentina (DOD 4172), Joaquin V. Gonzales/Argentina (DOD 50541), Antofagasta/Chile (IAGS 931870) and Calama/ Chile (IAGS 932070).

Tidal corrections could be neglected in the coastal area because of its small effect compared with the pronounced gravity anomalies observed. In the Puna and the eastern part of the Andes the tables of the "Annual Earth-Tides-Corrections", published yearly by Geophysical Prospecting, were used.

The large area and logistic problems did not always allow the determination of the drift of the gravity metres by repeating the measurements at each station. However, even if poor tracks were used by car, the drift of the LaCoste & Romberg instruments (models G and D) rarely exceeded 0.1 mGal per day (1 mGal = 10^{-5} ms^{-2}).

A reliablel determination of the geographic coordinates was not possible everywhere because a unique set of maps did not exist, e.g. at a scale of 1:50 000, covering the entire area. At worst, the positioning error amounts to approx. 0.5 km and corresponds to an error in the gravity anomaly of about 0.25 mGal, caused by the effect of imprecise latitude. Even worse was the problem of height determination at each station. Only 37% could be related directly to benchmarks, such as levelling lines, trigonometric heights, height points of water tubes, railways and spot heights. All others had to be determined with barometers using the benchmarks as base stations. To improve the quality of the barometric measurements, time-dependent drift corrections were calculated as it is usually done for gravity measurements, using as many benchmarks and repeated measurements as possible. Moreover, the profiles of several days were combined in order to eliminate systematic errors. Altimeter types Wallace & Tiernan FA181 and Thommen 3B4.01.2, were used. The scales of these instruments were calibrated on levelling lines with an altitude difference of about 2000 m. Error estimations showed that even in the worst case the accuracy was better than 20 m, giving an error in the Bouguer anomaly of about 4 mGal, which is less than 1%.

3.2 Reduction of the Absolute Gravity Data

The calculation of the gravity anomaly values was based on the following equations:

Bouguer anomaly

$$\Delta g'' = g_{abs} + \delta_h + \delta g_{top} + \delta g_{bou} - \gamma \qquad (1)$$

Free-air anomaly

$$\Delta g' = g_{abs} + \delta_h - \gamma \qquad (2)$$

where:

g_{abs} : absolute gravity at station (measured)
δ_h : normal gravity at station level h (calculated)
δg_{top} : topographic reduction (density: 2.67 g/cm^3)
δg_{bou} : Bouguer reduction (density: 2.67 g/cm^3)
γ : gravitational constant.

The normal gravity was calculated according to IGSN71 using the International Gravity Formula of 1967. For the topographic reduction, a method developed for gravity investigations in the Alps was used (Ehrismann et al. 1966), after adapting it to the special situation in the Central Andes. Calculations of topographic reduction were based on the DEM by B. Isacks, Cornell University (Isacks 1988).

3.3 Bouguer Anomaly

Map 2 shows the derived Bouguer anomaly map. As usual, in the offshore area the Bouguer anomaly is replaced by the free-air anomaly. Therefore one observes a high correlation of the trench topography and the gravity field. Onshore the gravity field drops down to a regional minimum of less than -400 mGal in the Central Andes, mostly related to crustal thickening due to isostatic compensation and tectonical processes (Götze et.al. 1991b and this Vol.).

3.4 Isostatic Anomaly

The effect of isostatic compensation of topography was calculated assuming the regional compensation model of Vening-Meinesz with the following parameters: crustal density, 2.67 g/cm^3, mantle density, 3.17 g/cm^3, water density, 1.03 g/cm^3, crustal thickness at sea level, 40 km. The gravity effect of this model was calculated using fast Fourier techniques and then subtracting the Bouguer anomaly (on- and offshore) at station level. The resulting anomaly serves as a residual field (Map 3) and is interpreted by Götze et al. (1991b and this Vol.).

3.5 Plotting Procedure

Both the original Bouguer gravity and the isostatic anomaly were interpolated onto a 5 km times 5 km grid using the method of weighted averages (corresponding to $1/R^4$) within an unlimited search radius (Mundry 1970). The isolines are not smoothed by automated numerical procedures.

3.6 Seismicity Data

Earthquake epicentre data, taken from ISC (1971 -1986) and USGS (1962-1987), are given in Map 2 (Bouguer anomaly). The number of detections (P-arrivals) for each event is ≥ 25. The plotted epicentres are divided into three classes of body wave magnitudes: $4.8 \leq mb \leq 4.9$; $5.0 \leq mb \leq 5.9$ and $mb \geq 6.0$, expressed by the increasing size of the various symbols which represent the different depth ranges: 0-70 km depth (circles); 71-140 km depth (triangles) and depths ≥ 140 km (squares). The events form a typical Wadati-Bennioff zone with a dip angle between 20°-30°. Concentrations of seismicity are observed at about 100-150 km and at 200 km depth. Between 63° and 64°W longitude very deep quakes from the depth range 470-630 km are detected.

References

General References

de Almeida FFM (1978) Tectonic map of South America, 1 : 5 000 000. DNPM - CGMW - UNESCO, Rio de Janeiro

Omarini R, Götze HJ (eds) (1991) Am Geophys Union, Int Lithosphere Program, Publ 192, 30 pp, 2 sheets

Schmidt S, Götze HJ (eds) (1992) Am Geophys Union, Int Lithosphere Program (in prep)

References to Geological Map: Argentine Territory

Alonso R, Gutierrez R, Viramonte JG (1984) Puna Austral - Base para el subprovincialismo geológico de la Puna Argentina. 9. Congr Geol Argentino 1: 48-49, with Mapa geológico dela Puna 1:1 250 000

Amengual R, Méndez V, Navarini A, Viera O, Zanettini JC (1979) Geología de la Región Noroeste, Republica Argentina, Provincias de Salta y Jujuy, 1:400 000, Dirección General de Fabricaciones Militares

Yacimientos Petrolíferos Fiscales, Gerencia de Exploración (Narciso, V, Fernandez C, Gebhard J; 1984): Mapa del Noroeste Argentino 1:500 000. 9. Congr Geol Argentino

References to Geological Map: Bolivian Territory

Marsh SP, Richter DH, Ludington S, Soria-Escalante E, Escobar-Diaz A (1992) Geological map of the Altiplano and Cordillera Occidental, Bolivia, 1:500,000. US Geological Survey Bulletin 1975

Servicio Geologico de Bolivia (Aranibar O, Urdinines M, Perez H, Villaroel C, Jarandilla J, Escobar A) (1979) Mapa Geológico de Tarija-Villazón, 1:100 000 (unpubl)

Servicio Geológico de Bolivia (Perez H) (1988) Mapa Geológico de Bolivia, Hojas Tarija-Villazón, 1:250,000 (unpubl)

Yacimientos Petroliferos Fiscales Bolivianos (Cherroni C, Cirbian M, Torrez E, Codime J, Flores A, Alarcon J) (1966-1970) Mapa Geológico del Altiplano y partes de las Cordilleras oriental y occidental, 1:250 000 (unpubl)

Yacimientos Petrolíferos Fiscales Bolivianos and Servicio Geológico de Bolivia (Pareja J, Vargas C, Suárez R, Ballón R, Carrasco R, Villarroel C) (1978) Mapa Geológico de Bolivia, 1:1 000 000, La Paz

Yacimientos Petrolíferos Fiscales Bolivianos (Tufiño D) (1988) Mapa Geológico de Bolivia, Sector Subandino y Llanura Sur, 1:250,000 (unpubl)

Yacimientos Petrolíferos Fiscales Bolivianos (Aranibar O, Jarandilla J) (1989) Mapa geológico compilado del Altiplano y parte de las Cordilleras Oriental y Occidental (Hoja 2), 1:250,000 (unpubl)

References to Geological Map: Chilean Territory

Bogdanic T (1990) Kontinentale Sedimentation der Kreide und des Alttertiärs im Umfeld des subduktionsbedingten Magmatismus in der chilenischen Präkordillere. Berl Geowiss Abh A123: 117 p (with a map 1:250 000 of the Sierra de Moreno, modified from Carta Geol Chile)

Ferraris F (1978a) Cordillera de la Costa entre 24° y 25° Latitud Sur, Región de Antofagasta. Inst Investig Geol, Santiago, Carta Geol Chile 26, 1:250 000, 15 pp

Ferraris F (1978b) Hoja Tocopilla, Región de Antofagasta. Inst Investig Geol, Santiago, Carta Geol Chile 3, 1:250 000, 32 pp

Ferraris F, Di Biase F (1978) Hoja Antofagasta, Región de Antofahasta. Inst Investig Geol, Santiago, Carta Geol Chile 30, 1:250 000, 48 pp

Gardeweg M Ramirez CF (1985) Hoja Río Zapaleri, II Región de Antofagasta. Serv Nac Geol Minería, Santiago, Carta Geol Chile 66, 1:250 000, 89 pp

Hervé M, Marinovic N (1989) Geocronología y evolución del batolito Vicuña Mackena, Cordillera del la Costa, Sur de Antofagasta (24-25°S). Rev Geol Chile 16: 31-49

Marinovic N, Lahsen A (1984) Hoja Calama. Serv Nac Geol Minería, Santiago, Carta Geol Chile 58, 1:250 000, 140 pp

Mpodozis C et al. (in prep.) Hoja Aguas Blancas, Región de Antofagasta. Serv Nac Geol Minería, Santiago, Carta Geol de Chile, 1:250 000 (unpublished draft)

Naranjo JA, Puig A (1984) Hojas Taltal y Chañaral, Regiones de Antofagasta y Atacama. Serv Nac Geol Minería, Santiago, Carta Geol Chile 62-63, 1:250 000, 140 pp

Ramirez CF, Gardeweg M (1982) Hoja Toconao, Región de Antogasta. Serv Nac Geol Minería, Santiago, Carta Geol Chile 54, 1:250 000, 121 pp

Ramirez CF, Huete C (1981) Hoja Ollagüe, Región de Antofagasta. Inst Investig Geol, Santiago, Carta Geol Chile 40, 1:250 000, 47 pp

Servicio Nacional de Geología y Minería (1982) Mapa Geológico de Chile 1:1 000 000, Santiago

Skarmeta J, Marinovic N (1981) Hoja Quillagua. Serv Nac Geol Minería, Santiago, Carta Geol Chile 51, 1:250 000, 63 pp

Vergara H and Thomas A (1984) Hoja Collacagua, Región de Tarapacá. Serv. Nac. Geol Minería, Santiago, Carta Geol Chile 59, 1:250 000, 79 pp

References to the Gravity Data Base

Ehrismann W, Müller G, Rosenbach O, Sperlich N (1966) Topographic reduction of gravity measurements by the aid of digital computers. Boll Geofisica Teor Appl 8 (29): 3-20

Isacks BL (1988) Uplift of the Central Andean Plateau and bending of the Bolivian orocline. J Geophys Res 93, B4: 3211-3231

Götze HJ, Strunk S, Schmidt S (1988) Central Andean gravity field and its relation to crustal structures. In: Bahlburg H, Breitkreuz C, Giese P (eds) The Southern Central Andes. Lecture Notes in Earth Sciences 17. Springer, Berlin Heidelberg New York pp 199-208

Götze HJ, Lahmeyer B, Schmidt S, Strunk S, Araneda M (1990) A new gravity data base in the Central Andes (20°-26°S). EOS Transactions, Am Geophys Union 71 (16): 401, 406-407.

Götze HJ, Lahmeyer B, Schmidt S, Strunk S, Araneda M, Chong G, ViramonteJ (1991a) The gravity data base of the transect compilation. In: Omarini R, Götze HJ (eds) Global Geoscience Transect 6: Central Andean Transect, Nazca Plate to Chaco Plains, SW Pacific Ocean, N Chile and N Argentina. Am Geophys Union, Washington DC, pp 20-23

Götze HJ, Meurers B, Schmidt S, Steinhauser P (1991b) On the isostatic state of the Eastern Alps and the Central Andes - a comparison. In: Harmon RS, Rapela CW (eds) Andean magmatism and its tectonic setting. Geol Soc Am Spec Pap 265, pp 279-290

Mundry E (1970) Zur automatischen Herstellung von Isolinienplänen: Beih Geol Jahrb, 98: 77-93

References to the Earthquake Data

ISC (1971-1986) Bulletin of the International Seismological Center. Newbury, UK

United States Geological Survey (USGS) (1962-1987) Preliminary Determinations of Epicenters. National Earthquake Information Center (NEIC), Denver, CO